Computational Neuroanatomy

Computational Neuroanatomy

Principles and Methods

Edited by

Giorgio A. Ascoli, PhD

*Krasnow Institute for Advanced Study
and Psychology Department
George Mason University,
Fairfax, VA*

Humana Press ✳ Totowa, New Jersey

© 2002 Humana Press Inc.
999 Riverview Drive, Suite 208
Totowa, New Jersey 07512

humanapress.com

All rights reserved. No part of this book may be reproduced, stored in a retrieval system, or transmitted in any form or by any means, electronic, mechanical, photocopying, microfilming, recording, or otherwise without written permission from the Publisher.

All authored papers, comments, opinions, conclusions, or recommendations are those of the author(s), and do not necessarily reflect the views of the publisher.

This publication is printed on acid-free paper. ∞
ANSI Z39.48-1984 (American Standards Institute) Permanence of Paper for Printed Library Materials.

Production Editor: Mark J. Breaugh.

Cover illustration: Composite map showing the 3D distribution of cholinergic cells projecting to four arbitrarily defined mediolateral sectors of the neocortex. *See* Fig. 3A on page 176.

Cover design by Patricia F. Cleary.

For additional copies, pricing for bulk purchases, and/or information about other Humana titles, contact Humana at the above address or at any of the following numbers: Tel.: 973-256-1699; Fax: 973-256-8341; E-mail: humana@humanapr.com; Website: http://humanapress.com

Photocopy Authorization Policy:
Authorization to photocopy items for internal or personal use, or the internal or personal use of specific clients, is granted by Humana Press Inc., provided that the base fee of US $10.00 per copy, plus US $00.25 per page, is paid directly to the Copyright Clearance Center at 222 Rosewood Drive, Danvers, MA 01923. For those organizations that have been granted a photocopy license from the CCC, a separate system of payment has been arranged and is acceptable to Humana Press Inc. The fee code for users of the Transactional Reporting Service is: [1-58829-000-X/02 $10.00 + $00.25].

Printed in the United States of America. 10 9 8 7 6 5 4 3 2 1

Library of Congress Cataloging-in-Publication Data

Computational neuroanatomy : principles and methods / edited by Giorgio A. Ascoli.
     p. cm.
  Includes bibliographical references and index.
  ISBN 1-58829-000-X (alk. paper)
  1. Neuroanatomy--Mathematical models. 2. Computational neuroscience. I. Ascoli,
Giorgio A.

QM451.C645 2002
573.8'33'015118--dc21
                                                                                       2001051865

Preface

The importance of computational modeling as a research approach in neuroscience is recognized today by most researchers in the field. Computational neuroscience is generally associated with simulations in electrophysiology and neural dynamics. Recently, an increasing number of neuroscientists have begun to use computer models to study and describe neuroanatomy, its subcellular bases, and its relationship with neuronal activity and function. Other researchers began importing accurate and quantitative descriptions of neuronal structure and connectivity into computer simulations of neuronal and network physiology. Perhaps owing to the broad range of scales spanned by these studies, from subcellular structures to very large assemblies of interconnected neurons, computational neuroanatomy literature is sparse and distributed among the many technical journals in neuroscience. Nevertheless, a common theme is easily recognized in all these research projects: the use of computer models, simulations, and visualizations to gain a deeper understanding of the complexity of nervous system structures. Neuroanatomy constitutes a central aspect of neuroscience, and the continuous growth of affordable computer power makes it possible to model and integrate the enormous complexity of neuroanatomy. It is not surprising that computational neuroanatomy research projects are stirring considerable interest in the scientific community. *Computational Neuroanatomy: Principles and Methods* is the first comprehensive volume discussing the principles and describing the methods of computational approaches to neuroanatomy.

Computational neuroanatomy is potentially as vast and diverse a field as neuroanatomy itself. In an attempt to capture this diversity, each chapter of this book is contributed by different authors. Each subject is presented and discussed by the experts who first defined the problems, implemented the methods to solve them, and formulated the principles underlying the solutions (brief biographies of the book's authors are provided at the end of the book). Principles and methods of computational neuroanatomy are explained through direct examples of recent or ongoing research. All chapters were peer-reviewed by the editor, by contributors of other chapters, and by "external" reviewers (who are acknowledged at the end of this Preface).

Most chapters are enhanced by electronic material included in the companion CD-ROM. Such material includes software packages used in computational neuroanatomy, step-by-step explanation of the algorithms implemented in such programs, and examples of data files. In addition, given the important contribution of computer graphics to neuroanatomical models, results reported in the book are further illustrated by animations and movies in the CD-ROM. While only black and white figures are reproduced in print, high-resolution color images are contained in the disk. Finally, the CD provides links to web sites containing updates and additional information.

Computational Neuroanatomy: Principles and Methods may be used as a back-to-back text by readers interested in learning the basic strategies, results, and language of computational neuroanatomy, or as a unique reference to consult for key material (both conceptual and technical) in these new areas of investigation. Active researchers and graduate students should be able to read the chapters as if they were published in a high-quality scientific journal. Advanced undergraduate students and interested non-academic thinkers with a background in neuroscience or computer science will also find this volume highly accessible.

The book was edited with particular attention to the expected diversity in background of the readership. A natural audience for this publication consists of all neuroanatomists interested in novel technology. The use of computers can aid neuroanatomical investigation and understanding, and the material of this book can be an inspiring source of research ideas as well as a basic guide to keep up to date with computational developments. As a rapidly growing field, computational neuroanatomy is of interest for the neuroscience community in general, and this book provides a review of many leading research paths. On the other hand, computer scientists and engineers are turning with ever deeper interest to biological architectures. Nervous systems are still remarkably superior to digital computers and artificial neural networks in a variety of computational and cognitive tasks, and a crucial reason is their structure. This book constitutes an intellectual bridge between information technology and neuroanatomy. Finally, the tremendous impact that computer graphics has had and will continue to have in education makes this material also useful for academic instructors involved with brain science, including neurologists, psychologists, biologists, and physicists.

Structural and functional human brain imaging and mapping is contributing enormously to the advancement of neuroscience. Neuroimaging is obviously anatomical in nature, and it involves a great deal of computational analysis and processing. However, most of the aspects of computational neuroanatomy described in this book revolve around the neuron as a fundamental brick of brain structure and function. Readers interested in the issues of computational neuroanatomy related to brain mapping should refer to the excellent recent publications specifically dedicated to neuroimaging.

Naturally, different research groups focus on different scales. Consequently, this book is organized in three main parts. Part One deals with single neurons and their internal structures, particularly dendritic morphology and its interaction with single-cell electrophysiology. Part Two discusses neuronal assemblies, axonal connectivity, and large-scale, anatomically accurate networks. Finally, Part Three tackles the major issues of integration of the massive knowledge necessary to describe (and generate) completely accurate neuroanatomical models at the system level. A detailed description of each chapter is beyond the scope of this preface. However, the first introductory chapter provides a review of several recent developments in computational neuroanatomy and introduces the subsequent chapters in this context. In addition, a summary of the contents is provided by the abstracts of each chapter.

Acknowledgment

I wish to extend my gratitude to all the people who made this book possible.

The publisher, Humana Press, demonstrated considerable courage in commissioning a book in such a novel and unexplored field. The help provided by the Humana support staff in all the phases of organization and editing was outstanding.

Preface *vii*

The time I spent wearing the Editor's hat really belonged to my wife, Rebecca, and to my two sons, Benjamin and Ruben. They were extremely patient and gave me much of the support and inspiration necessary for the completion of this book.

Rarely does a researcher have the fortune to meet and be mentored by such a nurturing advisor as Dr. James L. Olds. Without him this book would simply not exist.

The many authors of this book, of course, deserve the lion's share of the credit. Not only did they contribute chapters of great quality and substance, they also provided each other with invaluable feedback by cross-reviewing the manuscripts.

I am also deeply indebted to many external reviewers who read earlier versions of the chapters. They include Kim "Avrama" Blackwell (George Mason University), Jean-Marie Bouteiller (University of Southern California), Ann Butler (Krasnow Institute for Advanced Study), Mark Changizi (Duke University), Barry Condron (University of Virginia), Bard Ermentrout (University of Pittsburgh), Rebecca Goldin (University of Maryland), Kristin Jerger (Krasnow Institute for Advanced Study), Huo Lu (California Institute of Technology), Gianmaria Maccaferri (Emory University), Michael Moseley (Stanford University), Tay Netoff (Boston University), Kimberlee Potter (Armed Forces Institute of Pathology Annex), Bruce Rasmussen (George Washington University), and Charles Schroeder (Albert Einstein College of Medicine).

As pointed out by several contributors, the reviewers' insightful comments were instrumental in ensuring the highest quality of this book.

Giorgio Ascoli

Contents

Preface v
Contributors xi

Introduction

1 Computing the Brain and the Computing Brain
   Giorgio A. Ascoli 3

Part I

2 Some Approaches to Quantitative Dendritic Morphology
   Robert E. Burke and William B. Marks 27

3 Generation and Description of Neuronal Morphology Using L-Neuron:
   A Case Study
   Duncan E. Donohue, Ruggero Scorcioni, and Giorgio A. Ascoli 49

4 Optimal-Wiring Models of Neuroanatomy
   Christopher Cherniak, Zekeria Mokhtarzada, and Uri Nodelman 71

5 The Modeler's Workspace:
   Making Model-Based Studies of the Nervous System More Accessible
   Michael Hucka, Kavita Shankar, David Beeman, and James M. Bower 83

6 The Relationship Between Neuronal Shape and Neuronal Activity
   Jeffrey L. Krichmar and Slawomir J. Nasuto 105

7 Practical Aspects in Anatomically Accurate Simulations
   of Neuronal Electrophysiology
   Maciej T. Lazarewicz, Sybrand Boer-Iwema, and Giorgio A. Ascoli 127

Part II

8 Predicting Emergent Properties of Neuronal Ensembles Using a Database
   of Individual Neurons
   Gwen A. Jacobs and Colin S. Pittendrigh 151

9   Computational Anatomical Analysis
    of the Basal Forebrain Corticopetal System
    ***Laszlo Zaborszky, Attila Csordas, Derek L. Buhl, Alvaro Duque,
    Jozsef Somogyi, and Zoltan Nadasdy*** 171

10 Architecture of Sensory Map Transformations:
    *Axonal Tracing in Combination with 3D Reconstruction,
    Geometric Modeling, and Quantitative Analyses*
    Trygve B. Leergaard and Jan G. Bjaalie 199

11 Competition in Neuronal Morphogenesis
    and the Development of Nerve Connections
    Arjen van Ooyen and Jaap van Pelt 219

12 Axonal Navigation Through Voxel Substrates:
    A Strategy for Reconstructing Brain Circuitry
    Stephen L. Senft 245

13 Principle and Applications of Diffusion Tensor Imaging:
    A New MRI Technique for Neuroanatomical Studies
    Susumu Mori 271

Part III

14 Computational Methods for the Analysis of Brain Connectivity
    Claus C. Hilgetag, Rolf Kötter, Klaas E. Stephan, and Olaf Sporns 295

15 Development of Columnar Structures in Visual Cortex
    Miguel Á. Carreira-Perpiñán and Geoffrey J. Goodhill 337

16 Multi-Level Neuron and Network Modeling in Computational Neuroanatomy
    ***Rolf Kötter, Pernille Nielsen, Jonas Dyhrfjeld-Johnsen,
    Friedrich T. Sommer, and Georg Northoff*** 359

17 Quantitative Neurotoxicity
    David S. Lester, Joseph P. Hanig, and P. Scott Pine 383

18 How the Brain Develops and How it Functions:
    *Application of Neuroanatomical Data
    of the Developing Human Cerebral Cortex to Computational Models*
    ***William Rodman Shankle, Junko Hara, James H. Fallon,
    and Benjamin Harrison Landing*** 401

19 Towards Virtual Brains
    Alexei Samsonovich and Giorgio A. Ascoli 425

Index *437*

Contributors

GIORGIO A. ASCOLI, PhD • *Krasnow Institute for Advanced Study and Department of Psychology, George Mason University, Fairfax, VA*

DAVID BEEMAN, PhD • *Department of Electrical and Computer Engineering, University of Colorado, Boulder, CO*

JAN G. BJAALIE, MD, PhD • *Institute of Basic Medical Sciences, Department of Anatomy, University of Oslo, Oslo, Norway*

SYBRAND BOER-IWEMA, MS • *Krasnow Institute for Advanced Study, George Mason University, Fairfax, VA*

JAMES M. BOWER, PhD • *Research Imaging Center, University of Texas Health Science Center, San Antonio; Cajal Neuroscience Research Center, University of Texas at San Antonio, San Antonio, TX*

DEREK L. BUHL, PhD • *Center for Molecular and Behavioral Neuroscience, Rutgers University, New Brunswick, NJ*

ROBERT E. BURKE, MD • *Laboratory of Neural Control, National Institute of Neurological Disorders and Stroke, National Institutes of Health, Bethesda, MD*

MIGUEL Á. CARREIRA-PERPIÑÁN, PhD • *Department of Neuroscience, Georgetown University Medical Center, Washington, DC*

CHRISTOPHER CHERNIAK, PhD • *Committee on History & Philosophy of Science, Department of Philosophy, University of Maryland, College Park, MD*

ATTILA CSORDAS, MD • *Center for Molecular and Behavioral Neuroscience, Rutgers University, New Brunswick, NJ*

DUNCAN E. DONOHUE, BA • *Krasnow Institute for Advanced Study and Department of Psychology, George Mason University, Fairfax, VA*

ALVARO DUQUE, PhD • *Center for Molecular and Behavioral Neuroscience, Rutgers University, New Brunswick, NJ*

JONAS DYHRFJELD-JOHNSEN, MS • *Vogt Brain Research Institute, Heinrich University Düsseldorf, Düsseldorf, Germany*

JAMES H. FALLON, PhD • *Department of Anatomy and Neurobiology, University of California at Irvine, Irvine, CA*

GEOFFREY J. GOODHILL, PhD • *Department of Neuroscience, Georgetown University Medical Center, Washington, DC*

JOSEF P. HANIG, PhD • *Division of Applied Pharmacology Research, Food & Drug Administration, Rockville, MD*

JUNKO HARA, PhD • *Department of Information and Computer Science, University of California at Irvine, Irvine, CA*

CLAUS C. HILGETAG, PhD • *School of Engineering and Science, International University Bremen, Bremen, Germany*

MICHAEL HUCKA, PhD • *Control and Dynamical Systems, California Institute of Technology, Pasadena, CA*

GWEN A. JACOBS, PhD • *Department of Cell Biology and Neuroscience, Center for Computational Biology, Montana State University, Bozeman, MT*
ROLF KÖTTER, MD • *Vogt Brain Research Institute, Heinrich University Düsseldorf, Düsseldorf, Germany*
JEFFREY L. KRICHMAR, PhD • *The Neurosciences Institute, San Diego, CA*
BENJAMIN HARRISON LANDING, MD • *Deceased; Department of Pediatrics and Pathology, University of Southern California, Los Angeles, CA*
MACIEJ T. LAZAREWICZ, MD • *Krasnow Institute for Advanced Study, George Mason University, Fairfax, VA*
TRYGVE B. LEERGAARD, MD, PhD • *Institute of Basic Medical Sciences, Department of Anatomy, University of Oslo, Oslo, Norway*
DAVID S. LESTER, PhD • *Pharmacia Corporation, Peapack, NJ*
WILLIAM B. MARKS, PhD • *Laboratory of Neural Control, National Institute of Neurological Disorders and Stroke, National Institutes of Health, Bethesda, MD*
ZEKERIA MOKHTARZADA, BS • *Committee on History & Philosophy of Science, Department of Philosophy, University of Maryland, College Park, MD*
SUSUMU MORI, PhD • *Department of Radiology, Johns Hopkins University, Baltimore, MD*
ZOLTAN NADASDY, PhD • *Center for Neural Computation, The Hebrew University, Jerusalem, Israel*
SLAWOMIR J. NASUTO, PhD • *Department of Cybernetics, University of Reading, Reading, UK*
PERNILLE NIELSEN, MS • *Niels Bohr Institute of Astrophysics, Physics, and Geophysics, University of Copenhagen, Copenhagen, Germany*
URI NODELMAN, BS • *Department of Computer Science, Stanford University, Stanford, CA*
GEORG NORTHOFF, MD, PhD • *Department of Psychiatry, Otto von Guericke University, Magdeburg, Germany*
P. SCOTT PINE, MA • *Division of Applied Pharmacology Research, Food & Drug Administration, Rockville, MD*
COLIN S. PITTENDRIGH, PhD • *Center for Computational Biology, Montana State University, Bozeman, MT*
ALEXEI SAMSONOVICH, PhD • *Krasnow Institute for Advanced Study, George Mason University, Fairfax, VA*
RUGGERO SCORCIONI, BS • *Krasnow Institute for Advanced Study, George Mason University, Fairfax, VA*
STEPHEN L. SENFT, PhD • *Krasnow Institute for Advanced Study, George Mason University, Fairfax, VA*
KAVITA SHANKAR, PhD • *Division of Biology, California Institute of Technology, Pasadena, CA*
WILLIAM RODMAN SHANKLE, MD • *Department of Cognitive Science, University of California at Irvine, Irvine, CA*
FRIEDRICH T. SOMMER, PhD • *Department of Neural Information Processing, University of Ulm, Ulm, Germany*
JOZSEF SOMOGYI, PhD • *Department of Medicine, Flinders University, Adelaide, South Australia*

OLAF SPORNS, PhD • *Department of Psychology, Indiana University at Bloomington, Bloomington, IN*
KLAAS E. STEPHAN, MD • *Vogt Brain Research Institute, Heinrich University Düsseldorf, Düsseldorf, Germany*
ARJEN VAN OOYEN, PhD • *Graduate School of Neurosciences, Netherlands Institute for Brain Research, Amsterdam, The Netherlands*
JAAP VAN PELT, PhD • *Graduate School of Neurosciences, Netherlands Institute for Brain Research, Amsterdam, The Netherlands*
LASZLO ZABORSZKY, MD, PhD • *Center for Molecular and Behavioral Neuroscience, Rutgers University, New Brunswick, NJ*

Introduction

1
Computing the Brain and the Computing Brain

Giorgio A. Ascoli

ABSTRACT

Computational neuroanatomy is a new emerging field in neuroscience, combining the vast, data-rich field of neuroanatomy with the computational power of novel hardware, software, and computer graphics. Many research groups are developing scientific strategies to simulate the structure of the nervous system at different scales. This first chapter reviews several of these strategies and briefly introduces those that are expanded in the subsequent chapters of the book. The long-term end result of the collective effort by researchers in computational neuroanatomy and neuroscience at large will be a comprehensive structural and functional model of the brain. Such a model might have deep implications for scientific understanding as well as technological development.

1.1. INTRODUCTION

The modern scientific investigation of nervous systems started just over a century ago with the work of Santiago Ramon y Cajal (1). Cajal's "neuron doctrine" was revolutionary for two main reasons. On the one hand, it showed that, like all the other organs in the body, the brain is constituted by cells. On the other hand, it began to reveal the incredible complexity of the shape of brain cells (glia and, in particular, neurons) and their potential interconnectivity. These findings inspired the principal axiom of modern neuroscience: the key substrate for all the functions performed by nervous systems, from regulation of vital states, reflexes, and motor control, to the storage and retrieval of memories and appreciation of artistic beauty, lies not in some "magic" ingredient, but rather in the structure and assembly of neurons. Over the past hundred years, a series of fundamental discoveries about synaptic transmission, passive and active electric conductance, neurotrophic factors, structural and functional plasticity, development, and topographical representations (2) shaped the neurosciences into a highly interdisciplinary field, overlapping with chemistry, biology, physics, informatics, pharmacology, neurology, and psychology. Yet, anatomy remains the chief aspect in the investigation of the brain, embodying and framing the contributions of all other disciplines to our understanding of nervous systems. The ultimate, and arguably the hardest, challenge to human knowledge consists of understanding how organic matter gives rise to feelings, emotions, and logical thinking. The solution resides in those millions of millions of neurons and millions of billions of connections.

Understanding how cognition arises from the brain may seem like a hopeless goal, if by "understanding" a process we mean creating a mental model of its mechanism. In fact, the incredible complexity of neuroanatomy has so far prevented us from synthesizing the huge amount of collected experimental data into a complete and organic functional scheme. This grand task, however, might be achieved if we employ powerful computers to aid our mental model. In the last decade of the twentieth century, while the price of home computers has remained approximately constant (between $1000 and $2000), available personal computers have increased their speed by 100-fold (from 10 MHz to 1 GHz), their fast memory by 1000-fold (from 1 MB to 1 GB), their local storage capacity by 400-fold (from 100 MB to 40 GB), and their portable storage capacity by 1000-fold (from 720 kB to 650 MB) *(3)*. In other words, the evolution of microprocessors outperformed the 1965 "visionary" prediction by then-Intel chairman Gordon Moore that computational power would approximately double every 18 mo *(3)*.

It is unlikely that "Moore's Law" will hold indefinitely, due to chemical and physical limits of computing matter. However, progress has also been achieved through novel concepts and applications in computational technology (such as the Internet over the past 15 yr), and new such innovations are expected in the future. Critics of Moore's law predicted that a "plateau" in the development of computer hardware should have been already visible now (and it is not). On the opposite spectrum of opinions, several experts hold that the progress of computational power grows with a dual exponential rate. In other words, its speed of evolution would double at fixed intervals of time *(4)*. If this is the case, we should face an information "singularity" within a few decades: at that point, the rhythm of increase of computational speed and capacity will outpace the time scale of human reasoning (seconds to milliseconds), and the time of progress of information technology will "collapse". Whether we reach a plateau or a singularity, in the next few decades we will witness a historical transition in the evolution of science and computation.

Whatever the future holds, the present spectacular growing power of computers appears to be suitable to meet the challenge of a comprehensive description of the nervous system. The combination of neuroscience and computers has already produced dramatic results. In experimental research, computers have fostered progress by allowing quicker and more reliable setups and results. We have seen the emergence of a new field, computational neuroscience, which studies and develops computer-assisted models of neurobiological processes. Simulations of different neural structures at various scales, along with quantitative comparisons of model performance to biological data, have contributed extensively to our understanding of the functional connections between the system level (accessible by the study of behavior) and the microscopic level (accessible by molecular and cellular techniques). Examples of recent successful computational studies include a variety of subjects, such as visual cognition (figure–ground segregation *[5]*, object recognition *[6]*, attention *[7]*, but also stereopsis *[8]*), eye saccadic control *(9)*, memory encoding at the neuron *(10)* and network *(11)* levels, the role of inhibitory interneurons in the relationship among neuronal oscillations *(12)*, synchrony, physiological rhythms and pathological states (e.g., epileptic seizures [13]), and even the mechanism of pain and neurological rehabilitation (14). In all cases, computational modeling plays an integrating role in the study of the structure–function relationship by combining cellular and subcellular physiology, psychophysics, and mathematical analysis.

While computational neuroscience has a relatively long history in electrophysiological simulations (such as passive and active conductances implemented in "compartmental" models of neurons and networks [15,16]), its potential for neuroanatomy has only recently gained enthusiastic appreciation in the scientific community. Together with this enthusiasm, the availability of neuroanatomical databases and the most recent development of computer graphics have resulted in a plethora of high-level research projects focusing on computational modeling of neuroanatomy. These studies range from the description of dendritic morphology and the characterization of its relationship with electrophysiology to the analysis of the structural determinants of higher brain functions via the detailed mechanism of neuronal assembly into functional networks. Despite this wide range of scopes and scales, a considerable number of recent excellent scientific publications shared the common approach of using computational simulations to investigate neuroanatomy and its influence on neuroscience at large, thus virtually defining the new field of computational neuroanatomy.

1.2. COMPUTING THE BRAIN

Cajal's theory put the neuron center stage. Today, we know that neurons are themselves complex computational machines. Theories of dendritic, somatic, and axonal functions have matured well beyond the traditional scheme of "input–integration–output". Single neurons and their arbors are now considered sophisticated time filters (17), coincidence detectors (18), internally distributed devices of local memory storage (10,19), and dynamic metabolic assemblies with high internal spatial specificity (20,21), just to mention but a few examples (see also [22]). Not surprisingly, neuronal structure has been characterized as increasingly complex with each major discovery.

If neurons are not the elementary or "atomic" computational units of the brain, which substructures play this role? For a while, neuroscientists hoped that synapses would be the key to the mystery of nervous system computation. Further studies indicated that presynaptic and postsynaptic processes (mostly in the axons and in the dendrites, respectively) could independently modulate synaptic activity. Moreover, the simple early distinction between excitatory and inhibitory synapses has been put in a much broader and more complex perspective by the discovery of a large number of neuromodulatory neurotransmitters. More recently, much attention has been paid to the mutual influence between genetic expression in the soma and activity in axons and dendrites, involving issues of intracellular trafficking and communication. It is doubtful that any single piece of the puzzle will provide a complete answer.

An example of the issues raised by the ultrastructural investigation of neurons is provided by the characterization of spines, little mushrooms covering the dendritic surfaces of many neuronal classes in the central nervous system. Hypotheses on the function of dendritic spines have ranged from plastic loci of synaptic input (for a review, see [23]) to neuroprotection (24). Recently, electron microscopy has allowed an in-depth study of spine anatomy, revealing a great variability in size, shape, location, and clustering *(25,26)*. Further functional clues may derive from the quantitative description of other related subcellular structures, including actin cytoskeleton *(27)* and spine-specific protein complexes *(28)*. The quest for a complete structural (and functional) characterization of neurons must ultimately focus on proteins, their metabolism, regulation, and dynamics. For the time being, however, much still needs to be discovered about neuronal morphology, and computational studies have, so far, mainly considered dendrites as the "elementary" structural objects for modeling purposes. For the scope of this book, dendrites will constitute the first step in the bottom-up path towards an integrated structural model of the brain.

Neurons can be classified according to a variety of criteria, including location within the nervous system, main neurotransmitter(s) released, presence of specific protein markers, dendritic and axonal structure, and interconnectivity with neurons of other classes. Dendritic morphology is traditionally a fundamental criterion in neuronal classification, partly because it is immediately captured by optical microscopy under staining conditions discovered and optimized over a century ago *(1)*. Neurons in different morphological classes range widely in dendritic size (from a few micrometers to several millimeters of spatial extent) and complexity (from a handful to thousands of bifurcations per tree). This complex variability makes a large number of morphological classes easily recognizable by few peculiar characteristics, and the same neuronal class can be often recognized across species that are philogenetically very distant. Yet even within a given species and morphological class, no two neurons are ever exactly alike, and individual cells are remarkably different. This is analogous to different classes of botanical trees: a person can walk for miles and miles in an oak forest, seeing thousands of oak trees. These trees vary greatly in size, number of branches, bifurcation angles, etc. Yet each tree is clearly an oak tree and can be immediately recognized as different from a pine tree or a palm tree.

From a mathematical point of view, the interclass and intraclass variability of dendritic (and botanical) morphology can be captured statistically. Let us suppose to characterize trees by three geometrical parameters: the total surface area, the average amplitude angle at bifurcations, and the average diameter ratio of the two branches (thicker over thinner) in all bifurcations. In this scheme, every tree can be represented as a point in 3D, with each Cartesian axis quantifying one of three above measurements. If we analyze various neurons and plot them according to this classification, we obtain various clouds of points. It is very unlikely that two neurons will have identical measurements in these three parameters. However, neurons belonging to the same morphological classes will be closer to each other than to neurons belonging to different classes (Fig. 1).

Thus, it appears that neurons can be classified morphologically based on how they cluster together upon quantitative analysis *(29)*. In general, a much larger number of parameters is measured for morphological classification (in the example above, we have used only three parameters for convenience of graphical representation). Cluster analysis can be carried out precisely and quantitatively *(30)* and can help define major morphological classes as well as subclasses *(31)*: in the example above, one can imag-

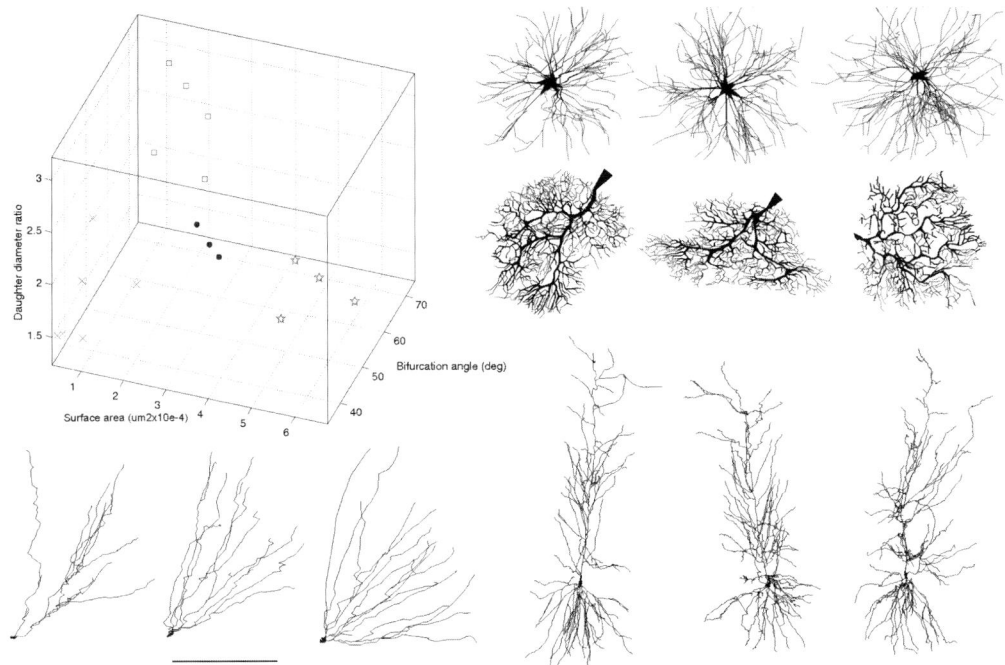

Fig. 1. 3D scatter plot of morphological parameters measured from neurons belonging to various morphological classes. Measured parameters are the average diameter ratio between the larger and the smaller branches at bifurcations (height), total surface area (width), and average amplitude angle at bifurcations (depth). Three examples of each neuronal class are illustrated: (clockwise) motoneurons (stars, 4 neurons measured), Purkinje cells (filled circles, 3 neurons measured), CA1 pyramidal cells (empty squares, 5 neurons measured), and Dentate granule cells (crosses, 6 neurons measured). Surface area of motoneurons has been reduced 10-fold in the 3D scatterplot. Scale bar: Granule cells and Purkinje cells, 200 μm; motoneurons, 1 mm; pyramidal cells, 300 μm.

ine analyzing many more neurons such that, "zooming in" on each cloud, several smaller clusters would become apparent.

Starting from the late 1970s, Dean Hillman published a series of influential papers proposing that a great deal of the intraclass and interclass variability of dendritic morphology could be captured by a restricted number of fundamental parameters of shape *(32,33)*. Hillman's description reflected important known or novel biophysical principles, and most of his parameters had important subcellular correlates, such as membrane metabolism and microtubules dynamics. Hillman's most important insight was that many morphological properties, which so peculiarly characterize different neurons, can in fact emerge from the complex interactions among the fundamental or basic parameters. Thus, different morphological classes could be characterized by different sets of statistical distributions describing the basic parameters, as measured from the experimental data. Within a given morphological class, the diversity of individual neurons would be reflected by the natural variability of the measured basic parameters.

Although we now know that Hillman's original formulation fails to capture all aspects of dendritic morphology quantitatively *(34–36)* (see also Chapter 3), these pioneering studies inspired many researchers to further explore, elaborate and refine similar ideas and principles *(35–39)*. With the advent of personal computers and the increasing availability of digital morphological data collected by semiautomatic tracing, many of these anatomical descriptions were "translated" into algorithms. The statistical characterization of neurons was then implemented as stochastic sampling (Monte Carlo method). Fundamental or basic (and now, algorithmic) parameters could be measured quantitatively and automatically from digital data. Algorithms would then sample random numbers within the appropriate statistical distributions describing the data.

Hillman's description aimed at capturing the statistical morphological properties of adult neurons and was almost exclusively based on local and intrinsic parameters, such as branch diameter. Other models focused on the growth process and incorporated environmental or global influences *(40–44)*. Most data on dendritic morphology available in the literature focus on the "dendrogram" properties, i.e., topology (pattern of branching) and internal geometry (length, diameters, and their combinations). Most of the models mentioned above aim at reproducing dendrograms. The 3D arrangement of dendrites in space, however, constitutes a difficult and important problem in the quantitative description of dendritic morphology. Few studies even define appropriate parameters for such a complex characterization *(45–47)*. Only recently have computational models attempted to incorporate 3D dendritic orientation in simulation designs *(48–50)*.

The study and parsimonious description of motoneuron morphology by Burke and coworkers in the early 1990s *(36)* constituted one of the first successful attempts to implement an algorithmic representation of dendrograms in a computer simulation. The same laboratory (which also contributed many important investigations in experimental, computational, and theoretical electrophysiology) was also among the first to attempt a quantitative description of the 3D architecture of dendrites *(45)*. In Chapter 2 of this book, Burke and Marks review these earlier studies and describe their ongoing efforts to integrate the description of dendrograms with those of the spatial orientation of dendritic branches and of packing density in a coherent stochastic formulation.

Recently, the editor's laboratory implemented several variations of Hillman's algorithm, as well as Burke's model, in a software package called L-Neuron *(39)*. L-Neuron "reads" statistical distributions of basic parameters measured from digitized experimental data and generates "virtual neurons". These virtual neurons can be quantitatively compared to the original real cells by measuring any parameter not used by the algorithm ("emergent parameters"). Any discrepancies between real and virtual cells may provide feedback on the anatomical rules underlying the algorithms. Further rules can be elaborated, implemented, and tested, until a better description is achieved (Fig. 2). Some of the algorithms implemented in L-Neuron proved satisfactory in the description of Purkinje cells and motoneuron morphology *(35)*. Chapter 3 presents a case study in which the first implemented algorithms fall short of an accurate description and shows the strategy illustrated in Figure 2 "in action". A beta version of L-Neuron (running under Unix and DOS) is released in the companion CD-ROM. The program to extract quantitative measurements from digital data (e.g., both basic and

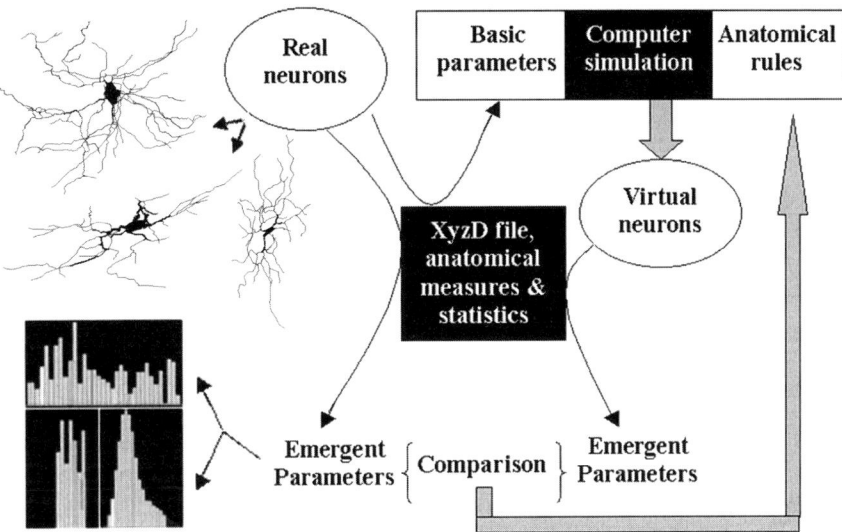

Fig. 2. Summary of the research strategy to simulate dendritic morphology (neurons are retinal ganglion cells).

emergent parameters), called L-Measure *(51)*, is also described in chapter 3 and is available in the CD-ROM.

Chapter 3 shows how the computer generation of dendritic morphology constitutes an exercise towards a complete and accurate anatomical description of single neurons. Although remarkable progress has been achieved since Hillman's seminal studies, the problem is still open, and different principles need to be "tried out". Numerous examples in the literature suggest that specific biophysical rules could be appropriate to describe neuritic branching *(52–55)*. In Chapter 4, Cherniak and colleagues review their studies indicating that some of the fundamental correlations underlying the "optimization" of dendritic structure reflect physical principles, such as energy minimization.

Although attempts to model dendritic growth and adult neuronal morphology have given encouraging results, in nature, anatomy and physiology can never be completely separated from each other. Neuronal structure and development are affected by the electrical activity of the growing cell as well as of neighboring neurons. While experimental neuroanatomy is the oldest branch of neuroscience, and computational neuroanatomy the newest, the field of electrophysiology has thrived with the parallel advancements of experiments and computer models. Experimental electrophysiological findings are framed within a quantitative model whenever possible, and computer simulations are most often based on experimentally measured data. Until recently, computational studies in electrophysiology almost entirely disregarded neuroanatomy, reducing neuronal structures to balls and sticks and concentrating on the biochemical and biophysical phenomenological aspects. In the last decade, however, investigators have started paying more attention to the real structures in their electrophysiological simulations. This has been made possible, in part, by the increasing availability of digital morphological data. Most importantly, this trend has been facilitated by the realization that, in order to study the relationship between structure and activity, and between structure and function, structure, indeed, needs to be taken into consideration.

The leading software packages for electrophysiological simulations, such as Genesis *(16)*, Neuron *(56)*, and, for large-scale parallel simulations, Neosim *(57)*, are nowadays fully compatible with a compartmental representation corresponding to real or realistic anatomy. Even more importantly, anatomy is becoming the representation of choice in the modeling community for integrating knowledge about compartmental properties including distribution of active ionic channels, synaptic receptors or metabolic networks, intracellular recording patterns, and calcium dynamics *(58)*. In Chapter 5, the developers of Genesis describe their workspace environment to integrate and represent information relevant to electrophysiological models.

If the biochemical machinery underlying electrical activity in neurons is selectively compartmentalized in specific subcellular locations, it should be apparent that anatomy must be considered when building biologically accurate computational models of neuronal activity. However, how can one study the specific influence of structural differences among neurons on their firing patterns and their function? Neurons are diverse both in their morphology and in their biochemical contents, and both sources of variability affect electrical activity. From the experimental standpoint, studying the structure–activity relationship in neurons directly is extremely difficult, but computer simulations constitute a powerful alternative. In classical computational models, anatomy is simplified or kept "constant", and the influence of various distributions of active and passive properties on neuronal firing is assessed. With a complementary approach, one can keep the biophysical model constant and implement it on different dendritic structures. In this way, investigators characterized the effect of morphological differences among different neuronal classes on their firing patterns *(59)* and on the dendritic back- and forward-propagation of action potentials *(60)*. These findings were recently extended by an analysis of topological influences of firing properties *(61)* and by studies of the electrophysiological effect of dendritic variability within the same morphological class *(62,63)*. In Chapter 6, Krichmar and Nasuto illustrate this strategy and review the experimental and computational evidence on the relationship between structure and activity in single neurons.

While an accurate anatomical representation of dendritic structure makes electrophysiological simulations more biologically plausible, it increases the complexity of the models. Single neurons can be represented by many thousands of compartments, and the choice of representation becomes both somewhat arbitrary and crucial for the outcome of the simulation. A few excellent publications have recently appeared that discuss technical aspects of compartmental modeling at length *(15,16,64)*. However, for research projects with a specific focus on neuronal anatomy, several practical issues should be considered when preparing electrophysiological simulations. Chapter 7 in this book is specifically dedicated to these aspects. This chapter also describes a series of software tools for spike train analysis *(65)* and morphological format conversion for electrophysiological simulations, several of which are included in the companion CD-ROM.

1.3. FROM NEURONS TO NETWORKS

Although individual neurons are extremely complex structures, it is important to maintain the perspective that the functions and computations carried out in nervous systems are generally supported by assemblies of neurons rather than by individual cells. How can we connect the levels of single neurons (and their substructures) with the level of networks of many neurons? While most aspects of single neuron morphology described above are being actively investigated in computational neuroanatomy, several researchers have started employing computer simulation and visualization tools to explore the even more complex systems of neuronal assemblies. Part II of this book is a collection of chapters representing a variety of these approaches.

The spatial location of a neuron in the nervous tissue strongly affects its connectivity in the network and, thus, the role it plays in the overall function of the system. This relationship between space and activity underlies the broad concept of brain functional maps, typically exemplified by the visual receptive fields characterized by Hubel and Wiesel *(66)*. In the primary visual cortex, neurons appear to be spatially arranged in columnar structures that are not separated from each other by any physical boundary, but rather are segregated functionally. Neurons in each column respond to the presence of lines or bars of a specific orientation and in a specific location of the visual field. Thus, the information represented in the cortex ("function") corresponds to (or "is mapped on") a specific location in the cortex. Starting from Hubel and Wiesel's discovery, much work has been focused on elucidating the functional mapping of the cerebral cortex in rodents and primates alike, using both electrophysiological techniques (as in the original work by Hubel and Wiesel) as well as modern imaging *(67)*.

In trying to connect the level of analysis and explanation considered in the first part of this book (dendritic morphology) with functional mapping, the question naturally arises of the relationship between dendritic structure and activity not just in single cells, but rather within a cell assembly. How do dendritic morphology and the spatial location of neurons interact in contributing to the system level activity of the nervous system? Unfortunately, the spatial resolution of functional imaging is far too limited to allow detailed investigation of dendritic activation in nervous tissues. Even in electrophysiological investigations, extracellular or somatic recording techniques can provide only indirect information on dendritic activity, while intradendritic recording is not yet practical enough to allow coverage of substantial space within the tissue. In any case, the experimental techniques currently available for the structural investigation accompanying electrophysiological studies (typically, staining and reconstruction) fall short of providing integrated information on both dendritic morphology and multiple neuronal locations.

Recently, Gwen Jacobs' laboratory used the cricket sensory system as the structure of choice to investigate the multiple interrelationships among dendritic morphology, relative spatial location of neurons, electrophysiological activity, and information encoding (function) *(68,69)*. The use of an invertebrate model system provided the necessary simplification to analyze in detail and integrate the many aspects underlying the interplay between structure, activity, and function. Chapter 8 reviews the mixed experimental and computational approach that allowed Jacobs and collaborators to reconstruct the morphology, electrophysiological activity, and spatial relationship of several cricket sensory neurons. The result is a system of detailed 3D probability maps representing (and predicting) the multicell patterns of dendritic activation and connectivity that correlate with a given sensory input ("information").

Vertebrate central nervous systems are too complex to isolate regions in which single neurons can be precisely mapped in terms of their detailed spatial location, relative orientation, dendritic morphology, connectivity, and activity. Even probabilistic characterizations usually focus on specific aspects (cell density, regional connectivity, etc.) without connecting the subcellular and multicellular levels. This is particularly true for subcortical areas of mammalian brains, where the functional organization of the maps (if any) is largely unknown. A typical example is provided by the basal forebrain, a region believed to be involved in a variety of crucial cognitive, emotional, and autonomic functions. The basal forebrain is characterized by an extreme chemical and anatomical heterogeneity at the neuronal level and by a remarkable complexity in the pattern of projections to the rest of the brain. For several years, the laboratory of Laszlo Zaborszky has used computational methods to assemble, analyze, and synthesize large amounts of sparse data on the basal forebrain *(70,71)*. In Chapter 9, they review their recent efforts, from the reconstruction of electrophysiologically identified neurons to the assembly of a cell population model.

Another example of the role of computational neuroanatomy in facilitating our understanding of brain areas whose fundamental organizational principles escape traditional anatomical approaches is provided by the studies of the brain stem in Jan Bjaalie's laboratory *(72,73)*. In parallel investigations of the cat auditory system and the rat cerebro-cerebellar pathways, Bjaalie and colleagues have adopted a variety of experimental and computational techniques, including axonal tracing and the use of specialized computer software to map, reconstruct, and visualize networks at the system level, with the goal to render massive anatomical data in 3D space. In Chapter 10, they describe both the methods and the principles of the analysis that led them to establish novel principles of sensory map transformation.

The research efforts of Zaborszky and Bjaalie show the great potential of computational neuroanatomy in the integration of the experimental data in highly complex systems. This approach is complementary to the attempt to simulate neuronal structure and activity with computer models constructed bottom-up, described in Part I of this book. The incompleteness of the data and the complexity of most brain regions makes a bottom-up approach to the system level not feasible at this stage. However, the fast (and increasing) rate of data acquisition and sharing in the neuroscience community, and the ability to integrate and efficiently interpret these data, could soon allow a neuroanatomical computer model of an entire brain region at the detailed level of single-cell morphology. In order to make such a future model functional in terms of electrophysiological activity, it will be essential to include the generation of suitable network connectivity and the description of the underlying axonal navigation.

Dendritic morphology and axonal navigation are naturally related to each other, and the attempts to model them separately are but coarse simplifications. Among the first researchers to employ computational methods to study neuronal anatomy, van Pelt and van Ooyen, have developed sophisticated models to simulate both dendritic geometry and axonal navigation and connectivity *(42,74)*. One of the main tenets of van Pelt and van Ooyen's approach is competition among growing neurites. In their models, dendrites compete for growth within trees or neurons, and axons compete for dendritic or somatic targets. This principle reflects the limited resources naturally available and also allows a population-based implementation, which constitutes an ideal computational tool to bridge single-neuron and network level neuroanatomical models. In Chapter 11, van Pelt and van Ooyen demonstrate how their simulations accurately capture a great amount of anatomically properties of real neurons, such as the topology of their branching patterns and the kinetics of neuritic elongation.

The regulation of dendritic and axonal growth by competition at the level of neuronal populations is implemented in Stephen Senft's software program ArborVitae *(43,48)*. In ArborVitae, cell bodies, spatially distributed in nuclei or layers, give rise to dendrites and axons that can grow according to internal rules as well as to distance-dependent repulsion or attraction from other structures. The program is flexible and complex and has been used to generate a variety of anatomically plausible neuronal networks. Given a specific algorithm for axonal navigation, the exact spatial location of the source and targets, shape of the tissue, and presence of barriers can drastically affect the connectivity of the network. For this reason, it is important to integrate the simulation algorithms with brain atlas data in order to use ArborVitae to develop neuroanatomical knowledge, novel hypotheses, and intuitions. In Chapter 12, Senft discusses the use of experimental 3D maps of the mouse brain gray and white matter as a substrate to model axonal navigation in a realistic environment, in the context of the ArborVitae implementation.

Computer simulations of single neuron morphology can be based on parameters extracted directly from the experimental data, and the generated virtual neurons can be quantitatively compared with their real counterparts. In future attempts to create anatomically accurate simulations of complete networks, where can the experimental data on axonal navigation be obtained? General information available from atlases does not contain enough detail to allow the construction of a single neuron-based bottom-up model. Single-cell axonal tracing experiments are time consuming, rarely complete, and usually lack the necessary "panoramic" view of the surrounding tissue regions. A possible approach is to utilize the maps developed and described by Bjaalie's group (Chapter 10). An alternative is provided by a relatively novel imaging technique, diffusion tensor imaging (DTI). In DTI, a magnetic resonance signal is used to track the diffusion of water molecules in the living tissue. Given the high anisotropy of fiber tracts in nervous systems, DTI can actually map axonal pathways at a remarkably detailed level *(75,76)*. In addition, DTI is noninvasive, thus allowing data collection from living humans under a variety of physiological and pathological conditions. In Chapter 13, Susumo Mori presents the principles of DTI and its potential applications to neuroscience and computational neuroanatomy.

1.4. THE COMPUTING BRAIN

The chapters in Part II of this book show the tremendous cellular complexity of nervous systems at the network-to-regional level. While computer simulations can be applied directly to investigate the structure–activity relationship at the single-neuron level, the same approach at the system level encounters difficult problems of detail and resolution that require the development of dedicated technical solutions. Biologically plausible models based on real data are now being developed and implemented for dendritic morphology, yet the assembly of multineuron networks with the same level of detail faces the challenge of data acquisition and organization. Current progress supports the hope that enough information will be collected "top-down" to allow the generation of a comprehensive bottom-up model of the brain. This model will be an exceptional tool for the investigation of the real brain and also a potential source of inspiration for the development of new machines. Will anatomy be an essential part of such a model? Absolutely. It might be present in a highly abstract form, where "anatomically realistic" shapes and connections are substituted by more compactly described functionally equivalent units. But the principles of anatomy will have to be part of a comprehensive model of the brain as long as we hold that structure underlies activity (and thus function).

The issue of knowledge integration and representation is thus encountered in system level computational neuroanatomy in at least two related strategic stages. First, a usable form of knowledge must be made available to the neuroscientist to make sense of the incredibly complex structural data of the brain and to allow the creation of data-based computer models at the network level. Second, the functional principles of the anatomical structure must be extracted, quantitatively formulated, and implemented in order to simulate the brain computational processes in a software model. In Part III of this book, knowledge integration and representation constitute the central focus in the characterization of the brain and its computing abilities.

If the internal structure of neurons were "modeled away" completely, the brain could be functionally represented as a gigantic connectivity network. The neuroscience literature and textbooks are full of examples of coarse connectivity knowledge. Typical examples are the statements that "retinal ganglion cells project the optic nerve fibers to the lateral geniculate nucleus of the thalamus," or that "dentate gyrus granule cells project the mossy fibers to the CA3 region of the hippocampus proper". This knowledge is coarse, because it represents a gross oversimplification of reality. The real situation is more complex, because almost every brain region is anatomically and functionally divided in a large number of subregions (lateral geniculate nucleus of the thalamus and the CA3 region of the hippocampus proper being no exceptions). Each of these subregions is typically heterogeneous with respect to morphological and chemical classes of the constituting neurons (see e.g., Chapter 9). In any case, even at the coarse level exemplified above, knowledge about major connectivity projections in the brain is still incomplete.

Let us assume that, after decades of progress, the knowledge about brain connectivity is "out there" in the neuroscience literature at whatever level of detail may be needed to construct a comprehensive functional model. Even in this hypothetical situation, the complexity of the system would prevent an intuitive understanding or representation. This fundamental problem in neuroanatomy has been recently approached by several groups with a computational perspective *(77–80)*. Typically, network connectivity among and within brain regions can be described and analyzed statistically as a graph model. Although many brain areas have been recently investigated at this level, the primate cortical visual system has become a "test model" for these techniques. In Chapter 14, Hilgetag and colleagues describe the basic mathematical concepts of cluster and graph analyses and review the main results recently obtained with these approaches. It is important to note that knowledge about brain connectivity is generally not absolute and uncontroversial. On the contrary, many hypotheses of brain connectivity are derived indirectly or are extrapolated from experimental results and may be controversial. A recent knowledge-based computational neuroanatomy project, called Neuroscholar, is addressing this issue at a variety of scales *(81)*. In Neuroscholar, neuroanatomical data is computationally represented to allow the explicit differentiation among alternative interpretations of the same data, while maintaining the source database of the original literature report and, where possible, links to available raw data.

The relationship between brain connectivity, activity, and functional mapping is still not well understood, but computational approaches have produced several interesting results where basic experimental data are available. As mentioned above, the primary visual cortex is organized in columns with respect to the represented object location and orientation in the visual field, as well as ocular dominance. These complex maps are formed during development based on the incoming electrophysiological activity from the retina via the thalamus. In Chapter 15, Perpinan and Goodhill present some of the computational principles underlying the formation of these maps. Their model successfully explains many of the emergent properties of visual columnar structure and is well suited to interface region level analysis with lower level principles of axonal navigation *(82)* (see e.g., Chapter 12). The columnar organization of the cerebral cortex allowed neuroanatomists to develop a surface-based mapping representation *(83,84)*. Under the assumption that the "canonical microcircuit" in the column is basically the same throughout the neocortex, the relationship between function and structure can be sought with a unique emphasis on location on the surface. In reality, the structure and function within the canonical column, and their variability throughout the cerebral cortex, are not yet fully understood, although recent experimental and computational progress have advanced this field rapidly *(85,86)*. The creation of a large-scale anatomically realistic model of the neocortex is, as of yet, premature. However, multiscale models with at least a plausible anatomical foundation are starting to appear in the neuroscience literature. The use of data-based knowledge allows, in certain cases, the connection between neuron level compartmental simulations (see Chapters 5–7) and function emerging from connectivity schemes and higher level models *(87)*. In Chapter 16 Rolf Kotter and colleagues present an original working example of a multiscale simulation of activity propagation in the primate visual cortex, using simplified compartmental models of neurons within a canonical microcircuit as the building units to provide network level insights into the functional anatomy of the visual system.

What are the necessary components of a complete neuronal level anatomical model of a brain region that can be used to simulate network connectivity and (with adequate computational power and parallel processing *[57, 88]*) electrophysiological activity? A first step could be the deployment of a very large number of detailed neuronal morphologies throughout virtual space representing tissue regions and their connectivity with intrinsic and extrinsic axonal pathways. Recently, a pilot study of this approach has been attempted for the hippocampus *(43)*. However, for the model to be accurate, it

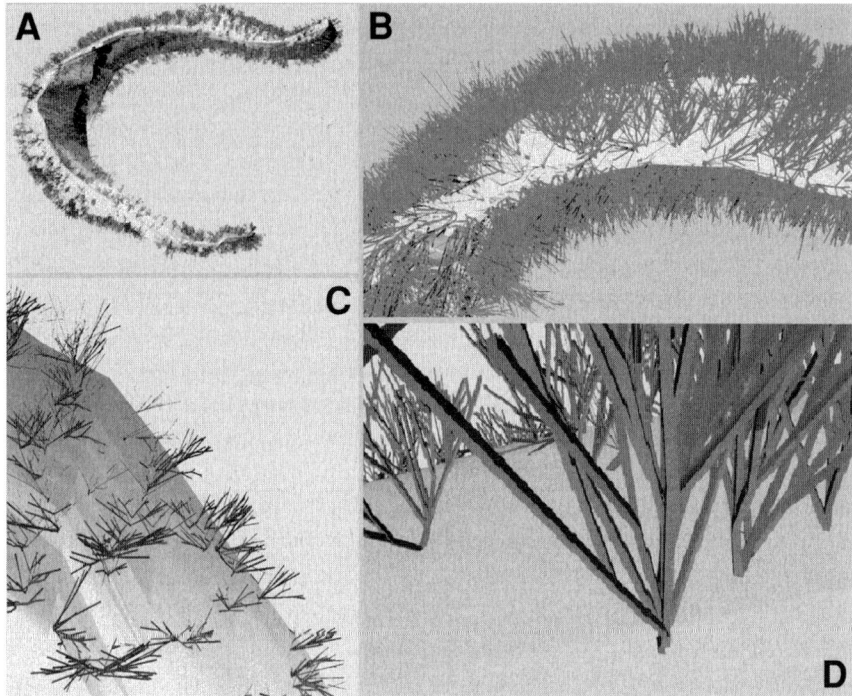

Fig. 3. A system level anatomical model of the rat dentate gyrus. The granular layer surface is reconstructed in 3D from experimental atlas sections. One million granule cells (20,000 replicas of 50 reconstructed neurons) are stochastically distributed within the layer, with primary axis oriented perpendicular to the surface. **(A)** A saggital view of the whole dentate gyrus, from the medial plane. Only one in a thousand cells is virtually stained ("pseudo-Golgi"). The canonical "C-shaped" slice of the dentate gyrus is perpendicular to the plane of the page. **(B–D)** Progressive zoom-in on the model. A color version of these images, as well as several animations, are included in the CD-ROM. Based on simulations by Ruggero Scorcioni.

is necessary to model or reconstruct the spatial boundaries of the given brain region from the experimental data (see also Chapter 10). One possibility is to assemble 2D images from a section atlas into a 3D virtual reality reconstruction. The editor's laboratory has produced such a surface for the granular layer of the dentate gyrus in the hippocampus *(89)*. Since granule cells are roughly perpendicularly oriented with respect to this surface, it is possible to computationally distribute reconstructed or simulated neurons in the virtual dentate gyrus in an anatomically accurate fashion (Fig. 3). The next step is to use reconstructed data of incoming axons from the entorhinal cortex, also digitally available *(90)*, to model the main extrinsic synaptic inputs from the perforant pathway. Although the sheer mass of data in such a realistic model makes progress extremely time-consuming, it would appear that this ambitious project can be, at least in principle, completed.

Surface reconstruction from histological sections is limited by tissue distortion or loss and by the lack of a systematic and accurate method to align sections (see also Chapter 12). An alternative is to use microscopic imaging such as *ex vivo* magnetic resonance imaging (microscopic MRI). In this technique, the whole brain of an animal (typically mouse or rat) is excised and, while maintained alive with perfusion of physiological solutions, extensively recorded for nuclear magnetic resonance signals. The very long possible exposure (12–24 h) allows resolutions comparable to those available in stained atlas sections *(91,92)*. This technique can thus be used to obtain "already reconstructed" 3D surfaces of brain regions whose boundaries are well delineated by water content contrast. The optical sectioning performed in the MRI scan avoids the problems of distortion and alignment typical of physically sectioned histological preparations. Microscopic MRI also has a great potential in the applied field of toxicology *(93)*: on the one hand, altered water content is a good marker for neuropathological states, while on the other hand, several drugs contain atoms such as ^{19}F that are not naturally present in the tissue and can be detected by nuclear magnetic resonance. In Chapter 17, Lester and colleagues review the computational principles and applications of microscopic MRI and other related imaging techniques to quantitative neurotoxicology.

If a comprehensive model of brain anatomy and physiology is to be used for the investigation of activity underlying information processing and ultimately cognitive functions, it is legitimate to consider the limitation of any investigation based on the brain of species other than human. High resolution *ex vivo* microscopic MRI is performed on small animals, and it requires their sacrifice. The resolution of noninvasive imaging techniques typically used on humans, such as positron emission tomography (PET) and functional or structural MRI, is still far from the neuron-based level considered in this book (for a recent quantitative comparison among imaging techniques, see [94]). The resolution of noninvasive imaging might eventually allow the collection of surface-based data that can be used in the construction of a bottom-up model of the brain (see also Chapter 13). In such a model, cellular data will most likely be a hybrid collection from different species. For the time being, neuroscientists are imaging human and nonhuman brains in parallel in search of a complete explanatory bridge between cellular structure, function, animal models, and human-level cognition *(67,95)*. It should be noted here that the human brain imaging community is very active in computational neuroanatomy research. The emphasis of this book is on bottom-up models with a strong cellular basis. For an extensive description of imaging-specific computational neuroanatomy principles and methods, the reader is referred to other recent publications (e.g., [67]).

An alternative to noninvasive human brain imaging in the structural investigation of the human brain is provided by the high resolution histological examination of *post-mortem* tissue. Nowadays, it is extremely difficult to obtain and destructively analyze enough human brains to allow for a meaningful statistical analysis of the anatomical inter-individual variability. However, large human *post-mortem* histological data sets collected in the past are available for reanalysis and mining. Using one such database, Shankle and colleagues have found evidence of continuous increase of the number of neocortical neurons throughout the first 72 mo of postnatal development *(96,97)*. In Chapter 18, they review their research results and present strategies to use the available human neuroanatomical data for computational modeling. Shankle's approach is particularly interesting because it involves developmental research. Classical neuroanatomical studies generally describe static structures, and computational neuroanatomical models fare no better. However, a large part (if not all) of the cognitive abilities of the human brain are linked to its plasticity. Thus, a truly realistic anatomical model should be 4D rather than 3D, describing the temporal evolution of the spatial structure. Multiple time scales should be included to describe and model developmental changes, as well as the fast morphological adaptation at the dendritic, axonal, and synaptic level underlying learning and memory.

Half a century ago, based on a novel yet limited understanding of neuronal function, many scientists and thinkers believed that artificial neural networks constituted an appropriate model of the brain and could reproduce its computational ability. From the original McCulloch-Pitts formulation, the "neurons" of neural networks became progressively more complex *(98)*, but still far less complex than our understanding of real neurons. Artificial neural networks are still today an active area of research, and they have provided many solutions for real world applications. However, most neuroscientists do not view artificial neural networks as a plausible model of brain function. There is something missing in all attempts conducted so far to model the human mind. Many cognitive modelers hold that neuroanatomical details are not important in the generation of a brain-based computer model capable of human-level intelligence. We claim that neuroanatomical details will be the most important aspects of such a model *(99)*. Certainly, there are elements of the computational ability of a human brain that are absolutely unmatched by present machines. We largely do not understand how these abilities arise from the neuronal structures in the brain. By reproducing the relevant structures in a computer model, we may obtain important insights in the computational process of the brain and, at the same time, inspire the next generation of artificial intelligence research projects.

A complete and detailed neuroanatomical model of the brain could potentially incorporate physiological features, thus allowing neuroscientists to test a large number of hypotheses with "virtual experiments", which are impossible to perform in real life (for technical limitations or in principle). In the extreme hypothetical case, if a computer model could reproduce all the aspect of the structure, biochemistry, and connectivity of the brain, down to the neuronal and subneuronal level, that model would also display brain-like behavior. Such a prospect would certainly boost scientific intuition and foster education, as well as raise ethical and epistemological issues quite different from those discussed today in the scientific community. The nineteenth and last chapter of this book discusses the prospect to build such a model and its potential impact on science.

1.5. CONCLUSIONS

Computational neuroanatomy promises to stir a long-term interest. Like no other field in neuroscience, neuroanatomy is extremely data rich and theory poor. One hundred years of experiments await interpretation in an organic mental model, and the computational resources to help build and analyze such a model have just become available. Computer tools and simulations are now used both to reproduce the structural and functional properties observed experimentally in nervous systems and to summarize, integrate, and represent the acquired knowledge. Still in its infancy, computational neuroanatomy research has already produced solutions to several outstanding problems in neuroscience and has also raised new questions in the study of the brain. Our hope and belief is that these are but the first conceptual and technical steps towards a comprehensive understanding of the most complex structure in the known universe.

REFERENCES

1. Ramon y Cajal S. Textura del Sistema Nervioso del Hombre y los Vertebrados. 1994 English translation by N and LW Swanson, Oxford University Press, New York, 1894–1904.
2. Squire L. (ed.) History of Neuroscience in Autobiography, Vol. 1–3. Academic Press, San Diego, CA, 1996–2001.
3. Polsson K. Chronology of Events in the History of Microcomputers from 1947 to 2000. Published online at http://www.islandnet.com/~kpolsson/comphist/ © 1995–2000 Ken Polsson.
4. Kurzweil R. The Law of Accelerating Returns. Published online at http://www.kurzweilai.net/articles/art0134.html © 2001 Raymond Kurzweil.
5. Li Z. Computational design and nonlinear dynamics of a recurrent network model of the primary visual cortex. Neural Comput 2001; **13**:1749–1780.
6. Riesenhuber M., Poggio T. Models of object recognition. Nat Neurosci 2000; Suppl:1199–1204.
7. Deco G., Zihl J. A neurodynamical model of visual attention: feedback enhancement of spatial resolution in a hierarchical system. J Comput Neurosci 2001; **10**:231–253.
8. Ohzawa I. Mechanisms of stereoscopic vision: the disparity energy model. Curr Opin Neurobiol 1998; **8**:509–515.
9. Findlay JM, Walker R. A model of saccade generation based on parallel processing and competitive inhibition. Behav Brain Sci 1999; **22**:661–674; discussion 674–721.
10. Poirazi P, Mel BW. Impact of active dendrites and structural plasticity on the memory capacity of neural tissue. Neuron 2001; **29**:779–796.
11. Schultz SR, Rolls ET. Analysis of information transmission in the Schaffer collaterals. Hippocampus 1999; **9**:582–598.
12. Wallenstein GV, Hasselmo ME. GABAergic modulation of hippocampal population activity: sequence learning, place field development, and the phase precession effect. J Neurophysiol 1997; **78**:393–408.
13. Traub RD, Jefferys JG, Whittington MA. Functionally relevant and functionally disruptive (epileptic) synchronized oscillations in brain slices. Adv Neurol 1999; **79**:709–724.
14. Britton NF, Skevington SM. On the mathematical modelling of pain. Neurochem Res 1996; **21**:1133–1140.
15. Koch C, Segev I. (eds.) Methods in Neuronal Modeling: From Ions to Networks. 2nd ed. MIT Press, Cambridge, MA, 1998.
16. Bower JM, Beeman D. (eds.) The Book of GENESIS: Exploring Realistic Neural Models with the GEneral NEural SImulation System. Springer-Verlag, New York, 1998.
17. Inoue T, Watanabe S, Kawahara S, Kirino Y. Phase-dependent filtering of sensory information in the oscillatory olfactory center of a terrestrial mollusk. J Neurophysiol 2000; **84**:1112–1115.
18. Pouill F, Scanziani M. Enforcement of temporal fidelity in pyramidal cells by somatic feed-forward inhibition. Science 2001; **293**:1159–1163.
19. Koch C, Segev I. The role of single neurons in information processing. Nat Neurosci 2000; Suppl.:1171–1177.
20. Gordon-Weeks PR. RNA transport in dendrites. Trends Neurosci 1988; **11**:342–343.

21. Newman EA. Calcium signaling in retinal glial cells and its effect on neuronal activity. Prog Brain Res 2001; **13**:241–254.
22. Stuart G, Spruston N, Hausser M. (eds.) Dendrites. Oxford University Press, New York, 1999.
23. Yuste R, Bonhoeffer T. Morphological changes in dendritic spines associated with long-term synaptic plasticity. Annu Rev Neurosci 2001; **24**:1071–1089.
24. Segal M. Dendritic spines for neuroprotection: a hypothesis. Trends Neurosci 1995; **18**:468–471.
25. Harris KM. Structure, development, and plasticity of dendritic spines. Curr Opin Neurobiol 1999; **9**:343–348.
26. Sorra KE, Harris KM. Overview on the structure, composition, function, development, and plasticity of hippocampal dendritic spines. Hippocampus 2000; **10**:501–511.
27. Capani F, Martone ME, Deerinck TJ, Ellisman MH. Selective localization of high concentrations of F-actin in subpopulations of dendritic spines in rat central nervous system: a three-dimensional electron microscopic study. J Comp Neurol 2001; **435**:156–170.
28. Pak DT, Yang S, Rudolph-Correia S, Kim E, Sheng M. Regulation of dendritic spine morphology by SPAR, a PSD-95-associated RapGAP. Neuron 2001; **31**:289–303.
29. Cannon RC, Turner DA, Pyapali GK, Wheal HV An on-line archive of reconstructed hippocampal neurons. J Neurosci Methods 1998; **84**:49–54.
30. Kaufman L, Rousseeuw PJ. Finding Groups in Data: An Introduction to Cluster Analysis. Wiley & Sons, New York, 1990.
31. Costa Ld, Velte TJ. Automatic characterization and classification of ganglion cells from the salamander retina. J Comp Neurol 1999; **404**:35–51.
32. Hillman DE. Neuronal shape parameters and substructures as a basis of neuronal form. In: The Neurosciences, Fourth Study Program (Schmitt F, ed.), MIT Press, Cambridge, MA, 1979:, pp. 477–498.
33. Hillman DE. Parameters of dendritic shape and substructure: intrinsic and extrinsic determination? In: Intrinsic Determinants of Neuronal Form and Function (Lasek RJ, Black MM, eds.), Liss, New York, 1988, pp. 83–113.
34. Ascoli GA. Progress and perspectives in computational neuroanatomy. Anat Rec 1999; **257**:195–207.
35. Ascoli GA, Krichmar JL, Scorcioni R, Nasuto SJ, Senft SL. Computer generation and quantitative morphometric analysis of virtual neurons. Anat Embryol 2001; **204**:283–301.
36. Burke RE, Marks WB, Ulfhake B. A parsimonious description of motoneurons dendritic morphology using computer simulation. J Neurosci 1992; **12**:2403–2416.
37. Tamori Y. Theory of dendritic morphology. Physiol Rev E 1993; **48**:3124–3129.
38. Nowakowski RS, Hayes NL, Egger MD. Competitive interactions during dendritic growth: a simple stochastic growth algorithm. Brain Res 1992; **576**:152–156.
39. Ascoli GA, Krichmar JL. L-Neuron: a modeling tool for the efficient generation and parsimonious description of dendritic morphology. Neurocomputing 2000; **32–33**:1003–1011.
40. Carriquiry AL, Ireland WP, Kliemann W, Uemura E. Statistical evaluation of dendritic growth models. Bull Math Biol 1991; **53**:579–589.
41. Li GH, Qin CD. A model for neurite growth and neuronal morphogenesis. Math Biosci 1996; **132**:97–110.
42. Van Pelt J, Uylings HBM. Natural variability in the geometry of dendritic branching patterns. In: Modeling in the Neurosciences from Ionic Channels to Neural Networks (Poznanski RR, ed.), Harwood Academic Publishers, Amsterdam, NL, 1999, pp. 79–108.
43. Senft SL and Ascoli GA. Reconstruction of brain networks by algorithmic amplification of morphometry data. Lect Notes Comp Sci 1999; **1606**:25–33.
44. Winslow JL, Jou SF, Wang S, Wojtowicz JM. Signals in stochastically generated neurons. J Comput Neurosci 1999; **6**:5–26.

45. Cullheim S, Fleshman JW, Glenn LL, Burke RE. Three-dimensional architecture of dendritic trees in type-identified alpha-motoneurons. J Comp Neurol 1987; **255**:82–96.
46. Jones CL and Jelinek HF. Wavelet packet fractal analysis of neuronal morphology. Methods 2001; **24**:347–358.
47. Claiborne BJ, Amaral DG, Cowan WM. Quantitative, three-dimensional analysis of granule cell dendrites in the rat dentate gyrus. J Comp Neurol 1990; **302**:206–219.
48. Senft SL. A statistical framework to present developmental neuroanatomy. In: Neural-Network Models of Cognition: Biobehavioral Foundations (Donahoe JW, ed.), Elsevier Press, New York, 1997.
49. Ascoli GA, Krichmar JL, Nasuto SJ, Senft SL. Generation, description and storage of dendritic morphology data. 2001; Philos Trans R Soc Lond B Biol Sci **356**:1131–1145.
50. Ascoli GA, Samsonovich A. Bayesian morphometry of hippocampal cells suggests same-cell somatodendritic repulsion. Advances in Neural Information Processing Systems 14. 2002; in press.
51. Scorcioni R, Ascoli GA. Algorithmic extraction of morphological statistics from electronic archives of neuroanatomy. Lect Notes Comp Sci 2001; **2084**:30–37.
52. Mitchison G. Neuronal branching patterns and the economy of cortical wiring. Proc R Soc Lond B Biol Sci 1991; **245**:151–158.
53. Condron BG, Zinn K. Regulated neurite tension as a mechanism for determination of neuronal arbor geometries in vivo. 1997; Curr Biol **7**:813–816.
54. Cherniak C. Neural component placement. Trends Neurosci 1995; **18**:522–527.
55. Cherniak C, Changizi MA, Kang D. Large-scale optimization of neuron arbors. Physiol Rev E 1999; **59**:6001–6009.
56. Hines ML, Carnevale NT. NEURON: a tool for neuroscientists. Neuroscientist 2001; **7**:123–135.
57. Goddard N, Hood G, Howell F, Hines M, De Schutter E. NEOSIM: Portable large-scale plug and play modelling. Neurocomputing 2001; **38–40**:1657–1661.
58. Mirsky JS, Nadkarni PM, Healy MD, Miller PL, Shepherd GM. Database tools for integrating and searching membrane property data correlated with neuronal morphology. J Neurosci Methods 1998; **82**:105–121.
59. Mainen ZF, Sejnowski T. Influence of dendritic structure on firing pattern in model neocortical neurons. Nature 1996; **382**:363–366.
60. Vetter P, Roth A, Hausser M. Propagation of action potentials in dendrites depends on dendritic morphology. J Neurophysiol 2001; **85**:926–937.
61. Duijnhouwer J, Remme MWH, van Ooyen A, van Pelt J. Influence of dendritic topology on firing patterns in model neurons. Neurocomputing 2001; **38–40**:183–189.
62. Krichmar JL, Nasuto SJ, Scorcioni R, Washington SD, Ascoli GA. Influence of dendritic morphology on CA3 pyramidal cell electrophysiology: a simulation study. Brain Research 2002; In press.
63. Nasuto SJ, Knape R, Krichmar JL, Ascoli GA. A computational study of the relationship between neuronal morphology and electrophysiology in an Alzheimer's disease model. Neurocomputing **38–40**:1477–1487.
64. De Schutter E. (ed.) Computational Neuroscience: Realistic Modeling for Experimentalists. CRC Press, Boca Raton, FL, 2000.
65. Nasuto SJ, Krichmar JL, Scorcioni R, Ascoli GA. Algorithmic analysis of electrophysiological data for the investigation of structure-activity relationship in single neurons. InterJournal Complex Syst **389**:1–7.
66. Hubel D, Wiesel T. Receptive fields, binocular interaction and functional architecture in the cat's visual cortex. J Physiol 1962; **160**:106–154.
67. Toga AW, Mazziotta JC, Frackowiak RSJ. (eds.) Brain Mapping: The Trilogy, 3 Vol. Set. Academic Press, San Diego, CA, 2000.

68. Paydar S, Doan CA, Jacobs GA. Neural mapping of direction and frequency in the cricket cercal sensory system. J Neurosci 1999; **19**:1771–1781.
69. Jacobs GA, Theunissen FE. Extraction of sensory parameters from a neural map by primary sensory interneurons. J Neurosci 2000; **20**:2934–2943.
70. Zaborszky L, Duque A. Local synaptic connections of basal forebrain neurons. Behav Brain Res 2000; **115**:143–158.
71. Zaborszky L, Pang K, Somogyi J, Nadasdy Z, Kallo I. The basal forebrain corticopetal system revisited. Ann NY Acad Sci 1999; **877**:339–367.
72. Leergaard TB, Alloway KD, Mutic JJ, Bjaalie JG. Three-dimensional topography of corticopontine projections from rat barrel cortex: correlations with corticostriatal organization. J Neurosci 2000; **20**:8474–8484.
73. Leergaard TB, Lyngstad KA, Thompson JH, Taeymans S, et al. Rat somatosensory cerebropontocerebellar pathways: spatial relationships of the somatotopic map of the primary somatosensory cortex are preserved in a three-dimensional clustered pontine map. J Comp Neurol 2000; **422**:246–266.
74. van Ooyen A. Competition in the development of nerve connections: a review of models. Network 2001; **12**:R1–R47.
75. Mori S, Barker PB. Diffusion magnetic resonance imaging: its principle and applications. Anat Rec 1999; **257**:102–109.
76. Stieltjes B, Kaufmann WE, van Zijl PC, et al. Diffusion tensor imaging and axonal tracking in the human brainstem. NeuroImage 2001; **14**:723–735.
77. Hilgetag CC, Burns GA, O'Neill MA, Scannell JW, Young MP. Anatomical connectivity defines the organization of clusters of cortical areas in the macaque monkey and the cat. Philos Trans. R Soc Lond B Biol Sci 2000; **355**:91–110.
78. Sporns O, Tononi G, Edelman GM. Connectivity and complexity: the relationship between neuroanatomy and brain dynamics. Neural Netw 2000; **13**:909–922.
79. Press WA, Olshausen BA, Van Essen DC. A graphical anatomical database of neural connectivity. Philos Trans R Soc Lond B Biol Sci 2001; **356**:1147–1157.
80. Burns GA, Young MP. Analysis of the connectional organization of neural systems associated with the hippocampus in rats. Philos Trans R Soc Lond B Biol Sci 2000; **355**:55–70.
81. Burns GA. Neuroscholar 1.00, a neuroinformatics databasing website. Neurocomputing 1999; **26–27**:963–970.
82. Goodhill GJ, Richards LJ. Retinotectal maps: molecules, models and misplaced data. Trends Neurosci 1999; **22**:529–534.
83. Dickson J, Drury H, Van Essen DC. 'The surface management system' (SuMS) database: a surface-based database to aid cortical surface reconstruction, visualization and analysis. Philos Trans R Soc Lond B Biol Sci 2001; **356**:1277–1292.
84. Dale AM, Halgren E. Spatiotemporal mapping of brain activity by integration of multiple imaging modalities. Curr Opin Neurobiol 2001; **11**:202–208.
85. Wang Y, Gupta A, Markram H. Anatomical and functional differentiation of glutamatergic synaptic innervation in the neocortex. J Physiol Paris 1999; **93**:305–317.
86. Gupta A, Wang Y, Markram H. Organizing principles for a diversity of GABAergic interneurons and synapses in the neocortex. Science 2000; **287**:273–278.
87. Kotter R. Neuroscience databases: tools for exploring brain structure-function relationships. Philos Trans R Soc Lond B Biol Sci 2001; **356**:1111–1120.
88. Howell FW, Dyhrfjeld-Johnsen J, Maex R, Goddard N, De Schutter E. A large-scale model of the cerebellar cortex using PGENESIS. Neurocomputing 2000; **32–33**:1041–1046.
89. Scorcioni R, Bouteiller JM, Ascoli GA. A real-scale anatomical model of the dentate gyrus based on single cell reconstructions and 3D rendering of a brain atlas. Neurocomputing, 2002; in press.

90. Tamamaki N, Nojyo Y. Projection of the entorhinal layer II neurons in the rat as revealed by intracellular pressure-injection of neurobiotin. Hippocampus 1993; **3**:471–480.
91. Dhenain M, Ruffins SW, Jacobs RE. Three-dimensional digital mouse atlas using high-resolution MRI. Dev Biol 2001; **232**:458–470.
92. Lester, DS, Lyon RC, McGregor GN, et al. 3-Dimensional visualization of lesions in rat brain using magnetic resonance imaging microscopy. Neuroreport 1999; **10**:737–741.
93. Lester DS, Pine PS, Delnomdedieu M, Johannessen JN, Johnson GA. Virtual neuropathology: three-dimensional visualization of lesions due to toxic insult. Toxicol Pathol 2000; **28**:100–104.
94. Nickerson LD, Martin CC, Lancaster JL, Gao JH, Fox PT A tool for comparison of PET and fMRI methods: calculation of the uncertainty in the location of an activation site in a PET image. NeuroImage 2001; **14**:194–201.
95. Chen JW, O'Farrell AM, Toga AW. Optical intrinsic signal imaging in a rodent seizure model. Neurology 2000; **55**:312–315.
96. Shankle WR, Romney AK, Landing BH, Hara J. Developmental patterns in the cytoarchitecture of the human cerebral cortex from birth to 6 years examined by correspondence analysis. Proc Natl Acad Sci USA 1998; **95**:4023–4028.
97. Shankle WR, Rafii MS, Landing BH, Fallon JH. Approximate doubling of numbers of neurons in postnatal human cerebral cortex and in 35 specific cytoarchitectural areas from birth to 72 months. Pediatr Dev Pathol 1999; **2**:244–259.
98. Haykin SS. Neural Networks: A Comprehensive Foundation. Prentice Hall, U. Saddle River, NJ, 1998.
99. Ascoli GA. The complex link between neuroanatomy and consciousness. Complexity 2000; **6**:20–26.

Part I

2
Some Approaches to Quantitative Dendritic Morphology

Robert E. Burke and William B. Marks

"If ye canna mak a model, then ye dinna understand it."
(Attributed to Lord Kelvin)

ABSTRACT

The availability of powerful desktop computers and of a large amount of detailed data about the morphology of a wide variety of neurons has led to the development of computational approaches that are designed to synthesize such data into biologically meaningful patterns. The hope is, of course, that the emerging patterns will provide clues to the factors that control the formation of neuronal dendrites during development, as well as their maintenance in the adult animal. One class of approaches to this problem is to develop quantitative computational models that can reproduce as many aspects of the original data as possible. The development of such simulations requires analysis of the original data that is directed by the model requirements, and their relative success depends on detailed comparisons between model outputs and the original data sets. Refinement of the models may require not only new experiments, as in other scientific disciplines, but also new ways of looking at the data already in hand. This chapter discusses some examples of this process, with emphasis on spinal motoneurons.

2.1. INTRODUCTION

Dendrites are critical to the processing of synaptic information in central nervous system neurons. Accordingly, there is considerable interest in their structure and function (1). Striking differences in neuron morphologies have been known for over a century (2,3). Such differences must be related to factors that govern the development and maintenance of their dendritic architectures. Over the past two decades, the development of intracellular labeling techniques using horseradish peroxidase (HRP) and biocytin (4,5), combined with computer-assisted methods for quantitative reconstruction of labeled neurons (6,7), have led to a large output of quantitative data about the morphology of dendrites. We now face the problem of how to reduce the mass of information, usually contained in multiple graphs and tables, into patterns that reveal their underlying meaning, without losing essential information. Hillman *(8)* called these patterns "fundamental parameters of form", but the underlying factors that produce them are not obvious upon inspection of the raw data. This chapter considers some approaches to this problem.

Two issues require some comment before proceeding. First, it has become clear in recent years that neuronal dendrites are not static structures; rather they can exhibit dynamic changes that presumably reflect functional changes in the nervous system *(9)*. Thus, the data about dendritic structure that are obtained by conventional neuroanatomical methods represent snapshots that may not be entirely representative. Second, there are a large number of practical difficulties inherent in gathering quantitative measurements of neuronal dendrites using conventional light microscopy (see *[10]*). These include factors such as tissue shrinkage, operator error, and the limited resolution of the light microscope, which is the only practical approach to reconstructing large neurons from serial sections. These sources of potential error must be kept firmly in mind when evaluating existing data, particularly when data from different sources are combined. Other options such as confocal microscopy of neurons filled with fluorescent tracers could, in principle, be more accurate and might even contribute an element of automation to the reconstruction process. However, technical problems, such as tracer bleaching, have confined most reconstruction efforts to more permanent forms of tracers, like HRP or biocytin, to be examined with conventional light microscopy.

With regard to data analysis, it is also important to remember that the process of quantitative reconstruction splits the continuous structure of the dendrite into discrete pieces, here referred to as "segments", each with a specified diameter and length. These discrete cylinders, plus information that identifies their positions within the dendritic tree, make up the usual computer data files. The position coding systems often vary between data sources, as does the presence or absence of information about the 3D location of each cylinder.

2.2. TWO-DIMENSIONAL ANALYSIS OF DENDRITES IN ISOLATED NEURONS

Quantitative analysis of neuronal dendrites began before the age of computers with the work of Sholl *(11)*, who plotted dendritic branching patterns of Golgi-stained cortical neurons in terms of distance from the soma. This straightforward approach guided many subsequent studies that used improved methods for intracellular staining of identified neurons, resulting in ever larger volumes of quantitative data about branch diameters and lengths, branching orders, and the locations of branching points and terminations, all considered as functions of distance from the soma (e.g., Fig. 1; see also *[12,13]*). An alternative approach, focused on dendritic branching patterns *per se*, concentrates on the topological complexity of trees from different types of neurons *(14–16)* (see Chapter 11). Neither approach takes account of the 3D tree structures.

Hillman *(8,17)* proposed that dendritic architecture can provide clues to the biological factors that control the multiplicity of neuronal shapes. Hillman's seven fundamental parameters that describe the morphology of a dendritic tree are: (*i*) stem diameter; (*ii*) terminal branch diameters; (*iii*) branch taper; (*iv*) branch lengths; (*v*) branch power (the relation between diameters of the parent branch with those of its two daughters);

Quantitative Dendritic Morphology

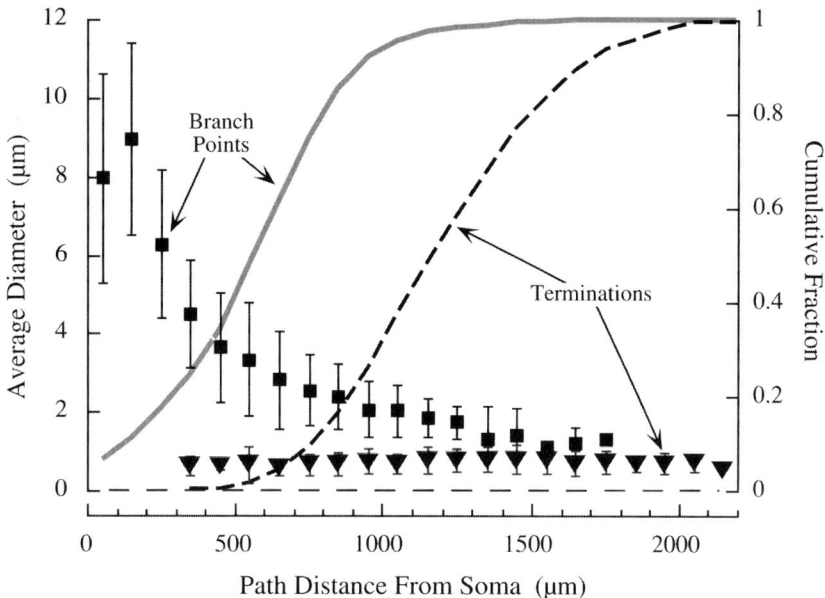

Fig. 1. Some features of cat α-motoneuron dendrites plotted as functions of the somatofugal distance along dendritic paths (abscissa). The solid and dashed lines show the locations of branching points and terminations, respectively, as cumulative fractions. The symbols indicate the average diameters (± one SD) of all branches that end in branching points (parent branches; solid squares) or terminations (solid triangles) within 100 μm bins of path distance. Data from 6 cat lumbosacral motoneurons reported in *(12)*.

(vi) the ratio of the two daughter branch diameters at each branch point; and *(vii)* the spatial orientation of branches. Hillman suggested that the cytoskeleton, particularly microtubule arrays, is critical to the control of dendritic architecture *(17)* (but cf *[18]*). The interrelations that he described implied that some combination of intrinsic factors could be used to simulate virtual dendrites that could be compared with real ones. It seemed possible that some correlation, which may be found among the descriptors, may arise as epiphenomena that depend on underlying mechanisms (see also *[19]*).

Building on this idea, we proposed an approach to this problem that began by analyzing correlations in quantitative data about completely reconstructed dendritic trees of a sample of cat α-motoneurons, in order to develop algorithms and appropriate data sets that might be used to construct virtual dendrite simulations *(20)*. We reasoned that a computational machine (algorithms plus parameters), which can construct virtual dendrites that reproduce not only the averages but also the variances of data from actual neurons, must contain all of the essential information inherent in that data set (see also *[21]*). There are a large number of correlations to choose from, and several different approaches were explored. All used a Monte Carlo simulation in order to generate the stochastic variations found in the observations.

The most successful algorithm was based on the relationship between the starting diameter and the length of dendritic branches, which are defined as beginning with the

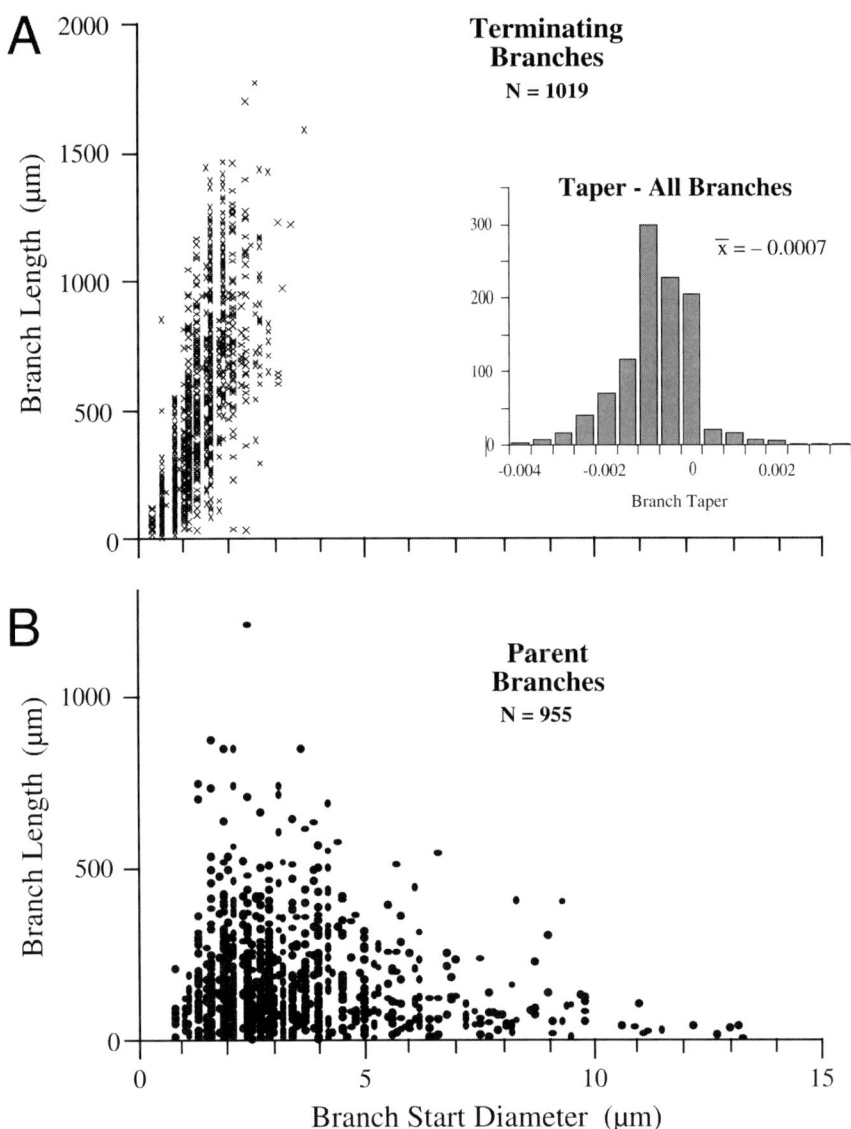

Fig. 2. Scatter plots of the lengths (ordinates) of terminating (**A**) and parent branches (**B**) in relation to their starting diameters (abscissae). The correlation between branch length and starting diameter was positive for terminating branches and negative for parents. The inset in panel **A** shows a histogram of branch taper, which had a mean of –0.0007 ± 0.0027 μm/μm. Adapted with permission from Figure 1 in *(20)*.

soma or a branch point and ending either at another branch point or at a termination (Fig. 2; see also Fig. 5 in *[8]*). As in most reconstructed neurons, the branches consisted of a sequence of segments, often with different diameters due to dendritic taper. Despite a great deal of scatter, the lengths of branches that ended at a branching point (called "parent" branches) varied inversely with their starting diameters, while the relation was direct in the case of terminating branches. It was also clear that the diam-

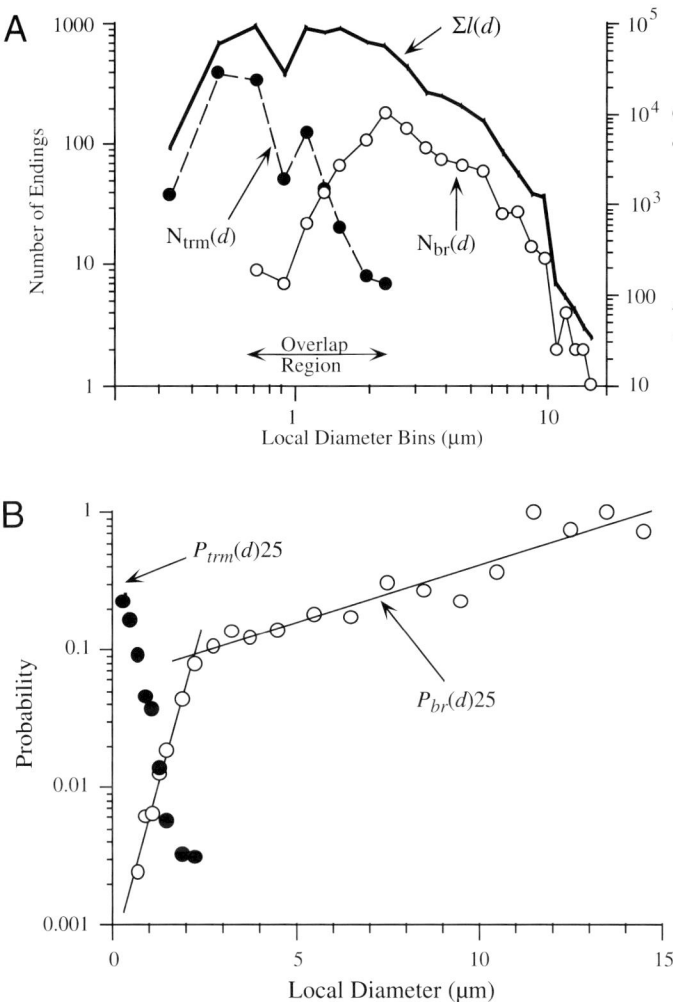

Fig. 3. Calculation of the probabilities of branching or termination as functions of local branch diameter [$p_{br}(d)$ and $p_{trm}(d)$, respectively]. **(A)** Log-log plot of the total length of dendritic material [$\Sigma l(d)$; solid line, referred to the right ordinate] and the numbers of branch segments giving rise to branching points [$N_{br}(d)$] or terminations [$N_{trm}(d)$], as functions of local diameter (d), binned by 0.25 μm for d ≤ 2.0 μm, 0.5 μm for d > 2.0 to ≤ 4.0 μm, and 1.0 for d > 4.0 μm. Segments with diameters between 0.8 and 2.2 μm (overlap region) could either branch or terminate. **(B)** Semilog plot of the ratios of the numbers of branch points or termination, divided by $\Sigma l(d)$, to give $p_{br}(d)$ and $p_{trm}(d)$, respectively. In this graph, $p_{br}(d)$ and $p_{trm}(d)$ are multiplied by Δl of 25 μm, which was the value of Δl used for simulations. Adapted with permission from Figure 2 in *(20)*.

eters of most individual branches became smaller in the somatofugal direction (i.e., they exhibited negative taper).

In order to simulate the wide scatter in branch lengths, which was as large for individual motoneurons as for the pooled data (Fig. 2), we used a stochastic (Monte Carlo) growth algorithm in which each increment in branch length, Δl, was controlled by the probability that it ended in a branching point or in a termination (p_{br} or p_{trm}, respectively). If neither was true, the branch continued to lengthen by Δl. The observations suggested that both p_{br} and p_{trm} depend in some way on branch diameter. In order to explore this possibility, we binned all of the branch segments by increments of diameter, d, and summed the total length $\Sigma l(d)$ in each diameter bin (Fig. 3A). The numbers of segments in each diameter bin that ended in either a branch point, $N_{br}(d)$ or a termination, $N_{tr}(d)$, were then divided by $\Sigma l(d)$ to give $p_{br}(d)$ and $p_{trm}(d)$ per unit length as functions of local segment diameter bins:

$$p_{br}(d) = \frac{N_{br}(d)}{\Sigma l(d)} \quad \text{and} \quad p_{trm}(d) = \frac{N_{trm}(d)}{\Sigma l(d)}$$

As shown in Fig. 3B, the resulting probabilities [$p_{br}(d) \Delta l$ and $p_{trm}(d) \Delta l$ with $\Delta l = 25$ µm] were well fitted by exponential functions of local diameter, d, with positive exponents for $p_{br}(d)$ and negative for $p_{trm}(d)$. Within the range of d where branches either branched again or terminated (overlap region; $d = 0.7$ to 2), the slope for $p_{br}(d)$ was much steeper and approx the inverse of the slope for $p_{trm}(d)$ (2.2 vs –2.9, respectively; see Table 1 in [30]).

To start the simulation process, the diameter of the first segment in the branch and a step length (Δl) are specified, and the computer compared p_{br} or p_{trm}, (multiplied by Δl and randomized as to which was tested first in order to eliminate bias) with a uniformly distributed random number, rnd between zero and 1. If this $p_x < rnd$, the branch ended appropriately. If neither event occurred, the branch extended by Δl, d_{i+1} changed by a selected value for taper ($\Delta d/\Delta l$), and the process was iterated until the branch either terminated or produced a branching point. The value of taper used was the only free parameter in the system, because it was difficult to specify a single value from the wide observed distribution (Fig. 2, inset). The algorithm produced parent and terminating branches with the observed length distributions (Fig. 2) with reasonable fidelity, given that a single value of taper was used for all branches. The value of taper that produced the least error in this simulation was –0.00125 µm/µm (Fig. 5 in [20]), which was comparable with the observed value (Fig. 2A, inset), given that taper is extremely difficult to measure with any accuracy.

Of course, this was only half of the process required to produce virtual trees; one must also specify the diameters for the daughter branches that are generated when a parent branch gives rise to a branching point. In real motoneurons, the average ratio between the sum of the daughter branch diameters (d_1 and d_2), raised to the 3/2 power, divided by the 3/2 power of the parent diameter (d_{par}), is slightly larger than 1.0 (actually 1.1), but shows wide variations (see Fig. 7 in [12]). Like most neurons, branch points in motoneuron dendrites give rise to only two daughter branches, and their diameters are negatively correlated (Fig. 4A). This relation was independent of both position in the tree and the end diameter, d_{par}, of the parent branch.

The distributions of d_1 and d_2 were Gaussian and exhibited similar means and standard deviations (SD). One of us (W.B.M.) devised a way to combine their values, using their observed correlation coefficient $\beta = -0.4$, into a single distribution, r, which was well approximated by a Gaussian with the same mean and SD (Fig. 4B). Drawing two numbers independently from r, multiplying this pair by the matrix {{1, –.2087}, {–.2087, 1}}, and then by d_{par}, produced a pair of daughter branch diameters that,

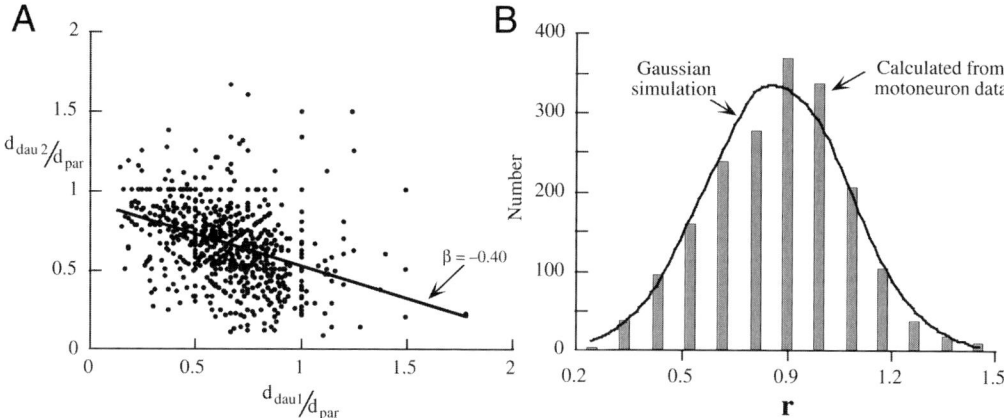

Fig. 4. Database for choosing diameters of daughter branches at simulated branch points. **(A)** Scatter plot showing the negative correlation between the diameters, d_1 and d_2, of the two daughter branches at 955 branching points, each normalized by the end diameter of the parent branch. The slope of the linear correlation, $\beta = 0.40$, was used to construct a distribution, **r**, that preserved the statistical relations between d_1 and d_2 (see [20]). The value of β was the same whether or not the data were shuffled. **(B)** The calculated **r** distribution (bars) were fit to a Gaussian function (solid curve) with the same mean and SD. The continuous function was used for generating virtual dendrites (see [20]). Adapted with permission from Figure 6 in (20).

when repeated, had the same means, SDs, and correlation coefficient as the observed daughter branch diameters.

A program was written that combined the branch growth algorithm discussed above with the algorithm for determining daughter branch diameters in order to construct virtual dendrites with a selected starting diameter, d_{stem}, that matched diameters of observed dendrites. The experimental database included 64 fully reconstructed dendrites from 6 α-motoneurons, with d_{stem} ranging from 2 to 18 μm. Each run of the program generated 64 virtual dendrites, using parameters based on the probabilities shown in Figure 3B and the algorithm for selecting daughter branch diameters (Fig. 4B). The program automatically calculated a wide variety of statistics about virtual dendrites that could be compared to their actual counterparts (e.g., Figs. 5 and 6).

Twenty simulations were run for several values of taper and each produced dendrite sets with different total numbers of terminations, indicative of the overall size of the simulated trees. There was a direct relation between taper and total termination numbers, in that more negative values of taper produced smaller trees. The value of taper used was adjusted empirically to produce dendrite sets with total termination numbers near that observed for the real motoneurons (n = 1974). A taper value of –0.0015 produced the closest approach to this number but individual runs varied rather widely.

This model system, referred to as "Model 1", produced virtual dendrites that matched many of the relations found in the actual database, including the distributions of total surface area and numbers of terminations in relation to d_{stem} for individual trees (see Fig. 7 in [20]) and the distributions of branch orders and diameters of parent and terminating branches as functions of distance from the soma (Fig. 5A, B, and D). None of

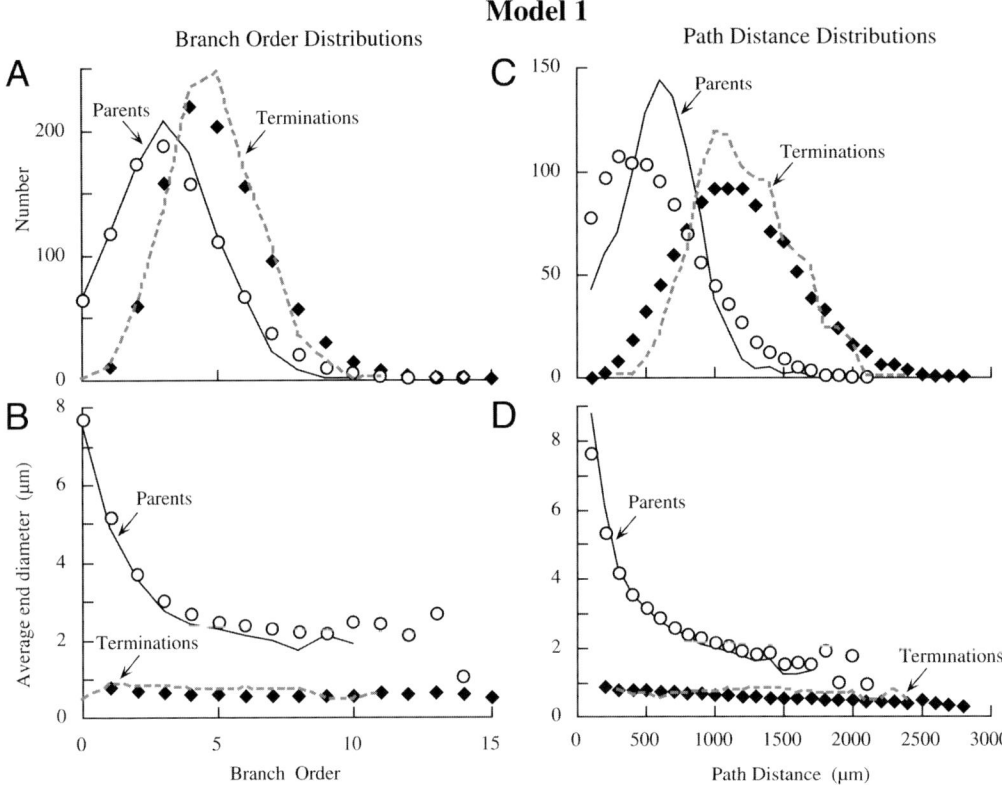

Fig. 5. Comparison of averaged data extracted from 64 simulated (symbols) and actual (lines) dendrites with the same set of stem diameters, using Model 1 simulations (see text). Note branches of order >10 in the simulations (**A** and **B**) and major discrepancies in the path distance locations of branch points and dendritic terminations (**C**). Adapted with permission from Figure 8 in (20).

these relations were built into the simulation algorithm; rather they are emergent properties that can serve as features for determination of goodness of fit. Although the path distance distributions of branch points and terminations were less satisfactory (Fig. 5C), this might be deemed good enough given the relative simplicity of the model.

This model result must be interpreted with caution. Although a growth model was used for these simulations in order to simulate the statistical variances in real dendrites, the results are a simulation of existing dendritic structures rather than of the dynamic processes that may have formed them. The computational machine and the parameters shown in Figures 3 and 4 should be thought of as a parsimonious description of the complex morphology of cat motoneuron dendrites that eliminates redundant information, rather than a model of how dendrites actually grow. Given this important distinction, the Model 1 results suggests that local branch diameter is a key factor that determines whether a given branch can or cannot maintain a branching point rather than terminating. This is consistent with the idea that the cytoskeleton, specifically the number of available microtubules (17), controls whether or not a given branch can give rise to a branching point, as well as how long parent and terminating branches can be.

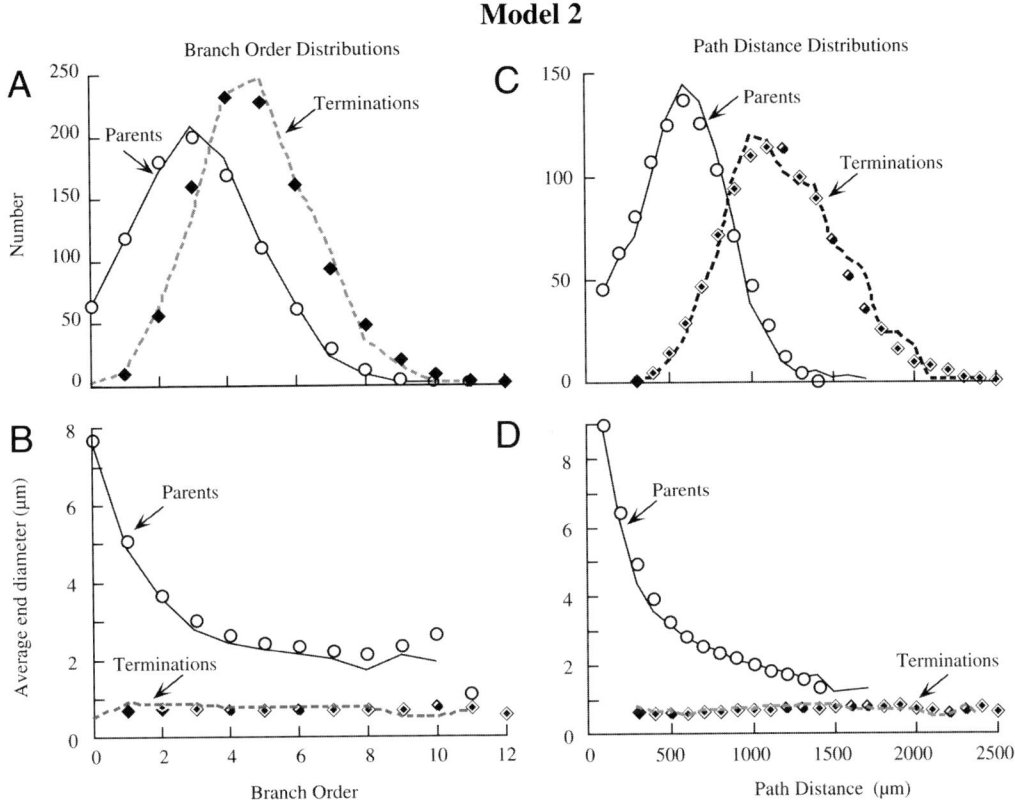

Fig. 6. As in Figure 5 but using Model 2 for simulations. Note lack of simulated dendrites with branch orders >11 (**A** and **B**) and much closer agreement between simulated and observed locations of branching points and terminations as functions of path distance from the soma (**C**). Adapted with permission from Figure 12 in *(20)*.

An alternative model, in which p_{br} and p_{trm} were functions of branch order alone rather than local diameter, produced poor fits to the actual data. Thus, observed correlations between branch order and dendritic architecture are probably epiphenomena.

2.3. HOW GOOD IS GOOD ENOUGH?

One of the more difficult questions in any computational model study is when to quit. The failure of Model 1 to completely reproduce the spatial distributions of branch points and terminations (Fig. 5C) suggested that it lacked some important factor. In addition, some runs of this model produced trees that were either much smaller or much larger than expected for the selected value of d_{stem}. The existence of "runaway" dendrites is evident in Figure 5B, which shows diameters of parent and terminating branches with branch orders >10 that are not found in real motoneurons. The richness of the database prompted us to explore other factors that might have accounted for these discrepancies.

The problem of trees that were too small or too large appeared to be caused by sequences of daughter branch diameters that, in rare instances, were either very small

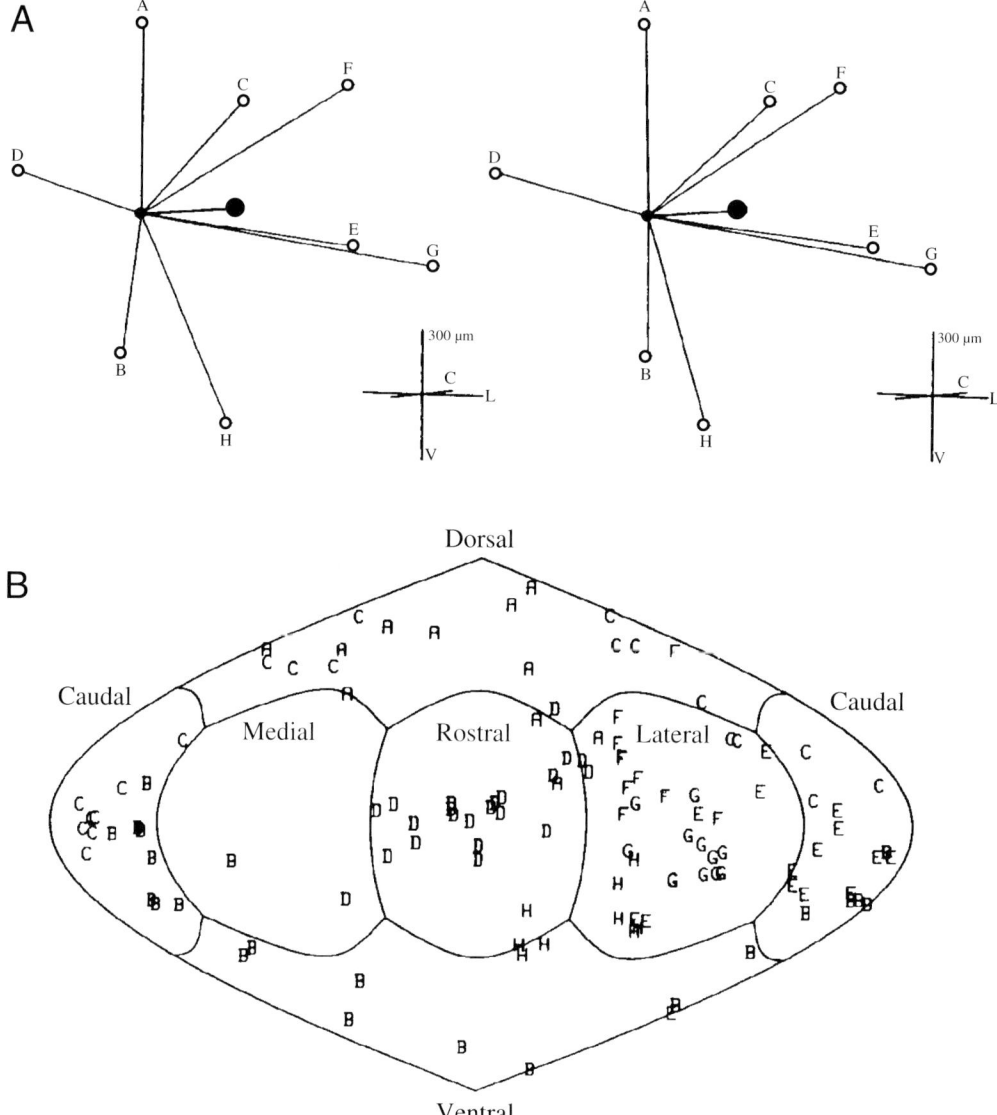

Fig. 7. Two simple approaches to quantitate the 3D morphology of a motoneuron. (**A**) Stereoscopic pair representation of the vectors of eight dendrites (labeled A through H) in terms of the centers of mass (open circles) of the membrane area in each dendrite, illustrating the projection of each dendrite away from the soma (small filled circles). The large filled circle represents the membrane area center of mass for the entire neuron. This was the most asymmetrical motoneuron of six similarly studied cells *(26)*. (**B**) 2D projection of the locations at which dendritic branches of the same cell as in panel **A** penetrate a spherical shell with radius 750 μm, centered on the soma. Individual branches of the different dendrites are labeled with the dendrite identification letter. The lines indicate the boundaries of six directional hexants, with the caudal hexant split in two. The mapping projection assigns approximately equal areas to each hexant region. Both panels adapted from Figure 6 in *(26)*.

or very large at successive branch points. This could occur by chance because the model had no memory of the preceding selections of d_1 and d_2. When we reexamined the experimental data, we found a small but significant dependence between the sum of d_1 and d_2 at a given branch point, normalized by the parent end diameter, d_{par}, and the starting diameter of the parent branch, normalized by the end diameter of its parent branch (i.e., the "grandparent" branch). This suggested the existence of a cytoskeletal or metabolic constraint on the size of the downstream subtrees. We implemented a "grandparent correction" based on the observed dependence, and this greatly reduced the occurrence of runaway trees, but neither this nor several other manipulations improved the discrepancy noted in Figure 5C (see [20] for details).

Another reexamination of the original data suggested that the values of $p_{br}(d)$ and/or $p_{trm}(d)$ might not be constant throughout the tree. Indeed, we found that both probabilities depended on the path distance, D, from the soma as well as on local diameter. Such a dependence could represent the metabolic cost of maintaining cytoskeleton at increasing distances away from the soma where the constituent proteins are generated. Estimation of this dependence was complicated by the problem of fitting smooth functions to the 3D surface described by the data points, now binned by both local diameter and path distance (see Fig. 10 in [20]). However, when equations for $p_{br}(d,D)$ and $p_{trm}(d,D)$ that fit the data were incorporated into the model, it produced trees that fit the observed data very well indeed (Fig. 6). It was necessary to analyze the data for dependence on d and D separately, because local diameters in individual branches were not strongly correlated with path distance, especially in the more distal parts of the trees (Fig. 1). These virtual dendrites also had branch topologies that fit those of actual motoneuron (Burke, Marks, and Ulfhake, unpublished). In this instance, all of the required parameters were intrinsic to the neuron itself. This may not always be true; in some cases external factors could be required in order to generate acceptable simulations (e.g., [22]).

The lesson for us in this work was that it is sometimes useful to extend a model that is reasonably good to one that is better, provided that the additions accurately reflect features that are in fact present in the original data. Each elaboration of the present model revealed factors that appear to be biologically relevant. The additional features were not at all obvious and emerged only after specifically tailored data extraction methods were employed. Indeed, the utility of quantitative biological models lies precisely in the fact that they force the investigator to search for relations that are hidden within the existing data or to guide the experiments that can generate the necessary new information.

2.4. NEURONS IN THREE DIMENSIONS

2D morphological data are relatively tractable for computational modeling, as exemplified by the discussion so far. However, it is considerably more difficult to extend such approaches to neurons as 3D entities *(17)*. The overall shape of neuronal dendritic trees have been analyzed by statistical methods, such as principle components *(23,24)*, and by a Fourier transform technique that can give concise information about the density of branches distributed in 3D space *(25)*. Cullheim and colleagues *(26)* introduced a simpler approach that tabulated the spatial distribution of the dendritic membrane area or branch volume within six "hexants", or subdivisions thereof, within an external coordinate system centered on the neuron soma and oriented by anatomical axes. These authors also used a variation of principle component approach to calculate the spatial locations of the centers of mass (COM) for membrane area or branch volume for individual dendrites, as well as for the neuron as a whole (Fig. 7A). On average, cat triceps surae motoneurons were found to be more or less radially symmetrical, although the dendrites projecting dorsally and ventrally tended to be slightly smaller than those that projected in the other directions. All of these methods provide ways to document the degree to which neuronal trees are polarized, either because of proximity to natural boundaries or, perhaps, to concentrated sources of synaptic input.

Another approach to the problem of analyzing the 3D spatial organization of dendrites is to map the spatial positions of their branches as they penetrate 2D spherical surfaces ("shells") located at different distances from the soma (Fig. 7B; *[26]*). Although such shell maps are on spherical surfaces, they are tractable for quantitation by spherical trigonometry. Questions such as the size of dendritic territories and how much they overlap can be approached by conventional nearest-neighbor or tessellation analyses to examine spatial clustering. Convex hulls (polygons with no internal angle <180° that encloses the target set of points) are computationally convenient, although they often include empty regions that properly belong to other dendrites. The disadvantage of such maps is that they do not lead to simplification of the data set or to identification of general principles that might be at work.

2.4.1. Building Three-Dimensional Dendrites

Renewed interest in computational neuroanatomy *(21)*, as well as the appearance of relevant software tools *(27,28)*, has stimulated the development of new approaches to quantitate neuronal morphology in 3D space that involve simulation. In an earlier section, we adopted the view that the simplest computational machine that can accurately reproduce a set of complex objects constitutes the most concise description of those objects. This philosophy predicts that new information may emerge if we can construct algorithms that can convert 2D dendrograms into statistically accurate 3D trees. At minimum, such a simulation requires extraction of two sets of data from the original morphological files: (*i*) the distributions of angles at which daughter branches emerge from branching points; and (*ii*) measurement of the degree to which individual branches meander (i.e., change in vectorial orientation) in space *(8)*. In principle, these data can be estimated from cells that have been digitized with sufficient spatial resolution.

As an example, we will consider here some possible approaches to measurement of daughter branch angles. The first decision is how to define the vectorial directions of daughter branches away a given branching point. There are at least three possibilities: (*i*) use the coordinates of the first digitized segment of each daughter branch (initial branch direction); (*ii*) use the coordinates of the point at which the branch ends (final branch direction); or (*iii*) use the least-squares fit to each meandering branch, perhaps weighted by local membrane area or segment volume (average branch direction). Because we were interested in the global shape of trees, we initially used the final branch directions.

The next issue is to define the frame of reference for measuring branch angles. An obvious choice is to calculate angular deviations from the parent branch direction.

Quantitative Dendritic Morphology 39

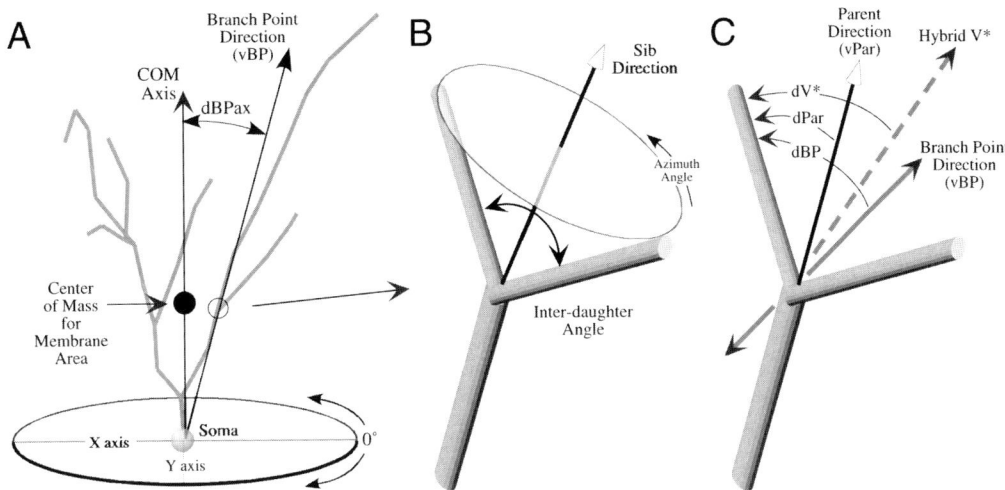

Fig. 8. Diagram to illustrate possible methods to calculate dendritic branching angles. **(A)** A motoneuron dendrite is shown after rotation into a Cartesian coordinate space, aligned in the Z (vertical) axis by the vector from the soma center to the COM for dendritic membrane area (filled circle). The direction of each branch point with respect to the COM axis can be specified by its dBPax away from the COM axis. Its horizontal position is described by an azimuth angle in the XY plane with respect to a reference direction (e.g., rostral). **(B)** The orientation of daughter branches at each branch point can be described by the angle between them (Interdaughter angle) and the vector midway between them (Sib direction). The Sib vector can be viewed as the axis of a cone around which the daughter branches can rotate in the perpendicular (azimuth) plane. **(C)** The direction of the individual daughter branches can be defined as the angle of deviation away from the branch point direction (dBP) or the direction of the parent branch (dPar). The most successful 3D simulations were obtained when daughter branch deviation angles (dV*) were referenced to a vector representing a linear combination of the branch point and parent directions (Hybrid V*; see text).

However, a 3D simulation algorithm, based on parent branch vectors, produced simulated trees that often had much larger lateral spread than natural cat motoneuron dendrites (Burke and Marks, unpublished), because such data has no relation to the shape of the tree as a whole. In fact, it can be shown that using only the parent branch direction leads to a 3D random walk. As discussed in the chapter by Ascoli (see also *[27,28]*), this can be overcome by introducing a spatial bias, or "tropism", to constrain tree growth in specified directions, but the underlying factors that produce such effects are unclear. We attempted to determine whether such biases are inherent in the statistics of the spatial disposition of branch points and path terminations in relation to a global frame of reference for a given tree.

In order to define a central axis for each individual dendrite, we chose to use the vector from the center of the soma to the COM for membrane area (Fig. 8A; *[26]*). The somatofugal COM axis was aligned with vertical (Z) axis of a Cartesian coordinate system, just as botanical trees are oriented with respect to gravity *(29,30)*. The location of each branch point (BP) in the tree was then specified by its angular deviation (dBPax)

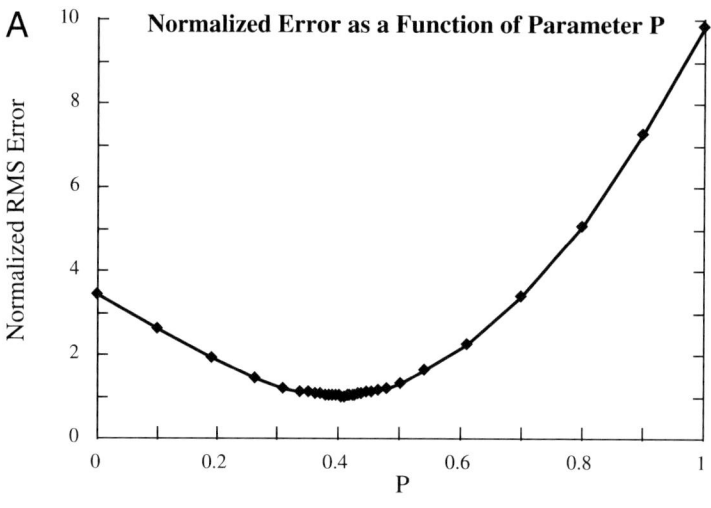
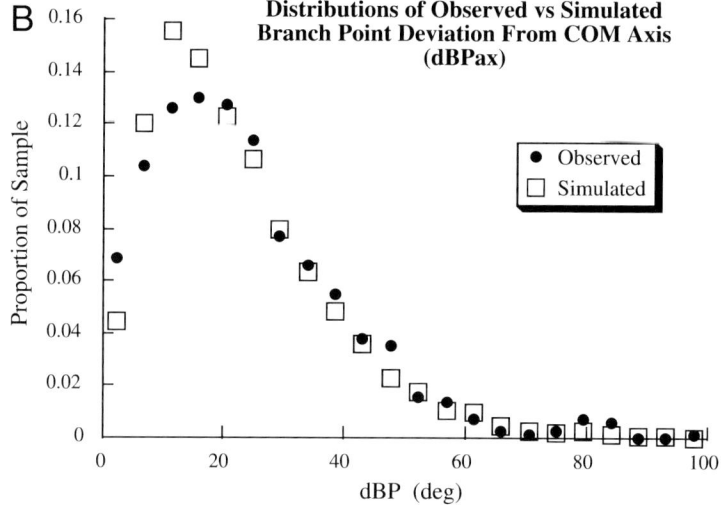
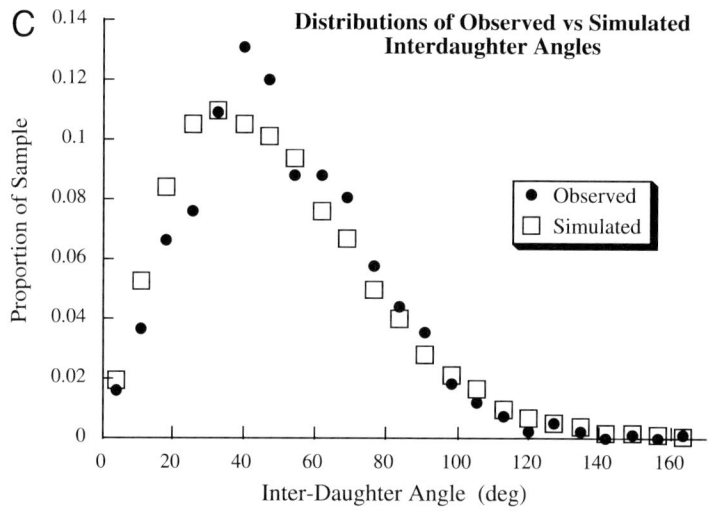

away from the COM axis and its azimuth angle in the XY plane (Fig. 8A). The distribution of dBPax angles provides a concise measure of the lateral dispersion of the dendrite's territory (see Fig. 9B).

We then needed a scheme for specifying the angular deviations of daughter branches at BPs that could serve as the basis for a simulation algorithm for 3D trees. Such a system should make maximum use of local frames of reference, as in the 2D models described above. We found it initially useful to define the direction of branching by calculating a "Sib" vector midway between the two daughter branches that represents the axis of a cone around which daughter branches, with any given interdaughter angle, can rotate into any azimuthal orientation (Fig. 8B). As with the azimuth angles of BPs, these orientations showed no rotational bias in motoneuron dendrites, so they can be evenly distributed in simulations. Still open is the question of what vector to use in measuring the Sib deviation. Using the parent branch direction alone again leads to a spatial random walk. However, using the direction of the branch point vector (vBP) defined in relation to the COM axis (Fig. 8A) preserves information about global tree structure. The azimuth of the Sib about this direction also turned out to be unbiassed. Thus, two distributions, one for the Sib deviation and the other for interdaughter angle, provided the basis for a 3D simulation algorithm. Both distributions were well fitted by Δ functions, each specified by two parameters, that are easily adapted for Monte Carlo simulations. Of course, the eventual COM axis of a simulated tree is unknown at the outset of 3D simulation, so we used the Z axis of the Cartesian frame as the reference vector.

We also explored a simpler algorithm that used only the distribution of deviations of the individual daughter branch directions (vDau) at each branching point as the basis for building complete trees. This approach was complicated by the fact that vDau was correlated with both vBP as well as with the direction of the parent branch (vPar) (Fig. 8C). The distributions of angular deviation between vDau and either vBP or vPar (dBP and dPar, respectively, in Fig. 8C) were both well-fitted by Δ distributions. As expected from the Sib data discussed above, neither vector exhibited any bias in azimuth orientation. Complete trees for 60 individual dendrites were simulated using either the dBP or the dPar distributions, based on length and diameter data from 60 real motoneurons. The total root mean square (RMS) error between a variety of angular measures

Fig. 9. *(facing page)* Results of simulating the 3D structure of 60 motoneuron dendrites. **(A)** Plot of RMS errors in 3D simulations based on using different proportions (P) of the branch point and parent directions [V* = (1 − P) vBP + P vPar] to calculate dV*. Simulations based entirely on the observed distribution of dBP (P = 0) produced less error than those based entirely on dPar (P = 1), but minimum error was found with P ~ 0.4. See text for details. **(B)** Comparison of the distributions of the angular deviation of branch points away from the COM axis (dBPax; see Fig. 8A) in 60 actual dendrites (filled circles) and averaged values from 10 repetitions of 3D simulation of the same 60 trees (open squares). Simulations used the distribution of daughter branch deviations, dV*, from the Hybrid vector, V*, with P = 0.4. The fit between the two distributions of the emergent property dBPax indicates that the real and simulated dendrites have the same (statistical) lateral spread. **(C)** As in panel **B**, but showing the comparison of real and simulated interdaughter angles. Interdaughter angle is also an emergent property, because the daughter branch directions are simulated separately and independently.

for simulated and real trees was smaller in the set of trees constructed using the dBP distribution (Fig. 9A; error = 3.4% with P = 0) as compared to the set simulated using the dPar distribution (Fig. 9A; error = 9.9% with P = 1). We then explored using a hybrid vector (V*) calculated as a linear combination of the branch point and parent directions (vBP and vPar, respectively)

$$V^* = (1 - P) \, vBP + P \, vPar$$

as the reference vector to calculate the deviation (dV*) for each daughter branch (Fig. 8C.). As with the Sib vector approach above, V* can be viewed as the axis of a cone that describes the locus of the distribution of daughter branch directions.

With simulations based on the dV* distribution, the overall RMS error for the 3D tree statistics were minimal with P ~ 0.4 (Fig. 9A). Furthermore, the distribution of angular deviations of branch points away from the COM axis (dBPax; Fig. 8A), as well as the interdaughter angles, for the simulated trees matched the observed data quite well (Fig. 9B and C). The azimuthal angles for daughter branch directions in the XY plane also matched those of real trees (not shown). These 3D statistical measures are emergent properties that are not specified by the simulation algorithm, so that the fits indicate that the simulation accurately reproduced the overall 3D structure of motoneuron dendrites with straight branches. The simulated trees exhibited the same range of constraint in lateral spread as real dendrites (estimated by dBP; Fig. 9B), and their overall shapes appeared appropriate to visual inspection. The final 3D algorithm used only three parameters, two to specify the Δ function that fits the daughter deviations from V*, plus the minimum error value of P = 0.4. This result suggests that the spatial distribution of daughter branches in motoneuron dendrites can be described by factors related to the central axis of the tree, which could reflect environmental constraints, and, to a lesser extent, on purely local factors related to the direction of the parent branch. We are investigating an analogous approach to simulation of the natural meander of individual branches.

A more difficult problem for simulation of dendritic trees in 3D is that multiple objects cannot occupy the same point in space. There may even be some active avoidance among branches from the same neuron *(31)*. Adjustment of the 3D positions of simulated dendrite branches would probably best be accomplished after the complete structure has been constructed. Ideally, such adjustments should be based on data from real dendrites that give information on the spacing between their components. To our knowledge, such analyses have not been made with *in situ* neuronal dendrite data, although some basic theoretical solutions to the problem have been suggested *(32)*.

In the same vein, simulation of the complete dendritic tree of a single multipolar neuron will require regional analysis of the positions of all elements, beginning near the soma. Although it is simple to arrange the COM vectors of simulated trees to project away from each other, the possibility of unnatural collisions exists after the first branch points. Any spatial adjustments to a given element made near the soma would propagate outward, presenting a potentially massive computational problem. It is tempting to sidestep this problem by simply accepting 3D virtual neuron simulations that subjectively appear "natural." However, it seems important for the field to develop objective assessments that could reveal important constraints that remain unresolved at present.

Quantitative Dendritic Morphology 43

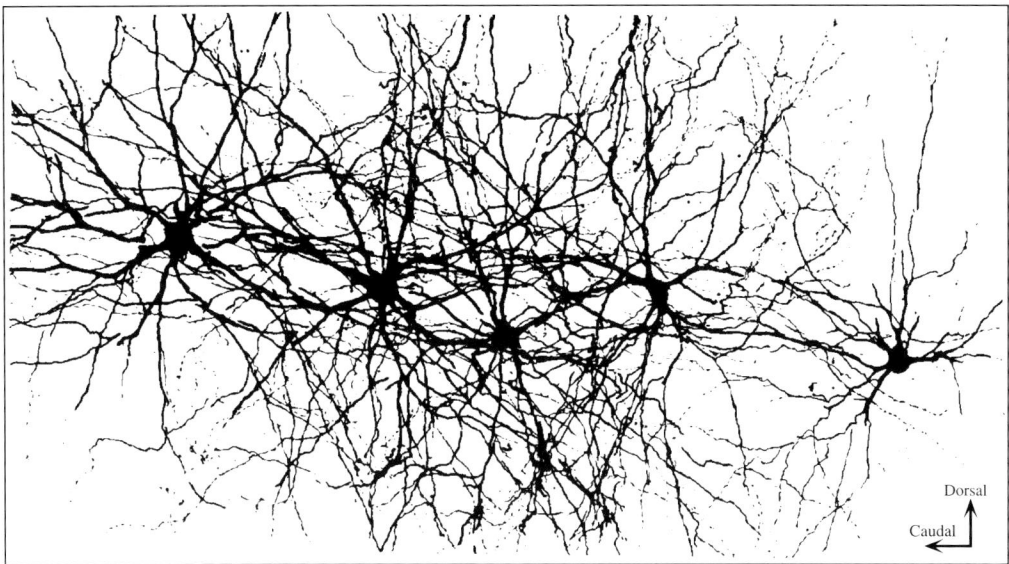

Fig. 10. Photomontage from three serial sagittal sections showing the somata and portions of the proximal dendritic trees of 5 filled α-motoneurons from a cat spinal cord. Note the complex interweaving of the trees from different cells. The somata are about 50 µm in diameter. Adapted from Figure 18 in *(41)*.

2.4.2. The Problem of Neuronal Packing

Clearly, accurate simulation of the 3D structure of individual neurons is a formidable problem. It is at least as difficult to devise quantitative approaches that can be used to measure how multiple neurons with overlapping dendritic territories are packed into the neuropil. The neuropil must provide space not only for somata, dendrites, and the synaptic boutons associated with them, but also for axons with and without myelin sheaths, glia, blood vessels, and extracellular space. Stereological methods can provide estimates of numbers of neurons *(33)* and the volume fraction occupied by these elements *(34)*. However, such data do not provide a clear picture of how individual neurons with extensive dendritic trees can share a given volume of neuropil. The complexity of this problem is illustrated in Figure 10, which shows the intermingled dendritic trees of just 5 HRP-labeled α-motoneurons that undoubtedly share this volume with many unlabeled cells.

We have looked at one approach to this problem using existing data to get estimates of the 3D volume fractions occupied by the dendritic trees of cat motoneurons, plus the synaptic boutons on them, as functions of radial distance from the soma. These estimates were used to explore the consequences of motoneuron packing density on the composition of the neuropil between the cells. Lumbosacral motoneurons in the central part of the ventral horn are, on average, radially symmetrical *(3,26)*, so that the same volume fraction function can be used for all directions. The average surface area and volume (expressed as percent) of the dendrites of 7 α- and 11 γ-motoneurons *(12,26,35)* were calculated within successive 100-µm-thick spherical shells centered on the soma

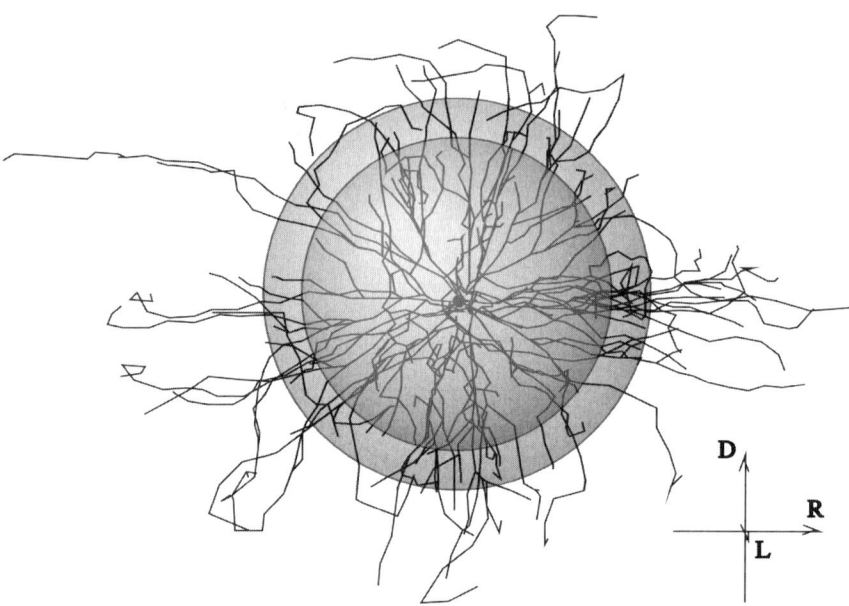

Fig. 11. 2D drawing of a completely reconstructed cat α-motoneuron superimposed on two spheres that represent different radial distances from the center of the soma. The volume between the spheres ("shell volume") contains elements of the dendritic tree, permitting calculation of the volume fraction occupied by elements of the neuron. The direction arrows are 500 µm long.

(Fig. 11). Similarly, we estimated the volume of the associated synaptic boutons in each shell, based on postsynaptic surface area distributions plus data about synaptic covering and bouton size data for cat α- *(36)* and γ-motoneurons *(37)*. The sum of dendritic and bouton volume, when divided by the total volume in each concentric shell, gave estimates of the average volume fraction within each radial shell that is occupied by each type of neuron plus its synaptic boutons (expressed as percent; symbols in Fig. 12A). The solid lines are fits to these data using the following equation:

$$V_{fr} = \gamma \left(\frac{\beta}{d} - 1 \right)^{\alpha}$$

where α, β, and γ are fitting parameters and d is radial distance from the center of the soma.

Assuming that a motor nucleus in the ventral horn includes 65% α- and 35% γ-motoneurons *(38)*, the fitted functions in Figure 12A were combined in those proportions to give the average volume fraction occupied by both types of motoneuron in the cat ventral horn. Because motoneuron dendrites are so extensive (up to 2000 µm from the soma), cells located at considerable distances from any given motoneuron can contribute to the neuropil in the center of the ventral horn. In order to evaluate the total volume fraction contributed by such overlapping dendrites and synaptic boutons, we assumed (for computational simplicity) that motoneurons are arranged in a cubic matrix with the separation (*S*) between somata as a free variable. A program was written in

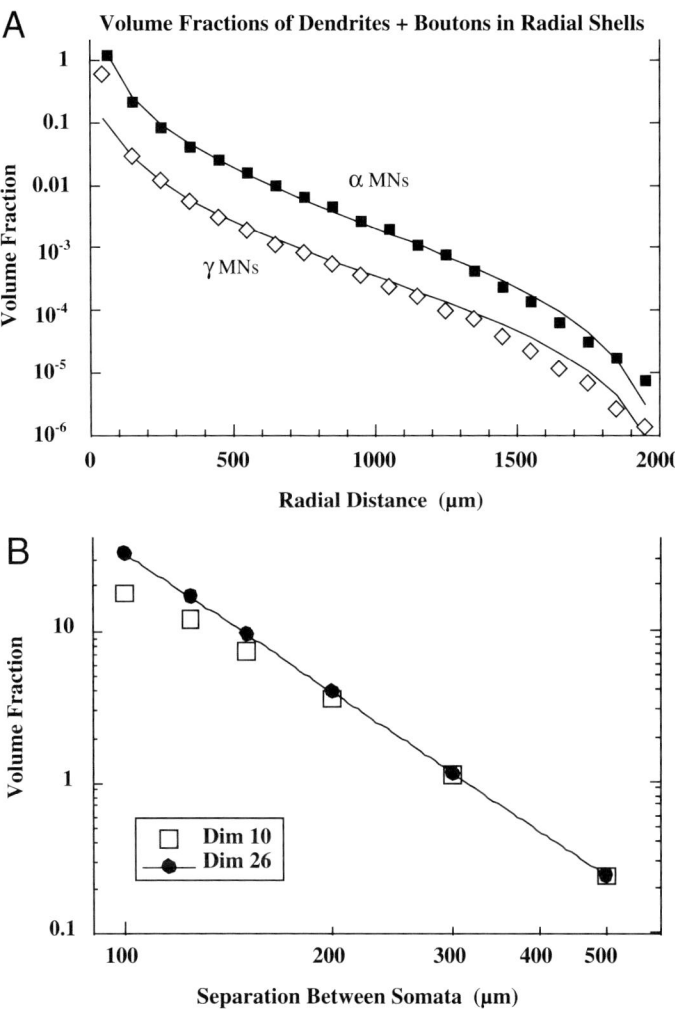

Fig. 12. Estimation of volume fraction occupied by motoneurons and associated synaptic boutons in the cat ventral horn. **(A)** Semilog graph of the estimated volume fractions (symbols) occupied by the dendrites and associated synaptic boutons of average α- (filled squares) and γ-motoneurons (open diamonds), calculated as described in the text. The calculated data were fitted with the equation given in the text. **(B)** Log-log graph showing the total volume fraction (ordinate, in percent) occupied by dendrites and boutons within the open central cube of a cubic array of ventral horn neurons with a uniform separation distance, S (abscissa), when calculated with different numbers of cells along each matrix edge (*Dim*). See text for further explanation.

MatLab® to calculate the average local volume fraction (V_{fr}) of dendrites plus boutons within the open cube at the center of the cell matrix. This was done by randomly sampling 100 locations within the central cube for various values of S. The log-log graph in Figure 12B illustrates how V_{fr} depends on S, as well as on the size (*Dim*) of the cell matrix, where *Dim* is the number of somata along each side of the cubic matrix.

When *Dim* was large enough so that the dendrites of cells along the edges could reach the central cube even with relatively small S (*Dim* = 26), V_{fr} (in percent) varied as

S^3. With smaller matrix dimensions ($Dim = 10$), V_{fr} departed from this power relation as S decreased, because the array limits were smaller than the dendrite extensions from the most peripheral cells. Although estimates of V_{fr} occupied by dendrites and boutons are not available for the cat ventral horn, such estimates from medial lamina VI in the rat spinal cord *(34)* suggest that dendrites and synaptic boutons occupy about 16–20% of the neuropil volume in lamina VI. The dimensions in the actual cat ventral horn are compatible with *Dim* between 10 and 26, so that this analysis suggests that V_{fr} would be between 11 and 18% for $S = 125$ μm (Fig. 12B). From the numbers and positions of motoneuron somata found in the lumbosacral ventral horn of the cat (Fig. 2 in *[38]*), we estimate that their average separation is about 125 μm (see also *[39]*). The fairly good agreement from these independent estimates suggests that this approach may be a valid way to get quantitative estimates of neuropil sharing when the required data are available. Of course, this calculation assumed an isotropic neuropil and other geometries would require more complex algorithms.

2.5. CONCLUDING COMMENTS

This chapter has dealt with some approaches to the problem of quantifying the morphology of individual neurons and of ensembles of neurons, using data from cat ventral horn motoneurons. The ability to mimic the statistical properties of cat motoneuron dendrites, viewed in terms of their 2D dendrograms, using a relatively simple growth model based on data extracted from the same data set, provides a parsimonious description of the original data, which separates factors that are determinative from those that are epiphenomena. The result suggests that local branch diameter, which in large measure depends on cytoskeleton *(8,17)*, is a key factor that maintains the architecture of mature dendrites. On the other hand, it is much more difficult to design computational engines that can reproduce the morphology of neuronal dendrites as 3D entities in ways that can be quantitatively verified. This is largely due to the difficulty of designing analytical tools that can adequately measure how these complex structures occupy space. One future direction for computational neuroanatomy is to solve this problem for individual neurons and then to apply the solution to multiple neurons that occupy the same region. This quest is more than an intellectual exercise, because the ability to simulate complex systems and to check the results of such simulation against the real thing has always led to deeper understanding of the biological world *(40)*.

REFERENCES

1. Stuart G, Spruston N, Häusser M. (eds.) Dendrites. Oxford University Press, New York, 1999, pp. 376.
2. Cajal S. Histology of the Nervous System of Man and Vertebrates. (translated by Swanson N, Swanson LW) Oxford University Press, New York, 1995.
3. Ramon-Moliner E. An attempt at classifying nerve cells on the basis of their dendritic patterns. J Comp Neurol 1962; **119**:211–227.
4. Cullheim S, Kellerth J-O. Combined light and electron microscopical tracing of neurons, including axons and synaptic terminals, after intracellular injection of horseradish peroxidase. Neurosci Lett 1976; **2**:307–313.
5. Snow P, Rose P, Brown A. Tracing axons and axon collaterals of spinal neurons using intracellular injection of horseradish peroxidase. Science 1976; **191**:312-313.

6. Glaser E, Van der Loos H. A semiautomatic computermicroscope for the analysis of neuronal morphology. IEEE Trans Biomed Eng 1965; **12**:22–31.
7. Johnson E, Capowski J. A system for the three-dimensional reconstruction of biological structures. Comput Biomed Res 1983; **16**:79–87.
8. Hillman D. Neuronal shape parameters and substructures as a basis of neuronal form. In: The Neurosciences. Fourth Study Program (Schmitt FO, Worden FG, eds.) MIT Press, Cambridge, MA, 1979, pp. 477–498.
9. Cline HT. Development of dendrites. In: Dendrites (Stuart G, Spruston N, Häusser M, eds.) Oxford University Press, New York, 1999, pp. 35–67.
10. Segev I, Burke RE, Hines M. Compartmental models of complex neurons. In: Methods in Neuronal Modeling (Segev I, Koch C, eds.) MIT Press, Cambridge, MA, 1997, pp. 93–136.
11. Sholl D. Dendritic organization in the neurons of the visual and motor cortices of the cat. J Anat 1953; **87**:387–401.
12. Cullheim S, Fleshman JW, Glenn LL, Burke RE. Membrane area and dendritic structure in type-identified triceps surae alpha-motoneurons. J Comp Neurol 1987; **255**:68–81.
13. Ulfhake B and Kellerth J-O. A quantitative light microscopic study of the dendrites of cat spinal α-motoneurons after intracellular staining with horseradish peroxidase. J Comp Neurol 1981; **202**:571–584.
14. Dityatev A, Chymykhova N, Studer L, Karamian O, Kozhanov V, Clamann H. Comparison of the topology and growth rules of motoneuronal dendrites. J Comp Neurol 1995; **363**:505–516.
15. van Veen M, van Pelt J. A model for outgrowth of branching neurites. J Theor Biol 1992; **159**:1–23.
16. Verwer R, van Pelt J. Descriptive and comparative analysis of geometrical properties of neuronal tree structures. J Neurosci Methods 1986; **18**:179–206.
17. Hillman D. Parameters of dendritic shape and substructure: intrinsic and extrinsic determination? In: Intrinsic Determinants of Neuronal Form and Function (Lasek RJ, Black MM, eds,) Alan R. Liss, New York, 1988, pp. 83–113.
18. Stevens J, Trogadis J, Jacobs J. Development and control of axial neurite form: a serial electron microscopic analysis. In: Intrinsic Determinants of Neuronal Form and Function (Lasek RJ, Black MM, eds.) Alan R. Liss, New York, 1988, pp. 115-145.
19. Tamori Y. Theory of dendritic morphology. Phys Rev E 1993; **48**:3124–3129.
20. Burke RE, Marks WB, Ulfhake B. A parsimonious description of motoneuron dendritic morphology using computer simulation. J Neurosci 1992; **12**:2403–2416.
21. Ascoli GA. Progress and perspectives in computational neuroanatomy. Anat Rec 1999; **257**:195–207.
22. Carriquiry AL, Ireland WP, Kliemann W, Uemura E. Statistical evaluation of dendritic growth models. Bull Math Biol 1991; **53**:579–589.
23. Brown C. Neuron orientations: a computer application. In: Computer Analysis of Neuronal Structures (Lindsey RD, ed.) Plenum Press, New York, 1977, pp. 177-188.
24. Yelnik J, Percheron G, François C, Burnod Y. Principle component analysis: a suitable method for the 3-dimensional study of the shape, dimensions and orientation of dendritic arborizations. J Neurosci Meth 1983; **9**:115–125.
25. Lindsey RD. Neuronal field analysis using Fourier series. In: Computer Analysis of Neuronal Structures (Lindsey RD, ed.) Plenum Press, New York, 1977, pp. 165–175.
26. Cullheim S, Fleshman JW, Glenn LL, Burke RE. Three-dimensional architecture of dendritic trees in type-identified alpha-motoneurons. J Comp Neurol 1987; **255**:82–96.
27. Ascoli GA, Krichmar JL. L-neuron: a modeling tool for the efficient generation and parsimonious description of dendritic morphology. Neurocomputing 2000; **32–33**:1003–1011.
28. Ascoli GA, Krichmar JL, Scorcioni R, Nasuto SJ, Senft SL. Computer generation of anatomically accurate virtual neurons. Anat Embryol 2001; **204**:283–301.

29. Ford R, Ford ED. Structure and basic equations of a simulator for branch growth in the *Pinaceae*. J Theor Biol 1990; **146**:1–13.
30. Honda H. Description of the form of trees by the parameters of the tree-like body: effects of the branching angle and the branch length on the shape of the tree-like body. J Theor Biol 1971; **31**:331–338.
31. Li GH, Qin CD, Wang ZS. Neurite branching pattern formation—modeling and computer simulation. J Theor Biol 1992; **157**:463–486.
32. Meyer F. Mathematical morphology: from two dimensions to three dimensions. J Microscopy 1992; **165**:5–28.
33. West MJ. Stereological methods for estimating the total number of neurons and synapses: issues of precision and bias. Trends Neurosci 1999; **22**:51–61.
34. Tredici G, Tarelli L, Cavaletti G, Marmiroli P. Ultrastructural organization of lamina VI of the spinal cord of the cat. Neurobiology 1985; **24**:293–331.
35. Moschovakis AK, Burke RE, Fyffe REW. The size and dendritic structure of HRP-labeled gamma motoneurons in the cat spinal cord. J Comp Neurol 1991; **311**:531–545.
36. Brännström T. Quantitative synaptology of functionally different types of cat medial gastrocnemius alpha-motoneurons. J Comp Neurol 1993; **330**:439–454.
37. Destombes J, Horchelle-Bossavit G, Thiesson D, Jami L. Alpha and gamma motoneurons in the peroneal nuclei of the cat spinal cord: An ultrastructural study. J Comp Neurol 1992; **317**:79–90.
38. Burke RE, Strick PL, Kanda K, Kim CC, Walmsley B. Anatomy of medial gastrocnemius and soleus motor nuclei in cat spinal cord. J Neurophysiol 1977; **40**:667–680.
39. Aitken JT, Bridger JE. Neuron size and neuron population density in the lumbosacral region of the cat's spinal cord. J Anat 1961; **95**:38–53.
40. Thompson D. On Growth and Form. Cambridge University Press, Cambridge, UK, 1942.
41. Burke RE. Motor units: anatomy, physiology and functional organization. In Handbook of Physiology, Sect. 1: The Nervous System, Vol. II. Motor Control, Part 1 (Brooks VB, ed.) American Physiological Society, Washington, DC, 1981, pp. 345–422.

3

Generation and Description of Neuronal Morphology Using L-Neuron

A Case Study

Duncan E. Donohue, Ruggero Scorcioni, and Giorgio A. Ascoli

ABSTRACT

L-Neuron is a software package that implements simple local anatomical rules to "grow" dendrites stochastically in virtual reality. This program can be used to obtain a compact description of dendritic morphology, to provide the substrates for physiological simulations, to aid neuroscience education, and to develop novel hypotheses about dendritic structure and development. Here, we explore the use of L-Neuron to model CA1 pyramidal cell morphology based on an archive of 24 real reconstructed rat hippocampal neurons. This chapter also describes the extraction of L-Neuron parameter distributions from digitized neurons by means of the companion program L-Measure. The quantitative comparison of virtual and real pyramidal cell dendrograms provides specific insights into neuronal structure and suggests possible avenues to improve the algorithm. Finally, we show how a remarkably accurate and complete spatial description of CA1 pyramidal cell dendritic morphology can be obtained from dendrograms by the addition of a very restricted number of model parameters.

3.1. INTRODUCTION

Historically, computational neuroscience has largely consisted of mathematical modeling of neuronal activity and electrophysiology. In multineuron simulations (or, in the extreme case, in artificial neural networks), neuroanatomy is typically disregarded altogether. Even in single-cell studies seeking to characterize and reproduce the natural physiology of neurons, cellular structure is rarely the focus of the investigation. From early theoretical studies, modelers have simplified anatomy away, attempting to reduce neuronal structure to one or few stylized spheres and cylinders representing the soma and the dendritic processes (for reviews, see [1,2]). However, neuronal shape is extremely important for neuronal function and activity. Dendritic and axonal structures underlie network connectivity and determine synapse numbers and positions. Even considering single neurons, the presence, variety, and complexity of dendritic active conductances and ionic dynamics make the reduction of dendritic structure to lumped equivalent cylinders a very rough, and often incorrect, approximation. The specific branching patterns of dendrites heavily affect signal integration and firing patterns *(3–6)* (see also Chapter 6 in this book). While it is the nature (and often the goal) of modeling to simplify a biological problem, neuronal models that do not take structure into account are severely limited in their ability to accurately represent biologically plausible neuronal behavior. Starting from the mid-1990s, computational neuroscientists have increasingly used real neuronal morphologies as the bases for their electrophysiological models *(4,7,8)*.

Anatomically accurate electrophysiological simulations are typically based on digital reconstructions of intracellularly injected neurons (see also Chapters 6 and 7 in this text). However, tissue preparation and neuronal tracing are extremely time-consuming processes: digitizing a single neurons can take several weeks of a skilled operator (for a discussion on morphological data acquisition, see *[9]*). Thus, the amount of time needed to trace enough neurons to build a large-scale anatomically realistic network model representing even a small brain region is staggering. Even for single-cell models, neuronal reconstructions only constitute raw experimental data. If the simulation results are to be correlated to the structural properties of the cells, many neurons must be used to constitute a representative sample, due to the typically large anatomical variability even within a given morphological class. Therefore, neuronal morphology needs to be characterized statistically.

What constitutes an appropriate characterization of dendritic morphology? Typical reports in the neuroscience literature quantify the geometry and topology of dendritic trees with a variety of scalar measurements, including the number of bifurcations or termination, dendritic surface or length, topological asymmetry *(10)*, and maximum distance from soma to tips. In addition, many of these parameters can be used for correlation measurements, such as the number of bifurcations vs distance from the soma (Sholl analysis). The number of different geometrical parameters that can be measured from neuritic trees is infinite, and of course, many such parameters are interdependent. The situation is further complicated by a lack of standard and by the intrinsic structural and functional complexity of dendrites. Even for a simple shape, for parameters such as "dendritic length", there are many possible measurements. Do we refer to the total length of the dendrites, or to the average length of a tree, or to the length of an individual branch of the tree? If we measure "by the tree", should we separate trees in different classes (e.g., apical and basal)? If we measure length by the branch, should stem, interbifurcation, and terminal branches be considered separately? Where exactly does the soma end and the stem branches begin? And length is relatively straightforward when compared with other typical measurements such as branch angles, fractal dimension, or tree spread.

Both problems of the difficulty to acquire experimental data and of the lack of its parsimonious characterization can be solved by computational modeling. Let us suppose to find a stochastic algorithm that generates digital dendrites (of a given class), whose morphology is statistically indistinguishable from that of real neurons (of the same class). By definition, the set of algorithmic parameters would constitute a complete characterization of this morphological class. In addition, this algorithm could generate any number of virtual neurons, thus allowing the creation of large-scale anatomical models. Computational neuroanatomy has achieved considerable progress towards the creation of such an algorithm *(11)*. Recently, we introduced a software package called L-Neuron *(12)*. L-Neuron (which is included in the CD-ROM) implements several existing algorithms for the generation of virtual dendrites, including variations of the description proposed by Hillman *(9,12–15)*, and the algorithm proposed by Burke and colleagues (see Chapter 2 in this text). The algorithms implemented in L-Neuron are mostly local in nature, i.e., each branch "grows" only based on its local properties (e.g., its diameter and the diameter and orientation of its parent), and independent of all others branches. Other algorithms have been proposed that take into consideration interbranch competition and other global factors based on topological or geometrical parameters. Examples of these algorithms are ArborVitae *(16,17)* (see also Chapter 11 in this text) and van Pelt's model, which is reviewed in Chapter 9 in this text.

Perhaps the greatest advantage of creating virtual neurons is in the knowledge and understanding of real dendritic morphology that is gained from the simulation process itself. Most of the parameters used in the L-Neuron algorithms (called basic parameters) can be measured directly from the real cells and correspond to subcellular structures and mechanisms (e.g., microtubule dynamics *[13]*). Virtual cells generated by L-Neuron can be compared to the real cells from which the basic parameters were extracted. This comparison is carried out by the statistical analysis of parameter not used directly in the algorithm (emergent parameters). If an algorithm lacks essential constraints (i.e., it is too simple), or imposes excessive constraints (it is too restrictive), or else is based on incorrect hypotheses, some of the emergent parameters measured from the virtual cells will have different average or variance compared to the real neurons. In this case, the analysis of the differences between real and virtual neurons can reveal novel biophysical principles or developmental insights (see Discussion and Fig. 2 in Chapter 1 of this book).

L-Neuron has been previously used to generate a variety of neuronal morphologies, including virtual Purkinje cells and motoneurons *(18)*. Here, we present a "case study" using L-Neuron to model the dendritic processes of CA1 pyramidal cells. We only consider one simple algorithm as a working example, but we discuss each step of the research strategy in depth. This chapter also describes the use of the computer program L-Measure *(19)* (also included in the CD-ROM) to extract basic parameters from the experimental morphology and to measure emergent parameters from both real and virtual neurons. L-Measure can be also used to carry out morphological analyses in studies correlating neuronal structure and activity (see Chapter 6 in this text). Together, L-Measure and L-Neuron constitute a powerful tool set for the computational neuroanatomist. L-Measure allows the fast and reliable quantitation of a large number of morphological properties of traced and virtual neurons. L-Neuron uses some of those measurements to generate an arbitrary number of nonidentical virtual neurons. Although the final goal of L-Neuron is to create cells that are anatomically indistinguishable from the real ones, the most useful outcome at this stage is the intellectual feedback from the simulation process.

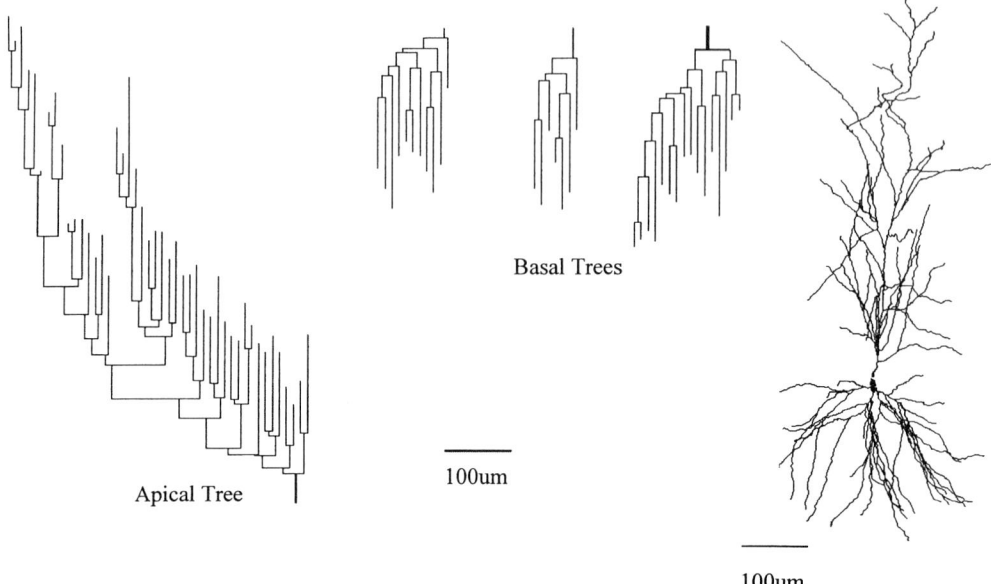

Fig. 1. Cell n421 from the Duke-Southampton archive and its dendrograms. Dendrogram to the left is the single apical tree, the other three are the basal trees. Scale bars are 100 μm. In the dendrograms, length is shown along the Y axis. The thickness of the lines denotes diameter.

3.2. METHODS

3.2.1. Experimental Data

Experimentally traced hippocampal neurons were obtained from the Duke-Southampton cell archive *(20)*. In particular, all of the currently available 24 CA1 pyramidal cells reconstructed in vivo from young rats were used (cells n400 to n423). Figure 1 shows an example of one such neuron and its dendrograms. Each neuron consisted of a morphological ASCII file in Duke-Southampton format (.swc) *(9,20)*. The reconstruction was represented by a series of points, or cylinders. Each cylinder was encoded as a line in the file. The line contained seven numerical fields corresponding to an identity (sequential integer), a type tag (1 for soma, 2 for axon, 3 for basal dendrites, 4 for apical dendrites), the X, Y, and Z positions of the cylinder ending point (in μm), the radius (also in μm), and the identity of the adjacent cylinders in the path to the soma (the "parent"). Examples of .swc morphological files are included in the CD-ROM. Several neurons had minor inconsistencies in the representation of one or more cylinders. Most of these "errors" consisted of an incorrect assignment of the type tag (for example, a few dendritic cylinders were assigned "basal" tag when they were located in the apical field, having apical parents and daughters). These problems possibly reflected human error in the semimanual reconstruction process. The appropriate assignment was generally obvious upon color-coded visualization of the structure, and files were thus corrected by manual editing. Seventeen of the 24 cells had at least one such inconsistency. Details about the changes and links to the edited files are available in the CD-ROM.

3.2.2. L-Neuron and the Modified Hillman Algorithm

The L-Neuron program is written in C++ and runs both under DOS and Unix systems. The CD-ROM contains the DOS executable of L-Neuron 1.07, as well as a link to the L-Neuron download page (http://www.krasnow.gmu.edu/L-Neuron/index.html), where Irix and Linux L-Neuron versions can be obtained. Since the program is under continuous development, the executables will be upgraded frequently. The executable included in the CD-ROM should be only considered as an example still in ß-release. Up to date versions should be downloaded from the L-Neuron Web page. The Delphi graphical user interface previously developed for L-Neuron *(12)* is no longer supported. A new Java-based interface is currently under development. The current version of L-Neuron can be executed with command lines both under Unix and DOS. Examples of executable DOS batch files (as well as of input and output files) are also included in the CD-ROM.

Several algorithms to generate virtual neurons are implemented in L-Neuron. These include Lyndenmayer rewrite rules *(11,12,21)*, several variations based on the original description by Hillman *(12–15)*, as well as the model by Burke and colleagues (see Chapter 2 in this text and references therein). The L-Neuron algorithms have been previously described in detail *(18)*. In this case study, we will only use a single Hillman-like algorithm, which is briefly described below (Fig. 2). The algorithm uses a set of basic parameters that are generally expressed as statistical distributions. Currently implemented distributions are Gaussian, gamma *(22)*, uniform, and delta (constant). Gaussian and gamma distributions can be further modified by a minimum and maximum constraint. In addition, linear mixtures of any number of distributions can be used in L-Neuron. When the algorithm uses a basic parameter, it samples a stochastic value from the appropriate distribution. The random seed is linked to the processor clock or set by the user as a simulation option.

Each dendritic tree is generated independently (the number of trees per neuron being the first basic parameter) and is attached to a spherical soma of given diameter. The simulation starts with sampling an initial stem diameter, taper rate, and branch length for each tree. Dendritic growth consists of an iterative process that crucially depends on branch diameter. Each branch (starting from the stem) bifurcates or terminates depending on whether or not its ending diameter (which is calculated from its initial diameter and the taper rate) is greater than a sampled diameter threshold. If the branch bifurcates, it stems two daughters whose initial diameters are calculated from a daughter diameter ratio and from the ending diameter of the parent using a modification of Rall's equation *(2,13)*: $PK \times d_p^{1.5} = d_1^{1.5} + d_2^{1.5}$, where PK is a sampled numerical param-eter (usually of value between 1 and 2). For each daughter, new values of branch length and taper rate are sampled, and the algorithm is repeated (twice). If the ending diameter of a branch is smaller than the sampled threshold, the branch grows by an additional "terminal" length and stops. The iterative algorithm is, therefore, active until all trees end in terminal tips, and it stochastically samples new values of branch length, taper rate, diameter threshold, and daughter diameter ratio and PK (or terminal length) for every branch. A precise definition of all these basic parameters (and how they are measured from the experimental data) is provided as additional electronic information in the CD-ROM.

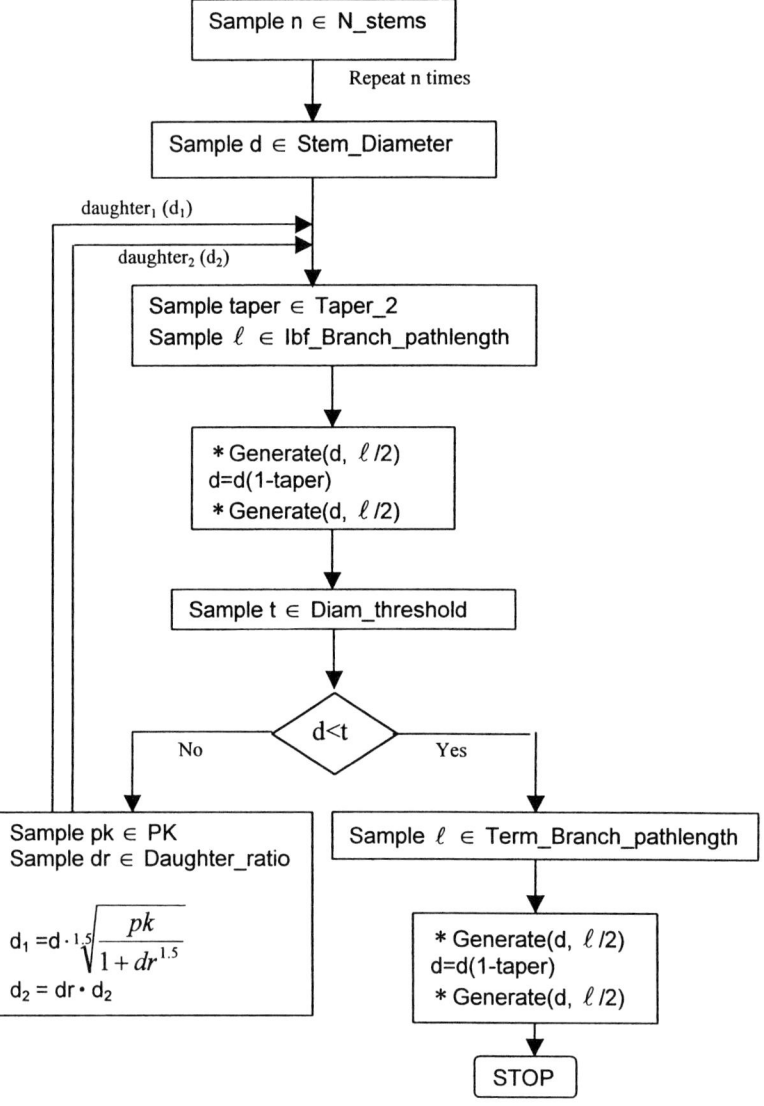

Fig. 2. The L-Neuron "Hillman/PK" algorithm flow chart.

The above algorithm actually only determines the topology and internal geometry (branch lengths, diameter, and connectivity) of the virtual neuron, and it thus generates a dendrogram rather than a complete neuron. L-Neuron can stochastically specify bifurcation angles and dendritic orientation along appropriately fragmented branches. The case study presented in this chapter, however, is limited to the dendrogram properties of neuronal morphology, which constitute the necessary and sufficient information used in anatomically accurate single-cell electrophysiological simulations (see Chapter 7 in this text). All aspects of the L-Neuron implementation concerning 3D geometry and dendritic orientation are described elsewhere *(9,12,18)*. The definition of the appropriate parameters is included in the companion electronic material. A further discussion of dendritic orientation is provided at the end of this chapter.

L-Neuron can output simulated morphology in a variety of formats, depending on simulation options set by the user. Available formats include the Duke-Southampton format (.swc), identical to the format of the real (reconstructed) neurons, a binary format (.vol) that can be interactively visualized and explored under DOS with a companion freeware program called L-Viewer (see Subheading 3.2.5), and various other graphical formats such as virtual reality markup language (.wrl), AutoCad (.dxf), and RayDream "persistency of vision" (.pov). A description of these formats is beyond the scope of the present chapter. The CD-ROM includes a complete list of the available output formats and appropriate command line instructions.

3.2.3. Extraction of Basic and Emergent Parameters with L-Measure

L-Measure is a software package specifically designed for the automatic extraction of morphological parameters from neuroanatomical files *(19)*. The current version of L-Measure (v1.6) runs under Windows and can be executed both from DOS command line and through a graphical user interface. The program is written in C++ and Java and is continuously upgraded. As for L-Neuron, the L-Measure executable included in the CD-ROM should only be considered a ß-release, and updated versions should be downloaded through the appropriate Web links.

L-Measure can read input files in a variety of formats (for a discussion of common morphological formats, see *[9]*), including Duke-Southampton (.swc), Neurolucida (.asc; http://www.microbrightfield.com), Eutectic (.txt; http://www.ls.huji.ac.il/~rapp/labpage.html), Nevins/Claiborne (.dat; http://cascade.utsa.edu/bjclab/), and ArborVitae (.seg; http://www.krasnow.gmu.edu/L-Neuron/index.html). L-Measure can convert any of these formats in the Duke-Southampton "standard". L-Measure recursively visits every cylinder in the neuron and uses the information (position, diameter, connectivity) to return a large variety of measurements. These include data regarding the whole neuron (e.g., total length) or each given branches (interbifurcation length) or individual point (cylinder length). Measurements can be returned as raw data (list of numbers), or as statistical summaries consisting of minimum, maximum, mean, and standard deviation calculated on a cell-by-cell basis or for groups of input cells. These options can all be set through the graphical user interface, as detailed in the electronic documentation.

One of the most useful features of L-Measure is the possibility to precisely specify dendritic subsets for which the information is required. For example, measurements could be selectivley taken from terminal apical segments of less than 0.4 µm in diameter. In addition, L-Measure can calculate any parameter distribution with respect to a second parameter (e.g., Sholl analysis of number of bifurcations vs distance from soma). In the present version, there are 38 functions that can be arbitrarily combined to return any measurement commonly used in the neuroanatomical literature, including lengths, diameters, topological asymmetry, angles, number and patterns of bifurcations, etc.

L-Measure was used to extract all basic parameters that appear in Figure 2 from the 24 real CA1 pyramidal cells taken from the Duke-Southampton archive. Apical and basal trees were analyzed separately, and axonal information (only available for a few neurons) was discarded. All measurements were extracted as raw data and further analyzed (see Subheading 3.2.4) to determine the appropriate statistical distributions. As an example of the basic parameter extraction process using L-Measure, we briefly illustrate here the case of interbifurcation path distance. The function *branch-pathlength* is chosen from the Function tab. To separate apical and basal dendrites, *type* is selected from the Specificity tab, the "=" button is selected, and either 3 or 4 is entered in the value field (in .swc files, type 3 and 4 correspond to apical and basal dendrites, respectively). In order to limit the measurements to interbifurcation branches, excluding terminal branches, *N-tips* is also selected from the Specificity tab, the ">" and the "and" buttons are selected, and 1 is entered in the data field. This will instruct the program to only return values for branches leading to more than 1 tip, therefore excluding terminal branches. In the Output tab, the *raw data* button is selected, and an output file name is entered. To start the extraction, the *go* button on the Go tab is selected. Similarly detailed procedures and definitions for all basic parameters are enclosed in the additional CD-ROM material.

L-Measure was also used to extract emergent parameters from both real and generated cells. For this case study, emergent parameters included the total number of bifurcations, total surface area, and total dendritic length. Finally, L-Measure was used to explore further morphological correlations in the real neurons (see Discussion).

3.2.4. Data Analysis

The raw measurements of basic parameters extracted from the real cells with L-Measure were imported in Microsoft Excel to determine the statistical function, or combination of functions, that best fit each distribution. The mean and standard deviation of the best Gaussian fit to the experimental distributions were calculated using the Excels *solver* function *(23)*. Gamma distributions are characterized by three parameters, α, β, and γ. Parameters α and β determine the shape of the distribution and are related to the mean (μ) and standard deviation (σ) of the data [$\alpha = \sigma^2/\beta^2$; $\beta = \sigma^2/(\mu - \gamma)$]. Parameter γ determines the offset of the gamma distribution and was usually zero. These parameters were also optimized using the Excel *solver* function. The range of uniform distribution fitting was simply taken from the minimum and maximum values of the raw data. A few (two in basal and one in apical) very large PK values of dubious biological plausibility were not included in the parameter distributions.

In order to quantify the quality of the fit (and to chose among different statistical functions or their mixtures when the "best fit" was not obvious by visual inspection), the following method was adopted *(24)*. For each statistical function, a number of points corresponding to the raw data was generated using the best fitting values of the function parameters (i.e., mean and standard deviation for Gaussian, α and β for gamma, minimum and maximum for uniform). Both the raw data and the generated distribution were then sorted by increasing order. The two ordered distributions were linearly correlated, and the corresponding Pearson's coefficient was taken as a measure of the fit quality.

3.2.5. Visualization of Neuronal Morphology and Dendrograms

Neuronal morphologies in Duke-Southampton formats (.swc) can be visualized with the freeware Java applet Cvapp *(20)* (www.compneuro.org). L-Neuron virtual cells can be generated in the same format (and thus visualized with Cvapp), or in a (more compact) binary format that can be visualized by L-Viewer *(12)* (download links included in the CD-ROM). Both Cvapp and L-Viewer allow users to move, rotate, and zoom neurons in a pseudo-3D environment. An additional software tool, called Dendro1, was used to convert morphological files in their dendrograms. Dendro1, which is also included in the CD-ROM, is written in Java and runs under Dos. Both input and output files are in .swc format. For all individual trees, Dendro1 reduces each branch to a single segment while conserving its total length, beginning, and ending diameter values. These segments are then vertically oriented (branch with greater number of daughters to the left) and connected with horizontal lines of arbitrary length (default 10 μm, but can be set by user) to conserve the original topology. The result is a 2D representation of each tree in the neuron, with all angle information removed (see Figs. 1 and 5 as examples). Usage documentation is included in the electronic material.

3.3. RESULTS

Basic parameters for the variation of the Hillman's algorithm described in Figure 2 were measured from the 24 CA1 pyramidal cells, keeping basal and apical dendrites separate. The number of data points varied greatly, from $N = 24$ for the cell-dependent parameters (such as the number of trees), to $N = 755$ and $N = 1172$, respectively, for the basal and apical parameters dependent on branches or bifurcations (such as interbifurcation length and daughter diameter ratio). Most of the measured parameters were fitted best by gamma curves. For instance, interbifurcation length, initial diameter, and diameter threshold, for both apical and basal dendrites, were all fitted by gamma curves with correlation of 0.9 or higher. As an example, Figure 3 shows the gamma fitting of the initial diameter for apical (Fig. 3A) and basal dendrites (Fig. 3B), together with the plots of the linear correlations between real data and data generated according to the fitting function.

A few parameters, such as basal and apical taper rate, and the number of basal trees, were best fitted by a Gaussian rather than gamma curve. Several of the parameters dealing with diameters showed around half of their measured data points at a single value. Specifically, many basal and apical bifurcations were symmetric as far as the diameters of the two daughter branches are concerned (i.e., $d_1 = d_2$, and the daughter diameter ratio equals one). Similarly, many branches were perfectly cylindrical from bifurcation to bifurcation (or termination), i.e., their initial and ending diameters were equal (taper rate of zero). Finally, when thin branches bifurcate (especially at high branching order), it was not uncommon that daughters had the same diameter of the parent, yielding a value of the constant PK of two. Instead of trying to fit these values into a gamma or Gaussian curve, these repeated values were treated as a separate constant (delta) distribution and were sampled according to their proportion of occurrence in the real cells. For example, the daughter diameter ratio had 65% of its basal points and 43% of its apical points of value 1. Only the remaining points were thus fitted to a

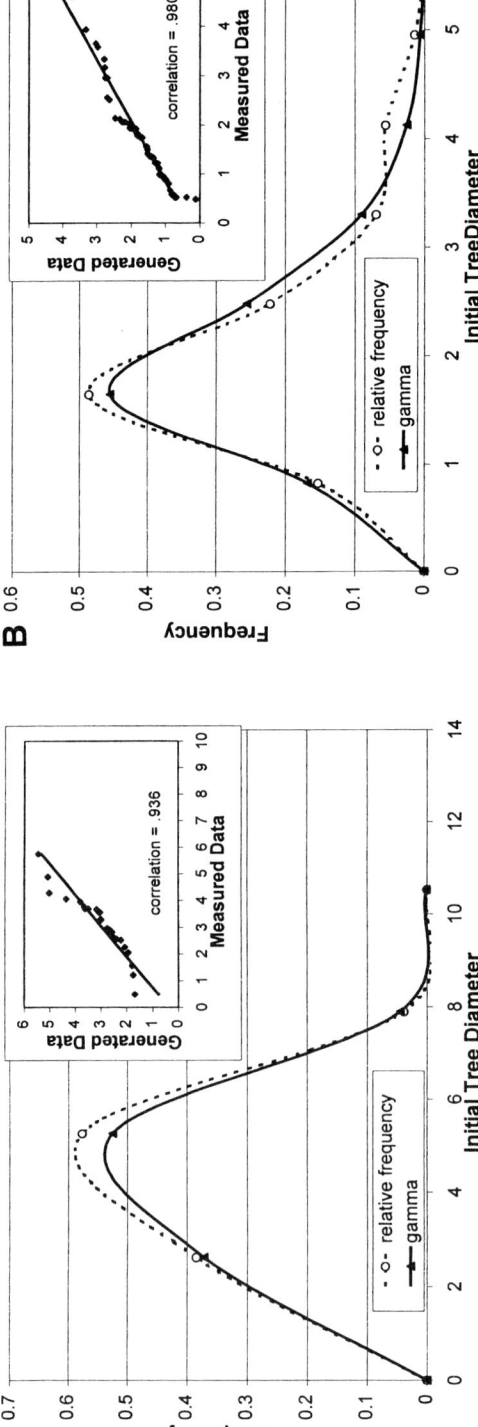

Fig. 3. Measured stem diameter frequency distribution. Diameter of first compartment of each tree are shown in dotted line. Data generated using best-fitting gamma distribution are shown in solid line. (**A**) apical; (**B**) basal. Inset: linear correlation (with Pearson's coefficient) between measured and sampled data.

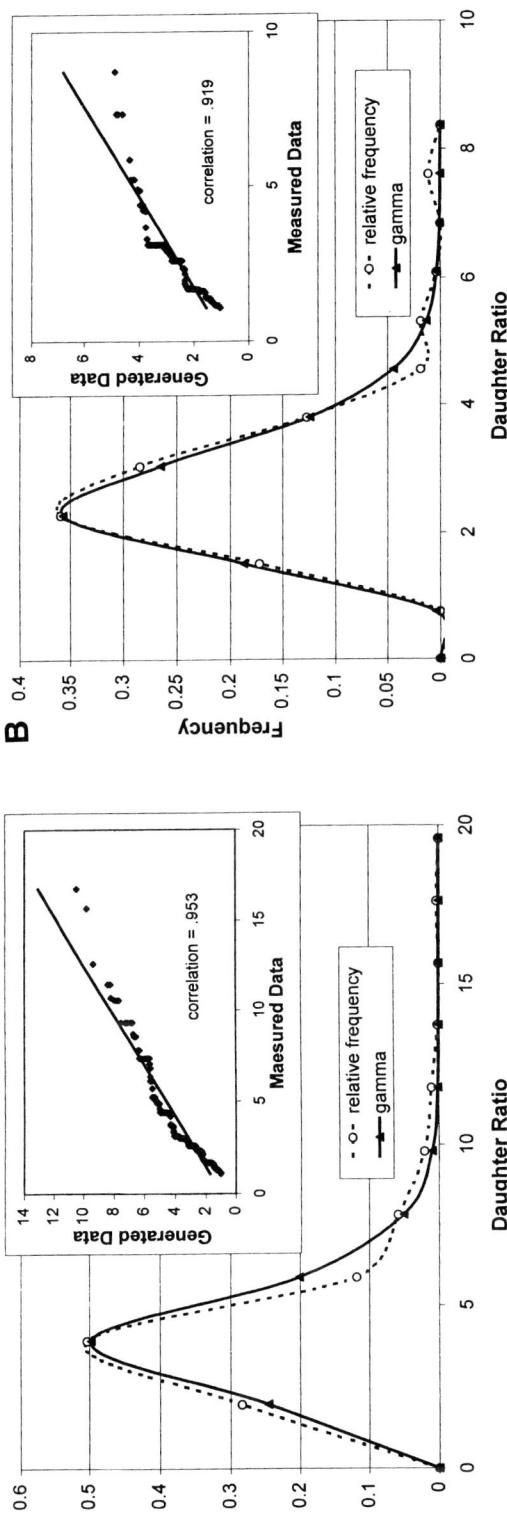

Fig. 4. Measured daughter diameter ratio frequency distribution. Measured data at each bifurcation (larger daughter diameter divided by smaller daughter diameter) are shown in dotted line. Data generated using best fitting gamma distribution are shown in solid line. (**A**) apical; (**B**) basal. Inset: linear correlation (with Pearson's coefficient) between measured and sampled data.

Table 1
Basic Parameters (Basal and Apical)

L-Neuron Parameter	Basal						Apical					
	Dist	%	Mean	Stdev	Min	Max	Dist	%	Mean	Stdev	Min	Max
Rall_Power	k		1.50				k		1.50			
Daughter_Ratio	y	0.35	2.30	1.07	1.00	8.70	y	0.57	3.18	2.01	1.00	16.75
	k	0.65	1.00				k	0.43	1.00			
Ibf_Branch_pahtlength (μm)	y		66.32	77.34	0.00	546.83	y		73.51	86.20	0.00	890.09
Stem_Diameter (μm)	y		1.56	0.88	0.48	4.61	y		3.10	1.11	0.48	5.74
Term_Branch_pathlength (μm)	k		61.42				k		50.73			
Diam_threshold (μm)	y		0.51	0.26	0.18	1.55	y		0.47	0.29	0.16	2.59
Taper_2	g	0.51	0.50	0.20	0.02	0.95	g	0.44	0.53	0.21	0.08	0.95
	k	0.49	0.00				k	0.56	0.00			
PK	y	0.45	1.53	1.57	0.10	9.88	y	0.62	1.35	0.95	0.17	6.49
	k	0.55	2.00				k	0.38	2.00			
N_stems	g		3.00	1.32	6.00		k	0.92	1.00			
							k	0.08	2.00			

Distribution type key: k, constant; g, Gaussian; y, gamma.
While means and standard deviations are given for clarity, distributions marked in y (gamma) were represented in the L-Neuron parameter input file in terms of α and ß. Means and standard deviation are calculated in this table as $\mu = \alpha\beta$ and $\sigma = \beta\alpha^{1/2}$.

gamma distribution. The graphs of these gamma curves and corresponding real points (all points that were not of value 1) are shown in Figure 4. The graphs and the linear correlations both indicate the high quality of the gamma fit for these basic parameter data.

The measured number of apical trees corresponded to a simple bimodal distribution: 22 out of the 24 cells had a single apical tree, while the remaining 2 cells had two apical trees. The corresponding parameter was thus set as a mixture of two constants, one at 1 and one at 2, with the appropriate relative proportions. Terminal path distance (used in the algorithm as an additional growth on top of the regular interbifurcation length) was also set at a constant determined by subtracting the mean interbifurcation path distance from the mean path distance, from bifurcation to termination, measured in terminal branches. Table 1 summarizes the results of the basic parameter extraction and curve fitting. The corresponding L-Neuron input files (.prm) are enclosed in the electronic material. Those files also contain several parameters concerning the spatial orientation of dendrites, which, though necessary to run L-Neuron, are not discussed here. The electronic documentation includes the definitions of these additional parameters and how they were extracted with L-Measure.

Fifty neurons were generated with seeds 1 to 50. Using the unedited data extracted from the 24 real cells (Table 1), only 3 of those seeds resulted in finished virtual cells. The remaining 47 simulations created neurons that kept growing beyond the bounds of the program (in this case set to approximately 3000 bifurcations). In order to limit infinite growth, the taper distributions used in the algorithm were recomputed disregarding the few (4% apical and 6% basal) negative values found in the real data. After excluding negative taper values from the basic parameter distributions, approximately 10% of the virtual cells still grew over the limit of 3000 bifurcations. In analogy with previous simulation studies (18), we thus imposed a maximum value of 2 for the PK distributions. This prevents daughters from being excessively larger than parents at bifurcation points. With this modified set of parameter distributions, all random seeds (1–50) resulted in 50 finished virtual cells. Of these, one had an apical tree with 2,735 bifurcations, over 36 standard deviations above the average of 45.14 bifurcations calculated over the remaining 49 cells. Figure 5 shows the dendrogram of one of these 49 cells.

Several basic and emergent parameters were measured from the basal and apical trees of real cells as well as of virtual cells generated using the raw parameter distributions, the distributions excluding negative taper values and PK values over 2, and this last set without the one extremely large outlier. Measured basic parameter included interbifurcation and terminal branch length and daughter ratio. Emergent parameters included bifurcation number, total length, and total surface area. Table 2 summarizes the data for these cell groups and parameters.

3.4. DISCUSSION

Table 2 indicates that the basic parameters extracted from the real neurons and used in the simulation are very close to those measured from the virtual neurons, in terms of both mean and variance. The emergent parameter measurements, however, show significant differences between real and generated neurons. When the raw basic parameters were used as extracted, most virtual neurons grew indefinitely. The few instances

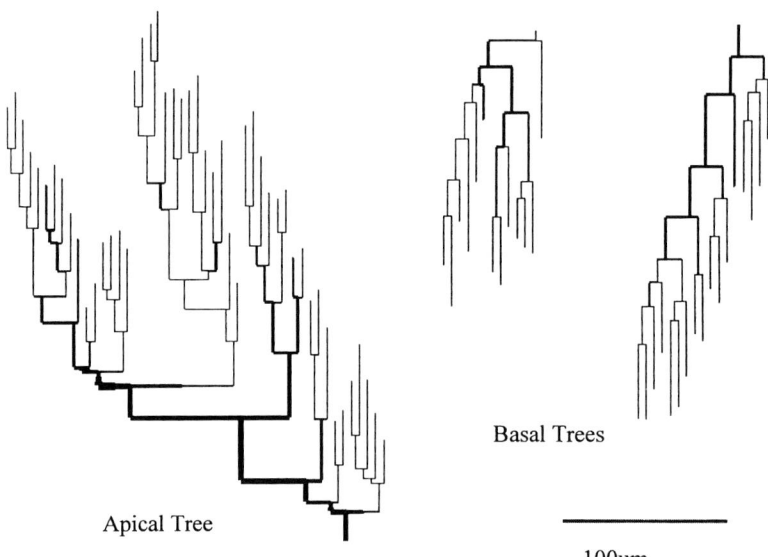

Fig. 5. Dendrogram of a virtual cell created from distributions disregarding negative tapers and having a maximum PK of 2 (random seed 34). Dendrogram to the left is the single apical tree, the two to the right are the basal trees. Scale bar is 100 μm. Virtual cells are included in the CD-ROM.

of finite growth were clearly biased towards excessively small size. When the statistical distributions of taper rate and PK were set limited maximum values (of 0 and 2, respectively), virtual neurons grew to much more plausible sizes. Emergent parameters measured from basal trees had means very close to those of the real neurons. A similar match was observed in apical trees once a single "hypertrophic" outlier was removed. However, even in these cases, the standard deviations of the emergent parameters are much higher in the simulated dendrites than in the real data. The problem of excessive variability in the emergent parameters of virtual neurons emphasizes the need for more constraints in the algorithms. Nevertheless, Figure 5 shows that this algorithm can produce instances of fairly realistic virtual CA1 dendrograms.

As most algorithms implemented in L-Neuron, the anatomical rules used in this study to model dendritic morphology are entirely local. This means that the growth of a branch at any point in the tree does not depend on any other event throughout the rest of the cell. The use of local rules has the advantage of simplicity, both in the extraction of parameter (that are measured directly from reconstructed neurons) and in the algorithm design: as the elongation of a branch is being processed, no information is required from any other structure. By and large, this assumption is also biologically plausible: there is no evidence that growth cones have detailed information about the branching of the rest of the cell. However, in nature, the overall size of the growing neuron is likely to have at least a coarse feedback effect on local branch elongation. In L-Neuron, such effect is not modeled, occasionally resulting in virtual neurons that are significantly larger or smaller than generally found in nature.

Table 2
Summary of Emergent Parameters

Tree Type		Apical				Tree Type		Basal			
Parameter		Real CA1	Unedited	Edited	No Outlier	Parameter		Real CA1	Unedited	Edited	No Outlier
n =		24	3	50	49	n =		24	3	50	49
Basic						**Basic**					
Ibf_Branch_pathlength (µm)	mean	73.51	74.52	78.48	80.71	Ibf_Branch_pathlength (µm)	mean	66.32	80.99	73.84	73.89
	stdev	86.20	50.47	74.94	77.22		stdev	77.34	59.75	67.61	67.65
Term_Branch_pahtlength (µm)	mean	124.24	108.40	108.50	110.98	Term_Branch_pahtlength (µm)	mean	127.88	79.62	108.54	108.57
	stdev	104.84	59.29	77.10	79.99		stdev	90.83	43.48	70.75	70.80
Daughter_Ratio	mean	2.25	1.94	2.29	2.29	Daughter_Ratio	mean	1.46	1.54	1.50	1.50
	stdev	1.87	1.42	1.62	1.60		stdev	0.89	0.67	0.87	0.87
Emergent						**Emergent**					
N_Bifs	mean	48.42	11.00	98.94	45.14	N_Bifs	mean	31.67	7.67	27.16	27.67
	stdev	15.61	9.54	387.25	73.24		stdev	9.53	5.51	47.83	48.19
Surface (µm²)	mean	18770.62	4318.46	43436.05	18561.08	Surface (µm²)	mean	11282.86	3648.81	9850.03	10039.35
	stdev	11032.09	4175.34	178516.43	30810.07		stdev	8051.04	2047.92	17026.20	17149.38
Length (µm)	mean	10451.52	2427.75	19667.44	9146.38	Length (µm)	mean	6866.79	1391.62	5465.35	5568.01
	stdev	3898.52	2102.72	75795.59	14653.52		stdev	2804.43	693.44	9082.06	9146.82

Column labeled *Real CA1* refers to the 24 traced hippocampal cells obtained from the Duke-Southampton archive. Column labeled *Unedited* refers to cells created using the raw extracted data. The *Edited* column refers to cells created when negative tapers were not included in the data analysis, and maximum PK value was set at 2. One of these 50 cells was several orders of magnitude larger than the others, and the *No Outlier* column summarizes the data on the remainder cells. See *Glossary* in the online material for descriptions of parameters.

The results of this case study confirm previous observations that virtual neurons generated with purely local algorithm tend to excessively vary in size compared to the real cells *(18)*. However, this study also exposed a potential "instability" of the algorithm that causes cells to grow indefinitely. This happens when the combination of sampled values for the parameters that control diameter (specifically, taper rate and PK) causes an initial "explosion" in the number of bifurcations, before the diameter drops below the threshold causing terminations. Once an excessive number of "growing" branches is present, it is likely that at least some of these will generate even larger daughters, causing a practically infinite loop. Our results demonstrate that limiting taper rate to nonnegative values and PK to values not larger than 2 (thus preventing the creation of daughters with much larger diameter than their parents) eliminates this instability. Other authors have also considered taper rate too sensitive a parameter, making it an (optimized) constant in their models (see e.g., Chapter 2 in this text). Similarly, previous implementations of the L-Neuron algorithms also limited possible PK values to 2 or less *(18)*. Why are these constraints necessary, given that basic parameters are directly measured from the real neurons?

In the current implementation of the algorithm, the values of many parameters are assumed to be uniformly distributed throughout the cell, but in reality they are not. This discrepancy between model and reality is, we believe, largely responsible for the instability of the algorithm that can lead to excessively large cells. Several parameters tend to depend on either diameter or distance from the soma. Taper rate, for example, is very large near the soma, where diameter falls off quickly. Distally in the dendrites, where diameters are small and generally almost constant, taper is mostly 0. If the statistical distribution of taper fitted from data extracted from the entire cell is used to sample taper values near the soma (e.g., at the beginning of neuronal growth), the probability to obtain a rapid initial shrinkage in the simulated dendrite will be significantly smaller than in real neurons. The top two graphs in Figure 6 demonstrate that a similar situation occurs for both the major parameters controlling diameter changes (taper rate and PK). In other words, in real neurons, the conditions (parameter values) leading to a branch diameters increase (negative taper and PK values greater than 2) are less frequent near the soma that distally. A corresponding tendency can be inferred from the plots of the average values of these parameters at different branching orders, in both basal and apical trees (Figs. 6C and D).

In order to address this issue, taper rate and PK could be made explicitly dependent on branch order or path distance from the soma in the algorithm. Alternatively, since diameter is generally monotonically decreasing with branch order and path distance from the soma, taper and PK could be made dependent on diameter itself, which would provide a useful feedback mechanism while maintaining the purely local approach. Whether these dependencies and mechanisms are in turn due to specific underlying developmental or cytoarchitecture properties of this cell class remains an open question. Implementation of these variations of the algorithm is currently underway.

The results of the present study can be compared to those obtained with the morphological simulation of motoneurons and Purkinje cells by using similar algorithms *(18)*. While the variability of emergent parameters was larger in virtual cells than in real cells for all the morphological cells, in the motoneuron models, such a difference in the

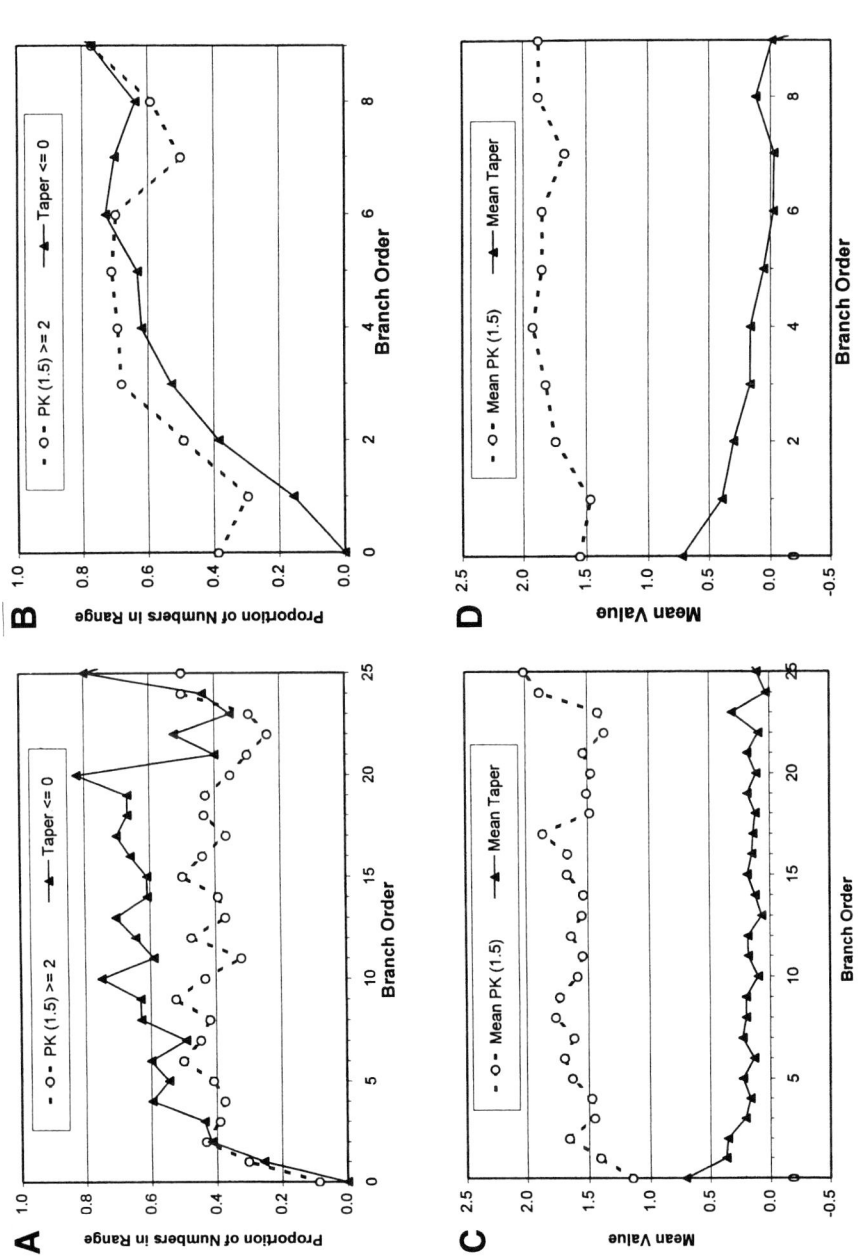

Fig. 6. Taper and PK vs Branch Order. Panels **A** and **B** show the proportion of values at different branch orders that would lead to a lack of decrease in diameter (taper values less than or equal to one, and PK values greater than or equal to 2). Panels **C** and **D** show the means for these two parameters by branch order. At an apical branch order greater than 25 and a basal branch order greater than 9, the amount of data drops drastically. Data from these distal tips is, therefore, not included in these graphs.

data variability was modest, and averages of all tested emergent parameters in virtual cells were remarkably accurate as compared to the real neurons. In Purkinje cells, the difference in data variability between real and virtual neurons was greater. Most emergent parameters were well reproduced, with the exception of the maximum branching order in Purkinje cells. Morphological analysis had connected that discrepancy with topological asymmetry, and, ultimately, with the daughter diameter ratio. The issue in the Purkinje cell case was that the daughter diameter ratio, which is assumed in these simple algorithms not to depend on the position of the bifurcation in the dendritic tree, had in fact a somewhat bimodal distribution, with great asymmetry (high daughter diameter ratio) close to the soma, and almost total symmetry (daughter diameter ratio close to one) distally *(18)*. A similar situation was encountered in the present case study with CA1 pyramidal cells. However, in this case, other parameters that crucially affect local diameter also heavily depended on distance from the soma (or diameter). Since all L-Neuron algorithms critically depend on local diameter in determining branching and terminating probabilities, and thus the size of the neuron, this observation explains the occurrence of "infinite growth" described in the Results section.

A further issue that emerged from this case study is that of stem diameter. The overall size of the generated neurons is highly dependent on the starting diameter, but in the traced neurons, the point where the soma stops and the dendrite starts is largely set arbitrarily. Since the decrease of taper rate with the distance from the soma is generally extremely steep in the very proximity of the soma, a possibly robust definition of dendritic stem would be linked to discontinuity in the taper rate.

3.5. SPATIAL ORIENTATION

A final item of discussion concerns the issue of the spatial orientation of dendrites. While dendrograms capture a great deal of morphological properties and are sufficient to run single-cell electrophysiological simulations, an important component of dendritic morphology is the occupation of space in three dimensions. In fact, this is one of the most important shape characteristics that neuroanatomists intuitively use in morphological classifications. L-Neuron tackles the problem of dendritic orientation at four levels *(12)*. First, trees stem out of the soma with a given orientation, typically determined in polar coordinates as elevation and azimuth. Second, each branch (between two consecutive bifurcations, between a stem and a bifurcation, or between a bifurcation and a termination) is fragmented in smaller segments, and each of these are also oriented, in polar coordinates, relative to their parent segment. Third, bifurcations are further characterized by an amplitude angle and a torque angle. As for the other basic parameters, these angles can be extracted (e.g., with L-Measure) from experimental morphology (the parameters, with their definitions and measurement procedures, are included in the L-Neuron files in the CD-ROM). Finally, all dendritic segments in the neuron can be pushed in one or more particular directions (tropism *[9]*), such as away from the soma, towards a plane, or along a fixed axis.

Recently, we introduced a simple model of dendritic orientation for hippocampal cells, including CA1 pyramidal cells *(25)*. In this model, only two tropic components were included: (*i*) a push away from the soma; and (*ii*) a push along the apparent preferential direction of growth (measured as the average orientation of all dendritic seg-

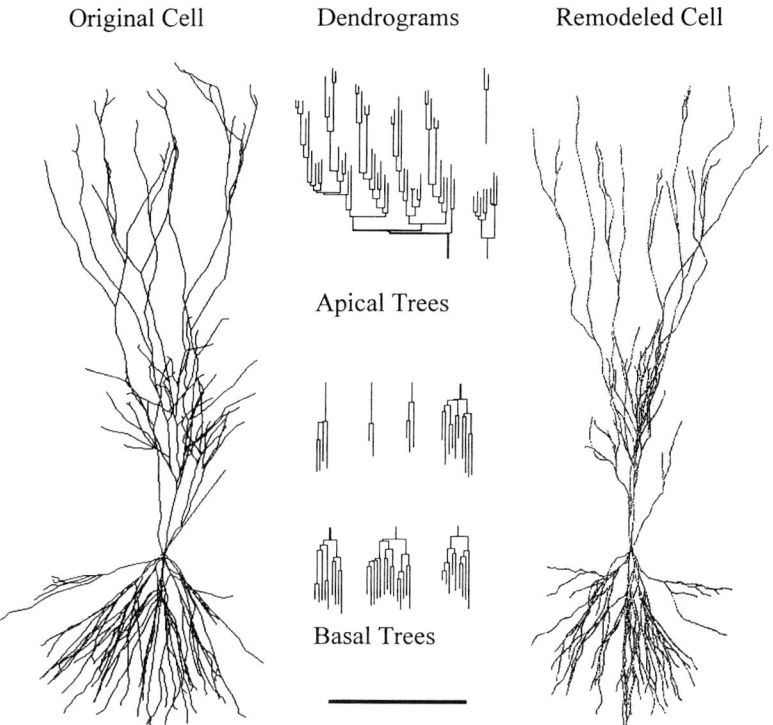

Fig. 7. Remodeling of a hippocampal pyramidal cell (cell c12866 from Amaral's collection). Scale bar is 200 microns for neurons and 400 microns for dendrograms. All angle information was eliminated from the experimental morphological file. Dendrograms were then reoriented according to a simple model that only includes stem direction, tendency to grow straight, repulsion from the soma, and noise *(25)*.

ments). Bayesian analysis demonstrated that the tendency to grow away from the soma is the dominant factor in hippocampal cells. In a further approximation, the two daughters of each bifurcation were treated as independent segments, so that no additional constraints were needed to describe bifurcations. Thus, remarkably accurate dendritic shapes were obtained with just four parameters characterizing spatial orientation: the azimuth and elevation of the tree stems, the "repulsion" from the soma (expressed relative to a tendency to grow straight, i.e., in the same direction as the parent segment), and a "noise" factor, accounting for random deviations (isotropic) in three dimensions. A complete discussion of this model is beyond the scope of the present chapter (see *[25]* for more details). However, it should be mentioned that the model could be used to "recreate" full neuronal geometry starting from the experimental "dendrograms" (Fig. 7). Thus, the same model of spatial orientation could be applied to the dendrograms generated with L-Neuron, such as those shown in Figure 5.

3.6. CONCLUSION

This chapter presented a "case study" of morphological simulation. We limited our analysis to one morphological class and, most importantly, to one simple entirely local algorithm. The goal of this study was not to find the perfect algorithm to reproduce all aspects of dendritic morphology. Such an algorithm will undoubtedly be fairly complex, as quantitative explanations of neuroscientific phenomena typically are. In contrast, the goal of this chapter was to illustrate the key steps of the research strategy underlying the search of such an algorithm. These include (*i*) the definition of the hypotheses (expressed algorithmically, as in Fig. 2); (*ii*) the measurement of the basic parameters from the experimental data; (*iii*) their fitting and representation as combinations of statistical distributions; (*iv*) the stochastic generation of virtual neurons; (*v*) the quantitative comparison of these simulated cells with the real ones by means of emergent parameters; and (*vi*) the morphological analysis to quantify the possible causes of the discrepancies (see examples in Fig. 6). This process can be iterated to improve the initial set of hypotheses by progressively implementing more refined algorithms (see also Fig. 2 in Chapter 1 of this text).

ACKNOWLEDGMENTS

This research was supported by Human Brain Project Grant No. R01-NS39600, funded jointly by the National Institute of Neurological Disorders and Stroke and the National Institute of Mental Health (National Institutes of Health [NIH]). This case study was made possible by the generous decision of Drs. G.K. Pyapali and D.A. Turner to share their digital morphological data through the Duke-Southampton archive. Authors are grateful to Mr. Philip O'Herron for technical assistance and to Dr. Alexei Samsonovich for providing the remodeled cell of Figure 7.

REFERENCES

1. Segev I, Rinzel J, Shepherd GM. (eds.) The Theoretical Foundation of Dendritic Function. MIT Press, Cambridge, MA, 1995.
2. Rall W, Burke RE, Holmes WR, Jack JJB, Redman SJ, Segev I. Matching dendritic neuron models to experimental data. Physiol Rev 1992; **72**:S159–S186.
3. Mainen ZF, Senjnowski TJ. Influence of dendritic structure on firing patterns in model neocortical neurons. Nature 1996; **382**:363–366.
4. Vetter P, Roth A, Hausser M. Propagation of action potentials in dendrites depends on dendritic morphology. J Neurophysiol 2001; **85**:926–937.
5. van Pelt J, Schierwagen A, Uylings HBM. Modeling dendritic morphological complexity of deep layer cat superior colliculus neurons. Neurocomputing 2001; **38–40**:403–408.
6. Krichmar JL, Nasuto SJ, Scorcioni R, Washington SD, Ascoli GA. Effect of dendritic morphology on CA3 pyramidal cell electrophysiology: a simulation study. Brain Research 2002; In press.
7. De Schutter E, Bower JM. An active membrane model of the cerebellar Purkinje cell. I. Simulation of current clamps in slice. J Neurophysiol 1994; **71**:375–400.
8. Migliore M, Cook EP, Jaffe DE, Turner DA, Johnston D. Computer simulations of morphologically reconstructed CA3 hippocampal neurons. J Neurophysiol 1995; **73**:1157–1168.
9. Ascoli GA, Nasuto SJ, Krichmar JL, Senft SL. Generation, description and storage of dendritic morphology data. Philos Trans R Soc Lond B Biol Sci 2001; **356**:1131–1145.
10. Van Pelt J, Uylings HBM, Verwer RWH, Pentney RJ, Woldenberg MJ. Tree asymmetry—a sensitive and practical measure fro binary topological trees. Bull Math Biol 1992; **54**:759–784.
11. Ascoli GA. Progress and perspectives in computational neuroanatomy. Anat Rec 1999; **257**:195–207.

12. Ascoli GA, Krichmar JL. L-Neuron: a modeling tool for the efficient generation and parsimonious description of dendritic morphology. Neurocomputing 2000; **32–33**:1003–1011.
13. Hillman DE. Neuronal shape parameters and substructures as a basis of neuronal form. In: The Neurosciences, Fourth Study Program. (Schmitt F, ed.) MIT Press, Cambridge, MA, 1979, pp. 477–498.
14. Hillman DE. Parameters of dendritic shape and substructure: intrinsic and extrinsic determination? In: Intrinsic Determinants of Neuronal Form and Function (Lasek RJ, Black MM, eds.) Liss, New York, 1988, pp. 83–113.
15. Tamori Y. Theory of dendritic morphology. Phys Rev E 1993; **148**:3124–3129.
16. Senft SL. A statistical framework to present developmental Neuroanatomy. In: Neural-Network Models of Cognition: Biobehavioral Foundations (Donahoe JW, ed.) Elsevier Press, New York, 1997.
17. Senft SL, Ascoli GA. Reconstruction of brain networks by algorithmic amplification of morphometry data. Lect Notes Comp Sci 1999; **1606**:25–23.
18. Ascoli GA, Scorcioni R, Krichmar, JL, Nasuto SJ, Senft SL. Computer generation and quantitative morphological analysis of virtual neurons. Anat Embryol 2001; **204**:283–301.
19. Scorcioni R, Ascoli G A. Algorithmic extraction of morphological statistics from electronic archives of neuroanatomy. Lect Notes Comp Sci 2001; **2084**:30–37.
20. Cannon RC, Turner DA, Pyapali GK, Wheal HV. An on-line archive of reconstructed hippocampal neurons. J Neurosci Methods 1998; **84**:49–54.
21. Prusinkiewicz P, Lindenmayer A. The Algorithmic Beauty of Plants. Springer-Verlag, New York, 1990.
22. van Pelt J, van Ooyen A, Uylings HBM. Modeling dendritic geometry and the development of nerve connection. In: Computational Neuroscience: Realistic Modeling for Experimentalist (De Schutter E, ed.) CRC Press, Boca Raton, 2000, Ch. 7.
23. Box GEP, Tiao GC. Bayesian Inference in Statistical Analysis. Wiley, New York, 1992.
24. Gilks WR, Richardson S, Spiegelhalter DJ (eds.) Markov Chain Monte Carlo in Practice. CRC Press, Boca Raton, 1996.
25. Ascoli GA, Samsonovich A. Bayesian morphometry of hippocampal cells suggests same-cell somatodendritic repulsion. Advances in Neural Information Processing Systems 14, 2002; in press.

4
Optimal-Wiring Models of Neuroanatomy

Christopher Cherniak, Zekeria Mokhtarzada, and Uri Nodelman

ABSTRACT

Combinatorial network optimization appears to fit well as a model of brain structure: connections in the brain are a critically constrained resource, hence their deployment in a wide range of cases is finely optimized to "save wire". This review focuses on minimization of large-scale costs, such as total volume for mammal dendrite and axon arbors and total wirelength for positioning of connected neural components such as roundworm ganglia (and also mammal cortex areas). Phenomena of good optimization raise questions about mechanisms for their achievement: the examples of optimized neuroanatomy here turn out to include candidates for some of the most complex biological structures known to be derivable purely from simple physical energy minimization processes. Part of the functional role of such fine-tuned wiring optimization may be as a compact strategy for generating self-organizing complex neuroanatomical systems.

4.1. INTRODUCTION

How well can combinatorial network optimization theory predict structure of invertebrate and vertebrate nervous systems? The working hypothesis explored here is that brain connections are singularly limited, both in volume and in signal–propagation times; therefore, minimizing costs of required connections strongly drives nervous system anatomy. Network optimization theory is the field in computer science that has developed formalisms of scarcity for expressing and solving problems of "saving wire." The question then is, how well do such concepts in fact apply to the brain? The main technique of these studies is computational experiments, the main hurdle the exponentially exploding computational requirements of optimization searches to evaluate connection-minimization of the neuroanatomy.

Good optimization findings focus attention upon possible biological mechanisms. Network optimization problems are among the most computationally intractable known; in general, only an exhaustive search of all possibilities can guarantee exact solutions. However, some "quick but dirty" probabilistic/approximation procedures developed for microcircuit design suggest candidate models for biological mechanisms of neuroanatomy optimization. In particular, we report positive results for neural optimization via genetic algorithms and via vector mechanical energy-minimization simulations. In fact, the latter models constitute an instance of self-organizing morphogenesis of highly complex biological structure directly from simple physical processes.

4.2. CONCEPTUAL BACKGROUND

The theoretical framework of this work grew out of methodological studies of prevailing models of the agent in microeconomic, game, and decision theory *(1,2)*. The basic finding was that these models typically presupposed agents with unlimited computational capacities, and more realistic bounded-resource models were then developed. Subsequently, the same approach was applied in computer science, to connectionist models of massively parallel and interconnected computation that were intended to be more neurally realistic than conventional von Neumann computational architecture (cf *[3]*); again, the models tended to overestimate available resources drastically—here, actual connectivity in the brain. At least initial connectionist models often tacitly assumed neural connections were virtually infinitely thin wires. In assembling the quantitative neuroanatomy necessary for evaluating neural feasibility of connectionist models, it became evident that a weaker but still discernible trend toward overestimation of resources then pervaded even some neuroanatomy *(4)*.

Thus, a bounded-resource philosophical critique of mind-brain science ("We do not have God's brain") focused attention on neural connections as a critically constrained neurocomputational resource. Through combinatorial network optimization theory, a positive research program emerged: if actual brain connections are in severely short supply, is their anatomy correspondingly optimized? The investigation thus falls in a Pythagorean tradition of seeking simple mathematical patterns in observed natural forms (e.g., *[5]*). In fact, minimum wiring interpretations of neuroanatomy can be traced back at least as far as Cajal's qualitative "laws of protoplasmic economy" *(6,7)* and have continued to receive attention (e.g., *[8]*).

The human brain is commonly regarded as the most complex physical structure known in the universe. In the face of such overwhelming intricacy, neuroanatomy traditionally tended toward "descriptive geography" of the nervous system, i.e., relatively low-level ad hoc characterization of individual neural structures. The abstractive power of concepts from computation theory would aid in coping with the unparalleled complexity of the brain. In particular, network optimization theory may provide a source for a "generative grammar" of the nervous system, some general principles that compactly characterize aspects of neuroanatomy. Of course, connection minimization is unlikely to be ubiquitous in the nervous system; indeed, given the many other competing desiderata driving design of a brain, the striking observation is that it should hold in even some conditions. The question, then, is characterizing where "save wire" does and does not apply.

For example, in the *Caenorhabditis elegans* ganglia case sketched below in Subheading 4.4.2., we reduced approximately one thousand pages of published anatomy diagrams *(9a–c)* to a 100-page database, which, in turn, was represented as a 10-page connectivity matrix (see Fig. 2 in *[10]*), which we then computationally verified to conform to connection-minimizing component placement optimization better than any of the nearly 40 million alternative possible layouts (see Fig. 1). If these types of result are confirmed, they constitute a predictive success story of recent quantitative neuroanatomy.

4.3. NETWORK OPTIMIZATION THEORY

The theory of NP-completeness emerged around 1972 *(11,11a)*; the key formal concept of a computational problem being NP-complete (nondeterministic polynomial-time complete) is strongly conjectured to be linked with a problem being intrinsically computationally intractable—that is, not generally solvable without exhaustive search of all possible solutions. Because the number of possibilities combinatorially explodes as the size of a problem-instance grows, such brute force searches are extremely computationally costly. For example, a 50 component system would have 50! possible alternative layouts, which is far more than the number of picoseconds since the Big Bang 20 billion years ago. Many of the most important real-world network optimization problems (e.g., the best known, Traveling Salesman) have been proven to be NP-complete or worse in computational complexity. Steiner tree and component placement optimization, problems examined here, are of this type, having been proven to be "NP-hard".

Steiner tree has been studied in its simplest form at least since the Renaissance *(12,13)*. The most relevant version of the problem is: given a set of fixed node loci, find the set of arcs (or branch segments) between those loci that interconnects all loci and has shortest total length. The resulting network will always constitute a tree. When it is permitted to have branch junctions only at node sites, it is a minimal spanning tree; when branch junctions may also occur at sites that are not nodes, it constitutes a Steiner tree. The total length of the Steiner tree for a set of nodes is equal to or less than the length of the minimal spanning tree for the nodes. For example, Figures 2A and B show, respectively, a minimal spanning tree and a Steiner tree for five nodes on a plane. The Steiner tree is about 4% shorter than the minimal spanning tree.

Since Steiner tree is a member of the class of NP-hard problems, it is not surprising that the largest unconstrained Steiner tree problems that can currently be solved have only approx 100 nodes (cf *[14]*). However, while minimal spanning trees are equal to or longer than corresponding Steiner trees, they are not at all computationally intractable; exact algorithms for them today perform well for quarter-million node sets. The basic question of goodness of fit of the Steiner tree concept to actual neuroanatomy is: do the dendritic and/or axonic arbors of a neuron form optimized Steiner trees interconnecting the cell body with a set of synaptic loci? The key idea needed for such applicability is that for real-world trees, living and nonliving, not all segments are equal: the concept of an optimal tree had to be extended to include variably weighted branches and trunks *(15)*.

Component placement optimization ("quadratic assignment problem"), the other wiring problem focused upon here, has received the most attention in computer science in connection with design of large scale integrated circuits *(16, 16a)*. The problem can be defined as: given the connections among a set of components, find the spatial layout

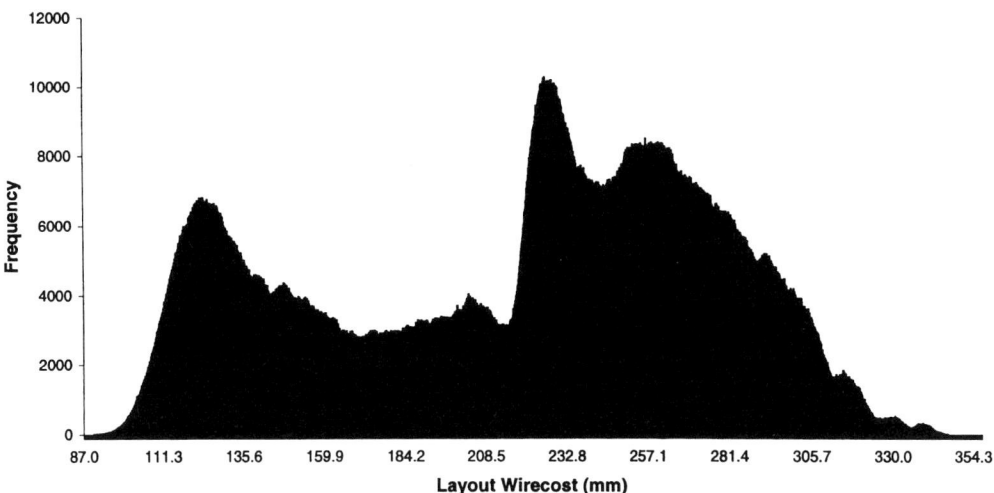

Fig. 1. Distribution of wirecosts (total wirelength) of all possible layouts of ganglia of *C. elegans*. A 10,000-bin histogram compiled from exhaustive search of all 39,916,800 alternative orderings of the 11 ganglia. Least costly and most costly layouts are rarest. In effect, the search approximates a simulation of the maximal possible history of the evolution of this aspect of the nervous system. The worm's actual layout (Fig. 2 in *[10]*) is in fact the optimal one, requiring the least total length of connecting fiber of any of the millions of possible layouts. For comparison, the last-place, "pessimal" layout would require about 4 times as much total connection fiber as the optimal one. (See *[7]*).

of components that minimizes total connection costs. The simplest cost-measure is length of connections (often represented as the sum of squares of the lengths); usually the possible positions for components are restricted to a matrix of "legal slots". As a simple example, Figures 3A and B diagram two of six possible configurations of components 1, 2, and 3 in slots A, B, and C; for the connections among the components, placement 3a requires the most total connection length and 3b the least.

Again, computation costs for the exact solution of component placement optimization problems are of a magnitude not encountered in most bioscience computing, outside of gene-sequencing, and constitute one of the principal technical impediments of this research. For n components, the number of alternative possible placements is n! (Size of this search space is generally unaffected by whether permissible component positions are located in 3, 2, or 1 dimensions.) Heuristic procedures that yield approx optimal solutions can be much more feasible, but their performance (e.g., how close to optimality are they likely to come) is not well understood.

Perhaps the most salient and daunting feature of nontrivial global optimization problems is the presence of local minima traps on the optimization landscape—that is, parameter values that yield least costs within a subregion of the search space, but not across the total space. For example, with regard to vector mechanical force-minimization treatments of the above two problems: (*i*) the dendritic tree of Figure 4C is suboptimal because of its topology, while Figure 4D shows the minimum cost topology; no vector mechanical tug of war re-embedding the suboptimal topology can ever transform it into the best topology; and (*ii*) similarly, Figure 5 shows a vector mechanical

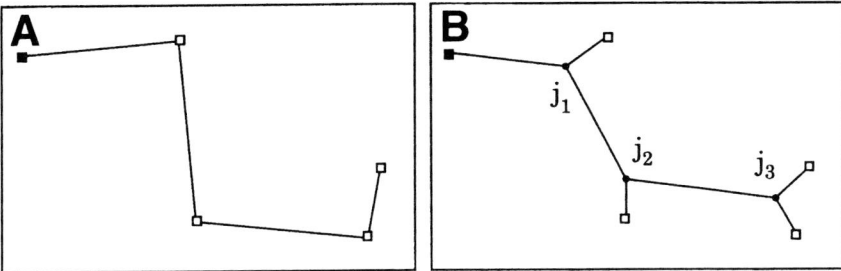

Fig. 2. Minimal spanning tree (**A**) and Steiner tree (**B**) for five nodes on a plane surface. The Steiner tree is shorter, but much more computationally costly to construct. Adapted from *(20)*.

local minimum trap for the roundworm ganglion component placement problem. The extensive modeling of cellular structures and processes in terms of compression–tension "tensegrity" by Ingber (e.g., *[17,17a]*) does not deal with local minima and, therefore, cannot account for such global optimization via evading such traps. Correspondingly, Van Essen's tension-based model of cortical folding, in terms of white matter tensegrity, also does not deal with local minimum traps and, so, will not suffice for global optimization problems of wiring minimization *(18–18b)*.

4.4. OPTIMIZATION MECHANISMS

From evaluating how well neural structures conform to minimum wiring principles of economical use of connections, we have gone on to seek biological mechanisms of the observed extremely fine network optimization. The emerging picture is that, corresponding to "Save wire" neuroanatomy optimization results, we have found neuroanatomical candidates for some of the most complex biological structures shown to be derivable purely from simple physical processes (cf *[19]*). This constitutes a further stage in developing an understanding of the generative rules that yield the highly complex anatomy of the nervous system.

4.4.1. Large-Scale Optimization of Dendrites and Axons

Some complex neuron arbor structure seems to be self-organizing, with no need of evolutionary mechanisms for its creation. While the key underlying pattern is network optimization, "Save wire" (in particular here, minimize total volume), the specific hypothesis in this case is, neuron arbor morphogenesis behaves like flowing water (see *[20]*). The volume minimizing fluid dynamic model yields two confirmed results: (*i*) it predicts diameters of branches at junctions; and (*ii*) from those diameters, branching angles and junction loci can then be predicted. The major methodological enterprise of the project centered on developing STRETCH—a package of algorithms for the computationally intractable (NP-hard) task of generating optimal trees *(13)* with variable branch-weights—as the gold standard against which to compare observed neuroanatomical trees.

Neural fluid mechanics is a simple fluid-dynamical model, for minimized walldrag of pumped flow through a system of pipes, that will predict branch diameters of some types of dendrites (e.g., of mammalian retinal ganglion cells) and axons (e.g., in rodent

Fig. 3. Component placement optimization: two alternative placements of elements 1, 2, and 3 in positions A, B, and C. For the given interconnections, placement (**A**) has greater total connection length than placement (**B**). Adapted from *(7)*.

thalamus) almost as well as it predicts configuration of nonliving tree structures such as river drainage networks. For neurons, the fluid dynamics falls in the laminar-flow vs turbulent regime.

Waterflow in branching networks, in turn, acts like a tree composed of a weights-pulleys-cords system (non-Hooke's Law), that is, vector mechanically; so also do the neuron arbors. As a result, the trees globally minimize their total volume to about 5% of optimum for interconnecting their terminals (see Fig. 4). One unanticipated moral that emerges is that, in a sense, "Topology does not matter", that is, the worst or "pessimal" connection pattern typically costs only relatively little more than the optimal pattern, compared to the wide corresponding possible range of costs for embedding a given topology. The conclusion here is only that this minimum volume configuration is the default neuron arbor structure, probably often modified in many complex ways (cf, e.g., *[21]*).

4.4.2. C. elegans *Ganglion Placement Optimization*

We have extended the above results on large-scale optimization of individual neuron arbors to the entire *C. elegans* nervous system. The basic picture is indeed that vector mechanics suffices for optimization of placement of the ganglia of *C. elegans*. As mentioned earlier, our prior research had found that the actual placement of the ganglia in the worm was optimal, in that it required the least total length for the animal's (approx 1000) interconnections, out of roughly 40 million alternative possible ganglion orderings (see Fig. 1). We had also reported a related set of optimization results for rat, cat, and macaque cortex, in terms of placement of connected Brodmann areas that conforms to an "Adjacency Rule" *(7,10,22,23)*. (Nonetheless, as we have noted *(22)*, the majority of connections in the actual worm are not to nearby components; therefore, merely positioning components so their connections tend to go to nearby components will not in itself suffice to yield the minimum wirecost layout.) As mentioned, if this 1-in-10-million type of result is replicated, it begins to approach some of the most precise confirmed neuroanatomy predictions (see *[24, 24a]*); hence, we sought convergent support by finding feasible mechanisms for such fine-grained optimization.

We have constructed Tensarama, a force-directed placement simulator, in which each of the worm's connections behaves like a microweight-and-pulley system (Fig. 5 shows a screendump). Analog hardware devices of this type have been used to solve simple (noncombinatorial) placement optimization problems for over a century *(25, 25a)*. (Similar "mesh of springs" simulations have become a focus of current modeling

Optimal-Wiring Models of Neuroanatomy

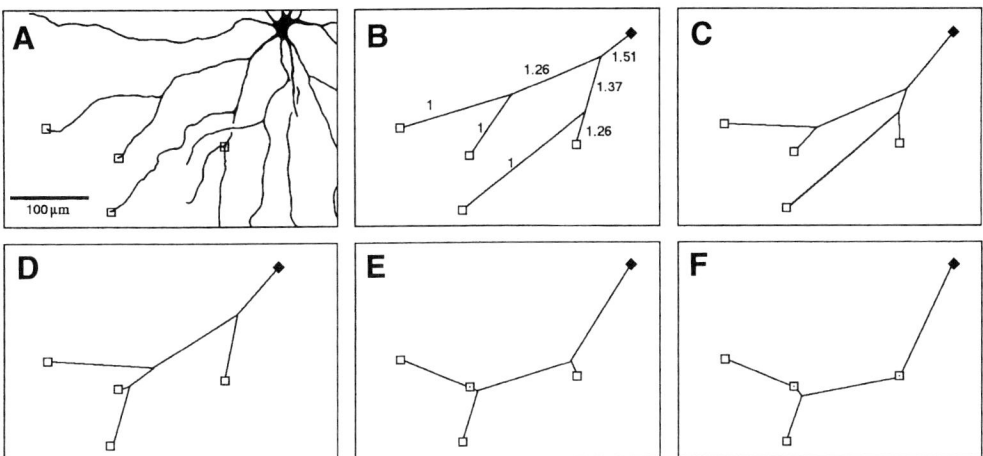

Fig. 4. Optimization analysis of a 5-terminal subtree from dendritic arbor of an α ganglion cell in rabbit retina. (**A**) A quadrant of the original camera lucida drawing containing the subtree (after *[35]*); soma is in upper right corner. "Leaf terminals" of the analysis are boxed (note that one of them is not a branch termination); "root-terminal" is at soma. (**B**) Wireframe representation of actual tree, with branch segments straightened between loci of terminals and internodal junctions. The labels give diameters assigned to the branch segments via the power law of the laminar-flow model. (**C**) Optimal re-embedding of the topology of the actual tree, with respect to total volume cost, via the STRETCH algorithm; this minimum volume embedding of the actual topology is 1.06% cheaper than the volume of the actual tree in panel **B**. (**D**) Optimal embedding of the optimal topology for the given terminal loci, with respect to volume-cost. It can be seen to differ from the actual topology of panels **A–C**; it is only 2.64% cheaper in volume than the actual topology in its actual embedding, in panel **B**. (**E**) Optimal embedding of the optimal topology, with respect instead to total tree surface area; actual vs optimal error is now 27.22%, much greater. (**F**) Optimal embedding of the optimal topology, with respect to total tree length; actual vs optimal error is now 60.58%, even greater. Thus, this dendritic arbor best fits a minimum volume model. Adapted from *(20)*.

of protein folding of amino acid chains *[26]*.) Over a wide range of initial input configurations of the ganglia, our vector mechanical net outputs the actual layout via tug-of-war, converging upon equilibrium at the actual minimum wirecost positioning of the ganglia—without major susceptibility to local minima traps. We have also constructed Genalg, a genetic algorithm (cf *[27]*) package that stably outputs the actual minimum wirecost placement (see Fig. 5); it is, in effect, a demonstration that evolutionary processes suffice for worm wiring optimality. (A caveat on interpretation of the vector mechanical models: while actual physical forces appear to drive neuron arbor optimization, it is likely in the case of nematode ganglion layout that the forces involved should instead be viewed more abstractly as governing natural selection processes; neuron somata need not, in fact, move during development of the individual organism. We have similarly argued *(7)* concerning the simplest neural component placement problem, of brain positioning, that the brain's sensory-motor connections, of course, do not behave literally vector mechanically over evolutionary history.)

But the bottom line here once more seems to be that, in a sense, "Physics suffices": since no genome is required for this self-organization, some interesting limits may thereby emerge on the Central Dogma of genetics. (Cf also the related picture regard-

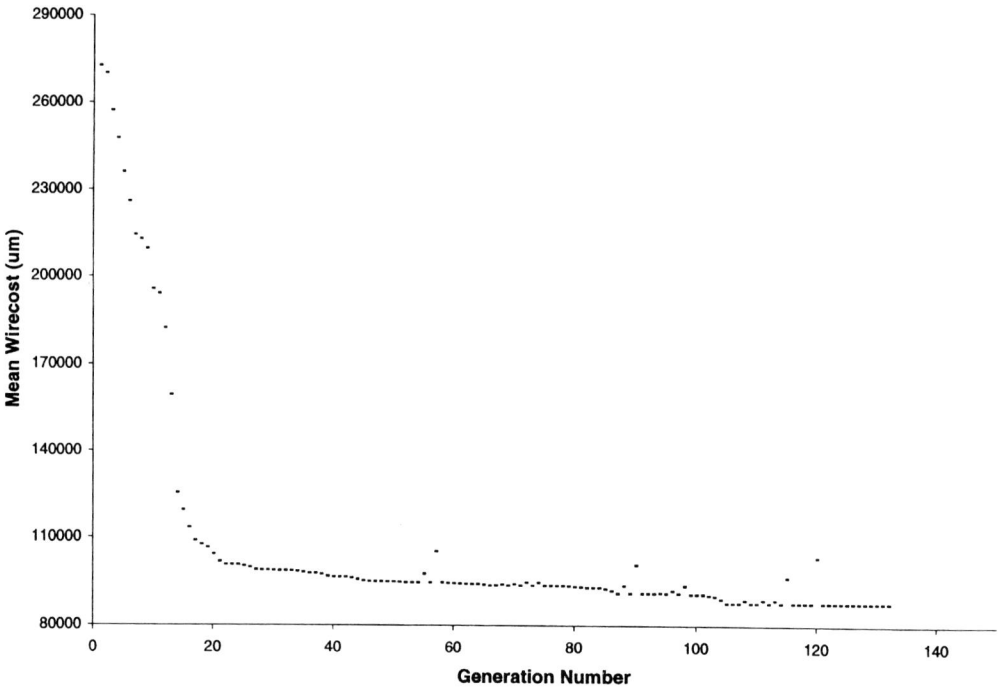

Fig. 5. GenAlg, a simple genetic algorithm, rapidly and reliably finds the optimal (minimum wirelength) layout of *C. elegans* ganglia. The initial population in this run is small, 10 individuals, each here with a reverse ganglion ordering of that found in the actual worm; the algorithm converges upon the minimum total wirecost layout (87,803 μm) in only 130 generations. The evolution of wirecost shows the usual pattern: a very rapid initial improvement of fitness (about 90% during the first 20 generations), followed by a much longer slower fine-tuning phase to optimality. Some of the random mutations cause the half-dozen brief "blips" of increased mean wirecost of the population during the later phase. The robust performance of this genetic algorithm, and also of our force-directed placement algorithm, is further converging support for the hypothesis that the actual layout of *C. elegans* is in fact perfectly optimized.

ing prions [28].) A discrete-state process like a genetic algorithm is not needed to generate highly complex types of biological structure. One rationale for nongenomic anatomy-generating processes, as well as for such simple generative rules as "Save Wire," is apparent in a dilemma that nature confronts: human brain wiring is among the most complex structures known in the universe, yet its layout information must pass through the "genomic bottleneck" of very limited DNA information representation capacity *(3,15)*. The harmony of neuroanatomy and physics suggested here would lower this hereditary information load by accomplishing network optimization without required participation of the genome.

Another observation of robustness worth further study is that, for both the global arbor and ganglion neural optimization problems, random noise injection (e.g., as in simulated annealing [29]) generally was not needed to evade local minima traps—unlike for typical network optimization problem instances.

Mapping "chaotic" optimization landscapes: we have found that both a genetic algorithm like GenAlg and a force-directed placement (FDP) algorithm like Tensarama

```
                                                Input: actual.mtx

                    T E N S A R A M A
Head                                                    Tail
0      0     1      1      2      2      3      3      4      4      5
0      5     0      5      0      5      0      5      0      5      0    Tetrons

   PH    (100.000000)
     AN    (300.000000)
       RNG    (440.000000)
         DO    (506.000000)
         LA    (564.000000)
            VN    (744.000000)
              RV    (1096.000000)
                        VCa    (2004.000000)
                                          VCp    (4004.000000)
                                                    PA    (4874.000000)
         DR    (854.000000)
           LU    (928.000000)

Final layout popped out after:   1,000,000 iterations
Tension Constant:    0.001000
Total Wirecost:   88485.250000 um
```

Fig. 6. Runscreen for Tensarama, an FDP algorithm for optimizing layout (minimizing total wirecost) of *C. elegans* ganglia. This vector mechanical simulation represents each of the worm's approx 1,000 connections (cf Fig. 2 in *[10]*) as a weight-and-pulley (non-Hooke's Law) element acting upon the movable ganglia "PH", "AN", etc. (Connections do not appear on the runscreen, nor do fixed components such as sensors and muscles). At each iteration, the program computes net horizontal force on each ganglion, and correspondingly updates its left/right position; the cycle is repeated a given number of times. (Ganglion locations are in "tetrons", or quarter-microns, to decrease round-off errors.) The most striking feature of Tensarama performance for the actual worm's connectivity matrix is its comparatively low susceptibility to local minima traps—unlike Tensarama performance for minor modifications of the actual connectivity matrix and unlike FDP algorithms, in general, for circuit design. However, the above screendump shows the final configuration of the system for an identified "killer" initial configuration input of the actual matrix: Tensarama has frozen in a local minimum with ganglia in positions (notably, ganglia DR and LU in head, rather than tail) that yield a final layout wirecost of 88,485.25 µm, about 0.8% more than the actual layout. The fatal initial layout here (ganglion left edges at 0 tetrons) differs only slightly from a quite innocuous initial layout (ganglion centers at 0).

perform notably well in optimizing ganglion placement for the actual connectivity matrix (Fig. 2 in *[10]*) of *C. elegans*. However, this good performance turns out to be interestingly narrow-tuned: (*i*) adding or removing as little as a single connection (of approx 1,000 total) in some cases can change the actual matrix into a "killer matrix" input that is highly prone to paralyzing an FDP algorithm in local minima traps; (*ii*) similarly, some "killer layout" initial input positionings of the ganglia of the actual matrix will paralyze the FDP algorithm (Fig. 6). Each of these instances of discontinuous, very sharply tuned performance prima facie suggests chaotic structure (e.g., *[30]*) and seems worthy of systematic exploration. Each exhibits a Butterfly Effect: some quite small changes of input conditions, but only in a limited range, yield drastic changes in behavior. We need to compare these natural neuroanatomical connection matrices with some typical benchmark micro circuits (e.g., [31]). Such studies entail basic mapping of the optimization terrain. For instance, the "neighborhood" around actual layouts (i.e., the subregion of nearby layouts that differ from the actual one by only a small number of component swaps) appears to be a particularly good one, which is richer in lower cost layouts than randomly sampled zones.

4.5. FUNCTIONAL ROLE OF NEURAL OPTIMIZATION

Finally, a larger question: if one takes seriously the above instances of distinctively fine-grained neural optimization, a larger question emerges: why is such extreme connectivity minimization occurring? Of course, "Save wire" has obvious fitness value as explained earlier—in reducing volume of a delicate metabolically costly tissue and in reducing signal propagation delays in a notably slow transmission medium. For instance, neuron volume minimization directly minimizes the significant metabolic "pumping" costs of maintaining ion concentrations across cell membranes *(32)*. Nonetheless, such optimization nearly to absolute physical limits is only rarely encountered in biology (for instance, human visual and auditory system amplitude sensitivities under good conditions and the silk moth olfactory system, which can detect single molecules of mating pheromones *[33–33b]*). The usual view (e.g., *[34]*) is that nature cannot afford to optimize, but instead, like any finite resource engineer, only satisfices with a compromise among competing desiderata that is "good enough". Natural selection almost never gets to begin with a clean slate, but instead must design organisms as a prisoner of prior evolutionary history.

Thus, the type of striking neural optimization we are observing, in itself, needs explanation regarding its functional role: it could be a clue about basic brain mechanisms that require such extraordinary connectivity minimization, and/or a sign of some unexpectedly feasible means of attaining such optimization, and/or, as mentioned above, part of an economical scheme for generation of complex structure. Neuroanatomical cases, where such optimization is not present, become as diagnostically significant as cases where it is present. "Why", thus, becomes as important as "how" here. Attention thereby turns to issues of neural function as well as structure—in any case, the two really mesh seamlessly. Just as a real brain does not consist of infinitely thin wires, its connections do not have virtually infinite signal propagation velocity. Hence, the methodological approach we started with for brain structure volume, and the issue of stringency of limits upon it, needs in turn to be recapitulated for brain function and its temporal constraints.

ACKNOWLEDGMENTS

We are grateful for suggestions of David Boothe. This work was supported by National Institute of Mental Health (NIMH) Grant No. MH49867. A version of this paper was presented at National Institute of Health/National Institute of Neurological Disorders and Stroke (NIH/NINDS) Colloquium, Bethesda MD, February 2001.

REFERENCES

1. Cherniak C. Minimal Rationality. MIT Press, Cambridge, MA, 1986.
2. Cherniak C. Philosophy and computational neuroanatomy. Philos Stud 1994; **73**:89–107.
3. Cherniak C. Undebuggability and cognitive science. Communica Assoc Comput Mach 1988; **31**:402–412.
4. Cherniak C. The bounded brain: toward quantitative anatomy. J Cognit Neurosci 1990; **2**:58–68.
5. Thompson D. On Growth and Form. Cambridge University Press. New York, 1917/1961.
6. Cajal S. Ramon y. Histology of the Nervous System of Man and Vertebrates, Vols. I & II. Swanson N, Swanson L., transl. Oxford University Press, New York, 1909/1995.
7. Cherniak C. Neural component placement. Trends Neurosci 1995; **18**:522–527.
8. Mitchison G. Neuronal branching patterns and the economy of cortical wiring. Proc R Soc Lond B 1991; **245**:151–158.
9. Albertson D, Thomson J. The pharynx of *Caenorhabditis elegans*. Philos Trans R Soc Lond B 1976; **275**:299-325.
9a. White J, Southgate E, Thomson J, Brenner S. The structure of the ventral cord of *Caenorhabditis elegans*. Philos Trans R Soc Lond B 1976; **275**:327–348.
9b. White J, Southgate E, Thomson J, Brenner S. The structure of the nervous system of the nematode *Caenorhabditis elegans*. Philos Trans R Soc Lond B 1986; **314**:1–340.
9c. Wood W (ed.) The Nematode *Caenorhabditis Elegans*. Cold Spring Harbor Laboratory Press, Cold Spring Harbor, NY, 1988. See also Achacoso T, Yamamoto W. *AY's Neuroanatomy of C. elegans* for Computation. CRC Press, Boca Raton, 1992.
10. Cherniak C. Component placement optimization in the brain. University of Maryland Institute for Advanced Computer Studies Technical Report, No. 91-98, 1991; J Neurosci 1994; **14**:2418–2427.
11. Garey M, Johnson D. Computers and Intractability: A Guide to the Theory of NP-Completeness. W.H. Freeman, San Francisco, 1979. Lewis H, Papadimitriou C. The efficiency of algorithms. Sci Am 1978; **238**:96–109.
11a. Stockmeyer L, Chandra A. Intrinsically difficult problems. Sci Am 1979; **240**:140–159.
12. Courant R, Robbins H. What is Mathematics? Oxford University Press, New York, 1941/1969.
13. Hwang F, Richards D, Winter P. The Steiner Tree Problem. North-Holland, Amsterdam, 1992.
14. Bern M, Graham R. The shortest-network problem. Sci Am 1989; **260**:84–89.
15. Cherniak C. Local optimization of neuron arbors. University of Maryland Institute for Advanced Computer Studies Technical Report, No. 90-90, 1990; Biological Cybernetics 1992; **66**:503–510.
16. Kuh E, Ohtsuki T. Recent advances in VLSI layout. Proc IEEE 1990; **78**:237–263.
16a. Sherwani N. Algorithms for VLSI Physical Design Automation, 3rd ed. Kluwer, Boston, 1995. Mead C. Analog VLSI and Neural Systems. Addison-Wesley, Reading, MA, 1989.
17. Ingber D. Cellular tensegrity: defining new rules of biological design that govern the cytoskeleton. J Cell Sci 1993; **104**:613–627.
17a. Ingber D. The architecture of life. Sci Am 1998; **278**:48–57.
18. Van Essen D. A tension-based theory of morphogenesis and compact wiring in the central nervous system. Nature 1997; **385**:313–318.
18a. Armstrong E, et al. Cortical gyrification in the rhesus monkey: a test of the mechanical folding hypothesis. Cereb Cortex 1991; **1**:426–432.
18b. Armstrong E, et al. The ontogeny of human gyrification. Cereb Cortex 1995; **5**:56–63.
19. Kauffman S. At Home in the Universe. Oxford University Press, New York, 1995.
20. Cherniak C, Changizi M, Kang D. Large-scale optimization of neuron arbors. University

of Maryland Institute for Advanced Computer Studies Technical Report, No. 96–78, 1996; Phys Rev E 1999; **59**:6001–6009. (http://pre.aps.org/).
21. Purves D, Lichtman J. Principles of Neural Development. Sinauer, Sunderland, MA, 1985.
22. Cherniak C. Reply to Letter to Editor. Trends Neurosci 1996; **19**:414–415. (www.wam.umd.edu/~cherniak/).
23. Young M. Objective analysis of the topological organization of the primate cortical visual system. Nature 1992; **358**:152–155.
24. Brush S. Feynman's success: demystifying quantum mechanics. Am Sci 1995; **83**:476–477.
24a. Kinoshita T. New value of the alpha^3 electron anomalous magnetic moment. Phys Rev Lett 1995; **75**:4728–4732.
25. Quinn N. The placement problem as viewed from the physics of classical mechanics. In: Proceedings of the 12th IEEE Design Automation Conference, 1975, 173–178.
25a. Francis R, McGinnis L, White J. Facility Layout and Location, 2nd ed. Prentice-Hall, Englewood Cliffs, NJ, 1992.
26. Berendsen H. Protein folding: a glimpse of the holy grail? Science 1998; **282**:642–643.
27. Mitchell M. An Introduction to Genetic Algorithms. MIT Press, Cambridge, MA, 1996.
28. Keyes M. The prion challenge to the 'Central Dogma' of molecular biology, 1965–1991. Stud Hist Biol Biomed Sci 1999; **30**:1–19.
29. Kirkpatrick S, Gelatt C, Vecchi M. Optimization by simulated annealing. Science 1983; **220**:671–680.
30. Thompson J, Stewart H. Nonlinear Dynamics and Chaos. Wiley & Sons, New York, 1986.
31. (http://vlsicad.cs.ucla.edu/~cheese/ispd98.html).
32. Laughlin S, de Ruyter van Steveninck R, Anderson J. The metabolic cost of neural information. Nat Neurosci 1998; **1**:36–41.
33. Cornsweet T. Visual Perception. Academic Press, New York, 1970.
33a. Green D. Introduction to Hearing. Lawrence Erlbaum, Hillsdale, NJ, 1976.
33b. Shepherd G. Neurobiology, 2nd ed. Oxford, New York, 1988.
34. Gould S. The Panda's Thumb. Norton, New York, 1980.
35. Peichl L, Buh, E, Boycott B. Alpha cells in rabbit retina. J Comp Neurol 1987; **263**:25–41.

5
The Modeler's Workspace

Making Model-Based Studies of the Nervous System More Accessible

Michael Hucka, Kavita Shankar, David Beeman, and James M. Bower

ABSTRACT

A realistic neuronal model represents a modeler's understanding of the structure and function of a part of the nervous system. The increasing number of such models represents a significant accumulation of knowledge about the structural and functional organization of nervous systems. However, locating appropriate models and interpreting them becomes increasingly more difficult as the number of online model and experimental databases grows. The central motivation for the Modeler's Workspace project is to address these problems.

The Modeler's Workspace is a collection of software tools being created to enable users to interact with databases of models and data. It will provide facilities for: searching multiple remote databases for model components based on various criteria; visualizing the characteristics of the components retrieved; creating new components, either from scratch or derived from existing models; combining components into new models; linking models to experimental data as well as online publications; and interacting with simulation packages such as GENESIS to simulate the new constructs. It is being written in Java for portability and extensibility. It is modular in design and uses pluggable components. To increase the probability that the Modeler's Workspace will be compatible with future databases and tools, we are using the XML, the eXtensible Markup Language, as the interchange format for communicating with databases.

5.1. INTRODUCTION

A structurally realistic neuronal model represents a modeler's understanding of the structure and function of a part of the nervous system. As the number of neurobiologists constructing realistic models continues to grow, and as the models become ever more sophisticated, they collectively represent a significant accumulation of knowledge about the structural and functional organization of nervous systems. But at the same time, locating appropriate models and interpreting them becomes increasingly more difficult as the number of online model and experimental databases grows. This

From: *Computational Neuroanatomy: Principles and Methods*
Edited by: G. A. Ascoli © Humana Press Inc., Totowa, NJ

is exacerbated by the fact that computational models developed by different researchers are often implemented using different software tools.

The central motivation for the Modeler's Workspace project is to address these problems *(1)*. Our goal is to develop a free open software environment that provides a variety of capabilities, including: searching multiple remote databases for model components based on various criteria; creating new components; combining model components together and translating them into formats suitable for simulation systems such as GENESIS *(2)*, NEURON *(3)*, XPP *(4)*, or NEOSIM *(5)*; managing personal databases of models and other information; and collaborating interactively with other researchers to work with models and simulations. The Modeler's Workspace is being written in Java *(6)* for portability and extensibility. It is modular in design and uses pluggable components. To increase the probability that the Modeler's Workspace will be compatible with future databases and tools, we are using the XML, the eXtensible Markup Language *(7)* as the interchange format for communicating with databases.

In this chapter, we describe a design for the Modeler's Workspace user interface, overerall architecture, model representation scheme, and the motivations behind the various design decisions. We believe that the Modeler's Workspace can provide integrated access to a wide variety of model databases and simulation tools. The system's modular and extensible design will make it possible for others to write new components, for example for handling new forms of model representations or interfacing to new databases. In fact, it is our hope that by providing a sound, open framework, and a collection of elements demonstrating its capabilities, others will be encouraged to contribute to its implementation and add new functionality to the environment, rather than create entirely new software tools or simulator-dependent model representations.

5.2. THE NEED FOR MODEL-BASED APPROACHES

The field of biology has made great advances over the last several decades in our ability to describe experimentally the organization of biological systems. However, with these technical advances has come an enormous increase in the amount of available data, in many cases already exceeding the ability of single investigators to keep up with what is known. For this reason neuroscience, and biology as a whole, is running the risk of becoming increasingly fragmented and disorganized.

The stress of information overload and the lack of appropriate conduits for information have lead to several national and multinational efforts to develop electronic databases *(8)*. Most of these projects are based on the idea that information in electronic form is more readily accessible and manipulated than traditional forms of publication. However, it is our view that in order for these efforts to be successful, they must provide enhanced functionality over traditional means of distributing information. To take one example, efforts at providing electronic forms of data delivery need to address the difficult issue of data validation. Data validation involves not only assessing the likely accuracy of the information stored, but also providing some measure of the relevance of the data to the current state of our understanding. In traditional forms of scientific data reporting, the peer review process provides the means of validating accuracy, while the discussion sections of published papers provide the opportunity to speculate on the functional significance of the results. It is not clear that electronic databases in their existing forms will provide these functions. Creators of electronic databases seem to assume that if the databases are available, interested parties will figure out how to use them appropriately.

5.2.1. The Role of Structurally Realistic Models

We believe that computational models provide a much more sophisticated, flexible, and powerful base for electronic storage and retrieval of information than do traditional on-line databases. Our work has, therefore, focused on developing tools for working with structurally realistic biological models. These are models whose first objective is to capture what is known about the anatomical structure and physiological characteristics of a neural system of interest.

In the case of a model of part of the brain, for example, this modeling approach typically starts with as detailed a description as possible of the relevant neuroanatomy. At the single-cell level, this usually means a description of the 3D structure of the neuron and its dendritic tree *(9)*. Most contemporary modeling approaches for realistic models of neural cells are based on compartmental modeling *(10)*. In each case, the overall compartmental structure of the model is first established on the basis of neuroanatomical details. Information about neuronal morphology used in computational models typically includes such parameters as soma size, length of interbranch segments, diameter of branches, bifurcation probabilities, and density and size of dendritic spines. At the neuronal network level, parameters include some description of the cell types found in the network and their connectivity *(11)*. The next stage of structurally realistic model building involves establishing the basic physiological behavior of the modeled structure, for instance by tuning the model to replicate neuronal responses to experimentally-derived data *(9,12)*. The final stage of model construction involves exploring the model's behavior to novel inputs, using it to generate new ideas about the neural system's function, as well as a guide for new experimental investigations.

As a result of their faithfulness to biological anatomy and physiology, structurally realistic models can be a means of storing anatomical and physiological experimental information. As model sophistication grows, structural models themselves become a form of information storage about the biological systems being studied.

5.2.2. Data Evaluation and Functional Assessment

Because of their nature, structurally realistic models also contain precise information about the relationships between known facts. For example, neurons display a wide range of dendritic morphologies, ranging from a single simple dendrite in retinal bipolar cells to compact but highly branched spiny and smooth arborizations of the cerebellar Purkinje cell. Detailed compartmental models of reconstructed neurons have become important tools for investigating how dendritic morphology and membrane biophysics interact *(13)* in integrating synaptic inputs *(14)* and propagation of action potentials *(15,16)*, with further implications for development and plasticity *(16)*.

Another issue related to the quality of stored electronic data involves the question of which data are most relevant and, therefore, useful to our current knowledge of a particular system. Just because data can be collected does not necessarily mean that this data will help expand our current understanding of a particular system. Modeling can reveal the importance of a particular data point as well as provide an immediate context for the data once obtained. To take one example, in one effort at modeling Purkinje cells *(9,17)*, the morphology of the models was based on a detailed light microscopic reconstruction of horseradish peroxidase-filled guinea pig Purkinje cells *(17,18)*, and all simulations were usually performed with a model based on the morphology of cell 1. However, when identical channel equations and densities were placed onto two additional Purkinje cells with morphologies labeled as cell 2 and cell 3 *(19)*, the details of the firing patterns were quite different for these three morphologies. The firing pattern of cell 3 with current injection was more similar to cell 1, but with a shift in the frequency-current curve. This turned out to be because of the small soma and short thin main dendrite, which caused smaller total potassium delayed rectifier and muscarinic conductances in the model of this cell *(9,17)*. This understanding would be difficult to achieve without the ability to perform "computer experiments" on structurally realistic models.

Finally, perhaps the most valuable contribution of a structural model is the way in which it can help to organize our understanding of the system being studied. In the large and growing electronic databases of the most common type in use today, it is not at all clear how the data will contribute to our understanding of function. Models, however, are specifically constructed to explore the functional implications of the data on which they are built. Models, thus, can serve not only as the point of entry for data; they can also serve as dynamic tools that can be used to understand its significance. As models become more sophisticated, so does the representation of the data. As models become more capable, they extend our ability to explore the functional significance of the structure and organization of biological systems. Thus, there is a direct link between the ultimate objective of acquiring the data and the data acquisition process itself.

5.3. OVERVIEW OF THE MODELER'S WORKSPACE

The Modeler's Workspace project is an effort to create a software environment that will make it easier for biologists to develop, use, and share structurally realistic models and, thereby, gain some of the benefits discussed above. As mentioned in the introduction, our goal is to provide the following kinds of facilities in the system:

1. Search and retrieval facilities for interacting with multiple databases of models and other information.
2. Facilities for creating, editing and visualizing models and their components.
3. Facilities for combining model components together and translating them into formats suitable for simulation systems such as GENESIS and NEURON.
4. Facilities for linking models to experimental data as well as online publications.
5. Facilities for managing a personal database where a user can collect models and other objects.
6. Collaboration facilities for connecting one or more users together, to allow them to simultaneously edit objects in a shared database and communicate with each other using chat facilities.

There are three main elements in the system: a user interface, a database server, and a global registry and repository called the Modeler's Workspace Directory. In this section, we summarize the overall organization and intended operation of the Modeler's Workspace from a user's perspective. In Section 5.4., we turn to the software architecture of the system, describing in more detail its extensible modular structure.

5.3.1. The Modeler's Workspace User Interface and Workspace Database

The Modeler's Workspace User Interface is being implemented as a stand-alone program written in Java that can run either as a separate application or as an applet from within a Web browser. All user interactions with the Modeler's Workspace take place through the User Interface component. The Workspace Database, the database component of the system, is a separate server program that can act as both a private repository for a user's work (where models, notes, and other objects are stored) and as a rendezvous point for collaborative activities. The User Interface communicates with the Workspace Database using network protocols.

A Workspace Database contains objects that represent different types of entities, such as models or bibliographic references. Each database object is structured according to a particular template. A template defines the format of an object, including its attributes and subattributes and the permitted types of data that can be stored in each attribute. Model attributes potentially can be of any type, including other models. Templates have names, and different templates are used to define different types of objects. Some of the basic types of objects predefined by the Modeler's Workspace are "Author", "Reference", "Data", "Neuron Model", and "Ion Channel Model." We discuss templates and representation issues in more depth in Subheading 5.5.

The separation of the two components allows the flexibility of connecting to a Workspace Database from different computers. The User Interface will be available for downloading from the Modeler's Workspace Web site and will run on any Java-enabled computer connected to the Internet. When the User Interface is started, it prompts for the network address of the Workspace Database, as well as a login name and password for accessing the database. This approach allows a roaming user to access his or her models and notes from anywhere on the network, using almost any kind of computer—from a public-access computer in a library to a laptop connected to the Internet via a modem.

5.3.2. Elements of the User Interface

There are three central regions in the User Interface: the Build pane, the Search pane, and the Connect pane. The first pane provides an interface for managing one's personal database of objects; the second, for searching other databases; and the third, for connecting to a specific Workspace Database for the purpose of browsing its contents and (optionally) engaging in collaborative activities with other connected users.

Figure 1 shows a prototype of the Build pane in the User Interface. The upper left region of the Build pane contains a list of the template types known to the system; the upper right region provides a listing of the objects in the user's Workspace Database that are based on a selected template; and the bottom half of the pane contains a window to the command line interpreter of a neuronal simulation package (in this example, GENESIS). The user may select objects in the Workspace Database list in the upper right and perform actions on them, such as editing them in an inspector window

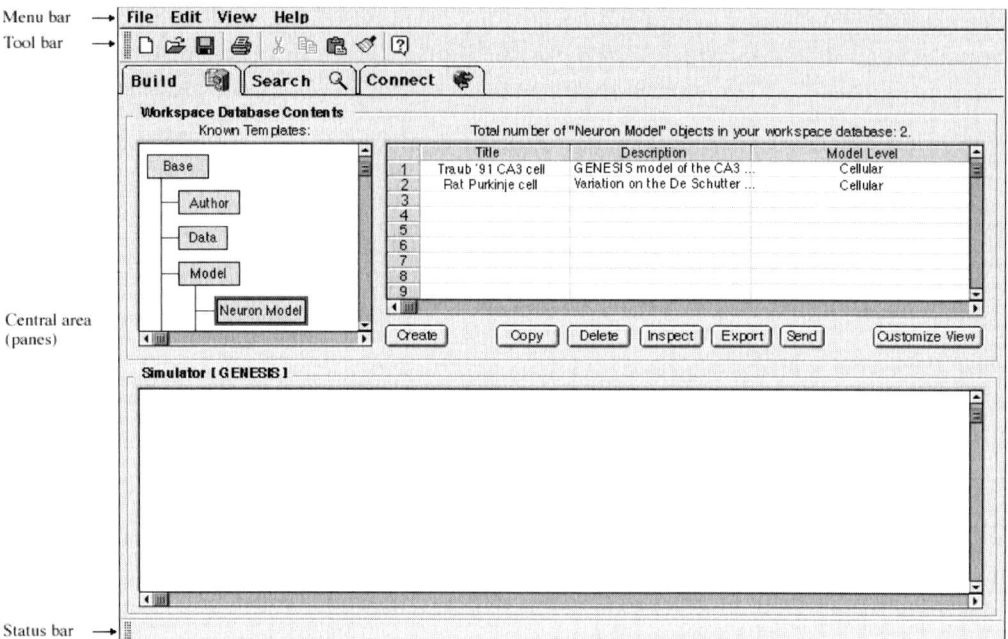

Fig. 1. A prototype of the Modeler's Workspace User Interface. It consists of a menu bar, a tool bar, a large central area, and a status bar at the bottom. This particular prototype shows the Build pane.

(described next) or sending them to the running simulation program. The columns displayed in the table can be changed through a dialog box accessed through the "Customize View" button.

Editing and viewing of objects in the Build pane takes place using graphical interface tools called inspectors. An inspector is simply a user interface module designed to let a user interact with information in a certain way. The default inspector in the Modeler's Workspace is called the Generic Inspector; it displays an object using a tree-structured table of attribute-value pairs and is used as the default editor/viewer for those types of objects that do not have their own specialized inspector.

Inspectors are incorporated into the Modeler's Workspace through a pluggable component architecture (with "plugin" software modules), so that new ones can be added dynamically to support new templates. Specialized inspectors may provide other modes of interaction besides the form-based approach of the Generic Inspector; for example, a single-cell inspector would provide a graphical, 3D tree viewer/editor for working with cell morphologies. A few different inspectors will be provided with the Modeler's Workspace. As others become available, they will be made available for downloading through the Modeler's Workspace Directory (see Subheading 5.6.).

The Search pane will provide the ability to search multiple databases for objects having specific characteristics. As shown in the prototype in Figure 2, the pane contains three main regions. The first region provides a pull-down list of known templates and a form. Once the user selects an object type from the list of templates, in order to indicate the type of object desired, the form is filled with attribute name-value slots

The Modeler's Workspace

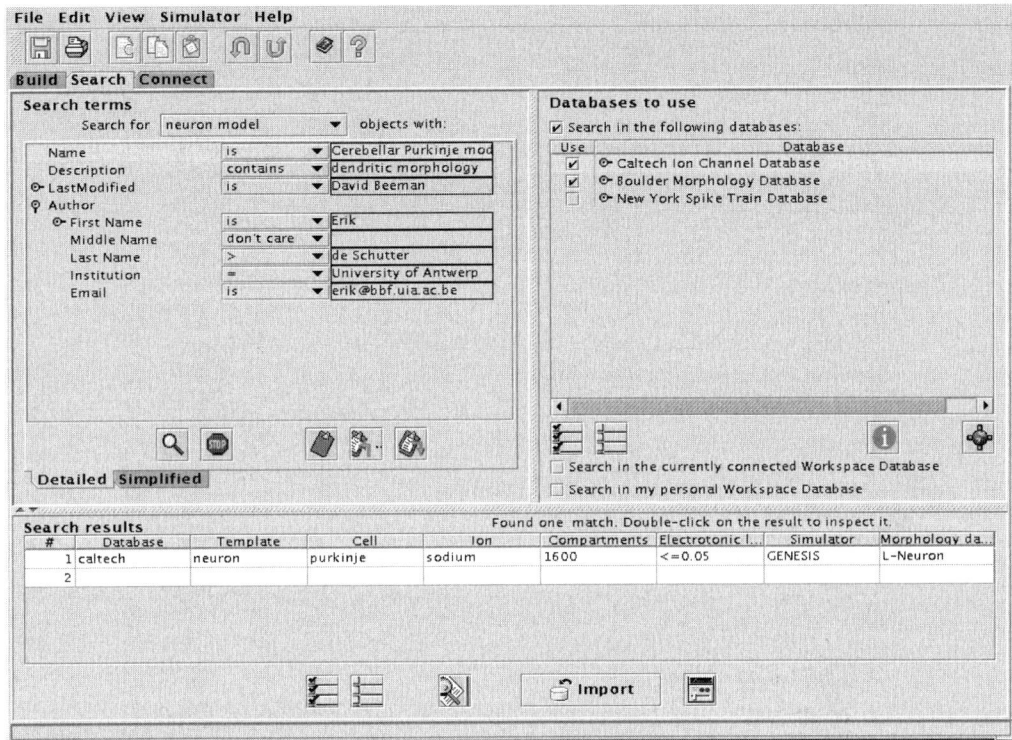

Fig. 2. A prototype of the Search pane. The upper left area allows the user to specify the object characteristics to search for; the upper right area allows the user to specify which databases should be used in a given search; and the bottom half provides a summary of the search results.

corresponding to the chosen template. This allows the user to specify the attribute values on which to search. The form is similar to that presented by the Generic Inspector mentioned above. The second region in the Search pane displays a list of databases. This allows the user to select which databases should be used for a search. The third region in the Search pane contains a table listing the results of the search. The columns in this table summarize the different attributes of the objects matched by the search. To obtain more detailed information about a given object, the user can double-click on a line in the table to view the object in an inspector window. More than one object can be inspected simultaneously, allowing, for example, multiple models to be examined side-by-side in separate windows. The user can also import objects from the search results into their personal Workspace Database.

In addition to searching multiple databases using the Search pane, a user may want to connect to a single Workspace Database and browse its contents. The Connect pane provides this one-database-at-a-time connection capability. It resembles the Build pane in organization, with a line at the top naming the currently connected database and a central region containing a table listing the contents of the database. As in the Build pane, the user can double-click on an entry to examine it in detail in an inspector window. If the user has write access to the Workspace Database, she or he can also edit

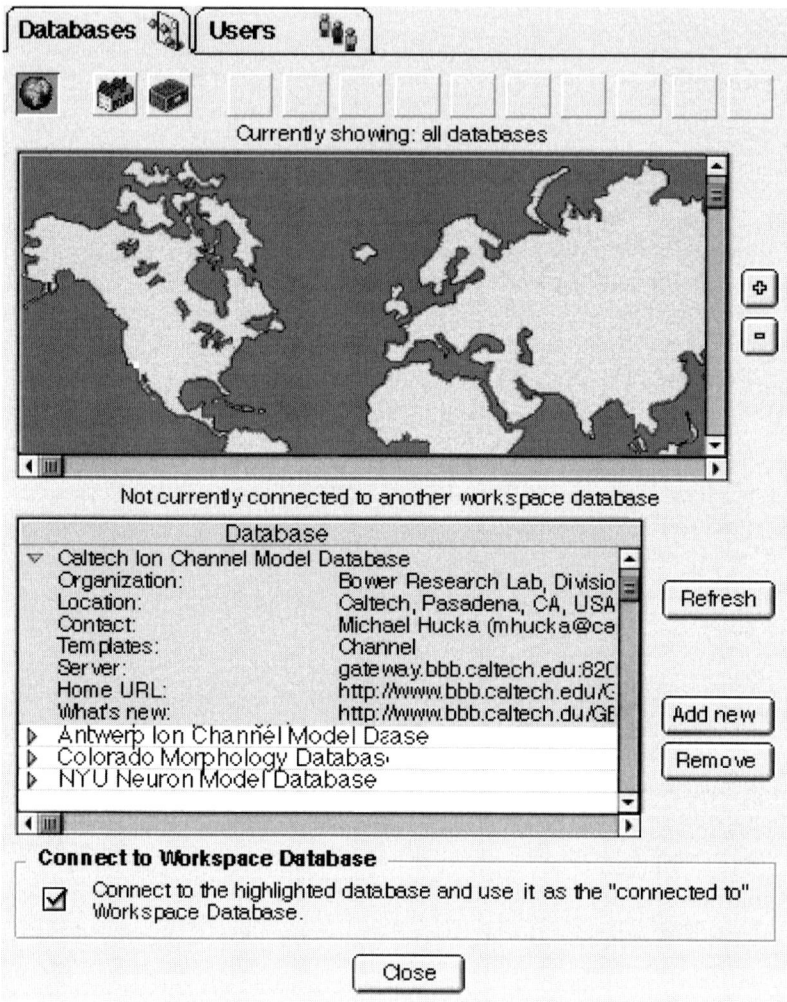

Fig. 3. Prototype of the Site Browser. This particular example screen displays the Databases pane, when the user is being asked to select a database for the Connect pane.

objects in this way. The Connect pane also provides the basis for engaging in collaborative activities with other Workspace users. Connecting to another user's Workspace Database (assuming that the owner has set appropriate permissions) allows one to view and possibly edit the objects in that database. The Workspace Database server is designed to allow multiple such simultaneous users to be connected. The User Interface implements a chat facility and a shared desktop viewing capability that allows all users connected to a particular Workspace Database to communicate with each other and see what each other sees on her or his computer screen. This is intended to make it easy for users to interact and work together on developing models.

5.3.3. The Site Browser

The need to select from a collection of databases or a collection of users occurs in several contexts in the Modeler's Workspace. Since databases and users are geographically scattered, a map-based interface, the Site Browser, will be presented to the user in these situations in order to provide cues that may help the user locate and remember the items involved. The Site Browser contains two tabbed panes, Databases and MW Users. The first pane displays a map of the world and a table of databases known to be publicly accessible. The map displays the geographic location of each database. In the MW Users pane, the same map-and-table interface is used, but here, the table contains a list of the users known to be actively using the Modeler's Workspace. The map in this case displays the geographic location of each user. The information in all cases is obtained by contacting the Modeler's Workspace Directory. Figure 3 shows a prototype of the Site Browser.

5.3.4. Access to Neural Simulation Packages

The final component of the Modeler's Workspace User Interface is the facility for communicating with simulation software. Interaction with simulators is supported through a plugin architecture, so that any simulation package can potentially be interfaced to the Modeler's Workspace once someone writes an appropriate plugin module and it is made generally available. Although this interface will be most useful for simulators such as GENESIS or NEURON, which construct structurally realistic compartmental models and can make use of the model representations that we provide, the programming interface is available for use by any simulator, including more general purpose ones such as Yorick or MATLAB. Interaction with the package takes place in the Build pane, which provides a window connected to the command-line interface of the simulation program (whether the simulator is GENESIS or another). A user can type commands directly to the simulator, and the output appears in the Build pane's interface window or in separate windows created by the simulator itself. The User Interface can also send model representations to the running simulation program; the plugin interface for the simulator translates objects from the representation used in the Modeler's Workspace to a format understood by the simulator.

5.3.5. An Example Usage Scenario

As an example of how one might use the Modeler's Workspace, consider the case of a neuroanatomist who is interested in the effects of dendritic morphology on the behavior of Purkinje cells. For example, this researcher may wonder whether a model with a simplified morphology might have sufficiently realistic behavior for use in a network model of part of the cerebellum. (We note that in the case of the Purkinje cell, the answer is likely to be "no" *[20].*) The user might begin by using the Search pane in the Modeler's Workspace to search for various Purkinje cell models in databases on the Internet. Figure 2 shows a possible search result that includes a hypothetical model based on that created by De Schutter and Bower *(9)*, but using a variant morphology that was generated by the L-Neuron program *(21)*.

The next step might be to use an inspector to examine the particular cell model in more detail. The user could do this by double-clicking the line containing the model of interest in the search results. The CD-ROM for this book contains additional documents and figures that could not be included in this chapter due to space limitations. There, the figure for the "Purkinje Cell Inspector" illustrates a mock-up of an inspector view with information about the geometrical and passive properties of the Purkinje cell model soma. In order to examine the specific channels used in the model in more detail, the user would employ a Channel Inspector, as illustrated in the figure "Purkinje Channel Inspector", which shows details of the channel dynamics used for the fast sodium channels in the model. Note that, as with the case of most of the inspectors, it is possible to see the actual equations that are used when the model is simulated. As well as providing further details of the model, the availability of these equations may give reassurance to modelers who distrust software that they did not write themselves.

As discussed in Section 5.4., the various templates used in the Modeler's Workspace contain a great deal of descriptive information about the model and the experimental data on which it is based. However, most information may be obtained if the model is actually used in a simulation, so that the behavior of different models may be compared under the same simulation conditions. In this case, users might want to import the model into the local Workspace Database, so that they can run it in a simulator. Importing a model can be done from the table of search results in the Search pane by highlighting the line containing the particular model desired and then clicking on the "Import" button. Once the model is copied into a user's Workspace Database, the user may switch to the Build pane to send the model to a simulator and compare current clamp simulations under the conditions described in *(9)*. The Build pane and the inspectors could also be used to create additional models, either by starting from and modifying an existing model, or by importing passive morphologies from elsewhere and populating them with channels taken from a Purkinje cell model database.

5.4. THE UNDERLYING ARCHITECTURE

The previous section makes clear that the Modeler's Workspace User Interface is the most essential part of the system. It must provide interfaces not only for interacting with model components, but also with databases and simulation tools. Because model representations, databases, and simulation/analysis tools will all change and evolve over time, the User Interface must itself be easily adapted and extended as the needs of users change.

We knew from the outset that the success of the project would be contingent on providing an open framework that could be taken and extended by other software tool developers. In designing the system, we identified the following characteristics as being crucial to meeting our needs:

1. Simplicity. The framework must be simple enough that interested developers can use it in their projects with a minimum amount of effort. The framework should not mandate the use of complex technologies such as CORBA *(22)*, although it should not prevent developers from using any particular technology in implementing an extension if they so desire.
2. Extensibility via plugin components. As new tools and methods are developed, it must be possible to implement them as pluggable modules that can be added to the existing framework without having to modify the framework itself. A plugin may either reimplement

The Modeler's Workspace

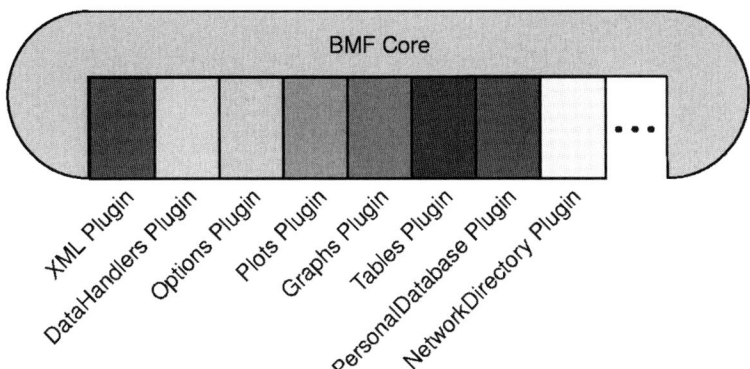

Fig. 4. The BMF Core and Core Plugins together constitute the foundations upon which applications are constructed. The BMF Core Plugins, shown here as shaded rectangles plugged into the BMF Core, are a set of essential modules provided with BMF and available to all applications built with BMF.

existing capabilities in new ways (for example, if someone develops an improved version of an existing module) or implement an entirely new capability.

3. Free distribution. All interested users must be able to obtain the system for free. Any software that is incorporated into the system and distributed with it, such as graphical user interface (GUI) widgets or object libraries, must itself be free of licensing fees or restrictions on redistribution.

4. Portability. Except for the Modeler's Workspace Directory server, the framework must be portable to at least Microsoft Windows (Win32) and Linux, with support for other varieties of Unix and MacOS X preferable as well. (The directory server exists only as a global server or multiple replicated servers, and runs separately from any user environments; therefore, it does not have the same restrictions on portability.)

We believe we have met these objectives by creating a layered, highly modular architecture. In the rest of this section, we discuss the high-level design of this architecture.

5.4.1. Layered Framework

We sought to maximize the reusability of the software that we developed for the Modeler's Workspace by dividing the Workbench infrastructure into two layers: the Modeler's Workspace itself and a lower-level substrate called the Biological Modeling Framework (BMF). The latter is a general software framework that provides basic scaffolding supporting a modular, extensible application architecture, as well as a set of useful software components (such as GUI tools) that can be used as black boxes in constructing a system. In fact, BMF is already being used to implement another tool, the Systems Biology Workbench *(23, 23a)*. Figure 4 depicts the core of BMF as a gray substrate holding a number of shaded blocks representing the basic plugins provided with BMF.

The Modeler's Workspace is a particular collection of application-specific components layered on top of BMF. These collectively implement what users experience as

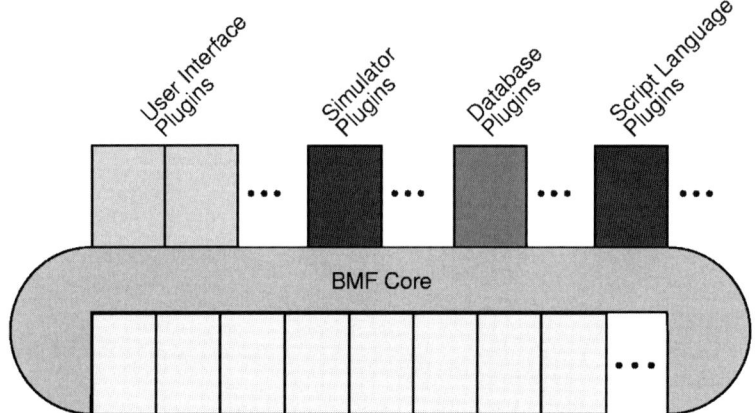

Fig. 5. Application specific plugins are added to the BMF Core and Core Plugins.

the "Modeler's Workspace." Some of these components add functionality needed for supporting the overall operation of the Workspace, such as the main screen of the User Interface described in Subheading 5.3.; other components implement the interfaces to databases and simulators. Figure 5 illustrates the overall organization.

As mentioned previously, the Workspace Database is a separate server program that can act as both a private repository for a user's work (where models, notebooks, and other items are stored) and as a rendezvous point for collaborative activities. The Modeler's Workspace system uses an XML-based model description language (discussed in Subheading 5.5.). The Workspace Database will accept data directly in XML format *(7)*, making it generic and capable of storing any information that can be encoded in XML.

5.4.2. Highly Modular, Extensible Architecture

The Biological Modeling Framework layer that underlies the Modeler's Workspace is implemented in Java and provides the scaffolding that (*i*) supports pluggable components, and (*ii*) provides a collection of basic components (such as an XML Schema-aware parser, file utilities, basic plotting, and graphing utilities, etc.) that are useful when implementing the rest of the Modeler's Workspace system.

Compared to many other frameworks, the Biological Modeling Framework is conceptually quite simple. The BMF Core consists of only one primary component, the Plugin Manager, and its operation is straightforward. Few requirements are placed on plugins themselves. For example, a plugin merely needs to be packaged in a Java JAR file, implement one specific interface (though it may implement others in addition), and obey a few rules governing behavior. There can be any number of plugins in the system, subject to the usual limitations on computer resources such as memory.

By virtue of the software environment provided by the Java 2 Platform, plugins can be loaded dynamically, without recompiling or even necessarily restarting a running application. This can be used to great advantage for a flexible and powerful environment. For example, the Modeler's Workspace can be smart about how it handles data types, loading specialized plugins to allow a user to interact with particular data objects on an as-needed basis. If the user does not already have a copy of a certain plugin stored on the local disk, the plugin could be obtained over the Internet, much like current-generation Web browsers can load plugins on demand. In this manner, plugins for tasks such as displaying specific data types or accessing third-party remote databases could be easily distributed to users.

5.5. REPRESENTATION OF MODELS AND DATA

One of the most difficult conceptual issues has been developing a strategy for describing models and their components. The Modeler's Workspace requires a representation language that abstracts away specifics of particular simulators, such as GENESIS, and also provides ways of interacting with existing neuroscience databases on the Internet.

Devising such a representation is difficult. The thorniest issue has been balancing the need for specificity in the representation (so that we can develop useful software tools for manipulating models) against the need for extensibility (so that as people's conceptualizations of neuronal characteristics change, the software does not need to be rewritten). It is not sensible to try to dictate every detail of how models and data are to be stored in databases. At the same time, some definitions of permissible data structures and formats must exist, so that we can proceed to develop software.

To begin addressing this problem, we first distinguish between a Modeler's Workspace Database, which is the database component of the Modeler's Workspace system, and a foreign database, meaning any other kind of database. In order to support some level of interoperability with foreign databases as well as neural simulators and allow users and software developers to evolve new representations and tools, we use a multifaceted approach having the following key aspects:

1. Model templates are organized following a simple object-oriented metaphor, with a base template serving as the root of all representations and new templates being derived from either the base or another existing template. We give users the ability to define new templates, but only through the addition of attributes; deletions are not allowed. This ensures that all models have at least a minimum set of common attributes that the Modeler's Workspace software can count on.
2. The Modeler's Workspace comes with a collection of default templates that can be used to describe the most common types of neuronal structures. These templates are part of a recently-established effort to produce NeuroML, a common exchange language for computational models in neuroscience *(24)*. We hope that these templates will act as de facto standards that will coalesce users around them and prevent the proliferation of many similar-yet-different representations for the most commonly used types of models.
3. Borrowing ideas from Burns (personal communication) and Gardner *(25)*, we require that the model templates used in a database be explicitly described and communicated by the database server when a client first contacts it. We also require that the templates used in publicly-accessible databases be made available separately through the Modeler's Workspace Directory (MWD). This permits the User Interface component to determine the structure of models in any given database by contacting either the MWD or the database in question.

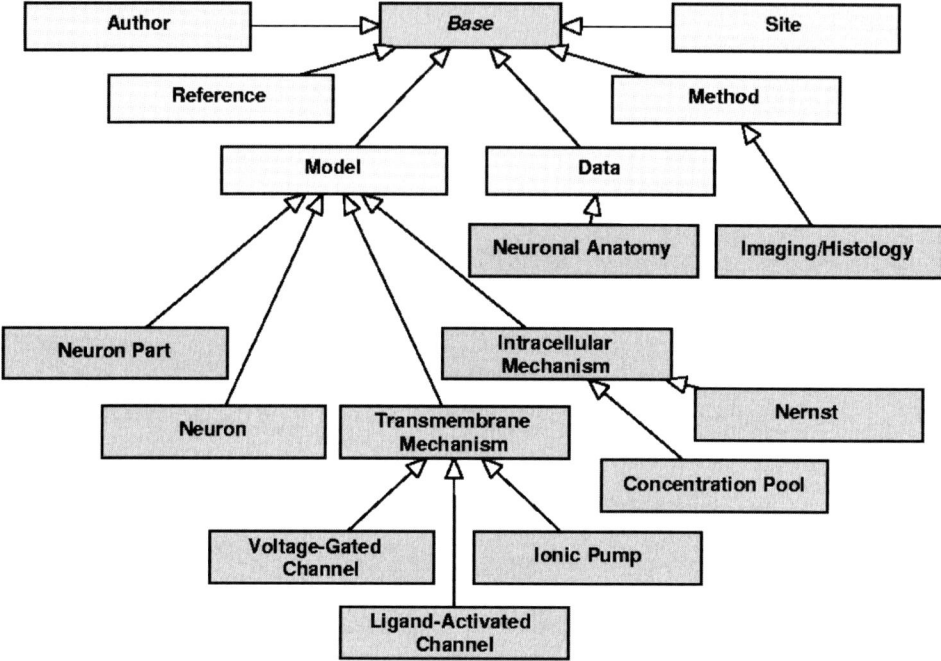

Fig. 6. All templates are derived from Base or another existing template. Open arrows indicate inheritance; for example, template Model inherits attributes from Base and adds its own new attributes. Additional templates derived from the six top-level tempates are shown with darker shading.

4. Interfaces to databases as well as simulation packages are mediated through software plugins. A database plugin for a particular foreign database provides the network interface and translation needed to interact with the database; a simulator plugin for a particular simulator handles translating commands and representations between the simulator and the Modeler's Workspace system.

5.5.1. Template Hierarchy

As mentioned above, models in NeuroML and the Modeler's Workspace are represented using a limited form of object-oriented description, in which each object in a database is defined according to either a root template called Base, or a template derived from it. Figure 6 shows the current hierarchy of templates used in the system.

The Base template contains a core set of attributes that are common to all main objects in the database. The first level of templates derived from Base consists of templates named Author, Reference, Method, Model, Data, and Site. The particular choice of first-level templates was taken from the work of Gardner et al. *(25)*. They are generic and do not contain any attributes that are specific to particular kinds of biological structures. The additional templates derived from the first-level ones then add specific templates for representing models of neurons and related structures. The following list briefly summarizes the templates defined at the time of this writing; the full definitions are available from the NeuroML web site (http://www.neuroml.org).

Base: This is the root of the template hierarchy. This template has only two attributes: id and version. The former places a unique identifier on every object; the latter allows the system to track the evolution of data objects.

Author: This template inherits attributes from Base and adds attributes for identifying a person by name, address, Web home page, and other characteristics. Using separate objects for author information allows users to enter into their databases the information about a given author once, then link to the author information from other objects (such as models and article references).

Reference: This is used to represent literature references. The attributes provided by this template are based on BibTeX records *(26)*. Author and editor information is represented in terms of links to Author objects.

Data: This is used for storing data in a Workspace Database or pointing to data stored in a remote database. It can record information describing the data, links to authors and references, and data (or pointers to data) grouped into data sets.

Method: This template is intended to capture information about experimental methodologies.

Site: This template is intended to capture information about such things as neuronal recording sites, brain regions, etc.

Model: This is intended to serve as a common starting point for all model template definitions. It is a generic structure, not specific to any particular kind of modeling. Specific kinds of models, such as for neuronal cells and intracellular and transmembrane mechanisms, are derived by starting from Model and adding new attributes.

Neuronal Anatomy: This template specializes the basic Data template to provide a container for anatomical information about neurons. It is primarily intended to be used for storing information about cell morphologies.

Imaging/Histology: This is a specialization of the Method template, used to describe methods of imaging and histology used in neuronal anatomy work.

Neuron: This is the main template for representing models of neurons. It extends the basic Model template with additional attributes for anatomical information, experimental information, the segmented cable structure of the model neuron, and other characteristics. It includes pointers to objects based on several other templates, in particular Neuronal Anatomy, Neuron Part, Transmembrane Mechanism, and Intracellular Mechanism.

Neuron Part: This exists to provide a way to construct reusable part models for Neuron objects. Portions of neuron models can be recorded in Neuron Part objects, allowing those portions to be reused in models by linking to them from within Neuron objects.

Transmembrane Mechanism: This template is intended to serve as a starting point for definitions of models of cell mechanisms such as ion channels, calcium concentration pools, etc.

Intracellular Mechanism: This is intended to serve as a common starting point for defining items such as calcium concentration pools.

Voltage-Gated Channel: As its name implies, this template can be used to represent models of voltage-gated ion channels. The representation is based in large part on that used by GENESIS and can handle channels not only of the common Hodgkin-Huxley variety, but also a number of variants.

Ligand-Activated Channel: This template can be used to represent models of channels that are activated by neurotransmitters.

Ionic Pump: This is used to represent various mechanisms for removing ions from a cell.

Concentration Pool: This template is used for models of intracellular ionic concentrations.

Nernst: This is one of several possible mechanisms for calculating changes in ionic equilibrium potentials.

The Modeler's Workspace uses XML Schemas to define templates. XML Schema is a recently introduced standard *(27–29)* for specifying the tags allowed in an XML data stream, how the tags can be organized into a hierarchy, and the data types of attributes delineated by the tags. The definition of the Base is expressed as a single XML Schema file. Each derived template (e.g., Model, Neuron, etc.) is similarly expressed using a separate XML Schema definition, making the organization very modular and easily extensible.

5.5.2. Advantages of the Approach

The object-oriented style of representation is useful for a variety of reasons. First, the existence of categorical templates allows the Modeler's Workspace User Interface to present the user with intelligent search forms. Specifically, the Modeler's Workspace search interface prompts the user to specify the type of object to search for (which is equivalent to specifying the template), and based on the user's choice, the system constructs a form using knowledge of the attributes defined by the template. The search form may include graphical elements specialized for the particular category of object involved. This allows the system to go beyond the usual fill-in-the-blanks search form and provide something more powerful and user-friendly.

A second reason is that, by choosing the search category appropriately, database searches can be made more or less specific. Because of the hierarchical relationships, a user can select a template in the middle levels of the hierarchy, and search operations can be designed to encompass all objects that are below it in the hierarchy. This means, for example, that a search using Model will encompass objects created from templates derived from it, such as Neuron class objects, Transmembrane Mechanism class objects, etc.

A final reason for the utility of the representational framework presented here is that software can be made modular and extensible. New software modules can be developed for the Modeler's Workspace alongside new templates, customizing the system to interact with new types of objects without redesigning or restructuring the whole system. For each representation derived from an existing template, all the software elements that worked with the parent template will also work with the derived templates. This is because the derived template can only add attributes, and while the existing tools will ignore the new attributes, they will continue to work with the attributes that were inherited from the parent template. Developers can write new software modules that interact with the additional fields in the new templates, and these software modules can be loaded into the Modeler's Workspace on demand, extending the software's functionality.

5.6. INTERACTING WITH DATABASES

The Modeler's Workspace design supports the ability for users to interact not only with theirs and other users' Workspace Databases, but with databases that were not designed specifically for the Workspace. In this section, we describe how the User Interface component of the system interacts with Workspace Databases and foreign databases.

5.6.1. Workspace Databases

Separating the Workspace Database from the User Interface, and making the former be a stand-alone server, is essential for providing the desired functionality in the Modeler's Workspace system. Not only does this approach support different access scenarios (described next), but it also has the advantage that when the user "starts running the Modeler's Workspace", they usually only need to start the User Interface—the Workspace Database typically will already be running, usually having been started up at computer boot time and waiting for connections over a network.

One of the access scenarios involves a one-to-one mapping between User Interface processes and Workspace Database processes. A certain user of a Workspace Database (typically, the user who establishes and configures it) is designated as its owner. In many cases, the owner may be an individual who simply wishes to use the Modeler's Workspace in private and may remain the sole user of the database. Whenever the user starts a copy of the Workspace User Interface, she or he only needs to supply the address of their Workspace Database server, and the User Interface connects over the network to this database process. The network-based nature of the database allows a user to roam anywhere on the Internet and still access their home Workspace Database. A user can connect, disconnect, and reconnect to their database from different computers.

Another access scenario involves multiple users connecting to a single Workspace Database server and simultaneously interacting with its contents. The owner of a Workspace Database has the ability to add other users to a list of designated permitted users—collaborators (possibly running at remote sites) who are allowed to connect to the Workspace Database and view and (possibly) edit its contents. This is the basis of the shared workspace facility of the Modeler's Workspace. The motivation for this facility is to permit various collaborative activities in the context of developing models. By allowing users to "meet" online, with the ability to see each other's Modeler's Workspace environments, we hope to encourage users to interact with each other and with their ongoing modeling work.

5.6.2. Foreign Databases

Foreign databases are those that were created by other groups prior to or without concern for the Modeler's Workspace. Because of the differences in interfaces and model representations, it is generally impossible to provide full interoperability between the Workspace and foreign databases. Models in such databases may be organized along different lines than those in Workspace Databases, and consequently, it may be impossible to provide more than the ability to search and retrieve models on the most basic characteristics (perhaps limited only to title, author, and similar fields). The degree to which a foreign database's representation can be mapped onto a Workspace representation must be determined on a case-by-case basis, and the interface must be implemented in the form of a plugin for each specific database.

Interaction with foreign databases occurs when the user performs search operations in the Search pane of the Modeler's Workspace User Interface. When the user chooses to search in selected foreign databases, each database plugin must do the work of translating the search request into the appropriate forms for the individual foreigns databases. Specifically, the database plugin performs the following functions: (*i*) engage the network communications protocol required by a particular foreign database (e.g., CORBA/IIOP *[22]*, HTTP, Z39.50 *[30]*); and (*ii*) translate back and forth between the Modeler's Workspace representation and search language and the corresponding elements of a foreign database.

A mediator must perform its translations by using the closest appropriate Modeler's Workspace templates. For example, a foreign database that stores models of ion channels would presumably be mapped to a representation based on the Transmembrane Mechanism template or perhaps even the more specific Voltage-Gated Channel template. In some cases, mapping foreign representations may only be possible in a partial way, with many attributes in the equivalent Workspace representation left blank. The database plugin must note which fields are missing in a particular model, so that the search process can distinguish between missing values and blank values. The treatment of missing attribute values during search can be handled specially and placed under control of a user preference setting; the user can elect to have the system either perform partial-matching (where missing fields are treated as "don't-care"), always-succeed-matching (where missing fields are treated as if they matched), or always-fail-matching (where missing fields are treated as if they failed to match).

Due to the lack of standardization in existing neuroscience databases, we expect that a custom mediator will need to be handwritten for each foreign database for which we want to provide interoperability. Initially, we expect that database plugins will be written by the Modeler's Workspace developers and distributed through the Modeler's Workspace Directory. In time we hope that other developers will also write database plugins for the system.

5.6.3. Template-Driven Search Interface

The existence of templates and the rules for exchanging them serve an important purpose in addition to structuring model representations: the definitions are used to construct search forms and inspector interfaces.

The search form in the Search pane (see Fig. 2) is constructed dynamically using the following approach. First, the names of all known model templates are collected together and presented in a pull-down list at the top of the Search pane (in the box in the line "Search for models having the following characteristics"). The user is required to first select one of the template names or else to select Any. The requirement to select a template is necessary to enable the Modeler's Workspace User Interface to present a search form that lists the attributes appropriate to that specific model category. The User Interface creates the search form by reading that class's XML Schema definition file.

5.6.4. The Modeler's Workspace Directory Server

The Modeler's Workspace Directory (MWD) is a global server located at a particular network URL. It acts as a global registry and repository supporting the community of users, making it possible for Workspace users all over the Internet to be able to learn about the databases that are available for public access. The MWD server will fulfill several roles:

1. It will maintain a list of all databases that are known to be available on the net and with which the Modeler's Workspace can interact. For each database, it will list information describing the contents of the database in terms of the corresponding Modeler's Workspace templates. If it is a foreign database, the MWD will include the appropriate mediator plugin which Modeler's Workspace clients can download if necessary.
2. It will maintain a list of all users of the Modeler's Workspace for those users who elect to be listed in the Directory. The list will be updated in real-time: whenever users start up a Modeler's Workspace process, they will be greeted with a request to allow the process to contact the MWD and register itself in the list of users. The users' locations can be viewed on the world map displayed in the MW Users pane of the Site Browser (see Subheading 5.3.).

The existence of a Modeler's Database Directory is important, not just to enable the Workspace to find out about available databases. It also means that the Workspace User Interface component does not need to contact all databases every time it is started. The alternative, having the Workspace query each known database directly at start-up time, would introduce a long startup delay. Centralizing the list of databases on a directory server will shorten startup time.

The MWD will run a simple program to update its contents daily by polling every database on its list and testing whether it is accessible. The MWD will record the last time that each database was known to be available, as well as download any new or changed template definitions from that database.

5.7. CONCLUSION

Structurally realistic neuronal models can serve as devices to collect, evaluate, and distribute information concerning the functional organization of nervous systems. As we have described in this chapter, the central goal of the Modeler's Workspace project is to provide the neuroscience community with a modular, extensible, and open software environment enabling neuroscientists to develop, use, and share structurally realistic models. The purpose of this chapter is to describe a design that we are just beginning to implement and to encourage others to participate in this effort. We hope that by providing common infrastructure for interfacing to databases and simulation packages, other software authors will gravitate towards this framework rather than developing entirely new software tools from scratch. Ultimately, we hope that others will be encouraged to contribute to the community new representations for models and new functionality in the form of plugins.

For the latest information on the status of this project, and the latest design documents for the the Modeler's Workspace, please visit the Web sites at (http://www.bbb.caltech.edu/hbp/) and (http://www.modelersworkspace.org). When working prototypes of the Modeler's Workspace become available for download, they will be announced on these sites. Further information about GENESIS may be obtained from (http://www.bbb.caltech.edu/GENESIS).

REFERENCES

1. Forss J, Beeman D, Eichler-West R, Bower JM. The Modeler's Workspace: A distributed digital library for neuroscience. Future Generation Computer Systems 1999; **16**:111–121.
2. Bower JM, Beeman D. The Book of GENESIS: Exploring Realistic Neural Models with the GEneral NEural SImulation System, 2nd ed. Springer-Verlag, New York, 1998.
3. Hines M, Carnevale NT. The NEURON simulation environment. Neural Computation 1997; **9**:1179–1209.
4. Ermentrout GB. XPP-Aut:X-windows PhasePlane plus Auto. (http://www.math.pitt.edu/bard/xpp/xpp.html). 2000.
5. Goddard N, Hood G, Howell F, Hines M, De Schutter E. 2001. NEOSIM: portable plug and play neuronal modelling. Neurocomputing, 2001; **38–40**:1657–1661.
6. Arnold K, Gosling J. The Java Programming Language. Addison-Wesley, Reading, MA, 1997.
7. Bosak J, Bray T. XML and the second-generation web. Sci Am 1999; **280**:89–93.
8. Koslow SH, Huerta MF. (eds) Neuroinformatics: An Overview of the Human Brain Project. Lawrence Erlbaum Associates, Mahwah, NJ, 1997.
9. De Schutter E., Bower JM. An active membrane model of the cerebellar Purkinje cell I. Simulation of current clamps in slice. J Neurophysiol 1994; **71**:375–400.
10. Segev I, Fleshman JW, Burke RE. Compartmental models of complex neurons. In: Methods in Neuronal Modeling, Ch. 3. (Koch C, Segev I., eds.) MIT Press, Cambridge, MA, 1989, pp. 63–96.
11. Wilson M., Bower, JM. Cortical oscillations and temporal interactions in a computer simulation of piriform cortex. J Neurophysiol 1992; **67**:981–995.
12. Bhalla US, Bower JM. Exploring parameter space in detailed single neuron models: simulations of the mitral and granule cells of the olfactory bulb. J Neurophysiol 1993; **69**:1948–1965.
13. Segev I. Single neurone models: oversimple, complex and reduced. Trends Neurosci 1992; **15**:414–421.
14. Johnston D, Magee JC, Colbert CM, Cristie BR. Active properties of neuronal dendrites. Ann Rev. Neurosci 1996; **19**:165–186.
15. Stuart G, Spruston N, Sakmann B, Hausser M. Action potential initiation and backpropagation in neurons of the mammalian CNS. Trends Neurosci 1997; **20**:125–131.
16. Vetter P, Roth A, Hausser M. Propagation of action potentials in dendrites depends on dendritic morphology. J Neurophysiol 2001; **85**:926–937.
17. Rapp M, Yarom Y, Segev I. The impact of parallel fiber background activity on the cable properties of cerebellar Purkinje cells. Neural Comput 1992; **4**:518–533.
18. Rapp M, Segev I, Yarom Y. Physiology, morphology and detailed passive models of cerebellar Purkinje cells. J Physiol (Lond) 1994; **474**:87–99.
19. Jaslove SW. The integrative properties of spiny distal dendrites. Neuroscience 1992; **47**:495–519.

20. Howell FW, Dyrhfjeld-Johnsen J, Maex R, Goddard N, De Schutter E. A large scale model of the cerebellar cortex using PGENESIS. Neurocomputing 2000; **32**:1041–1036.
21. Ascoli GA, Krichmar JL. L-Neuron: a modeling tool for the efficient generation and parsimonious description of dendritic morphology. Neurocomputing 2000; **32**:1003–1011.
22. Vinoski S. CORBA: integrating diverse applications within distributed heterogeneous environments. IEEE Communications Magazine, 1997, pp. 46–55. Specification available at (http://www.omg.org/).
23. Hucka M, Finney A, Sauro H, Bolouri H., Doyle J, Kitano H. The ERATO Systems Biology Workbench. In: Foundations of Systems Biology (Kitano H, ed.) MIT Press, Cambridge, MA, 2001.
23a. Systems Biology Workbench Development Group Homepage (http://www.cds.caltech.edu/erato), 2001.
24. Goddard N, Hucka M, Howell F, Cornelis H, Shankar K, Beeman D. Towards NeuroML: model description methods for collaborative modelling in neuroscience. Philos Trans R Soc Lond B 2001; **356**: 1209–1228.
25. Gardner D, Knuth KH, Abato M, et al. Common data model for neuroscience data and data model interchange. J Am Med Inform Assoc 2001; **8**:17–33.
26. Lamport L. LaTeX: A Document Preparation System. Addison-Wesley, Reading, MA, 1994.
27. Biron PV, Malhotra A. XML Schema part 2: datatypes. W3C Working Draft at (http://www.w3.org/TR/xmlschema-2). 2000.
28. Fallside DC. XML Schema part 0: primer. (http://www.w3.org/TR/xmlschema0/). 2000.
29. Thompson HS, Beech D, Maloney M, Mendelsohn N. XML Schema part 1: structures. W3C Working Draft at (http://www.w3.org/TR/xmlschema-1/). 2000.
30. ANSI/NISO Z39.50-1992 American national standard information retrieval application service definition and protocol specification for open systems interconnection. NISO Press, Bethesda, MD, 1992.

6
The Relationship Between Neuronal Shape and Neuronal Activity

Jeffrey L. Krichmar and Slawomir J. Nasuto

ABSTRACT

It has long been assumed that neuroanatomical variability has an effect on the neuronal response, but not until recently had research groups attempted to quantify these effects. In electrophysiological studies, the neuroanatomy is seldom quantified, and in neuroanatomical studies electrophysiological response is rarely measured. Computational techniques have the potential of bridging the gap between electrophysiology and anatomy, and testing the "morphology influences physiology" hypothesis. Computational techniques include modeling of neurophysiology, modeling and measurement of neuroanatomy, and mining publicly available archives of anatomical and electrophysiological data. In this chapter, we review studies that focused on the importance of neuronal shape's effect on neuronal function and describe the computational techniques and approaches used in these studies. We stress the need for developing a set of metrics that can quantify morphological shape and electrophysiological response and for including morphology in both experimental and simulation studies.

6.1. INTRODUCTION

In the preface to *Histology of the Nervous System of Man and Vertebrates* (*[1]*, originally published in Spanish in 1899 and 1904), the legendary anatomist Ramon y Cajal wrote: "…many theories, hypotheses, and simple guesses have been considered in an attempt to explain the functional role of the histological features associated with nerve cells and neural centers; we hope this will convince the reader that we are also attempting to create a conceptual science. In this we have been inspired by the old masters of anatomy, who believed that the goal of their work was physiology." In this age of computers, we are still inspired by the old masters of anatomy to understand relationship between the form and function of the nervous system. In the present chapter, we review computational approaches to this open area of research.

Neurons have a wide range of responses that may affect the functionality of brain regions *(2)*. The source of this variability is not obvious, but may involve factors such as synaptic connectivity, biochemical differences, type and distribution of active con-

From: *Computational Neuroanatomy: Principles and Methods*
Edited by: G. A. Ascoli © Humana Press Inc., Totowa, NJ

ductances, or morphological diversity. Differences in synaptic input, either in strength or connectivity, can alter the response of a neuron *(3,4)*. The neuronal response depends, in part, on the cell's concentration and distribution of ionic currents along its membrane. Recent studies suggest that cells adjust channel densities to accommodate differences in morphology *(5–9)*. This compensation may be the cause for the relative electrophysiological homogeneity in certain brain regions, such as hippocampal CA3, despite their diverse morphology *(10)*. In addition to synaptic connectivity and biochemical properties shaping neuronal electrophysiology, anatomical differences influence firing behavior and neuronal function. In the remainder of this chapter, we review experimental studies and computational approaches that investigate the effects of morphological variability.

6.2. EXPERIMENTAL STUDIES OF MORPHOLOGICAL VARIABILITY

Despite the difficulties, a few groups have recorded from neurons and accounted for 3D neuroanatomy within the same experimental preparation. Larkman and Mason investigated the relationship between soma/dendritic morphology and electrophysiology of visual cortex pyramidal cells in layer 2/3 and layer 5 *(11,12)*. After intracellular recording, they injected the cells with horseradish peroxidase (HRP) to obtain the soma and dendrite morphology. They divided these cells into three classes based on their qualitative and quantitative morphological differences: (*i*) layer 2/3; (*ii*) layer 5 with thick apical trunks; and (*iii*) layer 5 with slender apical trunks. The layer 5 pyramidal cells with thick apical trunks were generally larger than neurons in the other two classes. Of the three classes, only the layer 5 pyramids with thick apical trunks burst in response to current injections. This study showed that cell classes, differentiated by their morphology, could have different firing properties.

Dendritic branching influences the back propagation of an action potential into the dendritic tree. Williams and Stuart recorded from both the soma and dendritic tree of thalamocortical neurons and showed that action potentials initiated near the soma actively backpropagate into the dendrites and that action potentials attenuate more in a branched than in an unbranched region of a dendrite *(13)*. Moreover, the signal from dendrite to soma was more attenuated in highly branched dendritic regions. Thus, the shape of the dendritic tree could have an influence on synaptic integration.

Morphological studies in the hippocampus, which have been carried out on both a cellular level and a regional level, show that variations in neuroanatomy impact neuronal firing behavior. One study found bursting pyramidal cells predominantly on the borders of subfields (i.e., CA1a, CA1c, CA3a, and CA3c), while pyramidal cells in the medial areas, such as CA1b and CA3b, were predominantly characterized as spiking neurons *(14)*. In contrast, Bilkey and Schwartzkroin did not find this discrepancy and argued that firing differences may be more influenced by the cell's depth in the stratum pyramidale *(15)*. Cells with somata near the stratum pyramidale/stratum oriens border were more likely to burst. These firing properties may be influenced by the different morphology induced by the position in the hippocampal subfields. The depth of the soma location within the stratum pyramidale can influence how much volume the dendritic tree has to grow into before being compressed by a border. Additionally, dendritic fields of CA3c pyramidal cells bordering the dentate gyrus granular layer blades are more compact than the distal portion of CA3 near CA2 *(16)*. However, it is difficult to quantify the precise anatomical effect on firing properties experimentally because of the great morphological, biochemical, and electrophysiological variability, even within a subfield.

6.3. COMPUTATIONAL STUDIES OF MORPHOLOGICAL VARIABILTY

Computational approaches have the potential to test and quantify the effect morphological variability has on the firing properties of neurons. The task of neurophysiologists, who study the function of neurons by injecting current or voltage into a neuron and measuring the neuron's response, is painstaking and slow, and the number of neurons that can be measured in a given preparation is restricted. Often, the electrophysiological responses are recorded, but the neuromorphological data is not mapped. Alternatively, in a neuroanatomical study, the cell architecture is carefully mapped, but the electrophysiological data is not extensively recorded. The above situation follows from the difficulty in combining morphological and physiological measurements in a single preparation and still acquiring enough data points for a quantitative analysis. In computational neuroscience, modelers of single neurons have focused mostly on the biochemical details of the cell while ignoring the morphological detail. For example, Traub and his colleagues developed a biophysically detailed model of a CA3 pyramidal cell, that had either no branching structure *(17)* or a uniform branching structure *(18)*. Furthermore, individual cells in detailed network simulations usually have a uniform shape *(19–25)*.

In this section, we review computational approaches that have taken into account neuronal structure. We describe four types of computational tools that can be used for the neuronal structure–activity analysis: (*i*) neuronal modeling and simulation; (*ii*) sets of morphological metrics that can readily be applied to neuroanatomical data; (*iii*) electronic archives of neuronal data; and (*iv*) the artificial generation of neurons. We follow this discussion with a review of recent work by the authors and their colleagues in the Computational Neuroanatomy Group at the Krasnow Institute (http://www.krasnow.gmu.edu/L-Neuron) integrating these techniques.

6.3.1. Neuronal Modeling and Simulation

Neuronal modeling can be divided into two classes: passive models of neurons that have equations describing the flow of current, but no active currents, and active models that have equations describing active currents. Both types of models have been used to investigate the effects of morphology on neural processing.

The functional role of neuronal dendrites was first mathematically characterized by Wilfred Rall in his cable theory *(26,27)*, see *(28)* for review. Rall treated the dendritic tree as a passive (i.e., no active ionic channels) nerve cable. He derived partial differential equations to describe how current flows and voltage spreads across a dendritic tree in space and time. These cables had a membrane resistance, R_m, membrane capacitance, C_m, and an internal resistance, R_i. Rall showed that a cable divided up into smaller RC compartments could yield the same results as a single passive cable, but allowed for modeling of complex dendritic branching and nonuniformity along the dendrite. With this theory, Rall showed the functional importance of different spatiotemporal patterns of synaptic input, the effects of input to a single dendritic branch, and the effect dendritic branching has on the propagation of a signal.

Recently, London and colleagues compared the voltage transfer between uniform and nonuniform distributions of membrane conductance on a classic cable, and three classes (cerebellar Purkinje cell, layer V neocortical pyramidal cell, and hippocampal CA1 pyramidal cell) of reconstructed cells (29). In the nonuniform case, the membrane conductance increased linearly with the distance from the soma while keeping the overall conductance equivalent to the uniform case. In all the cell models tested, they found that monotonically increasing conductance increased the voltage response at the soma and decreased the electrotonic length of the cell. In fact, any membrane heterogeneity slows the decay of voltage and has important implications for the integration of synaptic signals. This study highlighted the importance of understanding the interplay between passive and active mechanisms in neural processing.

In order to analyze the spread of a signal within a complex dendritic morphology, Carnevale and colleagues developed a tool set, called the Electrotonic Workbench, and demonstrated that the electrotonic structure of a neuron is defined by the attenuation of voltage as it propagates toward or away from a reference point (30). Computing the attenuation, in both directions along each dendritic branch of the cell, transformed the cell from anatomic to electrotonic space. The electrotonic distance was defined as the natural logarithm of voltage attenuation. To use the Electrotonic Workbench, the user specified a file that contained morphometric data and the biophysical properties of the cell. A graphical interface depicted how the neuron would look after the electrotonic transform both for somatafugal (away from the soma) or somatopetal (towards the soma) signal flow. The attenuation could be calculated with DC and different frequencies of interest. Because of a recursive strategy and operating in the frequency range instead of the time domain, the attenuation for detailed hippocampal cells was calculated in less than 2 s. The NEURON simulation environment contains the Electrotonic Workbench and can be obtained via the World Wide Web (WWW) at (http://www.neuron.yale.edu) (31).

The Electrotonic Workbench analysis was applied to cells from hippocampal CA1, CA3, and dentate gyrus (30,32). Based on electrotonic analysis, it was predicted that cells with a long primary dendrite, such as CA1 or neocortical pyramidal cells, were more sensitive to a synapse's location for postsynaptic potential integration than cells without a long apical dendrite. Integration of postsynaptic potentials in neurons that lack a long primary dendrite, such as dentate gyrus granule cells, was less sensitive to synaptic location. In addition, pyramidal basal dendrites and dentate gyrus granule cells, which had similar dendritic morphology, had the same electrotonic properties and possibly similar functionality. The soma's sensitivity to synapse location may also be influenced by active processes in the dendrite. Recent work by Magee and Cook, measuring excitatory post-synaptic potentials (EPSPs) in CA1 pyramidal cells, showed that synaptic conductance increases with distance from the soma and may be responsible for normalizing the amplitudes of individual inputs (33). Clearly, combining tools such as the Electrotonic Workbench with active processes in dendrites and spines would help quantify this normalization effect.

Although passive models have been effective in describing neural processing, they cannot capture the dynamics of an action potential or a burst of action potentials. Here, we review modeling studies with active currents that included a description of neuromorphology.

Pinsky and Rinzel investigated the relationship between dendritic tree size and neuronal firing responses by developing a two-compartment (soma and dendrite) model reduced from the Traub CA3 pyramidal cell *(24)*. The soma had sodium and fast potassium currents, and the dendrite compartment had synaptic current, calcium current, after hyperpolarizing potassium current, and calcium activated potassium current. The morphological characteristics of the cell could be changed by adjusting the coupling strength between the soma and dendrite or by changing the ratio of the soma to dendrite surface area. Despite the simplicity of the model, it displayed a wide range of firing behaviors from regular spiking to bursting. Making the soma appear remote from the dendrite, by either increasing the soma to dendrite compartment coupling resistance or the ratio of somatic membrane area to dendritic membrane area, changed the Pinsky-Rinzel model from a spiking cell to a bursting cell *(34)*. This result was consistent with the Larkman and Mason cortical results described above.

Variations in dendritic morphology from different cell classes can have a qualitative effect on firing behavior. Using standard compartmental modeling techniques, Mainen and Sejnowski distributed identical ionic currents on the reconstructed dendritic morphology of a layer 3 aspiny stellate cell, a layer 4 spiny stellate cell, a layer 3 pyramidal cell, and a layer 5 pyramidal cell *(34)*. Their simulations showed that, given equal distributions of conductance parameters, smaller cortical cells (i.e., layer 3 aspiny stellate and layer 4 spiny stellate) tended to spike, whereas the larger cells (i.e., layer 3 and layer 5 pyramidal) tended to burst. In a recent simulation study, in which Vetter and colleagues examined eight different neuronal types, they found that dendritic branching patterns had a significant effect on the forward and back propagation of a signal *(35)*. For example, the distinctive branching found in Purkinje cells limited action potential backpropagation, whereas the dendritic geometry of dopamine neurons favored action potential propagation.

In contrast to the approach based on distributing ionic currents homogeneously across the dendritic tree, Migliore and colleagues changed the distribution of ionic currents in models of six morphologically varied CA3 pyramidal cells until they achieved similar electrophysiological behavior *(8)*. By trial and error, they changed the distribution of channels until all six cells were bursting. This was achieved by altering the Ca^{2+}-independent K^+ conductance within 100 μm of the soma. Although they claim that it was not difficult to achieve bursting in all the cells tested, they did not quantify the difference in morphology or its influence on the channel distribution.

6.3.2. Measuring Morphological Data

While it is obvious from the studies above that variations in dendritic structure affect neuronal function, it is still an open question as to which morphological characteristics are the most influential on neuronal function and to what degree. Before an investigation of the "neuromorphology influences neurophysiology" hypothesis can be undertaken, a set of morphometric parameters, which can be measured and compared, must be explored and defined.

Descriptions of the peculiar shape of neuronal arbors date back to the work of such pioneers as Purkinje *(36)*, Golgi *(37)*, Kölliker *(38)*, and Ramon y Cajal *(1)*. Quantitative descriptions of neuronal anatomy and morphology followed with the work of Lorente del Nò *(39)* and particularly Sholl *(40)*. Sholl introduced the use of diagrams summarizing centrifugal distribution of a variety of quantitative parameters, such as the distribution of dendritic branch diameters or branch lengths along the radial distance from the soma.

An important advance in the morphological analysis of neurons was made when computer technology was applied to reconstruct and render dendritic morphology *(41)*. Cells studied under the microscope could be acquired as computer files and displayed, rotated, and zoomed from different viewpoints. More sophisticated characteristics of dendritic shape could then be appreciated and evaluated. In a computer, dendritic morphology is generally stored as a list of dendritic points, each with 3D coordinates (x, y, z), a diameter, and a connectivity relationship. The "Cartesian" description of neuronal shape is suitable to characterize and display every detail of a dendritic tree.

We have identified a set of morphometric parameters, which are parameters useful in describing and comparing morphological differences between and within cell families *(42–44)*.

Scalar morphometrics, which are parameters that can be obtained directly from the 3D neuroanatomical description, include the dendritic size (average diameter, total length, total area), average path length (the distance from a terminal tip to the soma along the dendritic path), number of bifurcations, number of terminations, number of segments (number of branches between two bifurcations or a bifurcation and a terminal tip), and maximal branch order. Branch partition provides a quantitative measurement of how equally descendent branches are distributed on the two sides of a bifurcation *(45–47)*. It is defined for a parent branch as the relative and weighted difference between the numbers of terminal tips originating from the two child branches. Tree asymmetry is simply the average partition of all the branches and provides a characterization of the dendritic geometry *(48)*. The distribution of branch partitions may also be a critical morphometric parameter *(45,46)*. The overall tree size (height, width, and depth) and the interrelationship of these parameters characterizes the tree shape. For example, Claiborne and colleagues have identified ellipticity (width/depth) as a critical descriptor of Dentate Gyrus granule cell dendritic trees *(49)*.

Distribution morphometrics, are parameters derived from a histogram of one scalar morphometric vs a parameter characterizing spatial extent of an arbor, such as dendritic path, branch order, or distance from the soma *(11,12,43,44)*. For example, centrifugal distribution of branching points is derived by plotting the number of branch points vs the distance from the soma *(40,50,51)*. Other examples include the number of termination points vs branch order, the segment length vs branch order, or diameter size vs branch order *(50–52)*. Metrics describing the moments (e.g., mean or median, standard deviation *[50]*) and shape (e.g., kurtosis, skewness) of the resulting distribution can be used in the analysis. The program called L-Measure, developed in the Computational Neuroanatomy Group at the Krasnow Institute and described in Chapter 3, automated, integrated, and greatly extended the procedures for extraction of various scalar and distribution parameters. In particular, it enables a high flexibility and precision of defining distribution parameters. For example, it can extract scalar parameters at a given branch order, path, and Euclidean distance from soma, separately for basal or apical trees, or for branches of given topological type (e.g., terminal, internal, etc). Therefore, it is a very attractive and powerful tool for morphometric studies.

If we are to quantify the effect neuromorphology has on neurophysiology, a series of physiological measurements describing both spiking and bursting behaviors is also necessary. Spiking can be characterized by the current given at which the cell transitions from bursting to spiking and by the spike rate (recorded at current levels higher than the transition current). Bursting can be characterized by parameters such as burst rate, interburst interval, spike rate during burst, width of burst, burst shape, or burst variability. The burst shape can be measured by calculating the following ratio:

$$R = \frac{<\max\_ap> - <\min\_ap>}{<\max\_ap> - <\min\_ib>}$$

where <max_ap> is the average maximum voltage of an action potential within a burst, <min_ap> is the average minimum voltage between spikes within a burst, and <min_ib> is the minimum average voltage between bursts. R measures the "plateauness" of the cell's bursts: values of R approach 0 for plateau potentials, in which a train of action potentials stay above resting potential within a burst; they are close to 1 for trains of action potentials that return almost to resting potential within the burst. We devised a set of automatic procedures for extraction of features from the spike train data in order to speed up analysis and avoid introducing variation due to human factor *(53)*.

6.3.3. Archives of Neuroanatomy

Experimental data are continuously being accumulated and put into a digital format, but there are very few archives that are publicly available through the Internet to make this data readily accessible. One such database is the Southampton Archive of Neuronal Morphology (http://www.cns.soton.ac.uk/~jchad/cellArchive/cellArchive.html). The archive contains 3D morphological data of 124 hippocampal cells *(54,55)*. The archive makes available a Java applet, called the CellViewer, that allows a user to view cells in 3D space with pan, zoom, and tilt options; mark cell sections, such as axons, soma, or dendrites; and convert the cells into a format of a computational simulator, such as GENESIS or NEURON. The digitized cell reconstructions are in .swc format, consisting of a list of ASCII lines, each describing a small neuronal segment or "compartment" *(42,54,55)*. In this format, a compartment is approximated with a cylinder and described by a compartment number, a type (soma, axon, basal or apical dendrite), a position of the cylinder's end point (in x, y, and z Cartesian coordinates), a radius, and the number of the adjacent compartment in the path to the soma (thus, specifying the compartment interconnectivity). Anyone who converts their data into .swc format can store their data in the archive. To date, only Turner's group has contributed to the archive.

In contrast to other biological fields, a lack of standards and limited number of repositories has made data sharing among neuroscientists limited and difficult *(56)*. However, recent database development may be changing this trend. The NeuroScholar project's objective is to build a computational knowledge-based management system that can be accessed and queried by neuroscientists (http://neuroscholar.usc.edu; see also *57,58*). Currently, NeuroScholar holds neural connectivity data from the rat brain. The GENESIS group is developing a database in the form of the Modeler's Workspace holding objects ranging from ion channels to connectivity (http://www.bbb.caltech.edu/hbp, see also Chapter 5 in this volume). Additionally, the Virtual NeuroMorphology Electronic Database (http://www.krasnow.gmu.edu/L-Neuron/database) is a repository of both real and virtual neurons in .swc format *(59)*.

6.3.4. Artificially Generated Neurons

3D reconstruction of neurons is a painstaking and time-consuming endeavor. Therefore, there is a scarcity of data available for analysis by computational neuroscientists. An alternative approach to morphological databases of experimental data is to artificially generate a population of neurons for a given morphological class based on a set of fundamental parameters; parameters that are characterized by statistical distributions obtained from experimental data. The algorithms that generate "virtual neurons" sample values from these distributions stochastically. As a result, the population of algorithmically generated neurons will belong to the same morphological class, with no two virtual neurons being identical *(59,60)*.

The work in the Computational Neuroanatomy Group concentrated on two types of computational algorithms for the artificial generation and description of dendritic trees; namely local and global algorithms. Local algorithms, such as L-Neuron *(60)*, rely entirely on a set of local rules intercorrelating morphological parameters (such as branch diameter and length) to let each branch grow independently of the other dendrites in the tree as well as of its absolute position within the tree (see Chapter 3 in this volume). These algorithms are simple, intuitive, and their fundamental parameters can be measured directly from experimental data. Because of the small number of parameters, they are well suited to study structure–function relationship and the origin of emergent properties (i.e., anatomical parameters not explicitly imposed in the algorithm). In global algorithms, such as ArborVitae *(61)*, new dendritic branches are dealt "from outside" to competing groups of growing segments, also depending on their position in the tree (e.g., on their distance from the soma). Global algorithms are usually more flexible, but many of their fundamental parameters must be obtained through extensive and elaborate parameter searches. Global algorithms can be also extended to generate populations of interconnected neurons (networks), instead of single neurons (see Chapter 12 in this volume).

In another global approach, Winslow and colleagues stochastically generated dentate gyrus hippocampal granule neurons in order to analyze the response of these cells to stimulation applied to different perforant paths *(62)*. Their motivation was to investigate the relationship between cell response variation and neural shape. Winslow's approach was similar to global algorithms described earlier, because different tree features, such as number of branch points, branching pattern, or branch lengths were dealt "from outside" by using empirical probability distributions extracted from real morphological data. However, there were also fundamental differences between Winslow's tree generation method and algorithms such as implemented in Arbor Vitae and L-Neuron. In these local and global algorithms, one can argue that the process of virtual morphology generation mimics to some extent the development of dendrites, where structures are gradually expanding, meeting respective local and/or competitive constraints, until they reach their final form. In contrast, the process of dendritic tree construction used by Winslow and colleagues was very formal, that is, the constraints they used during dendrite generation, with the exception of the agreement of dendritic feature distributions with the corresponding empirical probabilities, were a few basic rules assuring conformity of the resultant structure with the overall properties of trees. In the algorithm, the number of branch points and their normalized distances along a predefined axis were first selected. Next, the number of primary branch points were selected and extended to a different proximal branch point. Consequently, every branch point is connected in the outward direction to the two nearest unconnected branch points (alternatively, an end point is created if no such branch points exist). The final stages of the algorithm involved spreading the trees in 3D by scaling their height, longitudinal, and transverse extent, as well as adding diameters; all using appropriate empirical distributions. Lastly, a "wiggle" was added to account for the curvature of dendrites.

6.3.5. Testing the "Morphology Influences Physiology" Hypothesis

We have approached the "neuromorphology effects neurophysiology" hypothesis by mining electronic archives for neuroanatomical data, utilizing computational modeling techniques to simulate neuronal firing behavior, and quantifying the effect by correlating the morphometrics described above with electrophysiological measurements of simulations. The general method consists of converting morphological measurements from 3D neuroanatomical data into a computational simulator format. In the simulation, active channels are distributed evenly across the cells so that the electrophysiological differences observed in the neurons would only be due to morphological differences. The cell's morphometrics and the cell's electrophysiology in response to current injections are measured and analyzed. In the following sections, we describe two case studies that analyzed the effects of neuronal shape on neuronal function: (*i*) comparing hippocampal CA3 pyramidal cells; and (*ii*) comparing normal cells with aged and Alzheimer's Disease (AD) cells in hippocampal CA1.

In a study to determine the effects of morphological variability on electrophysiology within a single cell type, 16 CA3 pyramidal cells were obtained from the Southampton archive *(44)*. The cells were converted to the GENESIS neural simulator format *(63)* by using the CellViewer tool from the Southampton archive. Equations for active currents, taken from the Traub CA3 model *(18)*, were described in GENESIS scripts. The distribution of the active currents was homogenous across all the 16 cells. In other words, any difference in the electrophysiological response to an identical stimulation was due to morphological variation. The stimulus response differences among the simulated neurons were quite dramatic (see Fig. 1). Cells responding to the same somatic current injection showed qualitatively different spiking patterns; regular spiking (Fig. 1A), regular bursting (Fig. 1B), or bursting with a plateau of action potentials (Fig. 1C).

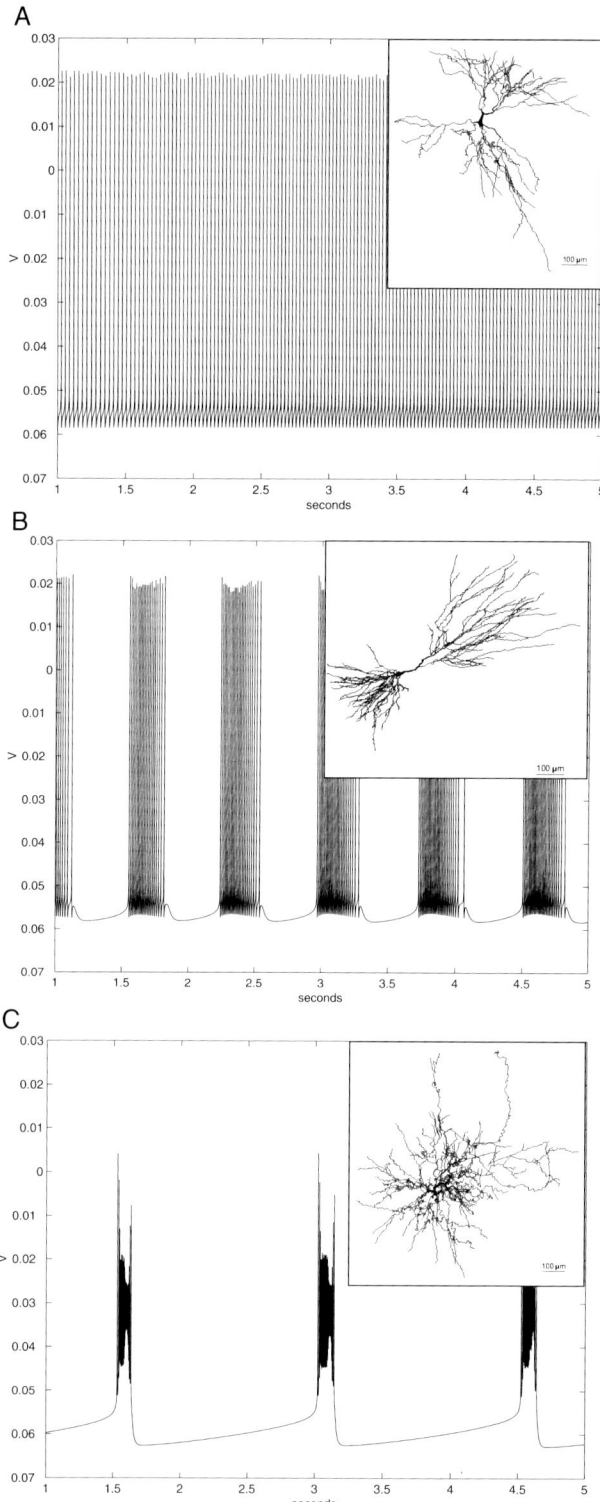

Fig. 1. Examples of different firing types for simulated CA3 pyramidal cells. **(A)** Regular spiking cell. **(B)** Regular bursting cell. **(C)** Plateau potential cell. All three simulated cells were injected with 0.2 nA. The insets to each voltage chart show a tracing of the cells.

A summary of the different firing responses at different current injections is shown in Figure 2. An analysis of the correlations between dendritic morphometrics and electrophysiological response showed that cells with smaller dendritic trees tended to be more excitable in both bursting and spiking. An analysis of the diameter as a function of path length showed that cells with a narrow diameter near the soma or cells in which the diameter decreased rapidly as the path length increased were more apt to be bursting cells (44). The degree to which the dendritic tree allowed current backpropagation was crucial in determining the duration and variability of a burst. These results implied that in addition to dendritic tree size, variations in the size and shape of the dendritic trunk could have a significant effect on firing behavior.

The above mentioned results are based on the investigation of the between-cell variability of the first order spike train parameters (features of spike trains in response to a fixed current injection). A natural question following this study is if the conclusions of this analysis hold true for different injection currents or if they are current-dependent. Two possibilities could arise that might limit the implications of this study. Either any relationship between morphology and first order parameters is injection current-specific and thus, epiphenomenal, or the relationship has a systematic but rather complex dependence on the input current.

We tested these possibilities by defining second order physiological parameters. The second order physiological parameters were defined with relation to the first order measurements. A second order parameter is a function describing the dependence of a first order parameter on the injection current. For example, the mean inter-burst-interval measured at a specific input current constitutes a first order spike train parameter. A function describing how the mean inter-burst-interval changes when the injection current is increased constitutes a second order parameter. Other examples of second order parameters include a change of the mean burst duration with the input current or a standard current–frequency curve. Thus, for a given cell, second order physiological parameters capture the dependence of the spike train waveform on the injected current.

In order to characterize a general tendency of the first order parameter change with respect to the level of current injection, we performed linear regressions for the plots of the first order parameters as a function of applied input current. The coefficients of the obtained linear trends, i.e., slopes and intercepts, characterized these second order parameters. These coefficients were subsequently used to investigate the relationship between morphological and second order electrophysiological parameters, analogous to earlier analysis using first order parameters. The significant correlations between morphology and first order physiological parameters were present for the corresponding second order parameters. Moreover, there were no additional significant correlations between second order parameters and morphometrics that would not follow from the previous analysis. Thus, in these studies the influence of morphology on physiology was indeed fully captured by the first order features of the spike train response.

As new metrics and more cells are put in the analysis, the parameter space has the potential of exploding and making analysis by conventional statistical methods insurmountable. For example, in the analysis of the above study, 7 scalar morphometrics and 1 distribution morphometrics were compared with 8 first order electrophysiological parameters for 16 different cells. Therefore, we recently explored visual data min-

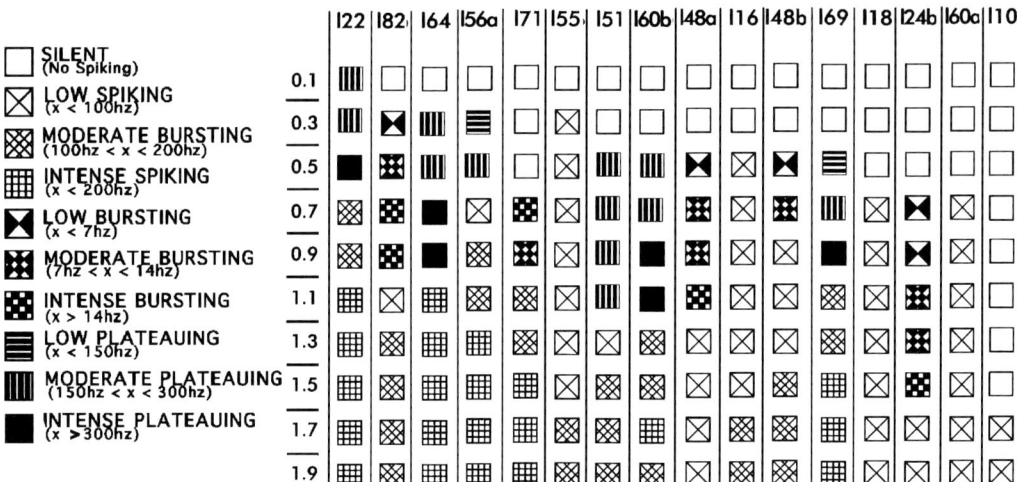

Fig. 2. Firing types at different current injections in nA. Each of the 16 CA3 pyramidal cells tested is listed by column and ordered from smallest (left) to largest (right) dendritic area. Each row represents the amount of current injected at the cell's soma. Each entry in the table denotes the qualitative response of a particular cell to a level of current injection. Adapted from *(44)* with permission.

ing techniques to detect structure in the data *(43)*. For our analysis, we used Xgobi *(64)*, a freely available data exploration package (http://www.research.att.com/areas/stat/xgobi/) for visual clustering and classification, through a technique called the brush-tour strategy. The brush-tour refers to a combination of brushing data points (i.e., graphically highlighting data points of interest) and performing a grand tour (i.e., examining the behavior of the highlighted data points as one tours through plots of different parameters and metrics) on the data. Touring through bivariate scatterplots of the morphological parameters revealed structure in the data. Cells with relatively small basal parameters near the soma and short path length from the soma to basal dendritic terminal tips, tended to burst with a plateau of action potentials (see Fig. 3). Figure 3 also revealed two outliers, cells l18 and cell l60a, which despite having a small diameter and short path length tended to spike. Cells l18 and l60a were brushed, and Xgobi was then used to investigate, by touring through the different morphometric values, whether these two cells had any extreme morphometrics. As the scatter plots in Figure 4 show, cells l18 and l60a had values in the upper range for total area and asymmetry. Although these results were preliminary, and the data set is small, the above example illustrates the power of exploratory data analysis. As the number of morphometric parameters and cells increases, so will the need for promising data visualization techniques such as Xgobi.

AD is among the leading causes of death in the United States, with a higher death rate than that of homicide or acquired immunodeficiency syndrome (AIDS). Most of the research on observable brain changes related to AD concentrates on how the accumulation of β-amyloid proteins affects the brain functioning. However, it is also known that AD involves morphological changes of the brain networks and single neurons *(65)*.

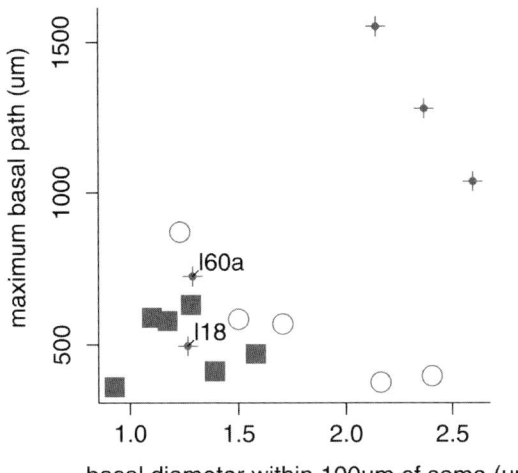

Fig. 3. Scatter plot of maximum basal path and basal diameter (within a distance of 100 μm of the soma) for the 16 cells tested. Bursts were brushed with a o, plateaus were brushed with a ■, and spikes were brushed with a +. The scatter plot reveals a relationship between firing type and the morphometric parameters. However, l18 and l60a appear to be outliers. Adapted from (43).

In order to investigate systematically the influence of AD-related morphological alterations on the electrophysiological response of neurons, it would be necessary to obtain the morphological and electrophysiological measurements from both healthy and AD affected subjects, and subsequently, to perform contrastive analysis between the AD and control groups. However, it is very hard to perform such studies using experimental methods due to the inherent problems in data acquisition. To our knowledge, there are no 3D morphological reconstructions of neurons from AD subjects. Nevertheless, this problem can be addressed by using corresponding data from an animal model of AD, e.g., Kainic Acid (KA) lesioned hippocampal neurons of a rodent (66). Although KA lesions are commonly used in animal models of other disorders such as epilepsy, they can be used as a model of AD, as it is believed that glutamate excitotoxicity may contribute to the pathological changes related to this disease (67). The above mentioned use of KA-based lesions for studies of the brain degenerative disorders makes our approach potentially interesting also to researchers investigating diseases other than AD.

Computational modeling is a viable methodology enabling us to address the question of a putative influence of AD morphological alterations on cell electrophysiological response. We downloaded morphological reconstructions of 18 neurons from CA1 area of rat hippocampus from the Southampton archive. The neurons were divided into 3 equal groups: normal (2 mo old), aged (24 mo old), and KA-lesioned neurons. We assumed an approach analogous to that described above for analysis of CA3 cells. Thus, we modified Traub's model of CA3 cells in order to account for relative differences in distribution of ion channels between CA3 and CA1 cells, and we kept all the physiological parameters constant across all the neurons (68). The electrophysiological response of cells was characterized simulating current clamp experiments, and the

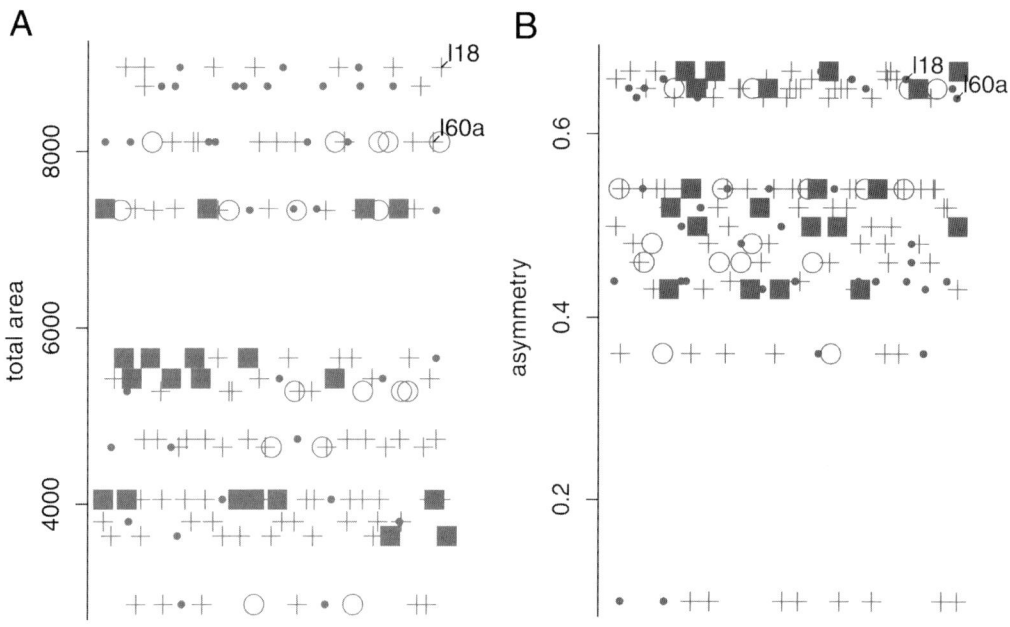

Fig. 4. (A) The outliers from Figure 3, l18 and l60a, were among the largest cells in terms of dendritic area **(A)** and the most asymmetric **(B)**. (Symbols are same as in Fig. 3.) Adapted from *(43)*.

results were used in analysis together with morphological characterization of all 3 groups.

To investigate the effect of AD on neuronal shape, normal, aged, and KA-lesioned groups of cells were compared with respect to parameters characterizing dendritic shape of basal and apical trees as well as the total cell morphologies. The most pronounced differences were observed between normal and aged cells (see Table 1). In general, the parameters suggested larger arbors in aged neurons than in normals. The size of the apical dendritic trees and the total dendritic length of basal trees were significantly larger for the aged than for the normal group. Both the apical and basal trees of the aged group had larger distal branch diameters than the normal group, as exemplified by the free term of fitted exponential relationship between the branch diameter and its distance from soma (diameter vs path length—Free Term). Although, KA-lesioned cells were less clearly distinguishable from the normal cells than the aged vs normal comparison, parameters characterizing apical tree size (the apical tree area, the number of apical branches, and the diameter of distal apical branches) did suggest that the KA-lesioned cells have larger than normal dendritic trees. This is consistent with the fact that only one parameter, the length of basal dendritic tree, did suggest a difference between KA-lesioned and aged cells. Despite the fact that no single parameter could differentiate all three groups of cells, combinations of parameters suggested differences between the three groups. Our studies confirmed results on morphological differences between normal, aged, and KA-lesioned cells reported in the literature *(69,70)*.

The three groups were compared with respect to first and second order physiological parameters. The first order parameters were dependent on the level of current injection.

Table 1
Differences Between Normal, Aged, and KA-Lesioned Cells with Respect to Scalar and Distribution Morphological Parameters

Parameters		Aged vs Normal mean ± SD		Normal vs KA mean ± SD		Aged vs KA mean ± SD	
Area (µm^2)	Total	**50842 ± 20986**	**31431 ± 12618**				
	Apical	**37097 ± 18367**	**21291 ± 10321**	*21291 ± 10321*	*32522 ± 16081.5*		
	Basal						
Number of Bifurcations	Total						
	Apical			*63 ± 35.6*	*93.8 ± 38.5*		
	Basal						
Number of Terminals	Total						
	Apical			*64.3 ± 34.8*	*96.2 ± 38.2*		
	Basal						
Branch Order	Total	*36 ± 3*	*31.2 ± 7.2*				
	Apical	*26.2 ± 1*	*21.7 ± 6.7*	*21.7 ± 6.7*	*27.33 ± 7.3*		
	Basal						
Diameter vs path length Free Term	Total						
	Apical	**0.491 ± 0.117**	**0.171 ± 0.247**	**0.171 ± 0.2471**	**0.484 ± 0.256**		
	Basal	**0.33 ± 0.25**	**0.15 ± 0.23**				
Diameter vs path length Decay Rate	Total						
	Apical	*1.07 ± 0.36*	*0.78 ± 0.38*				
	Basal	*0.94 ± 0.61*	*0.52 ± 0.41*				
Dendritic Length (µm)	Total	*25612 ± 12572*	*16061 ± 52171*				
	Apical	*17951 ± 10794*	*10828 ± 5337.8*				
	Basal	**7661 ± 2190**	**5233 ± 1276**			*7661 ± 2190*	*5490 ± 2254*

The significant differences in *t*-tests ($p < 0.05$) are shown in boldface, and the "trends" ($0.05 < p < 0.1$) are shown in italic. None of the parameters differentiates between all three groups of cells. The relationship between the diameter and the path length was assumed to obey an exponential function. Parameters of the fitted exponential expression (Decay Rate and Free Term) were used to represent this relationship.

At 0.03 nA, the mean inter-spike-interval during burst (ISI w/i Burst) was longest in lesioned cells (see Table 2). At 0.05 nA, the bursts were significantly longer in normal than in KA-lesioned cells (Burst Duration). The second order parameters showed a few more discernible differences between the groups. The most significant difference was in the second order parameter describing the relation between the mean inter-spike-interval during burst (ISI w/i Burst) and the input current. The inspection of the data suggested that, in all 3 groups, this relationship could best be described by a quadratic expression of the form $A(I_{input})^2 + BI_{input} + C$. Subsequently, the coefficients giving the best fit of this expression to the data were used in analysis. The dependence of the ISI w/i Burst on the input current for KA-lesioned cells was different than the corresponding relationship for normal cells. This relationship showed differences between normal and aged cells, and normal and KA-lesioned cells, but did not suggest significant differences between KA-lesioned and aged cells (Table 2). A few second order parameters showed significant differences between groups in some of their corresponding coefficients, but none could differentiate between all 3 groups of neurons.

A subsequent analysis confirmed a relationship between morphological shape and electrophysiological response in the entire population of 18 cells. However, in spite of this, analysis of morphological and physiological parameters separating the KA-lesioned from normal and aged cells did not reveal significant correlation between separation of the groups with respect to morphological and physiological parameters.

The high variability of morphological and electrophysiological data and the small number of available 3D reconstructions of KA-lesioned cells influenced the outcome of our analysis. Particularly, lack of clear separation between the groups with respect to their physiological responses, combined with the small number of cells per group, contributed to the relatively inconclusive assessment of the initial hypothesis. Larger cell sample size is needed in order to resolve the issue whether the lack of physiological separation observed in this study is an artifact of the sample size. However, the analysis did suggest the possibility of more complex or nonlinear relationships between morphology and physiology, and thus, it may be worthwhile to perform a sensitivity analysis to reveal a nonlinear dependence of physiology on morphological parameters.

The lack of 3D morphological reconstructions constitutes a fundamental problem for these type of studies. However, it could be overcome if researchers could generate artificial 3D dendritic trees, which nevertheless would possess all the characteristics typical of neurons of a given class. Availability of artificially generated models of normal and AD-lesioned cells could also aid tremendously in the analysis of dependence of parameters in this high-dimensional and complex problem. We hope that the algorithms for generating artificial biologically accurate neurons, described earlier in the chapter, will soon be able to successfully fulfill this important role.

6.4. CONCLUSIONS

In this chapter, we reviewed a number of studies that investigated the relationship between neuronal shape and neuronal function. In all cases, variations in shape, even within the same cell class, caused variations in neuronal response. Some of these studies have made an effort to isolate the effect of morphology from variations in physiology *(29,34,43,44,68)*, while other studies have done just the opposite and altered the physiology to overcome differences in morphology *(5–8)*. Now, with the advent of computational techniques, we have more tools at our disposal to understand the structure–function relationship. Computational modeling systems allow the user to specify the complex shape of a neuron in both single-cell and network simulations. Morphometric tools such as L-Measure and the Electrotonic Workbench *(30)* allow the user to measure experimental data efficiently. Methods to artificially generate neurons, such

Table 2
Differences Between Normal, Aged, and KA-Lesioned Cells with Respect to First and Second Order Physiological Parameters

Parameters	Aged vs Normal mean ± SD		Normal vs KA mean ± SD		Aged vs KA mean ± SD	
ISI within Burst (ms) I = 0.03 nA			35.14 ± 22.71	68.77 ± 47.87	30.06 ± 15.33	68.77 ± 47.87
ISI w/i Burst (var) I = 0.04 nA	*0.34 ± 0.17*	*0.2 ± 0.03*	**0.2 ± 0.03**	**0.29 ± 0.06**		
Burst Duration (ms) I = 0.05 nA			17.3 ± 16.72	3.66 ± 4.54		
Burst Duration (var) I = 0.05 nA			0.058 ± 0.055	0.017 ± 0.021	**0.049 ± 0.034**	**0.017 ± 0.021**
ISI w/i Burst (A)	−1688 ± 3782	29212 ± 2058	2921.65 ± 2058	−249.87 ± 1207		
ISI w/i Burst (B)	93.88 ± 139.8	−242.2 ± 256	−242.2 ± 256	20.95 ± 33.01		
ISI w/i Burst (C)			30.48 ± 8.04	13.39 ± 9.67	27.996 ± 16	13.39 ± 3.67

The significant differences in *t*-tests ($p < 0.05$) are shown in boldface, and the "trends" ($0.05 < p < 0.1$) are shown in italic. The variability of a physiological parameter (var) was defined as a ratio of its mean value to the standard deviation. Coefficients of the second order parameter quantify the relation between the mean inter-spike-interval during burst (ISI w/i Burst) and the input current. This relationship was best fitted with a quadratic expression of the form $A(I_{input})^2 + BI_{input} + C$.

as L-Neuron *(60)* and Arbor Vitae *(61)*, are fundamental for understanding the rules that govern a neuron's shape, but also have the potential of increasing the data set size in morphological studies. All of the tools need to be used in concert to fully understand morphology's effect on physiology.

In this chapter, we have only dealt with single-cell data. Most models that study the behavior of biological networks use a homogenous morphology to describe their neuronal elements. It is our belief that the variability in shape of neurons, connectivity patterns, and borders between regions are important and critical to understanding the function of the nervous system. Moreover, we believe the nervous system utilizes this variability, in a way that we do not yet understand, to increase its processing power. The only way to get at the heart of this problem is to combine experimental work with computational modeling. The tools and methods outlined in this chapter constitute a step in this direction.

ACKNOWLEDGMENT

JLK was supported by the Neuroscience Research Foundation which supports the Neurosciences Institute. SJN was supported by the Commonwealth of Virginia's Alzheimer and Related Diseases Research Award Fund (Award N. 00–1 to Giorgio Ascoli). The authors would like to thank the Krasnow Institute for Advanced Study for support of this work.

REFERENCES

1. Ramon y Cajal S. Histology of the Nervous System of Man and Vertebrates. Oxford University Press, New York, 1995.
2. Borg-Graham LJ. Interpretations of Data and Mechanisms for Hippocampal Pyramidal Cell Models. In: Cereb Cortex (Ulinski PSJ, EG, Peters A, eds) Plenum Press, New York, 1998, pp. 1–97.
3. Malenka RC, Nicoll RA. Long-term potentiation—a decade of progress? Science 1999; **285**:1870–1874.
4. Yeckel MF, Berger TW. Spatial distribution of potentiated synapses in hippocampus: dependence on cellular mechanisms and network properties. J Neurosci 1998; **18**: 438–450.
5. Andreasen M, Lambert JDC. Regenerative properties of pyramidal cell dendrites in area CA1 of the rat hippocampus. J Physiol 1995; **483**:421–441.
6. Andreasen M, Lambert JD. The excitability of CA1 pyramidal cell dendrites is modulated by a local $Ca(2+)$-dependent $K(+)$-conductance. Brain Res 1995; **698**:193–203.
7. Jensen MS, Yaari Y. Role of intrinsic burst firing, potassium accumulation, and electrical coupling in the elevated potassium model of hippocampal epilepsy. J Neurophysiol 1997; **77**:1224–1233.
8. Migliore M, Cook EP, Jaffe DB, Turner DA, Johnston D. Computer simulations of morphologically reconstructed CA3 hippocampal neurons. J Neurophysiol 1995; **73**: 1157–1168.
9. Turrigiano GG. Homeostatic plasticity in neuronal networks: the more things change, the more they stay the same. Trends Neurosci 1999; **22**:221–227.
10. Scharfman HE. Spiny neurons of area CA3c in rat hippocampal slices have similar electrophysiological characteristics and synaptic responses despite morphological variation. Hippocampus 1993; **3**:9–28.
11. Larkman A, Mason A. Correlation between morphology and electrophysiology of pyramidal neurons in slices of rat visual cortex I. Establishment of cell classes. J Neurosci 1990; **10**:1407–1414.
12. Mason A, Larkman A. Correlations between morphology and electrophysiology of pyramidal neurons in slices of rat visual cortex II. Electrophysiology. J Neurosci 1990; **5**:1415–1428.
13. Williams SR, Stuart GJ. Action potential backpropagation and somato-dendritic distribution of ion channels in thalamocortical neurons. J Neurosci 2000; **20**:1307–1317.
14. Masukawa LM, Bernardo LS, Prince DA. Variations in electrophysiological properties of hippocampal neurons in different subfields. Brain Res 1982; **242**:341–344.
15. Bilkey DK, Schwartzkroin PA. Variation in electrophysiology and morphology of hippocampal CA3 pyramidal cells. Brain Res 1990; **514**:77–83.
16. Ishizuka N, Cowan WM, Amaral DG. A quantitative analysis of the dendritic organization of pyramidal cells in the rat hippocampus. J Comp Neurol 1995; **362**:17–45.
17. Traub RD, Wong RKS, Miles R, Michelson H. A model of a CA3 hippocampal pyramidal neuron incorporating voltage-clamp data on intrinsic conductances. J Neurophysiol 1991; **66**:635–650.
18. Traub RD, Jefferys JGR, Miles R, Whittington MA, Toth K. A branching dendritic model of a rodent CA3 pyramidal neurone. J Physiol 1994; **481**:79–95.
19. Hasselmo ME, Wyble BP. Free recall and recognition in a network model of the hippocampus: simulating effects of scopolamine on human memory function. Behav Brain Res 1997; **89**:1–34.
20. Lumer ED, Edelman GM, Tononi G. Neural dynamics in a model of the thalamocortical system. II. The role of neural synchrony tested through perturbations of spike timing. Cereb Cortex 1997; **7**:228–236.

21. Lumer ED, Edelman GM, Tononi G. Neural dynamics in a model of the thalamocortical system. I. Layers, loops and the emergence of fast synchronous rhythms. Cereb Cortex 1997; **7**:207–227.
22. Menschik ED, Finkel LH. Neuromodulatory control of hippocampal function: towards a model of Alzheimer's disease. Artif Intell Med 1998; **13**:99–121.
23. Menschik ED, Finkel LH. Cholinergic neuromodulation and Alzheimer's disease: from single cells to network simulations. Prog Brain Res 1999; **121**:19–45.
24. Pinsky PF, Rinzel J. Intrinsic and network rhythmogenesis in a reduced traub model for CA3 neurons. J Comput Neurosci 1994; 39–60.
25. Traub RD, Jefferys JGR, Whittington MA. Simulation of gamma rhythms in networks of interneurons and pyramidal cells. J Comput Neurosci 1997; **4**:141–150.
26. Rall W. Theory of physiological properties of dendrites. Ann NY Acad Sci 1962; **96**:1071–1092.
27. Rall W. Theoretical significance of dendritic trees for neuronal input-output relations. In: Neural Theory and Modeling (Reiss R, ed.) Stanford University Press, Stanford, CA, 1964.
28. Segev I. Single neurone models: oversimple, complex and reduced. Trends Neurosci 1992; **15**:414–421.
29. London M, Meunier C, Segev I. Signal transfer in passive dendrites with nonuniform membrane conductance. J Neurosci 1999; **19**:8219–8233.
30. Carnevale NT, Tsai KY, Claiborne BJ, Brown TH. Comparative electronic analysis of three classes of rat hippocampal neurons. J Neurophysiol 1997; **78**:703–720.
31. Hines ML, Carnevale NT. The NEURON simulation environment. Neural Comput 1997; **9**:1179–1209.
32. Jaffe DB, Carnevale NT. Passive normalization of synaptic integration influenced by dendritic architecture. J Neurophysiol 1999; **82**:3268–3285.
33. Magee JC, Cook EP. Somatic EPSP amplitude is independent of synapse location in hippocampal pyramidal neurons. Nat Neurosci 2000; **3**:895–903.
34. Mainen ZF, Sejnowski T. Influence of dendritic structure on firing pattern in model neocortical neurons. Nature 1996; **382**:363–366.
35. Vetter P, Roth A, Hausser M. Propagation of action potentials in dendrites depends on dendritic morphology. J Neurophysiol 2001; **85**:926–937.
36. Purkinje JE. Bericht uber die versammlung deutscher naturforscher und arzte (Prag). Anat Physiologische Verhandlungen 1837; **3**:177–180.
37. Golgi C. Sulla fina anatomia del cervelletto umano. Istologia Normale 1874; **1**:99–111.
38. Kolliker AV. Die lehrer von den beziehungen der nervosen elemente zu einander. Anat Anz Ergazungshftr 1891; 5–20.
39. Lorente de No R. Studies on the structure of the cerebral cortex. II. Continuation of the study of the ammonic system. J Psychol Neurol (Leipzig) 1934; **46**:113–177.
40. Sholl DA. Dendritic organization of the neurons of the visual and motor cortices of the cat. J Anat 1953; **87**:387–406.
41. Capowski JJ. The reconstruction, display, and analysis of neuronal structure using a computer. In: The Microcomputer in Cell and Neurobiology Research (Mize RR, ed.) Elsevier, New York, 1985, pp. 85–109.
42. Ascoli GA Progress and perspectives in computational neuroanatomy. Anat Rec 1999; **257**:195–207.
43. Symanzik J, Ascoli GA, Washington SD, Krichmar JL. Visual data mining of brain cells. Comput Sci Stat 1999; **31**:445–449.
44. Washington SD, Ascoli GA, Krichmar JL. A statistical analysis of dendritic morphology's effect on neuron electrophysiology of CA3 pyramidal cells. Neurocomputing 2000; **32–33**:261–269.

45. van Pelt J, Uylings HBM, Verwer RWH. Distributional properties of measures of tree topology. Acta Stereol 1989; **8**:465–470.
46. van Pelt J, Uylings HBM, Verwer RWH. Centrifugal-order distributions in binary topological trees. Bull Math Biol 1989; **51**:511–536.
47. van Pelt J, Uylings HBM, Verwer RWH, Pentney RJ, Woldenberg MJ. Tree asymmetry—a sensitive and practical measure for binary topological trees. Bull Math Biol 1992; **54**:759–784.
48. van Pelt J, Schierwagen A. Impact of topological variability on dendritic geometry. J Biolog Syst 1995; **3**:1201–1210.
49. Claiborne BJ, Amaral DG, Cowan WM. Quantitative, three-dimensional analysis of granule cell dendrites in the rat dentate gyrus. J Comp Neurol 1990; **302**:206–219.
50. Larkman AU. Dendritic morphology of pyramidal neurones of the visual cortex of the rat: I. Branching patterns. J Comp Neurol 1991; **306**:307–319.
51. Larkman AU. Dendritic morphology of pyramidal neurones of the visual cortex of the rat: II. Parameter Correlation. J Comp Neurol 1991; **306**:320–331.
52. Uylings HBM, Ruiz-Marcos A, van Pelt J. The metric analysis of three-dimensional dendritic tree patterns: a methodological review. J Neurosci Methods 1986; **18**:127–151.
53. Nasuto SJ, Scorcioni R, Krichmar JL, Ascoli GA. Algorithmic statistical analysis of electrophysiological data for the investigation of structure-activity relationship in single neurons. InterJournal of Complex Systems 2001; **Report 389**.
54. Cannon RC, Turner DA, Pyapali GK, Wheal HV. An on-line archive of reconstructed hippocampal neurons. J Neurosci Methods 1998; **84**:49–54.
55. Turner DA, Li XG, Pyapali GK, Ylinen A, Buzsaki GJ. Morphometric and electrical properties of reconstructed hippocampal CA3 neurons recorded in vivo. J Comp Neurol 1995; **356**:580–594.
56. Chicurel M. Databasing the brain. Nature 2000; **406**:822–825.
57. Burns G. Neuroscholar 1.00, a neuroinformatics databasing website. Neurocomputing 1999; **26**:963–970.
58. Burns G. Knowledge Management of the Neuroscientific literature: the data model and underlying strategy of the NeuroScholar system. Philos Trans R Soc Lond B Biol Sci 2001; **356**:1187–1208.
59. Ascoli GA, Krichmar JL, Nasuto SJ, Senft S. Generation, description and storage of dendritic morphology data. Philos Trans R Soc Lond B Biol Sci 2001; **356**:1131–1145.
60. Ascoli GA, Krichmar JL. L-neuron: a modeling tool for the efficient generation and parsimonious description of dendritic morphology. Neurocomputing 2000; **32–33**:1003–1011.
61. Senft SL, Ascoli GA. Reconstruction of brain networks by algorithmic amplification of morphometry data. Lecture Notes in Computer Science 1999; **1606**:25–33.
62. Winslow JL, Jou SF, Wang S., JM, W. Signals in stochastically generated neurons. J Comput Neurosci 1999; **6**:5–26.
63. Bower JM, Beeman D. The Book of GENESIS: Exploring Realistic Neural Models with the GEneral NEural SImulation System. TELOS/Springer-Verlag, 1994.
64. Swayne DF, Cook D, Buja A. Xgobi: Interactive dynamics in the X Window system. J Comput Graph Stat 1998; **7**:113–130.
65. Coleman PD, Flood DG. Neuron numbers and dendritic extent in normal aging and Alzheimer's disease. Neurobiol Aging 1987; **8**:521–545.
66. Geddes JW, Monaghan DT, Cotman CW, Lott IT, Kim RC, Chui HC. Plasticity of hippocampal circuitry in Alzheimer's disease. Science 1985; **230**:1179–1181.
67. Shetty AK, Turner DA. Vulnerability of the dentate gyrus to aging and intracerebroventricular administration of kainic acid. Exp Neurol 1999; **158**:491–503.
68. Nasuto SJ, Knape R, Scorcioni R, Krichmar JL, Ascoli GA. Relation between neuronal morphology and electrophysiology in the Kainate lesion model of Alzheimer's disease. Neurocomputing 2001; **38–40**:1477–1487.

69. Pyapali GK, Turner DA. Increased dendritic extent in hippocampal CA1 neurons from aged F344 rats. Neurobiol Aging 1996; **17**:601-611.
70. Pyapali GK, Turner DA. Denervation-induced dendritic alterations in CA1 pyramidal cells following kainic acid hippocampal lesions in rats. Brain Research 1994; **65**:279–290.

7
Practical Aspects in Anatomically Accurate Simulations of Neuronal Electrophysiology

Maciej T. Lazarewicz, Sybrand Boer-Iwema, and Giorgio A. Ascoli

ABSTRACT

When computer simulations are employed to investigate mathematical models of electrophysiology, the details of the implementation can heavily affect the numerical solutions and, thus, the outcome of the simulations. In computational studies based on detailed dendritic morphology, relevant implementation details include, among others, the discretization of time and space. In particular, the anatomical representation of complex dendrites into isopotential compartments presents challenging issues (often overlooked in published reports) in the numerical approximation of the cable equation and its derivatives. Here, we discuss these issues using examples taken from variations of a model of CA3 pyramidal cell electrophysiology based on realistic anatomy and biophysics. In addition, we describe existing and novel procedures to produce model compartmentalizations that ensure stable numerical solutions, with references to popular simulation environments such as NEURON. Finally, we provide an overview of existing computational tools aiding the representation, conversion, and simplification of dendritic morphology for electrophysiological simulations.

7.1. INTRODUCTION

This chapter discusses some of the practical aspects that should be taken into consideration in computational studies of neuronal activity with detailed dendritic morphology. Only issues specific or related to the influence of the detailed anatomical representation on the electrophysiological simulation will be examined. For more general discussions of practical aspects in computational neuroscience, the reader is referred to the several excellent books that have recently appeared (e.g. [1–3]). Although particular attention should be paid to ensure that the complexity of the model is appropriate for the scientific question being investigated, in this chapter, we will assume that anatomical details (specifically, dendritic morphology) are important for the pursued research goal. We will use examples based on popular computational neuroscience software packages, such as NEURON [4] and GENESIS [5]. However, many

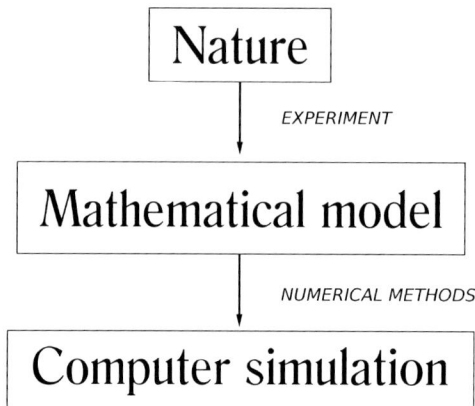

Fig. 1. Diagram representing the phases and components of the mathematical modeling process.

considerations are applicable to simulations implemented in other environments, such as Surf-Hippo *(6)*, NEOSIM *(7)*, and Catacomb *(8)*, as well as to simulations written in MatLab *(9)* or directly in FORTRAN or C++, possibly using libraries such as the Conical Library *(10)*.

Most of the aspects examined in this chapter concern the computational implementation of a neurobiological mathematical model (Fig. 1), i.e., the use of numerical methods to investigate the properties of the model. Thus, we will not discuss issues specific to mathematical modeling itself *(11,12)*, such as the construction of the model based on the experiments, or the general principles underlying the "experiments in computo" *(13)*. Specifically, we address issues related to the computational implementation of realistic models, as opposed to demonstration models *(14)*. Realistic models are based on the actual anatomy and physiology of the nervous system and are designed as tools to discover new ideas. In contrast, demonstration models do not necessarily attempt to reproduce the physical reality and are primarily intended to provide support for a particular preexisting theory.

One aspect of neurobiological models that is particularly relevant in computational neuroanatomy concerns the dendritic structure. Dendritic morphology, which is of course continuous in nature, must be discretized in the computer simulation. Typically, dendrites are represented as interconnected compartments that are assumed to have, at any time, uniform biophysical properties (membrane voltage, ionic concentrations, etc.). In the "compartmental" approach *(3)*, compartmentalization of the dendritic structure is performed at the level of the mathematical model itself: a given model contains a certain number of compartments, and changing the compartmentalization would imply a change of model. In contrast, the mathematical model derived from the cable theory *(15,16)* maintains the continuity of neuronal structure. In other words, the cable equation and its derivatives do not assume discretization and isopotential compartments. Compartmentalization is introduced during the implementation of this model as computer simulations used to approximate the solution by numerical methods *(17)*. Thus, changing the compartmentalization in cable theory-based models affects the approximation of the solution, but does not imply a change of the model itself.

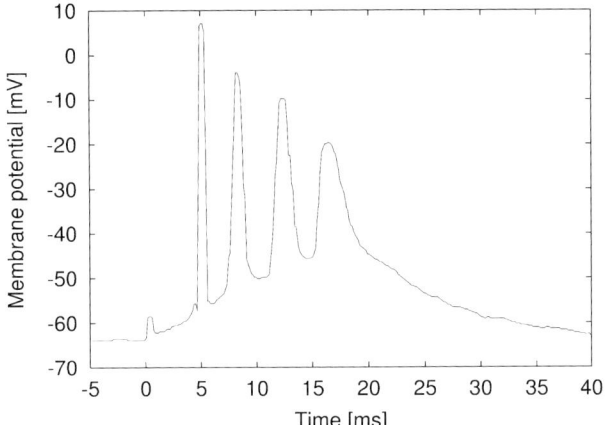

Fig. 2. Somatic membrane potential recorded from a CA3 pyramidal cell in response to somatic injection of a short (5-ms) current pulse (1 nA). Plotted from data in *(20)*.

A simple example of neurobiological modeling is provided by the following mathematical equation describing the passive electrical properties of a cell membrane *(2)*:

$$C \times \frac{\partial V(t)}{\partial t} + g_{pas} \times [V(t) - V_{rest}] = I_{inj} \quad \text{[Eq. 1]}$$

where C and g_{pas} are the membrane capacitance and conductance, respectively, $V(t)$ is the membrane potential at time t, V_{rest} is the resting potential of the membrane, and I_{inj} is the injected current. In this simple case, assuming the initial condition $V(0) = V_{rest}$, we can compute an analytical solution of Equation 1:

$$V(t) = \frac{I_{inj}}{g_{pas}} \times (1 - e^{-\frac{t}{R_m \times C_m}}) + V_{rest} \quad \text{[Eq. 2]}$$

Unfortunately, most realistic neurobiological models consist of complicated systems of nonlinear partial differential equations, for which analytical solutions are typically not known. For example, the following extension of the cable equation describes the voltage of a neuron as a function of position and time *(2)*:

$$\frac{\pi}{4} \frac{d^2(x)}{R_a(x)} \frac{\partial^2 V(x,t)}{\partial x^2} = \pi \times d(x) \times C_m(x) \times \frac{\partial V(x,t)}{\partial t} + \pi \times d(x) \times \frac{V(x,t) - V_{rest}}{R_m(x)} + I(x,t) \quad \text{[Eq. 3]}$$

where $V(x,t)$ is the membrane potential at location x and time t, $d(x)$, $R_a(x)$, $R_m(x)$, and $C_m(x)$ are the diameter of the neuron, cytoplasmatic resistance, membrane resistance, and membrane capacitance, respectively, at the location x, and $I(x,t)$ is the membrane current at location x and time t. This current includes contributions from voltage-gated and ligand-gated ion channels as well as current injections. The dependency of voltage-gated channel currents on voltage and time is usually described by systems of differential equations, such as Hodgkin-Huxley equations *(18,19)*. In more complex models, some of these current functions can additionally depend on calcium levels or other

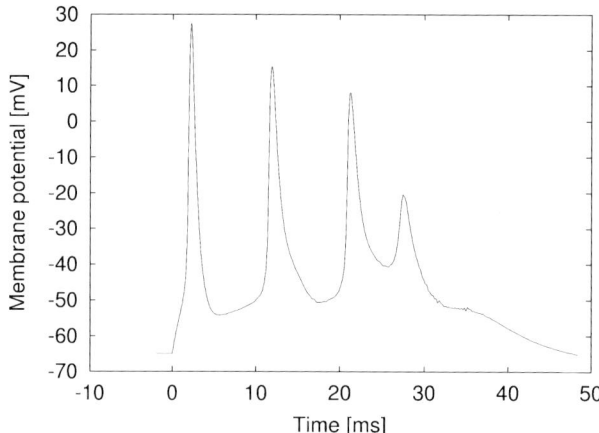

Fig. 3. Simulated CA3 pyramidal cell response to a 3-ms somatic injection of 1 nA of current. Based on a 385 compartment model *(22)* (see also http://senselab.med.yale.edu/senselab/ModelDB/).

factors. In general, the analytical solution of Equation 3 is not known. However, if a solution exists, it can be approximated by numerical methods using computer simulations.

In the following sections of this chapter, we use examples based on well-known models of the CA3 pyramidal cell. A characteristic electrophysiological behavior of this neuron is a train of spikes ("burst") in response to a short pulse of somatic current injection (Fig. 2). Such a behavior was fully described by Wong and Prince 20 yr ago *(20)*, and these experiments have been quoted in over 270 articles since then. A typical burst lasts about 50 ms, and it consists of 2 to 6 action potentials. Another peculiar feature is the progressive reduction of spike amplitudes during the burst *(21)*.

In this chapter, we examine several models of CA3 pyramidal cell electrophysiology deriving from the work of Migliore and colleagues *(22)*. The original model was based on accurate reconstructions of dendritic morphology, and it includes detailed calcium dynamics and nine voltage-gated ionic channels. The response of this model to a short pulse of somatic current injection is in good agreement with experimental data (Fig. 3).

7.2. COMPUTATIONAL IMPLEMENTATION

Although standard digital computers are deterministic devices, computer simulations should not be treated "literarily". The numerical implementation of mathematical models implies errors in the approximation of the solution, as well as errors in the very representation of numbers in the computer. In this section, we discuss some of these errors, which are intrinsic in numerical methods and computer implementations and are present in addition to any error inherent in the assumptions on which the mathematical model is based. At the end, although realistic simulations may help determine ranges of parameters underlying neuronal behaviors (e.g., quiescence, regular firing, or bursting), the occurrence of each spike in the simulation should not be necessarily considered with a precision of 1 ms.

In general, mathematical model equations (e.g., Equation 3) use continuous functions (and real numbers). Examples of these functions are membrane potential

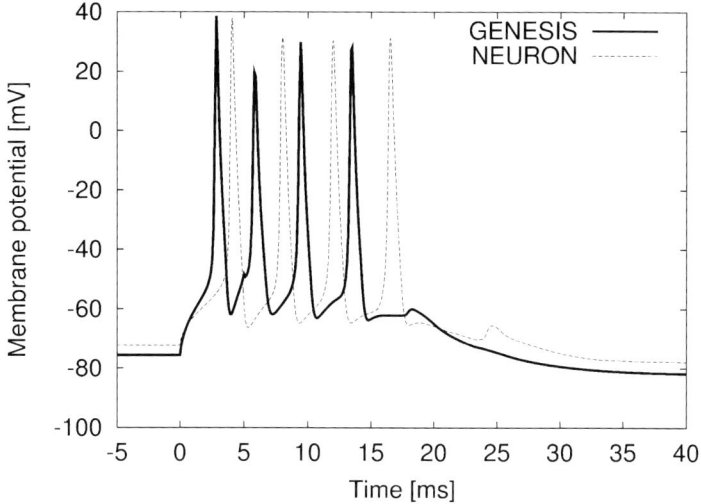

Fig. 4. Differences between the GENESIS and NEURON implementation of the Menschik model *(23)*. Somatic responses to somatic current clamp pulse (1 nA, 5 ms).

(depending on time and location within the neuron), dendritic diameter, passive membrane resistance, and ionic channel densities (all depending on location). In order to solve these equations numerically with a digital computer, these continuous functions and real numbers must be represented by bits. The precision of the representation can vary depending on the type of computer processor, operating system, and simulation software. Processor and operating system usually produce small discrepancies. Simulation software is more serious. Very often different software enviroments imply a different implementation of a given mathematical model. In C++ and FORTRAN, it is possible to implement almost any mathematical model. In contrast to that, high level software environments are particularly susceptible to portability issues. For example, NEURON and GENESIS contain several predefined basic strucures and objects. These elements are differently defined in the two environments, so it is usually necessary to adapt and change the mathematical model in order to port its implementation from NEURON to GENESIS or vice versa.

An example of that issue is provided by our attempt to port the GENESIS implementation of Meschik and Finkel's model *(23)* to NEURON. The Menschik model was built using the same 385 compartment cell of the Migliore model *(22)* and on similar active conductances. The structures and mechanisms controlling calcium concentrations and kinetics included four calcium shells and a calcium buffer, and are in principle identical in the Menschik and Migliore models. However, the implementation of calcium pumping in the Menschik model makes use of a GENESIS object specifically designed to describe the Michaelis-Menten steady state. NEURON does not have a predefined object for the Michaelis-Menten formalism. When we manually implemented the Michaelis-Menten equation for the Menschik and Migliore models in NEURON, the solution was numerically unstable. Thus, in our NEURON implementation, we described calcium pump starting from the reaction schemes (as in the original

Migliore model). Two models had a fairly different behavior although the difference in the two implementations is minor when considered in the complex context of the whole model. We found differences between firing frequency in response to somatic (steady or pulse) current injection, resting potential, subthreshold oscillations, spike shape and amplitude, and spike variability within bursts (Fig. 4).

Neurobiological models usually contain functions of time (e.g., Equation 3). Therefore, in anatomically accurate models, these functions depend both on time and spatial location, and the numerical approximation of the solution depends on the discretization of both variables. If the model only contains one compartment, Equation 3 can be reduced to Equation 1, making all functions $V(x,t)$, $I(x,t)$, $d(x)$, $R_m(x)$, $C_m(x)$, $R_a(x)$ constant over all values of x. Equation 1 can be also represented as:

$$\frac{\partial V}{\partial t} = f(v,t) \qquad [Eq.\ 4]$$

where the function f is typically nonlinear. If the initial condition $V(0) = V_0$ is known, Equation 4 can be resolved numerically by using the Taylor approximation:

$$V(t + \Delta t) \approx V(t) + \frac{\partial V}{\partial t} \times \Delta t \qquad [Eq.\ 5]$$

Equations 4 and 5 can be rearranged as:

$$V(t + \Delta t) \approx V(t) + \Delta t \times f(V,t) \qquad [Eq.\ 6]$$

In other words, in order to obtain an approximation of the solution of Equation 1, we divide the time domain into small intervals, or time steps (Δt). The use of an excessively big time step would result in an approximation of the solution so different from the "true solution", that it would not be scientifically meaningful. How small should the time step be? From the mathematical point of view, the smaller the step, the better the approximation. However, from the numerical implementation point of view, too small a time step could lead to round off errors, as well as unnecessarily slow execution of the simulation *(3)*. In general, the step should be considerably smaller then the time scale for the most rapidly occurring events in the model. For example, if action potentials typically rise to their maximal value in about 1 ms, a conservative choice for the time step is 0.01 ms. A practical approach to this issue is provided by the "rule of 1/2": if halving the time step causes substantial changes in the simulation outcome, the time step should be reduced until the approximation is "stable". Conversely, if doubling the time step results in no significant alteration of the simulation behavior, the time step should be increased. An example of the error introduced by an excessively large time step is shown in Figure 5, where the burst of the Migliore model *(22)* in response to somatic current injection is transformed in an irregular firing pattern.

The error of the Taylor approximation in Equation 6 is bounded by ΔV *(24)*. Therefore, in order to efficiently keep the maximum error below a given value, the time step in the simulation should depend on the neuronal activity itself (e.g., spiking vs quiescence): when the time derivative of voltage is bigger, a smaller time step Δt is required (Figure 6). Exploiting this relationship, NEURON implements the variable time step integration method CVODE *(24)*. The integrator dynamically adjusts the time step in

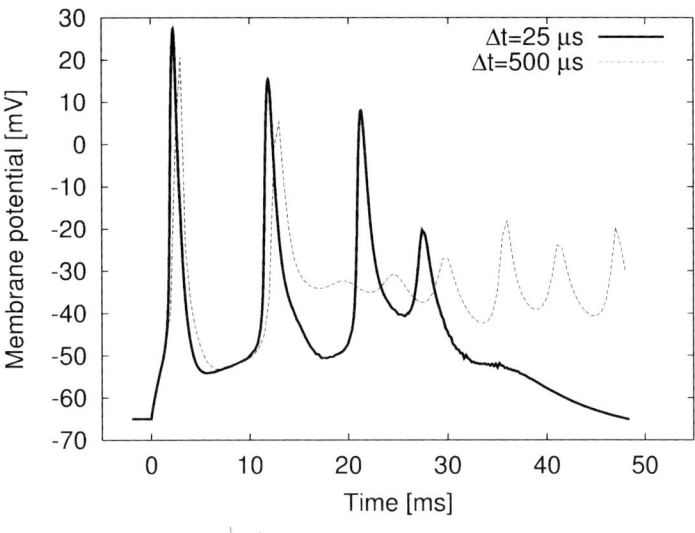

Fig. 5. Effect of time step on the simulation of the CA3 model used in Figure 1 (somatic injection of 1 nA, 3 ms).

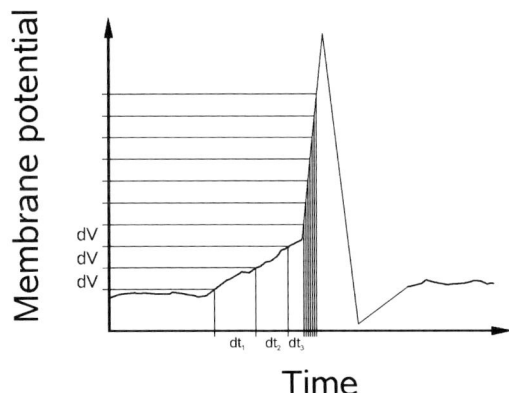

Fig. 6. An action potential requires smaller time steps then subthreshold oscillations in order to maintain the same accuracy in the computation of voltage.

order to maintain the estimated local error of each state variable below a specified maximum absolute error. Since different model variables are typically characterized by different ranges in their units, it is important to specify individual maximum allowable errors for each state variable *(25)*. Another issue with variable time step concerns discontinuous external events. If a current pulse is suddenly injected during a quiescent period of a simulation, the first time steps after the injection are likely to be too large. NEURON introduces a special mechanism [at_time()], to specify the occurrence of such external events *(25)*.

The issue of space discretization, or compartmentalization, in the computational implementation of anatomically accurate neurobiological models is in principle simi-

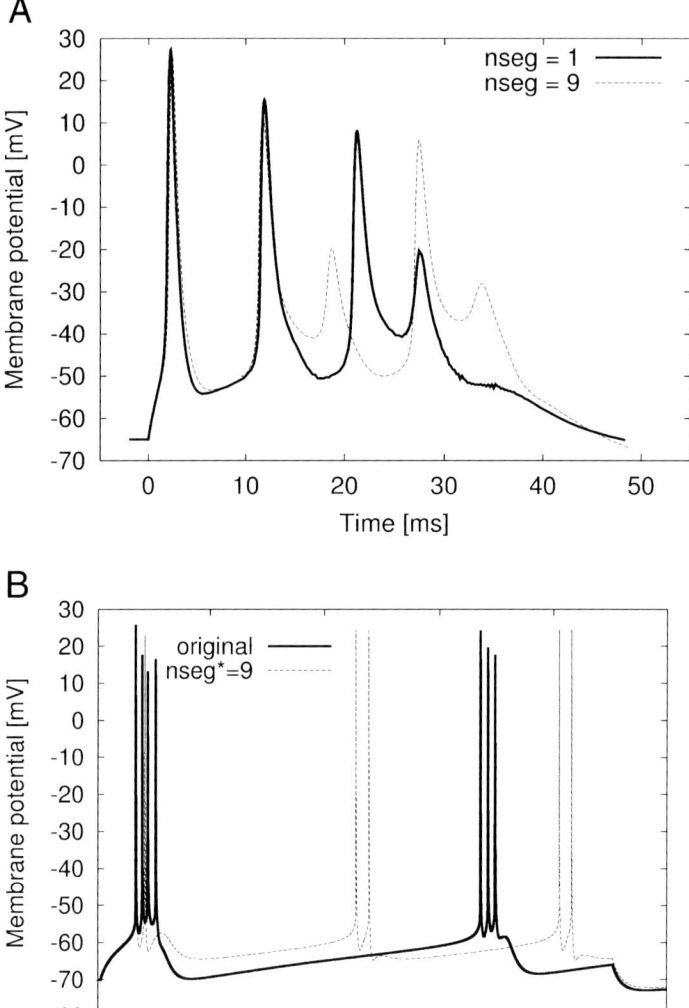

Fig. 7. (A) Effect of compartmentalization on the simulation of the CA3 model used in Figure 1 (somatic injection of 1 nA, 3 ms); nseg is the number of compartments in each dendritic branch. The CD-ROM contains mpeg animations displaying the variation of the simulation outcome with the number of compartments for various compartmentalizations of this and similar models. **(B)** Computer simulations based on the Mainen and Sejnowski model *(29)* with a 900 ms somatic current injection of 0.2 nA. The model (available at http://senselab.med.yale.edu/senselab/ModelDB/) was run with the original compartmentalization and by multiplying the number of compartments in each branch by 9 (further decrease of compartment size did not produce any significant difference).

lar to that of time steps. However, perhaps because neuroanatomy has been generally disregarded or "simplified away" in electrophysiological simulations until recently, numerical problems arising from compartmentalization are less known, and good general solutions, such as that of variable time step, have not yet been developed for spatial discretization. Even worse, as we discuss in the next section, the errors introduced in time and space discretization can interfere with each other in the numerical implementation, such that the best choice of compartmentalization can in practice prevent the use of variable time step.

7.3. ANATOMICAL REPRESENTATION

Equation 3 describes the mathematical model of the spatio-temporal distribution of membrane voltage of a neuron. The numerical approximation of the solution of this problem requires (in analogy to time discretization) space discretization, i.e., the division of the continuous dendritic branches in a finite number of compartments. Each of the state variables in Equation 3 is assumed constant within any one compartment and calculated only in the center of each compartment. Thus, for example, voltage and diameter are assumed to vary over x only between compartments (compartments are isopotential and cylindrical). The spatial discretization of the model implies that the accuracy of the approximation depends not only on the size of the time steps, but also on the "space steps", i.e., of the compartments.

For nonlinear equations such as Equation 3, it is very hard to assess the exact relationship between accuracy of a solution and integration steps. Theoretical results only provide an order of magnitude of the error. For example, in the implicit Crank-Nicholson integration method (which is the most common for neurobiological simulations) the local truncation error (for one time step) is of the order $O[(\Delta t)^3 + (\Delta t)(\Delta x)^2]$ and the global truncation error (cumulative, for the whole simulation) is of the order $O[(\Delta t)^2 + (\Delta x)^2]$ (26). Similarly to the case of time discretization, if we increase the number of compartments excessively, the simulation time will be unacceptably long, and the cumulative round-off error will become bigger then the error introduced by the discretization itself. The problem of compartmentalization, however, is even more difficult than that of choosing an appropriate time step, because the 1D cable equation is only a simplification of the 3D case. Such an approximation is only valid for compartment lengths that are not too small compared to the diameter (2,27). In addition, no variable discretization algorithm (analogous to that discussed in the previous section for time steps) is available for the spatial grid.

7.3.1. Rule of 1/3

In section 2 of this chapter, we described the "rule of 1/2" to empirically determine the appropriate time step for a simulation. A similar approach can be used to determine the appropriate size of compartments (28). We can call it "rule of 1/3", since it is based on the division of each existing compartment into three equal pieces (the use of an odd number guarantees that a "virtual electrode" in the middle of a compartment does not need to be repositioned upon the division). If the numerical solution changes significantly after this division, each compartment is again divided in three, until a satisfactory stability is reached. Surprisingly, several well-known models recently implemented have not been compartmentalized sufficiently. The simulation results change if each compartment is divided by 3 (see also Table 1). Examples include the Migliore model of CA3 pyramidal cell burst (Fig. 7A) and the Mainen and Sejnowsky model of cortical neurons *(29)* (Fig. 7B).

The rule of 1/3 is a very inefficient compartmentalization criterion, because the number of compartments grows exponentially with the power of 3. A possible alternative is the "length rule", requiring each compartment to be shorter then a given physical length (for instance, 1 µm), and then to reduce such a maximum length value until the simulation is stable. However, even the length rule falls short of an efficient compartmentalization criterion, because it does not take into account anatomical differences throughout the dendrites. In neurobiological models, time domain has a very simple topology: it is just an interval. In contrast, neuronal shape is generally characterized by complex tree architecture. The approximation of the solution of Equation 3 obtained by spatial discretization depends on the geometrical and electrophysiological properties of each compartment. As a consequence, a compartmentalization based on these properties is more effective than a uniform spatial compartmentalization.

For a given time step, the accuracy in computing voltage depends on the temporal derivative of the voltage (Fig. 6). Similarly, given the size of the compartment, the accuracy in computing voltage depends on the spatial derivative of voltage, i.e., how fast the potential is changing with the position along the dendritic axis (Fig. 8A). This means that an effective spatial discretization could be achieved with nonuniform compartment lengths, such that the voltage drop over each compartment is less than a maximum allowable value. Voltage drop is influenced among other factors by the diameter of the dendrite, passive properties, density of ion channels, presence of spines, synapses, etc. Transient signals propagating through the neuron, such as action potentials, further complicate the situation as they constitute spatial derivatives of voltage and calcium concentrations that are also variable in time (Fig. 8B).

7.3.2. Passive Case

How can we find an effective nonuniform compartmentalization? In the passive (linear) case, Rall defined a steady-state parameter, called space constant (λ_{DC}), representing the distance at which the voltage would drop to $1/e$ of its original value in an infinite cylinder with identical diameter, cytoplasmatic resistance, and specific membrane resistance of the neurite in question *(15,31)*. Rall demonstrated that

$$\lambda_{DC} = \sqrt{\frac{R_m d}{R_a 4}} \qquad [\text{Eq. 7}]$$

where R_m and R_a are the membrane and cytoplasmatic resistances, respectively, and d is the diameter.

A possible spatial discretization strategy is to use a small fraction (e.g., 5%–10%) of the DC space constant as the length of the compartments *(28)*. For convenience, length can be expressed in units of spatial constants (so called electrotonic length):

$$\text{EL}_{DC} = \frac{L}{\lambda_{DC}} \qquad [\text{Eq. 8}]$$

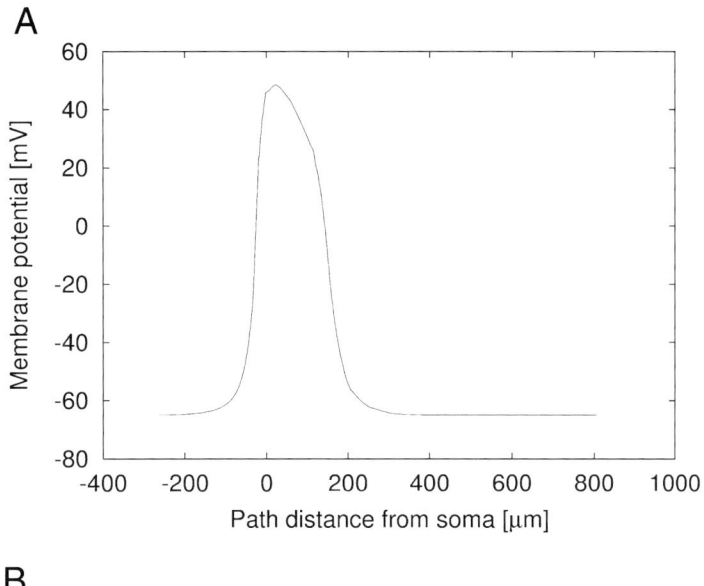

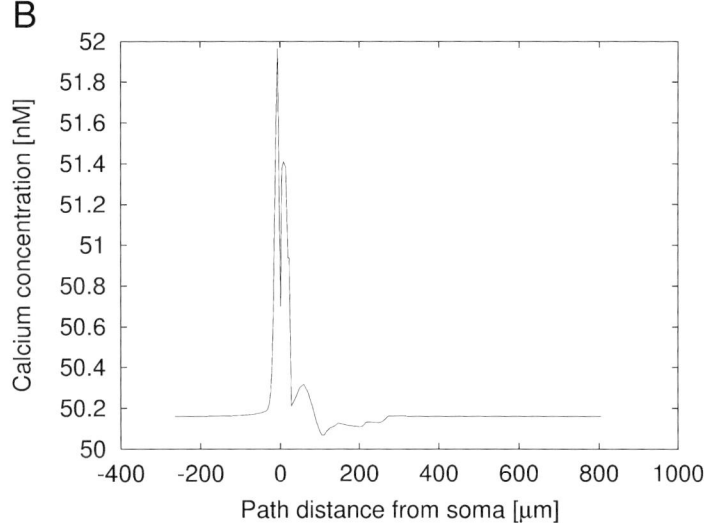

Fig. 8. (A) Spatial distribution of membrane potential and **(B)** internal submembrane calcium concentration from computer simulations of CA3 cell based on a modification of the Migliore model *(30)*. Neuron simulation files are included in the CD-ROM.

Thus a nonuniform compartmentalization algorithm can be implemented by sequentially dividing in 3 any compartment with an electrotonic length greater than a given threshold. In the passive case, it is even possible to calculate the approximation error produced by any electrotonic length threshold *(32)*. However, transient signals such as action potentials are subject to greater distortion and attenuation with distance by virtue of membrane capacitance and cytoplasmatic resistance than by membrane resistance and cytoplasmatic resistance. Thus the concepts of space constant and electrotonic length must be extended for nonsteady state inputs. For sinusoidal signals the space constant can be formulated as *(28)*

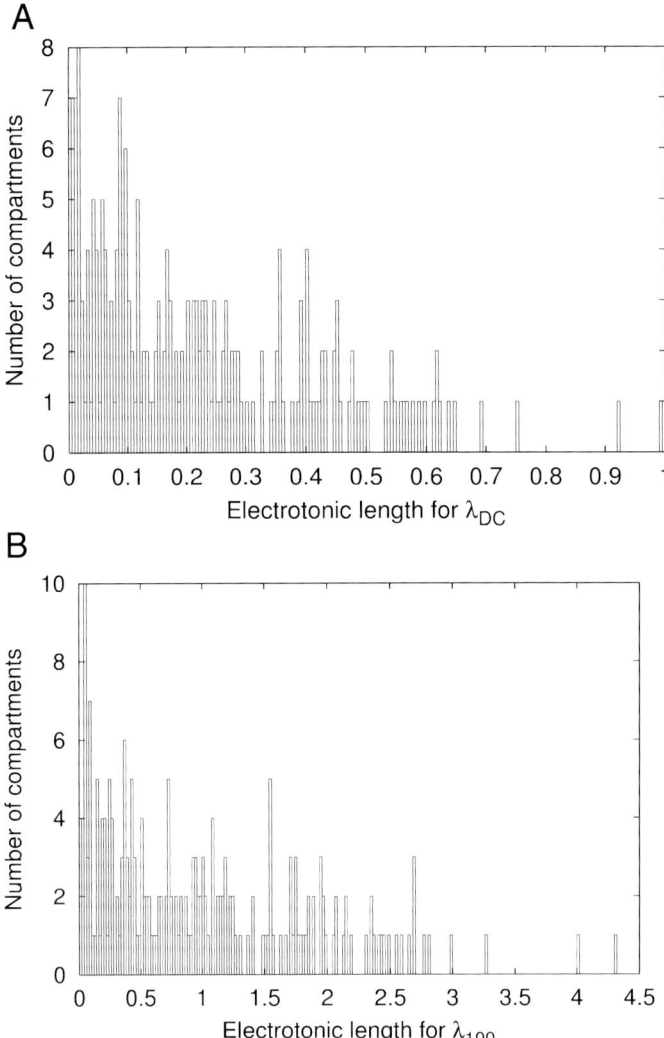

Fig. 9. Histograms of **(A)** EL_{DC} and **(B)** EL_{100} for the Migliore CA3 model *(22)*.

$$\lambda_f = \frac{1}{2}\sqrt{\frac{d}{\pi \times f \times R_a \times C_m}} \qquad \text{[Eq. 9]}$$

where f is the signal frequency, d is the diameter, R_a is the cytoplasmatic resistance, and C_m is the membrane capacitance. Accordingly, an alternative compartmentalization strategy consists of setting the threshold for the maximum allowed compartmental length to a fraction of λ_f[1].

---

[1]This is in fact one of the three possible mechanisms to compartmentalize models in Neuron: the NEURON Builder's command *d_lambda* sets the number of compartments in each branch, such that their electronic length for $f = 100$ Hz is less than a specified value (by default 0.1). Alternatively, one can set the number of compartments directly with parameter nseg. The third possibility in NEURON is to set the number of compartments with parameter *d_X*, which specifies the maximum allowed compartment length (by default 50 µm).

Figure 9 shows examples of the electrotonic length distributions for the DC and sinusoidal case (Equations 7–9) from the Migliore CA3 model. Given the large variability of space constants through the dendritic tree, the compartmentalization algorithm described in this subsection should be significantly more efficient than uniform algorithms (based e.g., on the 1/3 rule). This means that the number of compartments necessary to obtain the same level of accuracy is larger using the uniform algorithm than using the algorithm based on electrotonic lengths. In other words, given the same number of compartments, the approximation error introduced by the uniform algorithm is greater than that introduced by the algorithm based on electrotonic lengths.

In any case, it is important to stress that neither λ_{DC} nor λ_f take into account changes in the membrane conductivity produced by the activation of voltage-dependent channels, calcium dependent channels, or synapses. Thus, the above compartmentalization methods are not optimal for simulations involving action potentials.

7.3.3. Active Case

At the beginning of Subheadings 7.2. and 7.3., we have argued that efficient (spatial or temporal) discretization depend on the (spatial or temporal) derivative of transient signals. Therefore, the optimal compartmentalization should be based not only on the morphological characteristics of the model, but also on any other property influencing electric activity, such as densities and kinetics of ion channels and synapses, calcium pumping and buffering mechanisms, etc. For example, if a model produces high frequency ripples, its compartments should be very short in order to obtain a stable solution. In contrast, if the electrophysiological behavior of the model is relatively quiescent, a greater compartmental length will ensure the same accuracy. Similarly, a finer compartmentalization is required in the proximity of synapses with fast dynamics.

What is the optimal compartmentalization, given a model and its (expected) electrophysiological behavior? To the best of our knowledge, this general problem has not yet been solved theoretically. Based on the above discussion, we constructed the following empirical compartmentalization algorithm. From an initial compartmentalization of the model, we calculate the largest voltage drop in each compartment during the simulation. If the drop in any particular compartment is bigger then a given threshold, we divide that compartment into three compartments and then reapply the same procedure. This algorithm yields the most efficient spatial discretization of the model (Fig. 10). However, this strategy requires additional cycles of computationally intensive simulations. From our experience on CA3 pyramidal cell models, we found that the compartmentalization obtained with this "voltage drop" algorithm is only moderately more efficient than that obtained with the algorithm based on the space constant for sinusoidal signals (Equation 9). However, for complex models such as those discussed in this chapter, it is necessary to set the maximum compartmental length to as little as 0.04 electrotonic lengths (as opposed to the 0.3 electrotonic lengths recommended in the literature *[28]*). Such a constraint is fairly severe: the recent update of the Migliore pyramidal cell model *(30)* requires nearly 6000 compartments to be stable (compared to the approx 800 compartments obtained using the 0.3 electrotonic length threshold). The need for such a finer compartmentalization is due to the complexity of the model (11 active conductances, complex calcium dynamics) and bursting activity. A simpli-

Table 1
Examples of Recently Published Anatomically Realistic Models

Model Reference	Characteristic Action	{Number} and Distribution of Ion Channels and Calcium Mechanisms	No. Compartments Originally Used	No. Compartments Required to Obtain Stable Solution
Destexhe's thalamic relay cell (33)	Single spike	{3}, I_{Ca} distributed uniformly through dendrites, I_{Na} and I_K only in soma, calcium pumping	206	206
Migliore's CA3 pyramidal cell (22)	Burst, only in soma and main trunk	{9}, uniformly through dendrites (I_{Na} only in soma and main trunk), calcium diffusion, buffering, pumping	375	500
Mainen's cortical cells (29)	Burst, back propagation up to dendritic tips	{5}, uniformly through dendrites, calcium pumping	479	4500
Lazarewicz's CA3 pyramidal cell (30)	Burst, back propagation up to dendritic tips, dendritic generation of action potentials, forward propagation	{11}, nonuniformly hrough dendrites, t calcium diffusion, buffering, pumping	—	6000

fied model only containing a uniform distribution of I_{Na}, $I_{K(DR)}$, $I_{K(A)}$, and without calcium dynamics, only requires 500 compartments. In all of these cases, a constant time step of 25 µs is sufficient to obtain stable solutions.

In order to obtain the proper compartmentalization for an anatomically realistic model, we recommend using the algorithm based on the space constant λ_f starting from a threshold of 0.4 and decreasing this value (depending on the complexity of the model) until the solution becomes stable. We should also note here that the longitudinal diffusion of calcium (or of any other ion) is described by an equation similar to the cable Equation (1,2). Thus, the effects of compartmentalization on the solution of the calcium diffusion equation is similar to that on the cable equation. However, the spatial constant for calcium diffusion is 10 times smaller than the typical value for cable equation, the introduction of ionic longitudinal diffusion may require a 10-fold increase of the number of compartments necessary to ensure the same accuracy (2).

Since the optimal compartmentalization depends on neuronal activity, and neuronal activity is time-dependent, a time-variable compartmentalization should in principle be more efficient than a "static" algorithm. For electrophysiological simulations based on the cable equation, we are not aware of any attempt to implement a dynamical compartmentalization depending on transient activity. The only step in this direction consists of a dynamical lumping of axonal trees (34,35). This method consists of lumping different branches of the tree into larger equivalent cylinders during the simulation depending on the electric activity in the branches: silent branches are lumped, while active branches are unfolded. However, this method does not preserve tree morphology and has never been applied to complex models containing a variety of active currents.

7.3.4. Additional Practical Aspects on Compartmentalization

An important aspect of spatial discretization that is seldom taken into account regards models containing ionic channel densities distributed nonuniformly with respect to path distance. Since channel densities are discretized during the compartmentalization process, we have to consider two different types of approximation: the first one is connected with the numerical solution of the model's partial differential equations (as discussed in the previous sections); the second one is the discretization of the very distributions of the model's active mechanisms. As an example, let us consider a model with a dendritic distribution of $I_{K(A)}$ channel density that increases linearly with the distance from the soma (36). For a dendritic branch containing one compartment, this density is represented only by one number, i.e., the value taken from the position corresponding to the center of the compartment. If the branch is divident into three compartments, the density distribution in the branch is described by three different numbers, representing the centers of each compartment. In order to obtain the proper density distributions in any compartmentalization algorithm, it is, therefore, important to reload nonuniformly distributed properties after each change of compartmentalization. If the densities of mechanisms are calculated first, and then the compartments are further divided without redistributing nonuniform densities, these properties will be (incorrectly) constant across the divided compartments. In CA3 pyramidal cell models, we found that the spatial discretization of nonuniform distributions of mechanisms causes greater instability than the numerical approximation used to solve the partial differential equations. Models with uniform distributions require fewer compartments than models with nonuniform distributions in order to obtain a stable solution.

An extremely important practical note concerns NEURON simulations of complex anatomically accurate models (22,30). When a very fine compartmentalization is required to obtain a stable solution (i.e., thousands of compartments for a single neuron), fixed time step should be used instead of variable time step. In several cases, we found that the effect of variable time step on the approximation of the solution would prevent simulations from becoming stable even with very fine compartmentalization[2,3] (see CD-ROM for specific examples).

---

[2]This remark concerns the implicit numerical integration implemented in NEURON and might not apply to explicit numerical method (John Rinzel, personal comment).
[3]For simulations with realistic cell morphology using NEURON, we recommend the use of versions 5.0.0 or higher, because of problems connected with compartmentalization in previous releases (sometimes the area variable became zero after setting the nseg parameter).

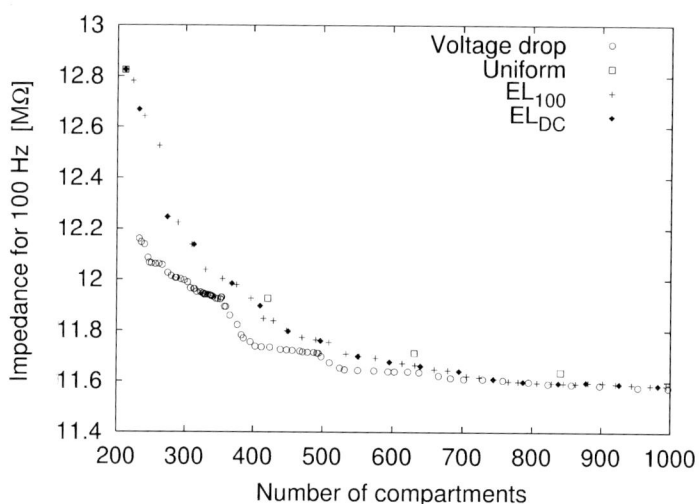

Fig. 10. Impedance (at 100 Hz) calculated for the updated CA3 model *(30)* using four different methods of compartmentalization: Voltage drop refers to the algorithm proposed in Section 7.3.3.; in the Uniform algorithm every section is divided into the same number of segments, EL_{DC} and EL_{100} refer to the algorithm described in Section 7.3.2., based on the spatial constants for λ_{DC} and for λ_{100}, respectively. Neuron simulation files are included in the CD-ROM.

In general, simulations based on realistic morphology are very computationally intensive. The compartmentalization algorithms described in the sections above, which should be run before the "real" simulations, may also be fairly time-consuming. Thus, it is useful to have clues on the (in)stability of the solution without the need to run an entire simulation. We found that input resistance and impedance are potentially useful parameters in this regard. Given a neuronal model, different compartmentalization can influence the numerical simulation so much as to affect the initial depolarization of the model in response to the same amount of current injection. This difference might only affect spiking frequency or delay, or it could change the results both quantitatively and qualitatively (Fig. 7). In the compartmentalization of the model, it is, therefore, convenient to assume the stability of the input resistance and impedance as a necessary prerequisite for the stability of the simulation (Fig. 10). Before performing any time-consuming simulation, an initial minimal number of compartment can be assessed by calculating the input resistance for different compartmentalizations. This procedure is particularly convenient in NEURON, where input resistance and impedance can be calculated directly with appropriate built-in functions.

7.4. FROM MORPHOLOGICAL RECONSTRUCTIONS TO ELECTRO-PHYSIOLOGICAL COMPARTMENTAL MODELS: TOOLS AND ALGORITHMS

The compartmentalization algorithms described in the previous sections can be used with any neuronal shape. The process of reconstructing neuronal morphology and digitizing it into a computer representation, however, is far from trivial *(37)*. At the end of this process dendritic trees are described in three dimensions as weighted graphs, with

Simulations of Neuronal Electrophysiology 143

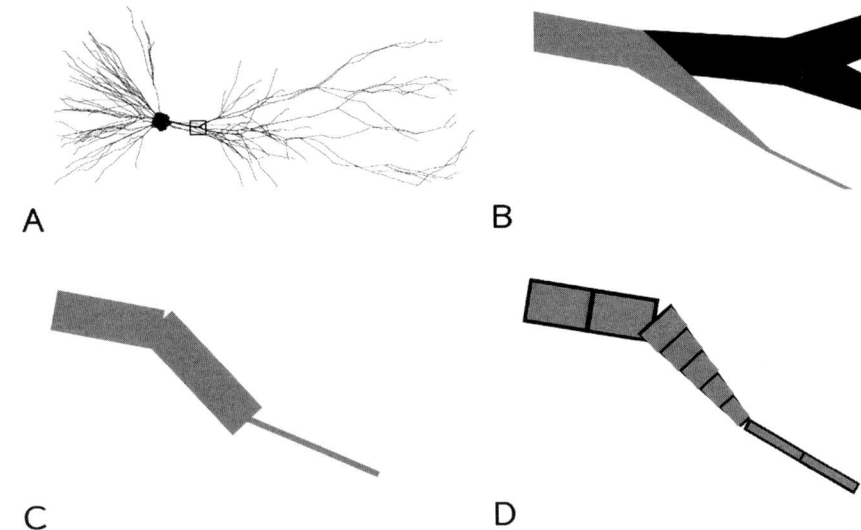

Fig. 11. (A) Computer representation of a CA3 pyramidal cell *(22)* with details (B) of branching dendrites. (C) Representation of dendritic branches as a sequence of cylinders. (D) Same branches from panel C represented in NEURON as three sections produced by the program swc2hoc (included in the CD-ROM). The first and the third sections are divided into two compartments, the second section in five.

labels describing diameters. In other words, the (continuous) branching structure of the tree is stored digitally as a connected series of cylinders (Fig. 11). The number of cylinders used in the representation depends both on the structural complexity of the neuron and on arbitrary or subjective factors intrinsic to the semimanual reconstruction system. It is important to point out that this spatial discretization is not related to the accuracy of the approximation of the solution of Equation 3. Thus, in order to use experimentally reconstructed neuronal morphology for electrophysiological analysis, it is necessary to redo the compartmentalization of the given structure, based on the needed accuracy of the solution approximation of Equation 3.

A considerable amount of reconstructed morphological data is available through the Internet. Examples of electronic morphological repositories for hippocampal neurons are the Southampton-Duke archive *(38)*, the Claiborne collection *(39)*, and the Guylas repository *(40)*. These cells are typically reconstructed from intracellularly filled (and usually sectioned) preparations with the Neural Tracing System (NTS or "Eutectic"), which is no longer supported, or the Neurolucida system *(41)*. A recently introduced plug-in for ImageJ *(42)*, called Neuron_Morpho, allows the semiautomatic reconstruction of neuronal morphology from stacks of confocal images *(43)*. Alternatively, neurons can be generated stochastically, i.e., the neuronal structure itself can be the product of computational modeling and simulations *(44,45)* (see also Chapter 3 in this book). In general, neurons reconstructed or generated with different software systems may have different digital formats. In the format adopted by the Southampton-Duke archive (called SWC), each cylinder is described by a (sequential) identification number, a tag (distinguishing soma, axons, apical, and basal dendrites, etc.), X, Y, and Z positions, as well as radius of the ending point, and the identification number of the (unique) adjacent cylinder in the path towards the soma. In all cases, a necessary step to perform electrophysiological simulations is format conversion.

Not all the morphological information of a neuronal reconstruction is used in Equation 3. The length, diameter, and connectivity of each cylinder, but not its spatial location, are relevant for electrophysiological simulations[4]. Therefore, the morphological representation can be simplified. For example, consecutive cylinders with the same diameter, and no intervening bifurcations, could be merged into a single section with length equal to the sum of the lengths of the original cylinders. Once again, the sections obtained in this way would have to be subsequently compartmentalized in order to obtain a stable numerical solution of Equation 3. The simplification of the morphological structure could be carried a step further by treating any sequence of cylinders between nodes (stems, branches, and terminations), independent of their diameter, as a single section (which must be compartmentalized before running the simulation). In this case, it is possible that a compartment would include in the end two original cylinders with different diameters. The diameter of the joint compartment may be set to conserve total surface (i.e., as the average of the original diameters weighted by the respective cylinder's lengths).

An alternative algorithm to simplify reconstructed morphology for electrophysiological simulations was developed by Borg-Graham *(46)* and is implemented in the Surf-Hippo simulation software. This algorithm lumps cylinders together preserving the topology, area, axial resistance, and the spatial location of the nodes. However, this algorithm affects dendritic length, volume, and diameter, and, in particular, it may produce implausibly huge diameters in distal dendrites. This kind of anatomical transformation can introduce artifacts in parameters such as calcium concentration, which depend on the surface to volume ratio. Moreover, due to the nonconservation of key geometrical properties, Borg-Graham's algorithm may cause problems with the localization of synaptic stimulation.

To maximize the anatomical accuracy of electrophysiological simulations, we created a novel algorithm that is more conservative (or less "invasive") than the above examples. This algorithm converts (experimentally reconstructed or computationally generated) morphological data files (SWC format) into NEURON simulation files (.hoc) by using the smallest possible number of sections without destroying information from the original morphology. The algorithm is implemented as an AWK script (called swc2hoc, and included in the CD-ROM) and runs under all UNIX systems. Two consecutive cylinders are merged if they have the same diameter and are not separated by a bifurcation. This method conserves the original diameter values as well as path length (and, thus, also surface and volume). In addition, when two or more cylinders are merged, swc2hoc stores the information about the original positions of their ending points as NEURON pt3d structures. This information can thus be used if needed (e.g. to distribute synaptic contacts accurately according to the appropriate anatomical 3D input patterns). Swc2hoc further creates frustums between consecutive cylinders with different diameters (and not separated by a bifurcation). This choice is justified by

---

[4]This is true only to the first approximation of neuronal behavior modeled in Equation 3. Modeling the effect of ephaptic interactions among dendrites would obviously require the explicit spatial representation.

the fact that the digital representation of neuronal morphology is only an approximation of the original structure (Fig. 11C). For example, consecutive points in the morphological reconstruction have finite diameter differences, whereas the physical reality is continuous. Ideally, the finest spatial discretization of the model should converge to the original shape. Thus, we linearly interpolate diameters at the points of discontinuity with a frustum made of three segments *(47)*. This architecture is designed to optimize the approximation of both the morphological structure (Fig. 11) and the electrophysiological process (Equation 3). The output of Swc2hoc is a NEURON (.hoc) file containing all the sections constituting the dendritic trees. These sections must be then compartmentalized using NEURON parameter nseg or any other algorithm described in the previous sections.

Several software tools are especially useful in anatomically accurate electrophysiological simulations. Cvapp is a visual editor for morphological files originally designed for the Southampton-Duke archive by Robert Cannon. Cvapp can read Neurolucida and SWC files and convert them to NEURON or GENESIS formats, as well as converting Neurolucida files into SWC. This program was recently modified by Steve van Hooser *(48)* to implement the compartmentalization algorithm based on a threshold proportional to the space constant λ_{DC} for the GENESIS environment. Additionally, this program can merge together consecutive cylinders depending on whether their diameter difference is smaller than a given threshold (set by the user). NeuroMesher is a similar java class tool that can convert Neurolucida and Eutectic files into GENESIS or NEURON format. NeuroMesher also implements compartmentalization mechanisms based on the same parameters as cvapp *(49)*. Finally, NTScable is another program to convert Eutectic files into NEURON format. Various compartmentalization algorithms are implemented in NTScable, based on parameters such as the minimum number of segments, the maximum absolute segment length, or the maximum segment length relative to the diameter *(50)*.

Buchs' Toolbox for Neural Modeling is a set of MatLab/JAVA tools including NeuroToolBox, NeuroTrace, and NeuroGenerator *(51)*. NeuroToolbox is designed to provide electrophysiological simulation mechanisms in MatLab. NeuroTrace extracts 3D morphological structures from stack of 2D images, producing either a cylinder representation of the neuron, or a smooth PostScript representation. NeuroGenerator integrates the morphological and electrophysiological components and builds the electrophysiological simulation in MatLab by inserting electrical properties from NeuroToolbox into morphology from NeuroTrace. In this program, compartmentalization is carried out based on λ_{DC}.

Dendritica is a useful application in the NEURON environment for relating dendritic geometry and signal propagation *(52)*. The software consists of three main components: (*i*) interactive morphological analysis and electrophysiological simulation of single neurons; (*ii*) automated batch simulation across a set of morphologies using the same simulation parameters; and (*iii*) automated analysis of batch simulation runs. Examples of the applications include simulations with voltage clamped waveforms such as somatic action potentials *(53)*, current clamp, synapse stimulation, or automatic calculation of I_{Na} density necessary for full back propagation of action potentials through all the dendritic terminal tips.

7.5. CONCLUSIONS

Computer simulation is only one of the stages in the process of modeling. Since the mathematical model is numerically approximated in this stage, precautions need to be taken to avoid that the errors produced by this approximation exceed the level of accuracy of the model itself. In particular, in anatomically accurate electrophysiological models, simulations require discretization of space and time. For models containing nonlinear elements, such as voltage-gated channels, there is no established method to assess the absolute error produced by a particular compartmentalization.

The major conclusion of this chapter is that compartmentalization algorithms should not be used blindly, but the effect of varying mesh resolutions on the simulation outcome should be systematically tested to establish the stability of the solution. Among the compartmentalization algorithms we discussed, uniform algorithms are less efficient than algorithms based on electrotonic length parameters. The algorithm we proposed based on voltage drop in the compartment is most efficient, but the differences with the algorithms based on electrotonic length are not dramatic.

REFERENCES

1. De Schutter E, Cannon RC. (eds.) Computational Neuroscience: Realistic Modeling for Experimentalists. CRC Press, Boca Raton, FL, 2000.
2. Koch C. Biophysics of Computation: Information Processing in Single Neurons. Oxford University Press, New York, 1999.
3. Bower JM., Beeman D. (eds.) The Book of GENESIS: Exploring Realistic Neural Models with the GEneral NEural SImulation System, 2nd ed. Springer-Verlag, New York, 1998.
4. (http://www.neuron.yale.edu).
5. (http://www.bbb.caltech.edu/GENESIS/genesis.html).
6. (http://www.cnrs-gif.fr/iaf/iaf9/surf-hippo.html).
7. (http://www.neosim.org).
8. (http://www.compneuro.org/CDROM/catacomb/).
9. (http://www.mathworks.com).
10. (http://www.strout.net/conical/).
11. Murthy DNP, Page NW, Rodin EY. Mathematical Modelling. A Tool for Problem Solving in Engineering, Physical, Biological and Social Sciences. Pergamon Press, Oxford, UK, 1990.
12. Gershenfeld NA. The Nature of Mathematical Modeling. Cambridge Univ. Press, Cambridge, UK, 1999.
13. De Schutter E. Using realistic models to study synaptic integration in cerebellar Purkinje cells. Rev Neurosci 1999; **10**:233–245.
14. Protopapas AD, Vanier M, Bower JM. Simulating large networks of neurons. In: Methods in Neural Modeling (Koch C, Segev I, eds.), 2nd ed. MIT Press, Cambridge, MA, 1998, pp. 461–498.
15. Rall W. Membrane time constant of motoneurons. Science 1957; **126**:454.
16. Norman RS. Cable theory for finite length dendritic cylinders with initial and boundary conditions. Biophys J 1972; **12**:25–45.
17. Langtangen HP. Computational Partial Differential Equations, Numerical Methods and Diffpack Programming, Lecture Notes in Computational Science and Engineering, Vol. 2. Springer-Verlag, New York, 1999.
18. Hodgkin AL, Huxley AF. A quantitative description of membrane current and its application to conduction and excitation in nerve. J Physiol 1952; **117**:500–544.

19. Hodgkin AL, Huxley AF. Currents carried by sodium and potassium ions through the membrane of the giant axon of Loligo. J Physiol 1952; 116:449–472.
20. Wong RK, Prince DA. Afterpotential generation in hippocampal pyramidal cells. J Neurophysiol 1981; 45:86–97.
21. Ranck JB Jr. Studies on single neurons in dorsal hippocampal formation and septum in unrestrained rats. I. Behavioral correlates and firing repertoires. Exp Neurol 1973; 41:461–531.
22. Migliore M, Cook EP, Jaffe DB, Turner DA, Johnston D. Computer simulations of morphometrically reconstructed CA3 hippocampal neurons. J Neurophysiol 1995; 73:1157–1168.
23. Menschik ED, Finkel LH. Cholinergic neuromodulation of an anatomically reconstructed hippocampal CA3 pyramidal cell. Neurocomputing 2000; 32–33:197–205.
24. Cohen SD., Hindmarsh AC. CVODE User Guide. Lawrence Livermore National Laboratory technical report UCRL-MA-118618, 1994.
25. Hines ML, Carnevale NT. Expanding NEURON's repertoire of mechanisms with NMODL. Neural Comput 2000; 12:995–1007.
26. Mitchell AR. Computational Methods in Partial Differential Equations. John Wiley & Sons, London, UK, 1969.
27. Rall W. Distributions of potential in cylindrical coordinates and time constants for a membrane cylinder. Biophys J 1969; 1509–1541.
28. Hines ML, Carnevale NT. NEURON: a tool for neuroscientists. Neuroscientist 2001; 7:123–135.
29. Mainen ZF, Sejnowski TJ. Influence of dendritic structure on firing pattern in model neocortical neurons. Nature 1996; 382:363–366.
30. Lazarewicz MT, Migliore M, Ascoli GA. A new bursting model of CA3 pyramidal cell physiology suggests multiple locations for spike initiation. 4th International Neural Coding Workshop—NCWS'2001. Plymouth, UK, 2001. To appear in Biosystems.
31. Segev I, Rinzel J, Shepherd GM (eds.) The Theoretical Foundation of Dendritic Function: Selected Papers of Wilfrid Rall with Commentaries. MIT Press, Cambridge, MA, 1994.
32. Eichler-West RM. On the development and interpretation of parameter manifolds for biophysically robust compartmental models of CA3 hippocampal neurons. University of Minnesota Doctoral Dissertation, 1996.
33. Destexhe A, Neubig M, Ulrich D, Huguenard J. Dendritic low-threshold calcium currents in thalamic relay cells. J Neurosci 1998; 18:3574–3588.
34. Manor Y, Gonczarowski Y, Segev I. Propagation of action potentials along complex axonal tree: model and implimentation. Biophys. J. 1991; 60:1411–1423.
35. Manor Y, Koch C, Segev I. Effect of geometrical irregularities on propagation delay in axonal trees. Biophys J 1991; 60:1424–1437.
36. Migliore M, Hoffman DA, Magee JC, Johnston D. Role of an A-type K^+ conductance in the back-propagation of action potentials in the dendrites of hippocampal pyramidal neurons. J Comput Neurosci 1999; 7:5–15.
37. Ascoli GA, Nasuto SD, Krichmar JL, Senft SL. Generation, description, and storage of dendritic morphology. Philos Trans R Soc Lond B Biol Sci 2001; 356:1131–1145.
38. (http://www.cns.soton.ac.uk/~jchad/cellArchive/cellArchive.html).
39. (http://cascade.utsa.edu/bjclab/).
40. (http://www.koki.hu/~gulyas/ca1cells/cellfiles.htm).
41. (http://www.microbrightfield.com/prod-nl.htm).
42. (http://rsb.info.nih.gov/ij/).
43. (http://www.maths.soton.ac.uk/staff/D'Alessandro/morpho/).
44. Ascoli GA, Scorcioni R, Krichmar JL, Nasuto SD., Senft SL. Computer generation and quantitative morphological analysis of virtual neurons. Anat Embryol 2001; 204: 283–301.
45. (http://www.krasnow.gmu.edu/ascoli/CNG/l-neuron/).

46. Borg-Graham L. Interpretations of data and mechanisms for hippocampal pyramidal cell models. In: Cerebral Cortex: Cortical Models, Vol. 13 (Ulinsky, ed.) Plenum Publishing Corporation, New York, NY, 1999, pp. 19–138.
47. Hines ML, Carnevale NT. The NEURON simulation environment. Neural Comput 1997; **9**:1179–1209.
48. (http://www.compneuro.org/CDROM/docs/cvapp.html).
49. (http://www.science-renaissance.org/rogene/Software/index.html).
50. (http://www.cns.fmed.ulaval.ca/alain_demos.html).
51. (http://www.cns.unibe.ch/~buchs/diplomas.html).
52. (http://dendrite.physiol.ucl.ac.uk/software.html).
53. Vetter P, Roth A, Hausser M. Propagation of action potentials in dendrites depends on dendritic morphology. J Neurophysiol 2001; **85**:926–937.

Part II

8
Predicting Emergent Properties of Neuronal Ensembles Using a Database of Individual Neurons

Gwen A. Jacobs and Colin S. Pittendrigh

ABSTRACT

Neurobiologists have known for decades that to understand the computational properties of a neural system, it is necessary to understand the relationships between the physiological properties of individual neurons and their anatomical structures. We have addressed this problem by developing an approach for analyzing the relationships between structure, function, and computation within a network of neurons. We have developed a suite of visualization and analysis tools, called NeuroSys (http://www.cns.montana.edu/projects/NeuroSys), that allows the investigator to reconstruct the anatomical features of many neurons and store them in a database that preserves their correct spatial relationships in the nervous system. This ensemble reconstruction can then be used as a precise anatomical template on which to predict connectivity patterns and image the functional properties of the network. The visual format of the database is a probabilistic atlas, which preserves the spatial relationships between all objects within the nervous system. The database can be queried for information regarding structural, functional, and relational attributes of the objects and be used to predict functional properties of the neural system. We have used NeuroSys to predict the connectivity relationships between neurons in a model sensory system and to predict the steady state response patterns of the ensemble of neurons to sensory stimuli. This technique has been extended to incorporate the dynamic patterns of activity of the sensory neurons to predict the spatial and temporal aspects of the response patterns.

8.1. INTRODUCTION

A grand challenge in neuroscience is to understand the biological basis of information processing at the cellular and network levels. This challenge is being approached from many fronts, some of which are purely technological in nature. One major technological front involves the creation of informatics and analysis tools to explore the structural organization of complex neural circuits and to discover and understand the emergent properties of neural ensembles. For the last several years, research in our laboratory has been focused on determining the cellular mechanisms through which sensory information is represented and decoded by ensembles of neurons. In the course of these studies, we have developed a number of experimental approaches that combine anatomical and physiological data with computational and visualization tools. The goal was to create an interactive problem solving environment for testing ideas and hypotheses about structure function relationships in the nervous system. In this chapter, we will describe the system we have developed and illustrate how we have used it to learn more about basic sensory processing mechanisms in a model sensory system: the cricket cercal sensory system.

Three major challenges face neuroscientists interested in understanding the computational properties of neural networks: (*i*) to understand the relationships between spatio-temporal activity patterns in neural ensembles and the information they convey; (*ii*) to understand how the spatio-temporal patterns are decoded by cells at the next processing stage; and (*iii*) to understand how computations (e.g., pattern recognition) are carried out on that decoded information *(1)*. Many studies focus on the algorithms and biological mechanisms through which spatio-temporal patterns of activity in a mapped sensory system are decoded by postsynaptic neurons imbedded within the map. There are now numerous examples of how information represented within these patterns can be decoded by nerve cells by virtue of the shape and/or location of their dendrites. The postsynaptic target interneurons in a variety of sensory systems derive their stimulus sensitivities from the global location and the finer-scale shape of their dendrites within these maps *(1–3)*. For example, within the primary visual cortex, axons are aligned along the axis of preferred orientation; in the cricket cercal system, the shape and position of sensory interneuron dendrites are tightly linked to their directional tuning characteristics, and in the auditory system, the length of an axonal aborization and its position within the nucleus creates a delay line that affects the relative timing of inputs to postsynaptic cells. For modalities such as the olfactory system, where the relevant input features cannot be represented as values within some continuous parameter space, the temporal dynamics of the ensemble response patterns appear to play a much more significant role in information representation. Within some olfactory systems, the presentation of a particular odor results in a temporal pattern of activity among an ensemble of cells that changes over time, thus encoding the odor quality as a temporal pattern of activity *(1)*.

Given the complexity of neuronal interactions at the structural and biophysical levels, the challenge of understanding how neural ensembles represent and encode information is a daunting one. One of the most promising techniques involves recording the activity of large numbers of neurons with multiunit electrode arrays or with optical imaging techniques. These approaches enable the investigator to analyze the response properties of many neurons simultaneously and to study how these global activity patterns may encode information about a stimulus or motor output. We have taken a different complimentary experimental approach to study the characteristics of the ensemble activity. We have developed methods and tools to predict ensemble activity based on measurements of the response properties recorded from a large sample of individual neurons. We have collected anatomical and physiological data from a large sample of individual sensory neurons and interneurons and used it to create a functional atlas of the system. A key aspect of this approach has been the ability to collect anatomical data from a large number of individual neurons and scale and align this data to a common 3D coordinate system, thereby preserving the anatomical relationships between individual neurons in the network *(4,5)*.

To create the functional atlas, we developed a relational database and a set of computational and visualization tools, which were combined into a single suite of software tools called "NeuroSys". Our goals were to meet basic storage and retrieval capabilities for anatomical and physiological data, and to go beyond those capabilities in several fundamental respects. First, as well as containing information about the anatomy and functional attributes of the neurons, the database was structured to manage information about measured and hypothetical relations *between* neurons. The visual format of the graphical user interface (GUI) includes a 3D atlas, which preserves the spatial relationships between all objects within the nervous system. Second, software was developed to enable the formulation of queries on multiple data types (e.g., on the basis of spatial location of objects within the atlas and by the physiological properties of the objects calculated from time-series data). The query responses could be returned as graphical predictions of dynamic ensemble activity patterns that represent both the spatial and temporal aspects of the patterns. Third, we developed means to integrate the database environment with a variety of data analysis and simulation tools. Through the use of these query and analysis tools, it was possible to formulate queries which predict dynamic activity patterns of large neuronal ensembles across multiple processing stages within the animal's nervous system.

Another goal of creating the suite of programs in NeuroSys, was to make these programs accessible to neuroscience researchers with little expertise in computer science or programming. All interactions with the system take place within the GUI. The user can query the database for individual neurons based on their anatomical or physiological attributes using text based query screens within the GUI. For example, to examine a group of neurons that are sensitive to similar type of stimulus, the user can make a query to retrieve the structures of "all neurons with peak sensitivity to stimulus A." The data files matching this query are loaded into the GUI, where the user can view the structures of the neurons in three dimensions. The user can examine their morphology, the spatial relationships between the neurons, or calculate anatomical overlap between the cells.

Data collected from both anatomical and physiological experiments was loaded into the database with user defined data entry screens. These data entry screens allow the user to pick from a set of attributes (data collection parameters, age of animal, stimulus parameters, etc.) and link them to the data set he or she wants to store. Users can also create new experimental attributes to associate with their data. The data and attributes are then loaded into the database by the database administrator. Our long term goal is to automate this process, so that the user can add new data directly to the database by using the data entry screens. The system was designed to accommodate the storage and retrieval of any physiological property of a neuron, such as transmitter phenotype or developmental age. Thus, the NeuroSys database and tools can be used to examine both the structural and functional relationships between different populations of neurons. In the following sections, we will describe how we have used NeuroSys to study the cellular mechanisms underlying information processing in the cercal sensory system.

8.2. A MODEL SENSORY SYSTEM FOR STUDYING ENSEMBLE ENCODING OF SENSORY INFORMATION

The cercal sensory system of the cricket has emerged as a powerful model system for studies of neural development, sensory information processing, neural coding, and computation *(6–9)*. This sensory system provides information about the direction and dynamics of air current stimuli and is involved in a number of orientation, escape, and mating behaviors *(10–15)*. It is implemented around a representation of air current direction and dynamics that demonstrates the essential features of neural maps found in more complex systems, including mammalian visual and auditory systems *(6,16)*. As in other mapped sensory systems, primary sensory interneurons in this system derive their sensitivity for one stimulus parameter (i.e., stimulus *direction*) from the placement of their dendrites within the neural map *(6,8)*. Stimulus direction is represented as a pattern of activity within the map, and interneurons are tuned to stimulus direction via the architecture of their dendritic arbors. Unlike more complex mammalian systems, the accessibility and relatively small number of cells in this system (2000 receptors, 50 local interneurons, 20 primary projecting interneurons) allow an exhaustive detailed analysis of the anatomy, physiology, and synaptic interconnectivity of the constituent neurons.

Over the last several years, we have collected anatomical and physiological data from a large population of sensory neurons and interneurons in the cricket cercal system. This data is stored in the NeuroSys database and has been used to make predictions about the ensemble activity patterns within the network. In the following sections, the physiological properties of these cells will be summarized followed by a description of their anatomical characteristics.

8.2.1 Physiological Characteristics of Neurons in the Cercal System

Sensory receptor neurons. The receptor organs for the cercal system are two antenna-like appendages called cerci at the rear of the abdomen (Fig. 1A and B). Each *cercus* is covered with approximately 1000 filiform mechanosensory hairs, and each hair is innervated by a single spike-generating mechanosensory receptor neuron. These sensory neurons display directional and dynamical sensitivities that are derived directly from the mechanical properties of the hairs *(17–22)*. In particular, the amplitude of the response of each sensory neuron to any air current stimulus depends upon the direction of that stimulus, and these "directional tuning curves" of the receptor afferents are well described by cosine functions, as shown in Figure 1C *(18)*. The median frequency of each sensory neuron's frequency tuning curve is strongly correlated with the length of its associated hair. Receptors innervating long mechanoreceptor hairs (>900 μm) are most sensitive to low frequency air currents (<150 Hz); receptors innervating medium length hairs (500 – 900 μm) are most sensitive to frequency ranges between 150 to 400 Hz *(20–22)*. Receptors innervating the shortest hairs (50 – 500 μm) respond to frequencies up to 1000 Hz.

Primary sensory interneurons. The sensory neurons synapse with a group of approximately 30 local interneurons and approximately 20 identified projecting interneurons that send their axons to motor centers in the thorax and integrative centers in

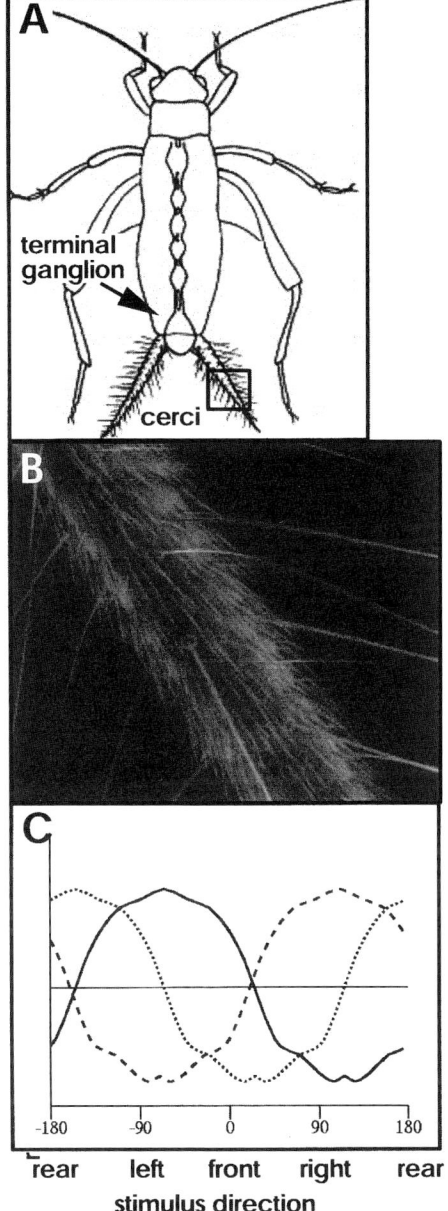

Fig. 1. Functional organization of the cricket cercal sensory system. (**A**) Schematic diagram of the common house cricket, *Acheta domestica,* showing the location of the abdominal nerve cord. The cerci are two abdominal appendages projecting from the rear of the animal's body. Both cerci are covered with mechanosensory hairs, each of which is innervated with a single sensory neuron. The axons of the sensory neurons project into the terminal abdominal ganglion, located at the caudal end of the abdominal nerve cord. (**B**) Scanning electron micrograph of a portion of the cercus showing the filiform sensory hairs. Note the different lengths of the hairs and the cuticular sockets supporting each hair. (**C**) Directional tuning curves of three primary sensory afferents (plotted in solid, dashed, and dotted lines), plotted as relative response amplitude vs stimulus direction. The center horizontal line indicates baseline activity level. Each cell increases or decreases its activity level according to the stimulus direction. The response curves are approximately sinusoidal and were derived from physiological measurements *(18,19).*

the brain. Like the sensory neurons, these interneurons are also sensitive to the dynamics and the direction of air current stimuli *(8,23–26)*. We have measured stimulus-evoked neural responses in several projecting and local interneurons, using several different types of air current stimuli *(24–26)*. Each of the interneurons studied so far has a unique morphology and also a unique set of directional and dynamic response characteristics.

8.2.2. Anatomical Characteristics of Neurons in the System

The axons of all sensory neurons project in an orderly array into the terminal abdominal ganglion to specific locations according to their directional tuning characteristics *(6,16,28)*. This projection pattern forms a continuous representation (i.e., *neural map*) of the direction of air currents in the central nervous system. The synaptic terminals from sensory neurons with similar peak directional sensitivities arborize in adjacent areas, and the spatial segregation between arbors increases as the difference in their directional tuning increases *(16)*. Interneurons have large complex dendritic arbors that overlap extensively with a large number of primary sensory neurons. The anatomical characteristics of both sensory neurons and interneurons are highly conserved from animal to animal, both in terms of the shapes and sizes of the arborizations and their location within the nervous system.

3D anatomical reconstructions of neurons in the system. The NeuroSys database contains over 250 3D reconstructions of individual neurons in the cercal system. These reconstructions were collected with the use of a computer-aided 3D reconstruction system developed in our own laboratory *(4)*. Individual cells were filled with a dye and traced manually using the reconstruction system. Our system is very similar to commercially available systems such as the Neurolucida System developed by Microbrightfield Ltd. Each reconstructed cell was represented as an ascii file of points describing: (*i*) the x, y, and z coordinate endpoints of each segment of the cell's branching structure, and (*ii*) the diameter of that segment. A set of common fiducial features were also recorded in every data file. These fiducial features were used for scaling and aligning the data to a common coordinate system. An image of a reconstructed interneuron and three afferent terminal arbors (indicated with arrows), in their correct spatial relationships, is shown in Figure 2A. This figure is also shown in color in the accompanying CD-ROM (Figure2_color_image). The cells were collected from different animals, but all were scaled and aligned to one another using the fiducial features. Note that commercially available reconstruction systems operate in an equivalent manner and yield data files that are compatible with our database, analysis tools, and viewing software.

Probabilistic representations of neural ensembles. The neurons in the NeuroSys database are a representative sample of a much larger population of neurons. The goals of our recent experiments involve predicting the response properties of a large population of neurons, so we developed a probabilistic representation of this population from our representative sample. We developed a statistical representation of the arborization patterns of neurons within their synaptic neuropil areas as probability density clouds *(16)*. A probability density cloud represents the 3D spatial location and distribution of membrane surface area for the synaptic varicosities of a sensory neuron. Each density

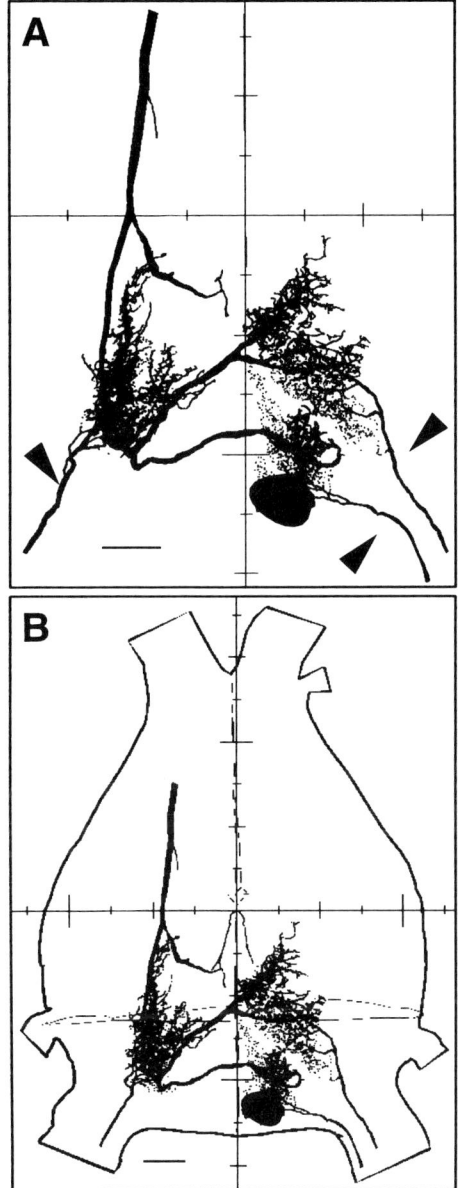

Fig. 2. 3D reconstructions of individual neurons. (**A**) An image from the database, showing a dorsal view of four different reconstructed neurons. Three of the neurons are primary sensory neurons whose axonal arborizations overlap extensively with the dendrites of a primary sensory interneuron. The axons of the sensory neurons have been marked with arrow heads. The large primary sensory interneuron sends an axon anterior to higher centers of the nervous system. Note all neurons have been scaled and aligned to the database and are shown in their correct spatial relationships to each other. (**B**) Inset shows the four neurons shown in panel **A** with an outline of the terminal abdominal ganglion. An image of this figure in color has been included in the companion CD-ROM (Figure2_color_image). Scale bar, 40 μm.

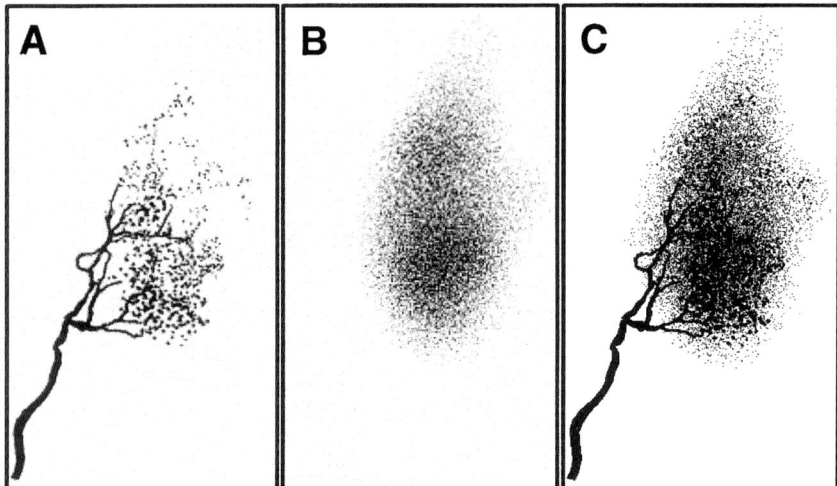

Fig. 3. Density cloud representation for a primary afferent neuron. The left image shows a 3D reconstruction of a primary afferent in dorsal view. The middle image shows the density cloud of the synaptic varicosities. The right image shows the density cloud in register with the neuron.

cloud represents a statistical mean derived from multiple examples of the same identified sensory neuron. The density cloud representation has been used to measure the anatomical overlap between pairs of sensory neurons *(16,28)*, or between a sensory neuron and an interneuron *(28)*. For an interneuron, a probability density cloud represents the 3D spatial location and distribution of membrane surface area associated with its dendrites. This type of representation permits calculations of the spatial relationships between neurons. In the cercal system, the shapes and 3D positions of the terminal arborizations of all cells are highly conserved across animals, allowing the derivation of reliable statistical estimates of dendritic and axonal terminal arbor locations within the cercal glomerulus.

A probability density cloud is derived for an sensory neuron arborization as follows. Each reconstructed afferent arborization contained many varicosities (approx. 500), which are presumed to be the sites at which the afferents make their presynaptic contacts with target cells. The diameter and specific x, y, z location in space of each varicosity of the afferent was measured. The contribution of each varicosity to that afferent's probability density cloud was considered to have three essential characteristics: (*i*) the magnitude of a varicosity's contribution to the net density cloud is proportional to its surface area; (*ii*) each varicosity's total contribution to the net density cloud is distributed throughout the local volume of neuropil according to a Gaussian function centered on the varicosity, rather than being concentrated at any point or surface; and (*iii*) the net density at any point in the cloud can be estimated by a linear sum of the contributions from *all* varicosities in the local neighborhood. Figure 3 shows a single sensory neuron terminal arbor in the two different formats: a 3D reconstruction and the probability density cloud. The magnitude of a sensory neuron's density cloud at any given point, therefore, corresponds to the local surface area of neuronal membrane per unit volume surrounding that point.

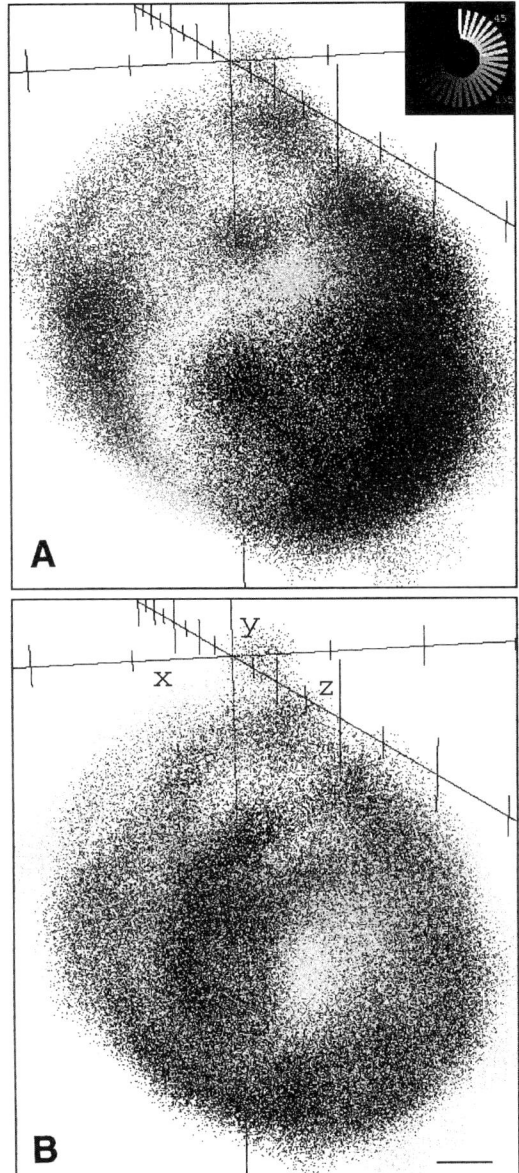

Fig. 4. Functional representation of two different stimulus parameters within a neural map. **(A)** Combined image of the entire ensemble of sensory afferents which innervate the filiform mechanoreceptor hairs on the left cercus shown in parasagittal view. Each cloud represents a population of sensory neurons tuned to a specific air current direction in body coordinates. Inset: the gray scale wheel corresponds to the peak directional tuning in body coordinates. White indicates air currents directed at the animal's head, light gray at 45° indicates the animal's upper left. **(B)** Same population of sensory neurons as shown in panel **A**, gray scale-coded according to frequency tuning. Sensory neurons tuned to low frequencies are colored light gray, and neurons tuned to higher stimulus frequencies are colored dark gray. Note the lack of anatomical segregation between the two populations of sensory neurons. Scale bar, 40 μm. A color version of this figure is included in the CD-ROM accompanying this volume (Figure4_color_image).

8.3. USING NEUROSYS TO STUDY EMERGENT PROPERTIES OF NEURONAL ENSEMBLES

The general approach we have used to study emergent properties of the cercal sensory system is to query the database for a set of neurons and their attributes and then to use the computational and visualization tools in NeuroSys to test predictions about system function. In all cases, these predictions cannot be made by studying individual neurons in isolation, but emerge when the attributes of many neurons are studied in combination. In the following sections, this process will be illustrated, and some of the scientific results obtained will be described.

8.3.1. Neural Maps of Direction and Frequency in the Cricket Cercal System

In most sensory systems, certain functional parameters of the constituent neurons are represented as neural maps. For example, the visual cortex contains a retinotoptic map of the visual field, and the somatosensory cortex contains multiple maps of the body surface. In the cricket system, sensory neurons are sensitive to two independent parameters of air currents: direction and frequency. We chose to examine these two parameters, because they are the most important physiological parameters that define the functional properties of the system. We used NeuroSys to determine whether both of these parameters were mapped continuously within the ensemble of sensory afferents. Figure 4 shows the results of these studies. These images are best viewed in color in the CD-ROM accompanying this volume (Figure4AB_color_image). In the black and white version of the images, Figure 4A shows density cloud representations of a large ensemble of sensory cells. Each density cloud, shown in gray scale, represents a population of neurons. The gray scale level of each cloud represents the peak directional tuning of that population of sensory neurons. The inset in Figure 4A shows the gray scale spectrum; white indicates stimulus directions aimed at the animal's head; light gray indicates directions aimed at the animals left (e.g., 45°) and darker grays indicate stimulus directions aimed at the animal's right (e.g., 315°). The density clouds overlap extensively forming a continuous 3D contour. The directional tuning of the sensory neurons changes continuously along the contour, thus creating a continuous map of air current direction. Note, although these are 3D structures, all images shown are 2D projections of these 3D structures. The color version of this figure (Figure4A_color_image) shows the ensemble of density clouds color coded according to stimulus direction.

Figure 4B shows the same ensemble of sensory neurons, now gray scale-coded according to the frequency tuning characteristics of the sensory neurons. Neurons sensitive to low frequencies are shown in dark gray, and those sensitive to higher frequencies are shown in light gray. The spatial relationships between these two populations of neurons can be seen in the color version of the figure where sensory afferents sensitive to low frequency air currents are shown in red, and those sensitive to higher frequencies are shown in green (Figure4_color_image.). Although there are regions of segregation between these two populations of sensory neurons, there is no consistent pattern or axis of segregation indicating a continuous representation of stimulus frequency within the ensemble. The continuous representation of direction is conserved in the

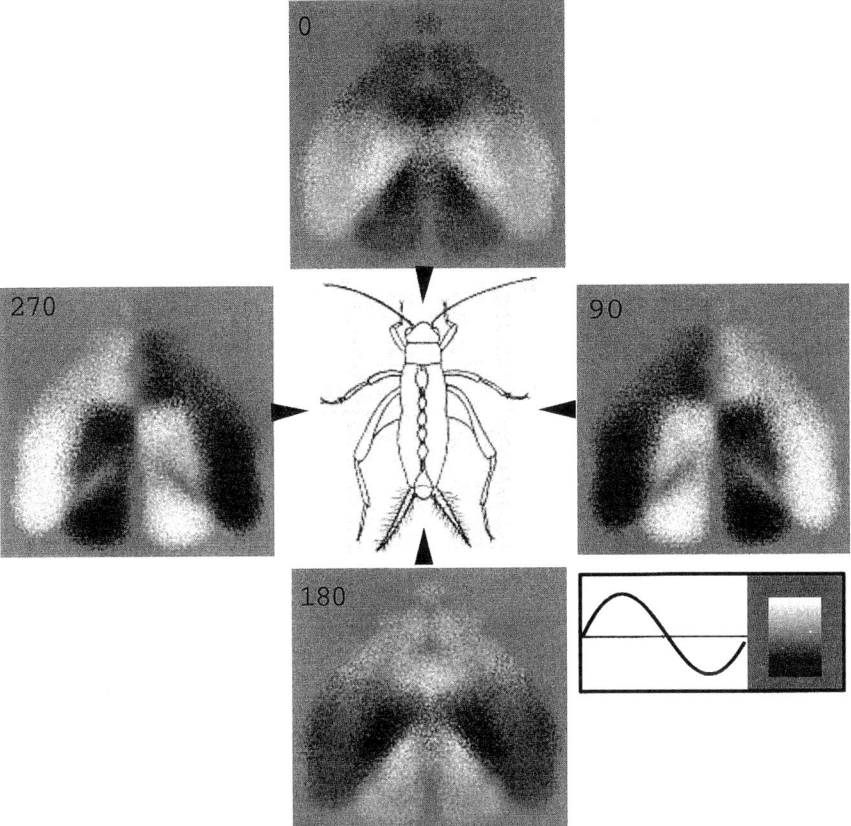

Fig. 5. Predictions of the spatial patterns of activity that would be elicited within the neural map by unidirectional steady-state air currents from 4 different directions. This is a dorsal view showing sensory neurons from both cerci; the relative level of activity within the population is indicated by a gray scale. The activity level of each afferent in the ensemble will be modulated up or down from its baseline level, as a function of the stimulus direction, resulting in a unique activation pattern for each different stimulus angle. The direction of the air current is indicated in the upper left corner of each image; arrows indicate these directions in the body coordinate system. The maximum level of activity is indicated as white, baseline activity as mid-gray, and a decrease below baseline activity is indicated as dark gray to black. The inset shows the gray scale aligned with a cosine function to represent an afferent directional tuning curve.

combined projection pattern of these populations, yet there does not appear to be a systematic representation of stimulus frequency through any dimension.

8.3.2. Predicting Spatio-Temporal Patterns of Activity within an Ensemble of Sensory Neurons

Within any mapped sensory system, the response of an ensemble of neurons to a sensory stimulus will be a spatio-temporal pattern of activity. The characteristics of the pattern will depend on the anatomical structure of the map and the response properties of the constituent neurons. In the cricket system, any air current stimulus, regardless of its direction, will deflect all of the mechanosensory hairs on both of the two cerci. The tuning curves of the sensory neurons are cosine functions spanning the entire 360° range in the horizontal plane *(18,19)*, therefore, the firing pattern of each neuron will be modulated up or down as a function of the angle between the air current stimulus and the peak of the cosine tuning curve *(18,19)*. Depending on the angle of the stimulus, some neurons will increase their firing rate in response to the stimulus, where as others tuned to opposing directions will decrease their firing rate. As a result of the continuous mapping of stimulus direction among the population, these differential firing patterns should result in a different spatial pattern of activity for each stimulus direction.

We tested this hypothesis using NeuroSys by predicting steady state spatial patterns of activity within the network by combining the anatomical data with the directional tuning curves of the sensory neurons. A stimulus direction was selected, and the response of the system was calculated at each voxel in the 3D voxel space. The contribution of a single sensory neuron to each voxel was determined by scaling the measured physiological response of its associated sensory neuron by the local varicosity density at that voxel in the coordinate system. The physiological response of the sensory neuron was calculated as the cosine of the difference in angle between the peak directional tuning of the afferent and the angle of the stimulus with respect to the animal's body. The response of the entire system of sensory neurons was then calculated as the sum of responses from all neurons in the database at each voxel. Note that these predicted patterns of activity are calculated by combining two independent characteristics of the sensory neurons, their anatomical structures, and their directional tuning characteristics. The patterns represent a prediction of the global activity pattern of the entire ensemble of neurons synthesized from these two parameters, not a visualization of previously stored data.

The predicted response patterns were imaged as changes in the levels of activity in the neural map relative to the baseline level of activity. A gray scale was used to indicate relative levels of activity, with white indicating maximum activity and black indicating minimum activity. The baseline activity level in the sensory neurons was represented as mid-gray (see Fig. 5, inset). In this manner, images could be generated to predict the relative response levels throughout the map for any given stimulus.

Figure 5 shows the predicted spatial patterns of activity for four different stimulus directions. The inset in Figure 5 shows the gray scale next to a tuning curve of a sensory neuron. If the sensory neuron was stimulated maximally by an air current directed along its peak tuning axis, its activity level would be indicated as white. If the angle of the wind stimulus is 180° away from the peak tuning of the afferent, the activity in the sensory neuron would be indicated as black. Thus, each neuron will respond differently to the stimulus depending on its directional tuning. Each stimulus results in a unique response pattern, and the patterns vary continuously as a function of the stimulus direction. Note that stimuli that are similar in direction (i.e., less than 90° different) elicit activity in contiguous regions of the map, whereas stimuli from opposite directions are represented by patterns that are inverses of each other. Each spatial pattern has approx the same net amount of excitatory area, and this area changes in shape with a change in air current direction *(2)*.

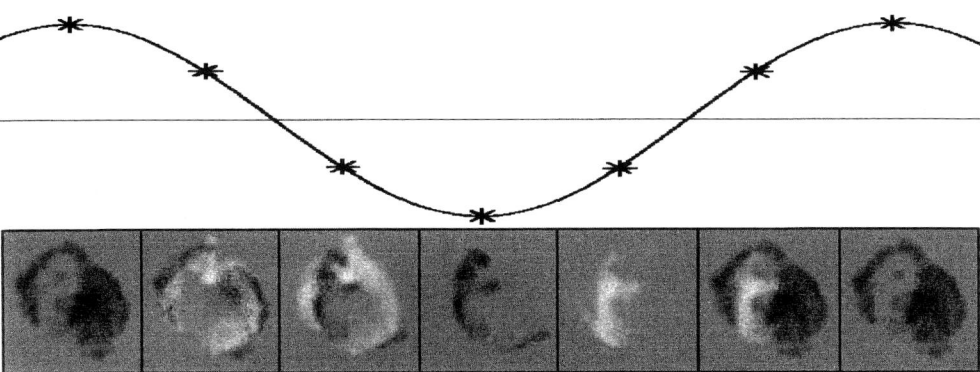

Fig. 6. Predictions of dynamic patterns of activity within the neural map. This series of images shows a sequence of activity patterns (from left to right) within the map in response to a predicted 100 Hz sine wave air current, which alternates direction back and forth across the animal's body. The top panel shows the sine wave stimulus with asterisks indicating the time when each image occurs with respect to the stimulus. The direction of the stimulus is encoded by two different activity patterns: a horseshoe shaped pattern when the stimulus is directed at the animal's left (when the sine wave stimulus is above the baseline) followed by a "C" shaped pattern when the air current changes direction towards the animal's right (when the sine wave stimulus is below the baseline). Each image shows the relative contribution of sensory neurons tuned to low frequencies (white clouds) and sensory neurons tuned to higher frequencies (black clouds). Note that, at some phases of the activity pattern, both sets of afferents tuned to the same directions are activated together (second image from the left), and at other phases cells tuned to opposite directions are activated together (second image from the right).

8.3.3. Predicting Spatio-Temporal Patterns of Activity within Neural Ensembles

Most sensory stimuli are dynamic in nature. The resulting activity patterns within the nervous system should exhibit both spatial and temporal characteristics. Our previous work indicates that different stimulus directions will evoke different steady state spatial patterns of activity within the ensemble of sensory neurons. An oscillating air current that moves back and forth across the animal's body should, therefore, evoke a spatial pattern of activity that changes with the direction of the stimulus. Since the population of sensory neurons contains cells with different frequency tuning characteristics, the spatio-temporal pattern may also change as a function of the frequency of the oscillation *(31)*. Different sensory neurons have very different peak response amplitudes, depending on the lengths and movement axes of their associated hairs. Different neurons with the same directional sensitivities may be activated as much as 180° out of phase with one another, depending on the lengths of their associated hairs. As a corollary phenomenon, we predict that there will be frequencies at which sensory neurons of opposite directional sensitivity will be firing synchronously.

In collaboration with John Miller and Sharon Crook, we are developing a computational tool called the DynamicAtlas, which is a combined query–computational–visualization tool for studying the dynamic response properties of cell ensembles *(32–34)*. The DynamicAtlas is part of the suite of computational tools in NeuroSys. It uses the same graphical user interface and displays the dynamic activity patterns within the viewing environment as animated sequences of images. This tool requires queries on multiple data types, integration of the database with analysis tools, and display of the results with dynamic multidimensional visualization tools. The basic approach is as follows:

1. A subset of neurons to be studied is selected from the database (e.g., "select all afferents having peak directional sensitivities within 30° of the rear of the animal and median frequency sensitivity of less than 200 Hz").
2. Computational tools operate on the physiological stimulus–response data stored for those cells in the database to predict the firing pattern of each individual neuron in the set in response to a user-defined stimulus waveform.
3. The predicted response patterns of all indicated cells are animated simultaneously, as dynamic sequences in which amplitude of the response (as a function of time) is used to set the color or intensity of the density clouds of individual neurons. This creates a global dynamic predicted activity pattern across the entire ensemble of selected cells.

Note that the dynamic animation created by this procedure is *not* a model-based graphical simulation, but rather is a graphical report of the results of a complex query on multiple data types. Specifically, the stimulus–response properties measured from real neurons are used to calculate responses to user-defined stimuli, and the calculated time-series responses are "painted" onto anatomical reconstructions of those neurons. The only aspect of this process that is not strictly a query on existing data is the calculation of the predicted activity patterns. In our previous experiments, this has been carried out through a "white noise analysis" *(20,29,35)*. Specifically, a Wiener kernel expansion was carried out on stimulus–response data sets stored in the database. A set of functions, called kernels, were derived from this analysis. The first order kernel is the first order cross-correlation of the spike train and the stimulus waveform and allows a linear approximation of the cell's response to any stimulus. The second order kernel characterizes nonlinear relationships between the stimulus and response and was combined with the first order kernel to yield a more accurate approximation of the response. These kernels are used in the *DynamicAtlas* program to predict the activity patterns of all cells as a function of time and in response to any specified stimulus waveforms *(34)*. Figure 6 illustrates these results. The top panel of Figure 6 shows the sine wave stimulus with an asterisk indicating the time at which a specific spatial pattern of activity would be produced by the ensemble. These spatial patterns are shown below the stimulus waveform. By inspecting the changing patterns shown in the images from left to right, the relative contributions of low frequency sensory neurons (shown in black) and high frequency neurons (shown in white) can be seen. At some phases of the stimulus, either the low frequency or high frequency afferents are active in isolation (from the left, images 1, 4, 5, and 7). At all other phases of the stimulus, both sets of sensory cells are active simultaneously. As the direction of the stimulus changes, the shape of the activation pattern also changes, e.g., at the peak of the sine wave (images 1 and 7) vs the trough of the sine wave (image 4). The computational techniques used to create these images were described in detail in *(9)*. These dynamic patterns of activity can be viewed as a color animation in the CD-ROM accompanying this volume (Figure6_color_animation). In this animation, low frequency sensory neurons

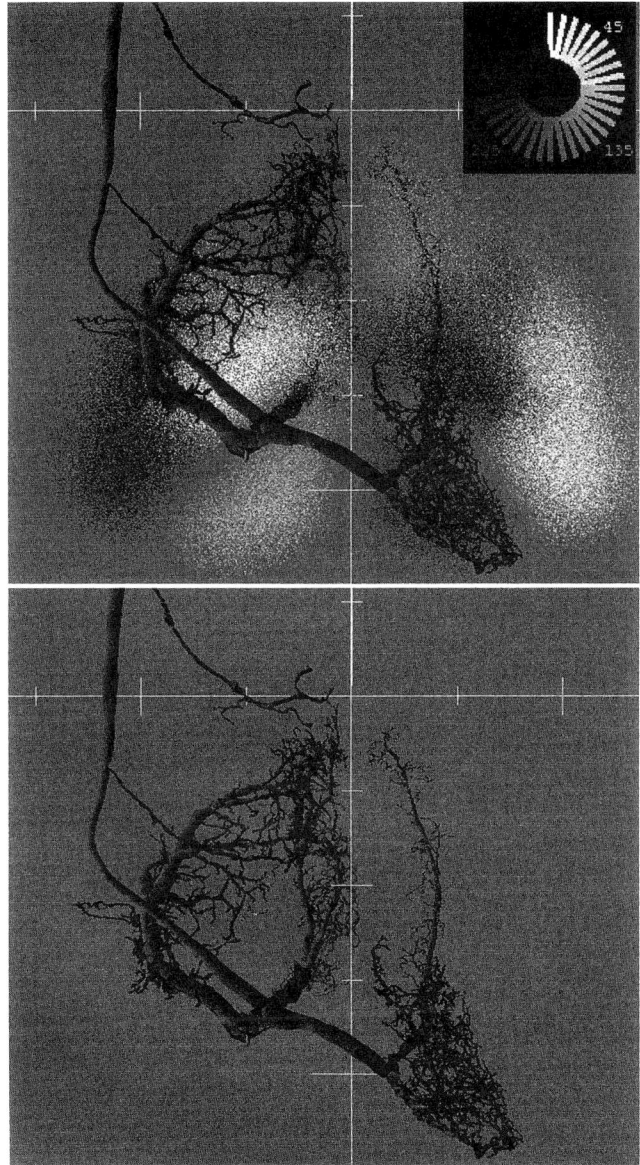

Fig. 7. Prediction of excitatory inputs to an identified interneuron. The images show a primary sensory interneuron and its spatial relationship to the map of air current direction. In the top panel, the interneuron has been superimposed over the map, illustrating the spatial location of its dendrites with respect to specific regions of the neural map. In the bottom image, the interneuron's dendrites have been gray scale-coded according to their location within the map and the distribution of excitatory inputs to the dendrites. The images represent a quantitative prediction of the spatial distribution of excitatory inputs to the interneuron, gray scale-coded according to the peak directional tuning of the afferents. A color version of this figure has been included in the companion CD-ROM to this volume (Figure7_color_image).

are colored red and high frequency neurons are colored green. By viewing the animation, the reader can gain an appreciation for the complexity of these patterns of activity and the relative contributions of the two populations of neurons to the overall pattern.

8.4. TRANSFER OF INFORMATION BETWEEN ENSEMBLES OF NEURONS

Information contained within these spatial patterns of activity must be extracted and decoded by postsynaptic interneuron "target" cells at the next processing stage. To study the anatomical basis for information transfer between these two ensembles of neurons, we developed methods to map the functional properties in one ensemble of neurons onto another ensemble. For example, consider a query whose goal is to obtain a first order prediction of the distribution of excitatory synaptic inputs from an ensemble of afferent terminal arborizations onto the dendritic branches of a particular interneuron. The process, carried out by another tool in the NeuroSys system called the Functional_Masker, was as follows. First, the database was queried for a specific neuron and loaded into the graphical user interface. A segment of the interneuron's dendrite was selected, and its spatial location within the atlas was identified (as the set of voxels through which that segment passes.) Second, the net local varicosity surface density for the selected subset of afferents within those same voxels was retrieved from the database. Third, the total surface area of the interneuron segment within those voxels was multiplied by the afferent surface area density within those same voxels. This yields a value proportional to the probability with which those two sets of objects overlap. Fourth, the functional attribute associated with those afferents (e.g., the mean angle of their peak directional sensitivity or their mean peak frequency sensitivity) was retrieved. Both values (i.e., the overlap probability and the associated functional attribute) were stored for return as part of the graphical query response. This 4-step procedure is repeated for every segment of the interneuron's dendritic structure. The image shown in Figure 7 depicts a sensory interneuron with its dendrites shaded in gray scale according to the peak directional of the primary afferents that provide excitatory synaptic input. This image is best viewed in color, where a color spectrum was used to indicate peak directional tuning; please see Figure 7 in the accompanying CD-ROM (Figure7_color_image). Note that this operation is conceptually equivalent to masking the entire ensemble afferent map structure with a spatial filter in the shape of an interneuron. The details of the algorithm, and an extensive set of queries using that algorithm, are presented in detail elsewhere *(2)*.

These tools can be used to predict the levels of synaptic input from the ensemble of sensory neurons onto individual postsynaptic interneurons. The goal of these calculations was to correlate predictions of the spatial patterns of activity within the sensory neuron ensemble with the known physiological response properties of sensory interneurons. One set of predictions is shown in Figure 8. First, the spatial pattern of activity within the afferent population in response to four specific stimuli was predicted as shown in the top row of images of Figure 8. The relative level of afferent activity was then computed for all locations along the dendrite of the interneuron, and a gray scale value (as described above) was assigned to each dendritic segment (Fig. 8, middle row set of images). This mapping from the afferent activity pattern onto the interneuron

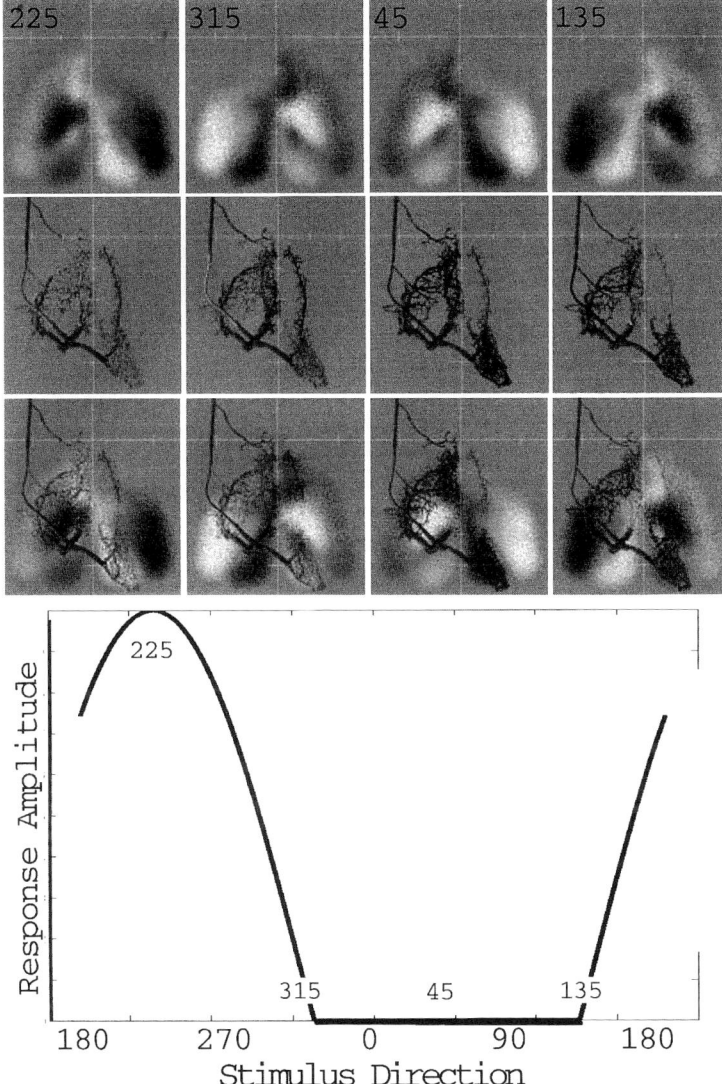

Fig. 8. Prediction of the relative level of excitatory input to a primary sensory interneuron in response to air currents from four orthogonal directions. The top row of panels show predicted spatial patterns of activity in response to air current from different directions. The middle row of panels show the interneuron with its dendrites gray scale-coded, according to the predicted level of excitatory input from the population of afferents. The third row of panels show the interneuron superimposed over each spatial pattern of activity. For each direction, the spatial pattern of activity elicited within the map is different, and thus, the activation pattern masked onto the interneuron's dendrites appears different. The maximum level of activation occurs for 225°, the minimum level for 180° opposite, at 45°. These directions correspond to the peak and trough of the cell's directional tuning curve, respectively. The response amplitude in the cell to directions 315° and 135° is the same, however the distribution of excitatory input to the cell is quite different for these two stimulus directions. The cell's directional tuning curve is presented in the bottom panel for reference. Note that this cosine shaped curve is truncated at the x axis to indicate the change from the very low level of baseline activity in the cell.

dendrites represents a first order prediction of the relative level of excitatory input to each dendritic region of the interneuron during the activation of that afferent response pattern.

We predicted that, since each interneuron in the cercal system has a unique arborization pattern, each should be sensitive to a specific subset of activity patterns within the map. Decoding information about air current direction is thus achieved primarily by a spatial matching function that compares the activity pattern in the afferents to the shapes of the dendritic arbors of the interneurons. Each interneuron should be sensitive to a specific range of spatial patterns, specifically those patterns that match the locations of its dendrites. Just as the spatial pattern of activity in the map changes as a function of stimulus direction, so does the level of excitatory input to the interneuron's different dendritic regions. Figure 8 shows the predictions of the level of excitatory input to a representative interneuron for 4 orthogonal stimulus directions: 45°, 135°, 225°, and 315°. These images are shown in comparison to the directional tuning curve of the interneuron. For stimuli at 225°, (the peak of the interneuron's tuning curve), a large portion of the dendritic arbor is activated maximally. At the orthogonal direction (45°), the level of excitation is suppressed below baseline activity over most of the dendritic tree. At intermediate directions (135° and 315°), the level of excitation is less than at the peak direction, yet the spatial distribution of that input to the interneuron is unique for each direction. The relative levels of activity within the dendritic arbor correlate well with the level of spiking output in the interneuron as measured physiologically (shown in the tuning curve in Fig. 8, bottom panel). Thus the interneuron was activated maximally when the spatial pattern of excitatory activity matches the location of its dendrites, thus providing the greatest amount of excitatory input. Conversely, for stimuli that do not activate the interneuron, the spatial pattern of excitatory activity does not overlap with the dendrites of the interneuron *(2)*.

8.5. GENERAL APPLICATIONS OF NEUROSYS

Although the NeuroSys database and tools were developed to investigate mechanisms of sensory information processing in the cricket cercal system, the tools and general approach can be used in a variety of different systems to address important questions in neuroscience. The only major requirement is that anatomical data from different experiments can be registered in a common coordinate system. For example, one could adapt the system to studying dynamic patterns of gene expression in a developing embryo, or study the distribution of membrane channels, receptors, and organelles in a specific neuron type. The NeuroSys environment was designed to allow the investigator to explore ideas and test hypotheses about system function by studying the interactions of the constituent elements. Despite the technical challenges involved, this systems level approach holds great promise for understanding the computational mechanisms of the brain.

REFERENCES

1. Laurent G. A systems perspective on early olfactory coding. Science 1999; **286**:723–728.
2. Jacobs GA, Theunissen F. Extraction of sensory parameters from a neural map by primary sensory interneurons. J Neurosci 2000; **20**:2934–2943.

3. Schwartz EL. Computational studies of the spatial architecture of primate visual cortex: columns, maps, and protomaps. In: Primary Visual Cortex in Primates, Vol. 10 of Cerebral Cortex (Peters A, Rocklund K, eds.) Plenum Press, New York, 1994.
4. Jacobs GA, Nevin R. Anatomical relationships between sensory afferent arborizations in the cricket cercal system. Anat Rec 1991; **231**:563–572.
5. Troyer TW, Levin JE, Jacobs GA. Construction and analysis of a data base representing a neural map. Microsc Res Tech 1994; **29**:329–343.
6. Bacon JP, Murphey RK. Receptive fields of cricket (*Acheta domesticus*) are determined by their dendritic structure. J Physiol (Lond) 1984; **352**:601–613.
7. Bialek W, Rieke F, deRuyter van Steveninck RR, Warland D. Reading a neural code. Science 1991; **252**:1854–1857.
8. Jacobs GA, Miller JP, Murphey RK. Cellular mechanisms underlying directional sensitivity of an identified sensory interneuron. J Neurosci 1986; **6**:2298–2311.
9. Warland D, Landolfa MA, Miller JP, Bialek W. Reading between the spikes in the cercal filiform hair receptors of the cricket. In: Analysis and Modeling of Neural Systems (Eeckman FH, ed.) Kluwer Academic Publishers, Boston, 1991, pp. 327–333.
10. Altman J. Sensory inputs and the generation of the locust flight motor pattern: from the past to the future. In: Biona Report 2 (Nachtigall W, ed.) Gustav Fischer, Stuttgart, 1983, pp. 127–136.
11. Boyan GS, Ashman S, Ball EE. Initiation and modulation of flight by a single giant interneuron in the cercal system of the locust. Naturwissenschaften 1986; **73**:272–274.
12. Camhi JM. The escape system of the cockroach. Sci Am 1980; **243**:144–157.
13. Gnatzy and Heusslein Digger wasp against crickets. I. Receptors involved in the antipredator strategies of the prey. Naturwissenschaften 1986; **73**:212–215.
14. Fraser P. Cercal ablation modifies tethered flight behavior of cockroach. Nature 1977; **268**:523–524.
15. Heinzel HG, Dambach M. Traveling air vortex rings as potential communication signals in a cricket. J Comp Physiol A 1987; **160**:79–88.
16. Jacobs GA, Theunissen F. Functional organization of a neural map in the cricket cercal sensory system. J Neurosci 1996; **16**:769–784.
17. Kamper G, Kleindienst H-U. Oscillation of cricket sensory hairs in a low-frequency sound field. J Comp Physiol A 1990; **167**:193–200.
18. Landolfa M, Jacobs GA. Direction sensitivity of the filiform hair population of the cricket cercal system. J Comp Physiol A 1995; **177**:759–766.
19. Landolfa M, Miller JP. Stimulus/response properties of cricket cercal filiform hair receptors. J Comp Physiol A 1995; **177**:749–757.
20. Roddey JC, Jacobs GA. Information theoretic analysis of dynamical encoding by filiform mechanoreceptors in the cricket cercal system. J Neurophysiol 1996; **75**:1365–1376.
21. Shimozawa T, Kanou M. Varieties of filiform hairs: range fractionation by sensory afferents and cercal interneurons of a cricket. J Comp Physiol A 1984a; **155**:485–493.
22. Shimozawa T, Kanou M. The aerodynamics and sensory physiology of range fractionation in the cercal filiform sensilla of the cricket *Gryllus bimaculatus*. J Comp Physiol A 1984b; **155**:495–505.
23. Kanou M, Shimozawa TA. Threshold analysis of cricket cercal interneurons by an alternating air-current stimulus. J Comp Physiol A 1984; **154**:357–365.
24. Miller JP, Theunissen FE, Jacobs GA. Representation of sensory information in the cricket cercal sensory system. I. Response properties of the primary interneurons. J Neurophysiol 1991; **66**:1680–1689.
25. Theunissen FE, Miller JP. Representation of sensory information in the cricket cercal sensory system. II. Information theoretic calculation of system accuracy and optimal tuning curve widths of four primary interneurons. J Neurophysiol 1991; **66**:1690–1703.

26. Theunissen FE, Roddey JC, Stufflebeam S, Clague H, Miller JP. Information theoretical analysis of dynamical encoding by four identified primary sensory interneurons in the cricket cercal system. J Neurophysiol 1996; **75**:1345–1364.
27. Clague H, Theunissen F, Miller JP. Effects of adaptation on neural coding by primary sensory interneurons in the cricket cercal system. J Neurophysiol 1997; **77**:207–220.
28. Paydar S, Doan CA, Jacobs GA. Neural mapping of direction and frequency in the cricket cercal sensory system. J Neurosci 1999; **19**:1771–1781.
29. Marmarelis PZ, Marmarelis VZ. Analysis of Physiological Systems. Plenum Press, New York, 1978.
30. Borst A, Haag J The intrinsic electrophysiological characteristics of fly lobula plate tangential cells: I. Passive membrane properties. J Comput Neurosci 1996; **3**:313–336.
31. Osborne LC. Biomechanical Properties Underlying Sensory Processing in Mechanosensory Hairs in the Cricket Cercal Sensory System. PhD Thesis. University of California, Berkeley, 1997.
32. Hodge K. NAPA: The Neural Activity Pattern Animator. Masters Thesis. Montana State University, 1998.
33. Hodge K, Starkey JD, Jacobs GA NAPA: The neural activity pattern animator. Proceedings of CGIM IASTED Halifax, Canada, 1998, pp. 15–18.
34. Jacobs GA, Crook SM. Hodge K, Roddey C, and Paydau S. Dynamic patterns of activation in a Neural Map in the cricket cercal sensory system. Computational Neuroscience Annual Meeting; Pittsburgh, PA, Abstract #10.
35. Roddey JC, Girish B, Miller JP. Assessing the performance of neural encoding models in the presence of noise. J Comput Neurosci 2000; **8**:95–112.

9
Computational Anatomical Analysis of the Basal Forebrain Corticopetal System

Laszlo Zaborszky, Attila Csordas, Derek L. Buhl,
Alvaro Duque, Jozsef Somogyi, and Zoltan Nadasdy

ABSTRACT

The basal forebrain (BF) is comprised of a neurochemically heterogeneous population of neurons, including cholinergic, GABA-ergic, peptidergic, and possibly glutamatergic neurons, that project to the cerebral cortex, thalamus, amygdala, posterior hypothalamus and brain stem. This multitude of ascending and descending pathways participate in a similarly bewildering number of functions, including cognition, motivation, emotion, and autonomic regulation. Traditional anatomical methods failed to grasp the basic organizational principles of this brain area and likened it at best to the organization of the brain stem reticular formation. Our studies, using various computational methods for analyzing the spatial distribution and numerical relations of different chemically and hodologically characterized neuronal populations, as well as fully reconstructed electrophysiologically identified single neurons, began to unravel the organizational principles of the BF. According to our model, the different cell types form large-scale cell sheets that are aligned to each other in a specific manner. Within each cell system, the neurons display characteristic discontinuous distributions, including high density clusters. As a result of nonhomogeneity within individual cell populations and partial overlapping between different cell types, the space containing the bulk of cholinergic neurons comprises a mosaic of various size cell clusters. The composition, dendritic orientation, and input–output relationships of these high density cell clusters show regional differences. It is proposed that these clusters represent specific sites (modules) where information processed in separate streams can be integrated. Via this BF mechanism a topographically organized prefrontal input could allocate attentional resources to cortical associational areas in a selective self-regulatory fashion.

9.1. INTRODUCTION

The term basal forebrain (BF) refers to a heterogeneous collection of structures located close to the medial and ventral surfaces of the cerebral hemispheres. This highly complex brain region has been implicated in attention, motivation, and memory as well

From: *Computational Neuroanatomy: Principles and Methods*
Edited by: G. A. Ascoli © Humana Press Inc., Totowa, NJ

as in a number of neuropsychiatric disorders such as Alzheimer's disease, Parkinson's disease, and schizophrenia *(1–3)*. Part of the difficulty in understanding the functions of the BF, as well as the aberrant information processing characteristic of these disease states, lies in the anatomical complexity of the region. BF areas, including the medial septum, ventral pallidum, diagonal band nuclei, substantia innominata, and peripallidal regions contain cell types different in transmitter content, morphology, and projection pattern *(4,5)*. Among these different neuronal populations, the cholinergic corticopetal neurons have received particular attention in numerous functional and pathological studies.

Recent interest in BF research was prompted by discoveries showing that a specific population of neurons in this region, namely those that use acetylcholine as their transmitter and project to the cerebral cortex, are seriously compromised in Alzheimer's disease *(6–9)*. However, cholinergic projection neurons represent only a fraction of the total cell population in these forebrain areas, which also contain GABA-ergic, peptidergic, and possibly glutamatergic neurons *(10,11)*. According to our unpublished estimations in one hemisphere of the rat brain, in the cholinergic BF areas, 20,000 cholinergic corticopetal cells are intermingled with other neurons, including about 35,000 calbindin, 26,000 calretinin, and 24,000 parvalbumin-containing neurons. These calcium-binding proteins are used to characterize different nonoverlapping populations of non-cholinergic BF neurons.

A quasi 3D representation of the cholinergic cell bodies (Fig. 1A) or the dendritic arborizations of their neurons (Fig. 1B) does not appear to show any recognizable architectural features, confirming a classical view in the literature that arousal is supported by a diffuse reticular activating system, including core brain stem structures, the BF, and the so-called nonspecific thalamic nuclei *(12,13)*. On the other hand, careful monitoring of the behavioral effects of lesions in the BF using an immunotoxin selective for cholinergic neurons, suggests that compartments of the BF, together with their specific cortical target areas, may participate in different cognitive operations *(14)*.

If the BF participates in different operations, we would expect that this may be reflected both in the local, as well as in the large-scale structural organization of its constituent neuronal populations. For example, one would expect that the BF would be constituted of repetitive building blocks (modules) as found in many other areas of the central nervous system (CNS), including the cortex, striatum, hypothalamus, brain stem, or the spinal cord *(15-18)*. The modular structure in various brain regions is the prerequisite structural basis for parallel, distinct operations *(19)*. Other structural features, like anisotropic dendritic orientation or segregation of various afferents and efferents, can also be taken as evidence for selective information processing *(20-26)*. In the past several years, we systematically investigated the 3D spatial organization of the various BF neural populations, including their dendritic organization and input–output relationship with the aim of uncovering the organizational principles of the BF, in particular within areas that are most heavily populated by cholinergic corticopetal neurons. This review is an attempt to summarize how anatomical features may constrain information processing in this brain area. The chapter is divided into several sections, each with subheadings indicating the special methods used. Following the main body of the text, the reader can find an Appendix with a detailed explanation of the data acquisition and methods analysis presented.

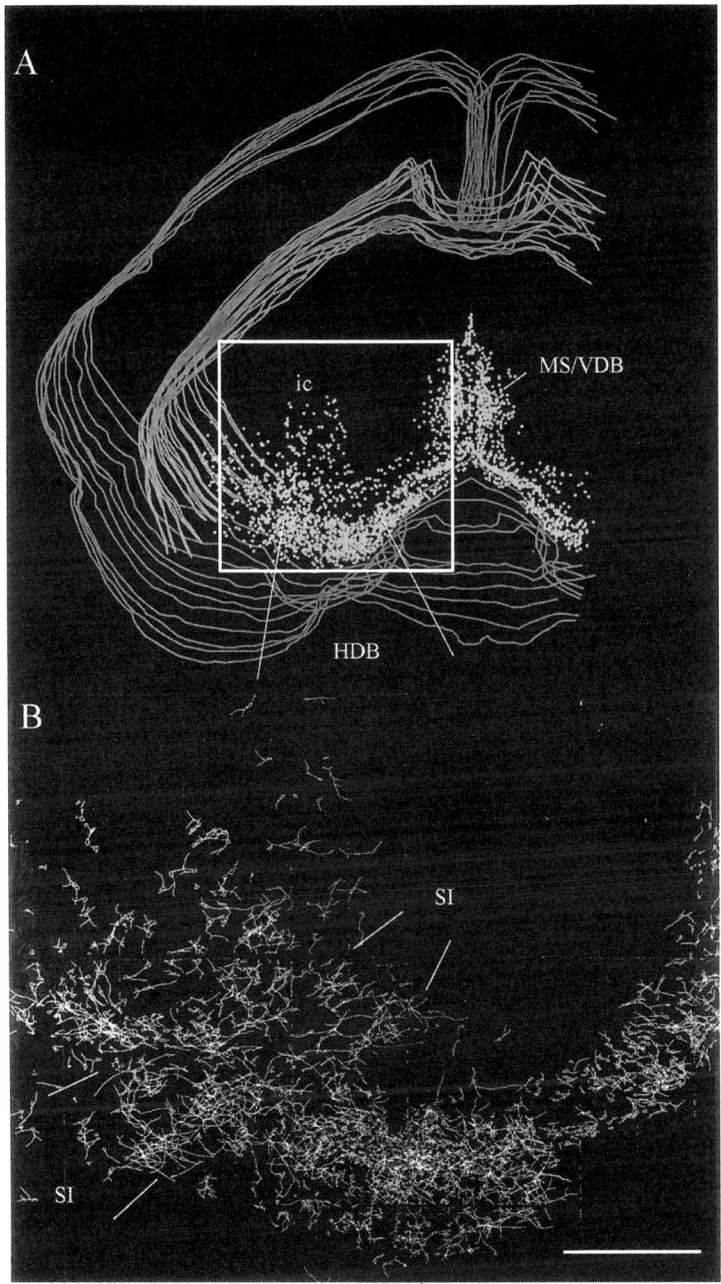

Fig. 1. (A) 3D wireframe diagram showing the distribution of cholinergic neurons in the BF. Cholinergic cells (dots) were mapped from 12 sections, approx 300 µm apart. The contours of the corpus callosum and the section outlines are marked. **(B)** Composite map illustrating the dendritic architecture of the BF cholinergic system. The location of panel **B** corresponds to the enclosed box in panel **A**. Dendrites of approximately 1300 cholinergic neurons were traced from 7 coronal sections. Diagonal white lines delineate the approximate location of the corresponding major forebrain areas. HDB, horizontal limb of the diagonal band; ic, internal capsule; MS/VDB, medial septum/vertical limb of the diagonal band; SI, substantia innominata. Scale bar, 1 mm (applies only to panel **B**). A color version of this figure is enclosed in the CD-ROM.

9.2. ASSOCIATION AND SEGREGATION OF DIFFERENT HODOLOGICALLY IDENTIFIED NEURAL POPULATIONS

9.2.1. Overlap Analysis

Although there is considerable species variation in the precise locations of cholinergic projection neurons in the BF, the efferent projections of these cells follow basic organizational principles in all vertebrate species studied. Thus in rodents, neurons within the medial septum and nucleus of the vertical limb of the diagonal band provide the major cholinergic innervation of the hippocampus; cholinergic cells within the horizontal limb of the diagonal band project to the olfactory bulb, piriform and entorhinal cortices; cholinergic neurons located in the ventral pallidum, sublenticular substantia innominata, globus pallidus, internal capsule, and nucleus ansa lenticularis, collectively termed as nucleus basalis, project to the basolateral amygdala and innervate the entire neocortex according to a rough mediolateral and anteroposterior topography *(27–35)*. Similarly, in primates, including humans, corticopetal cholinergic cells are subdivided according to the topography of their projections *(36)*.

It is unclear, however, what the functional equivalent of this topography is, especially in light of a study in rat, showing that neighborhood relationships in the BF projection neurons do not correspond to near neighbors in the representational areas of sensorimotor cortices, thus arguing against a simple functional organization *(37)*. Knowing the importance of the cholinergic BF system in modulating cortical activity *(38)*, we asked whether the organization of the basalocortical system can, in any sense, be related to the distributed and hierarchical organization of corticocortical connections, as proposed by Felleman and Van Essen *(39)*. Figures 2 and 3 display cases of overlapping and segregated projection neurons from a study aiming at a comprehensive reevaluation of the basalocortical projection (Csordas and Zaborszky, in preparation). Figure 2A is from a case in which two different retrograde tracers were injected into two cortical areas that were in the same mediolateral topographical register but they differed in their rostrocaudal location. This 3D image suggests that the two neuron populations (marked by different symbols[1]), projecting to two cortical areas, are, at least in the rostral part of the BF, intermingled. Using an overlap analysis program described recently *(21,25,40)*, Figure 2C shows that a substantial population of the two types of projection neurons are, indeed, located in overlapping voxels[2] (the method is briefly described in Appendix 9.9.4.). Figure 2B shows another case in which the two retrograde tracer injections were in different mediolaterally located cortical areas. As can be seen from this 3D rendering, there is little overlap in the location of the neurons projecting to these two cortical target areas. Figure 2D, using the overlap analysis program, supports the subjective impression that no overlap exist between these two distinct cell populations.

9.2.2. Isodensity Surface Rendering

Figure 3A shows a 3D rendering of the distribution of four BF cell populations that project to four arbitrarily defined mediolateral sectors of the neocortex reconstructed

---

[1] A color version of this and other figures are available in the companion CD-ROM file.
[2] Voxel is a 3D pixel that is the spatial unit of our analysis. See also Appendix 9.9.4. A mathematical description of the definition of voxels can be found in *(41)*.

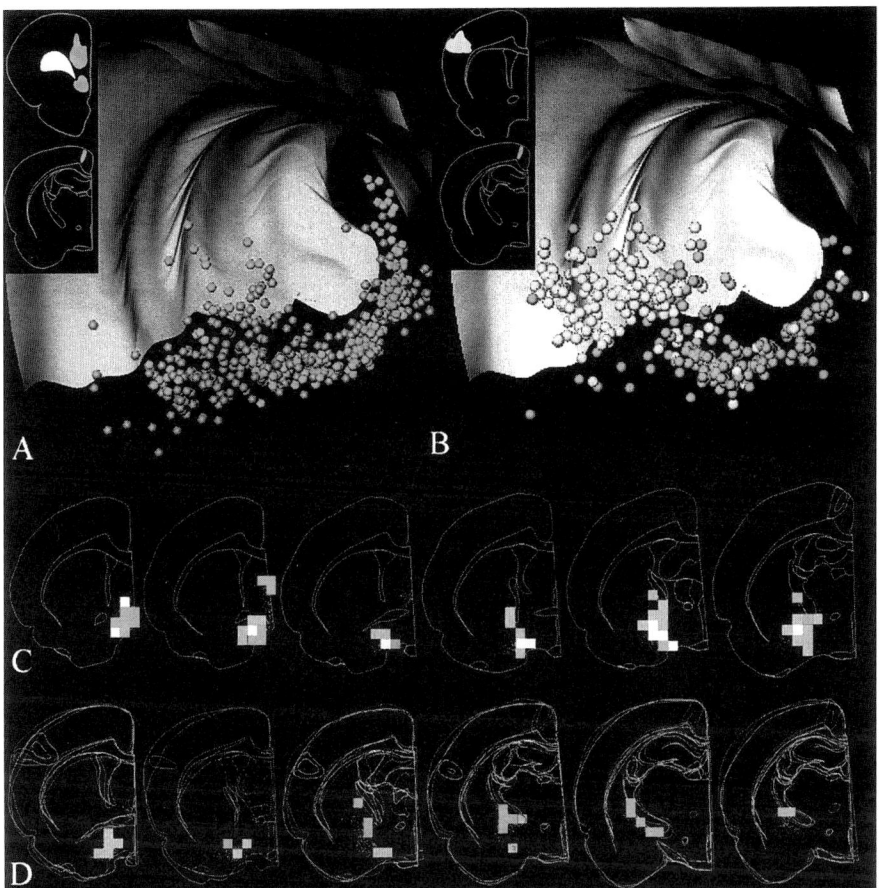

Fig. 2. (**A**) Distribution of non-cholinergic neurons projecting to the medial prefrontal cortex (light) and the border region between M1 and M2 region (dark). Note the substantial overlap of light and dark cells in the rostral (right hand side of the model) BF. Fluoro-Gold was injected into two sites in the prefrontal cortex and Fast Blue into the border of the M1/M2 regions (upper left insets). (**B**) Distribution of cholinergic neurons projecting to the somatosensory (light) and the M1/M2 association region (dark). Note the apparent minimal overlap between the light and dark symbols in the basal forebrain. (**C**) Overlap analysis from selected sections of case shown in panel **A**. (**D**) Overlap analysis from the case depicted in **B**. For panels **C** and **D**, each section was subdivided into $500 \times 500 \times 50$ μm voxels, and the number of cells from each of the two populations (populations "1" and populations "2") was counted in each voxel. Voxels containing at least 3 cells of either population are marked with light gray and dark gray, respectively; those containing at least 3 of both marker types are marked in white. Note the substantial overlap in panel **C**, as indicated by the white voxels. In panel **D**, no white voxels are detected indicating no overlap in this case. Note that the gray scalings (colors in the companion CD-ROM file) of the voxels here represents population "1" and/or population "2" and does not correspond to the coding in panels **A** and **B**. The corpus callosum is rendered by double gray/white surfaces around the cingulum bundle in the 3D models. A color version of this figure is enclosed in the CD-ROM.

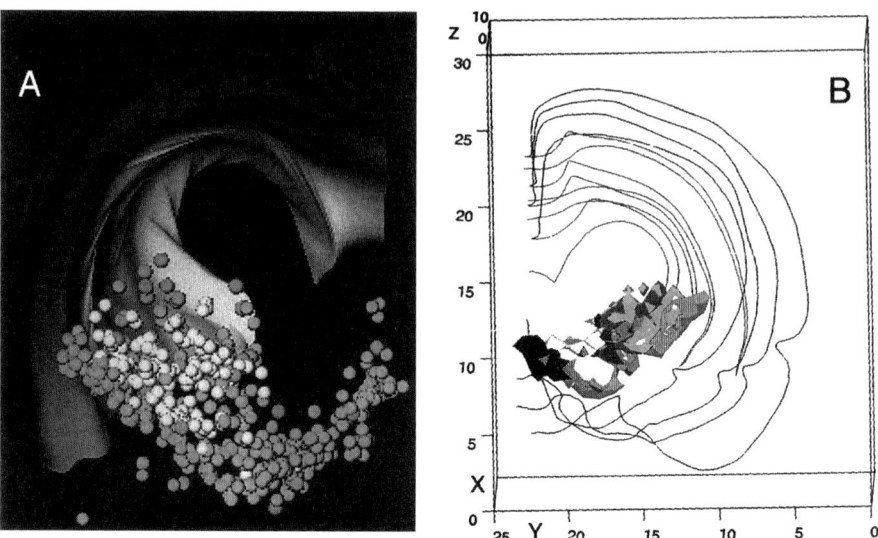

Fig. 3. (**A**) Composite map showing the 3D distribution of cholinergic cells projecting to four arbitrarily defined mediolateral sectors of the neocortex. In the color version of this figure (accompanying CD-ROM), cells projecting to different regions are color-coded (medial, red; intermediary sector, blue and yellow; and lateral parts of the neocortex, green). Note the relatively ordered rostromedial to caudolateral distribution of cells that project to mediolaterally located cortical areas. Dark (red) symbols in the lower right side of the model are rostral. Medial is right, lateral is left. (**B**) Isodensity surface rendering to show the major organizational features in the BF. Unit space: 400 × 400 × 50 µm, density threshold ≥2 cell/voxel. For appreciation of the different cell groups see the color version of this figure where dark blue surface covers unit spaces that contain cholinergic cells projecting to the posteromedial (M1/M2) cortex; yellow, medial prefrontal cortex; red, barrel cortex; green, posterior insular-perirhinal; light blue, agranular insular-lateral orbital; magenta, lateral frontal (motor) cortex. The isorelational rendering of panel **B** is placed into the wireframe of the section outlines and the corpus callosum to show their real position in the original brain. Note that the view in panel **B** is a mirror image of panel **A**. Here and at the rest of the 3D representations, the numbers along the z axis are the layers (sections), and the x and y values correspond to the voxel indices. A color version of this figure is enclosed in the CD-ROM.

from eight individual experiments. Since the overlap analysis is limited to the simultaneous comparison of only two cell populations and only in two dimensions, in order to appreciate the overall projection pattern in the 3D space, we developed an algorithm that renders a surface around voxels of similar cell densities (Appendix 9.9.4. and *[41]*). Since cells are replaced by densities, and densities are rendered around by surfaces, the simultaneous 3D visualization of multiple cell populations is feasible. Figure 3B is a 3D composite of the isodensity maps of six different cell populations, suggesting that the bulk of each cell population projecting to the six cortical targets is separated in the BF. Unfortunately, when isosurfaces of different cell types are combined, the larger surface area may have included isosurfaces of other cell types. Therefore, separate renderings of the individual cell populations and pairwise overlap analysis have to be considered *(41)*.

A detailed overlap analysis of 9 cases, each with paired injections, and some 30 computer-generated combinations of these cases (Csordas and Zaborszky, in preparation) suggest that corresponding mediolaterally located frontal and posterior cortical areas receive their input from a partially overlapping area in the BF. On the other hand, topographically noncorresponding frontal and parieto-insular areas receive their projections from nonoverlapping areas of the BF. Since the location of overlapping voxels in the BF is highly specific for the injection sites that represent cortical associational columns, these data suggest that the BF cholinergic input is transferred via specific corticocortical nodal points toward hierarchically related frontal cortical areas.

9.3. INHOMOGENEOUS DISTRIBUTION OF CHEMICALLY IDENTIFIED CELL POPULATIONS

9.3.1. Differential Density 3D Scatter Plot

Figure 4A, using a differential density 3D scatter plot (for a brief description of this method, see Appendix 9.9.4.) shows that the density of cholinergic cells is not uniform (see also Fig. 2 in [4]). Cholinergic cells often form clusters consisting of 3–15 tightly packed cell bodies. The saliency of these clusters, nonetheless, depends on the density threshold setting. For example, when using a relatively low threshold ($d \geq 5$ cells per $250 \times 250 \times 50$ μm voxel size), these clusters seem to be diffusely distributed. In contrast, when using a relatively high threshold ($d \geq 15$ cells/voxel), the clustering of cholinergic cells seems to deviate from a random distribution. Interestingly, in primates, in comparison to rodents, a proportionally higher percentage of cholinergic cells are located in clusters (4), suggesting that increasing clustering in the phylogeny of BF cell populations might be related to the increased specialization of the cortical areas they project to. Similarly, the location of other cell populations in rat, including calretinin, calbindin, and parvalbumin-containing neurons, suggests inhomogeneous distributions (Zaborszky, Buhl, Pobalashingham, Somogyi, Bjaalie, and Nadasdy, in preparation). Figure 4B shows a similar type of differential density scatter plot of parvalbumin-containing neurons, where dots represent the cell bodies, and large filled circles represent high density spots.

9.3.2. Isorelational Surface Rendering

Since the simulateneous visualization of more than two populations using differential density scatter plots is difficult, we applied another surface rendering algorithm that uses both density and spatial relational constrains (Appendix 9.9.4. and [41]). Figure 4C shows the isorelational surfaces (dark solid) rendered around regions where the density of both cholinergic and parvalbumin cells met two criteria: (i) density is at least five for each cell type within the voxel ($250 \times 250 \times 50$ μm); and (ii) the ratio of cholinergic to parvalbumin cell counts is at least 0.5. In other words, the voxels covered by the dark surface contain at most twice as many parvalbumin as cholinergic neurons. With the isorelational surface rendering, we introduced double constraints, density and relational, which led to a further simplification of our model. Comparing the locations covered by such isorelational surfaces with the scatter plot distribution of the corresponding two-cell populations clearly shows that these surfaces form a central core, consisting of high density cells from both cell populations that is flanked on all

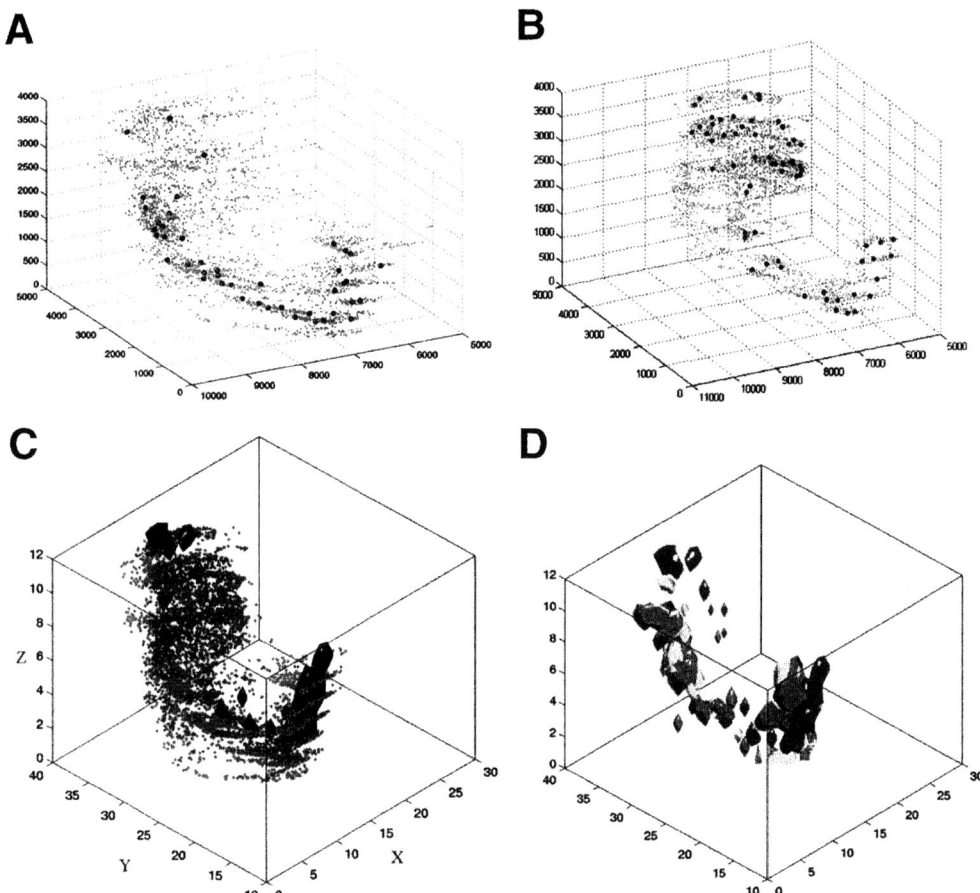

Fig. 4. Differential density scatter plots and isorelational surface mapping. (**A and B**) represent the spatial distribution of cholinergic (dots in panel **A**) and parvalbumin (dots in panel **B**) cells from the same brain showed separately. Filled circles mark the high density locations where the density of cholinergic or parvalbumin cells is higher than 15 cells in the unit space (250 × 250 × 50 μm). (**C**) The scatter plots of both cholinergic (red in the color version of this figure) and parvalbumin (green) cells are superimposed on the isorelational surface (dark solid area; violet in the CD-ROM file) where the density of both the cholinergic and parvalbumin cells is >5 and the ratio of cholinergic–parvalbumin cells is at least 0.5 or higher. (**D**) Merging the cholinergic–parvalbumin, cholinergic–calbindin, and cholinergic–calretinin isorelational surfaces (using cell density ≥5 in the unit space) into one scheme reveals that the cholinergic "column" can be parcellated into clusters of different sizes. Different shading of surfaces cover the spaces where the relationship of cholinergic cells to parvalbumin (green in the color version), calretinin (yellow), and calbindin (blue) neurons is similar (0.5 or higher). A color version of this figure is enclosed in the CD-ROM.

sides with single-cell populations of gradually decreasing densities (Fig. 4C). Merging the three pairwise isorelational surfaces (cholinergic–parvalbumin, cholinergic–calretinin, and cholinergic–calbindin) into one scheme suggests that the cholinergic cell "column" can be parcellated into several smaller clusters or larger amalgamations, in which cholinergic cells are mixed with the other three cell types in a specific fashion

(Fig. 4D). Using a section-by-section analysis of the overlap, as shown in Figures 2C and D, one can get a fairly good idea about the composition of the mixed clusters. The advantage of the combined 3D isorelational surface rendering of Figure 4D is indeed in the totality of this image. Comparing similar types of renderings from different brains, a similar global pattern emerges suggesting that the configuration of the isorelational surfaces is not by chance and that the high density clusters in the individual cell populations may correspond to the zones where the different cell populations overlap with each other. The location of these overlapping zones may be determined during ontogenesis.

9.4. CHOLINERGIC CELL GROUPS SHOW REGIONALLY SELECTIVE DENDRITIC ORIENTATION

9.4.1. Mean 3D Vector of Dendritic Processes

Since the geometry of axons and dendrites imposes constraints on their connections, in order to understand how information is handled in the BF, it is important to determine how the shape of the axonal and dendritic arborizations could influence regional connectivity patterns. Cholinergic cell bodies give rise to 2–5 primary dendrites radiating in all directions. The relatively straight primary dendrites bifurcate in an iterative fashion, and the sum of the lengths of the daughter branches is usually larger than that of the mother branch. The dendrites of adjacent cholinergic neurons often constitute overlapping fields. The dendrites are freely intermingled with passing myelinated fiber bundles within which they are embedded. Thus, the dendritic organization of the cholinergic BF neurons resembles that of the isodendritic type of neurons of the reticular formation *(42–44)* or the so-called interstitial neurons characterized by Das and Kreutzberg *(45)*. The total length of the dendrites of individual cholinergic neurons in rats is about 4 mm, arborizing in a box of about 0.1 mm^3, filling, however, only a fraction of its spatial domain. According to our estimation, one cholinergic cell dendritic domain might share its space with 50–80 other cholinergic neurons, depending on its location in the BF. Although a particular orientation of cholinergic dendrites could be noticed upon inspecting areas where the density of dendrites is low (see Fig. 1 in *[46]*), it is not possible to appreciate dendritic orientation with certainty in regions where the cell density is high, as can be judged from Figure 1.

We assumed that the cholinergic cell clusters, beyond their spatial segregation, must fit into the functional network of their input–output connections. In other words, we assumed that cholinergic cell clusters develop under the constraints that link together functionally related output (neocortical) and input (brainstem and telencephalic) pathways, and this input selectivity, we reasoned, must be reflected by the anisotropic dendritic orientation of the putative cell clusters.

In order to correlate regional differences of dendritic orientation to the spatially distributed population of neurons, we developed a method of representing the main dendritic branches of individual neurons with 3-D vectors and embedded them into the 3D coordinate system of the cell bodies. The origin of a vector represents the position of the neuron, its orientation represents the dominant orientation of the dendritic tree (for details, see Appendix 9.9.4.), and the length of vector represents the average length of the dendritic branch. The main orientation vectors of 750 individual cholinergic

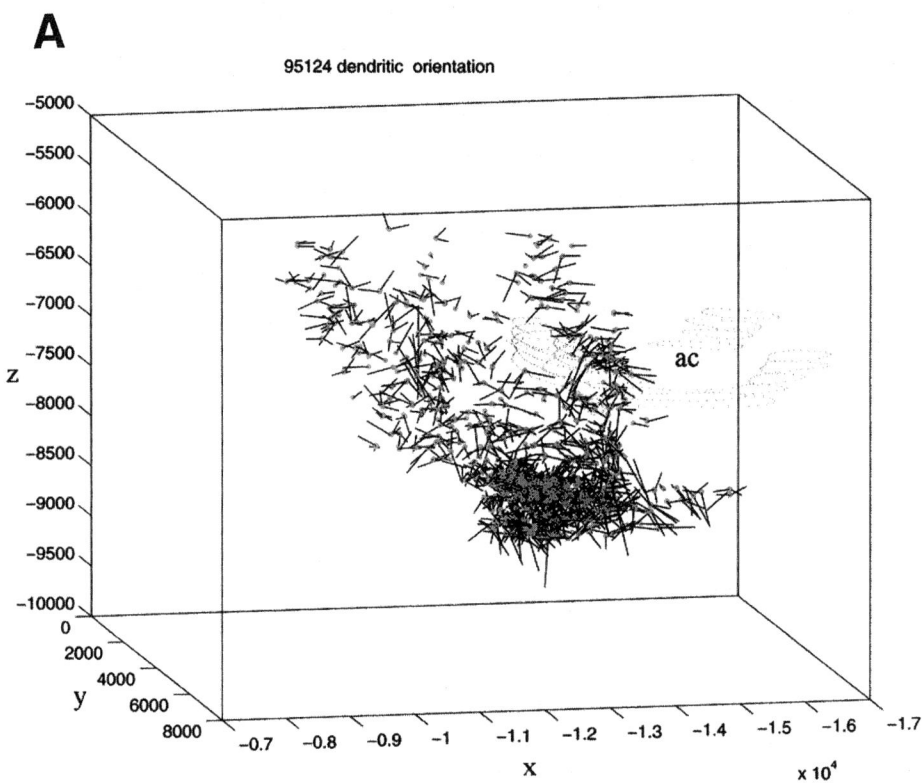

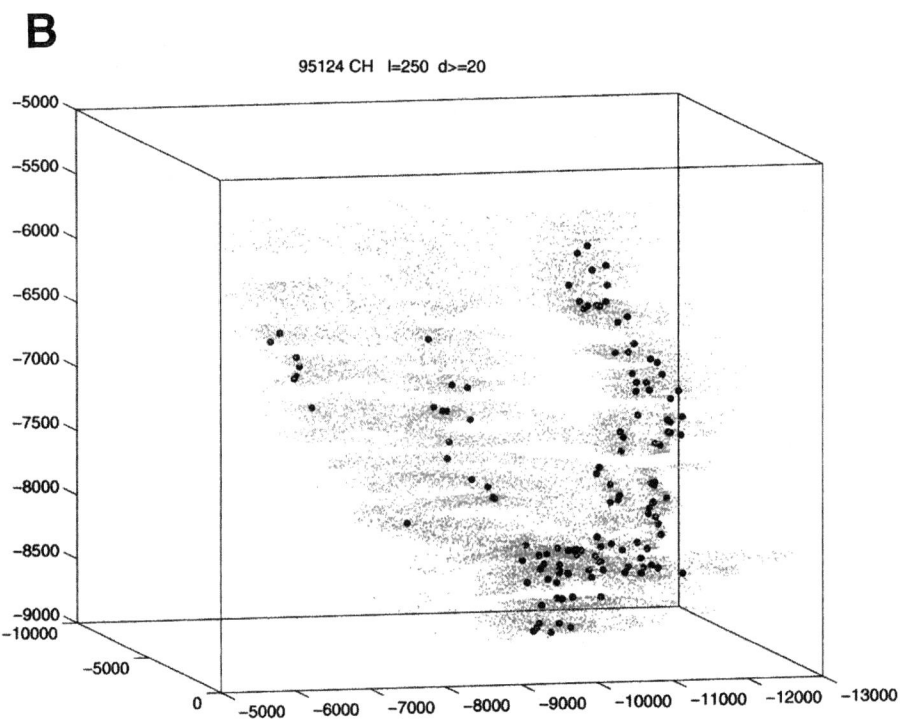

cells, selected from a population of about 15,700 cholinergic cells, are shown in Figure 5A. Rotation and navigation in the 3D plot made it possible to gain insight into the vector orientation even in the denser cell clusters. Comparison of Figure 5A with the differential density scatter plot of the same dataset, shown in Figure 5B, suggests a tendency of iso-orientation of dendrites within a given cholinergic cell cluster.

9.4.2. 2D Dendritic Stick Analysis (Polar Histogram)

To obtain a quick qualitative characterization of the directional distribution of dendritic growth projected onto the plane of sectioning, the polar histogram is the method of choice (Neurolucida® software package; see also Appendix 9.9.4. and *[47]*). In essence, using only the x and y coordinates of the traced dendritic segments, where individual segments are composed of pairs of adjacent points, the algorithm pools together all the segments around a center but preserves their length. The total range of angles is then binned to equal sectors, and the program calculates the sum of segments in each bin. The radial length of a filled sector is proportional to the total length of the dendritic branches of that specific orientation, thus the contributions of segment lengths and segment counts of that specific orientation are inseparable. In other words, a few long segments can add up to the length of many short dendrites. Depending on the choice of binning interval, the angle discrimination can be finer or broader. Analyzing BF cholinergic dendritic orientation by polar histograms suggests a regional orientation preference (Zaborszky, Nadasdy, and Somogyi, in preparation).

In contrast with the polar histogram method, the vector representation preserves the neuronal identity of dendrites, instead of pooling them together, and still provides an overall view of orientation of dendrites. The main advantage, however, is that vectors relate the dendrites to the spatial distribution of the neurons in a simplified and meaningful fashion. To compare the regional dendritic orientation derived from polar histograms with the orientation vectors calculated for individual cells, we selected a subspace of the septal area where a more detailed analysis of subpopulations of cholinergic neurons was available. The Neurolucida program allows one to outline and select cell populations from any number of sections in a series and to construct a polar histogram of the dendrites pooled together from the selected cell populations. Figure 6A represents a portion of the septal area from Figure 5A viewed from a sagittal direction. Figure 6B displays cholinergic cells whose dendrites were traced from a series of sections cut in the sagittal plane. The selection areas of the four polar histograms of Figure 6C–F are indicated by boxes of various sizes in the upper right diagram. Comparing the dendritic orientation obtained from the polar histograms of pooled dendrites with the mean 3D vector of the dendritic branches indeed suggests that subpopulations of

Fig. 5. (*facing page*) (**A**) Mean orientation of dendritic branches. The initial segments of dendrites are represented by dots. The outlines of the anterior commissure (ac) are indicated by small dots. (**B**) Differential density scatter plot of the same database. Dots represent cholinergic cells (n = 15,700), filled circles mark the high density locations where the density of cells is ≥20 per unit space (250 × 250 × 100 μm). Flakes are due to the section steps along the z axis. Cells and their dendrites were mapped from 34 consecutive horizontal sections stained for choline acetyltransferase. The comparison of panels **A** and **B** suggests the iso-orientation of dendrites in the high density cell cluster. A color version of this figure is enclosed in the CD-ROM.

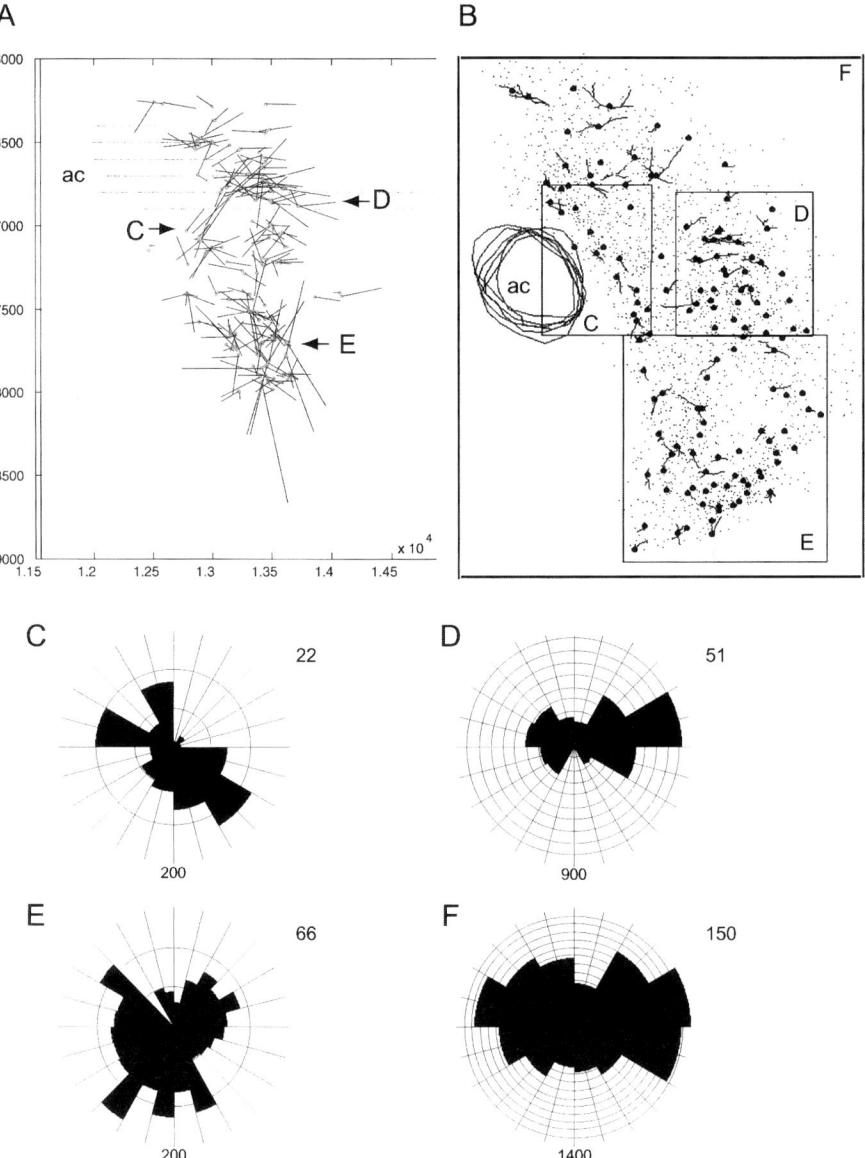

Fig. 6. Comparison of the dendritic orientation derived from polar histograms with the orientation vectors. **(A)** Part of the septal area from Figure 5A as viewed from the sagittal direction. Letters C, D, and E with arrows point to regions that may correspond to the same cell populations as selected for the polar histograms from sagittal sections of a different brain as shown in panel **B**. **(B)** Cholinergic cells (137) (filled circles) were selected from a stack of sagittal sections comprising the septal region (n = 2266 cholinergic cells, dots) for dendritic tracing. Letters C, D, E, and F mark boxes that were used to select dendrites for the orientation analysis. In both panels **A** and **B**, ac indicates the location of the anterior commissure. Despite slightly different orientation of the sagittal sections, one can recognize the same cell groups as seen in the 3D rendering. **(C–F)** Polar histograms representing dendritic orientation from indicated areas. Numbers at upper right indicate the number of dendritic segments in the sample. Numbers along the circles within the polar histograms mark distances in micrometers from the origin (see section 9.9. for explanation). A color version of this figure is enclosed in the CD-ROM.

cholinergic cells can be delineated based upon their density and main orientation of their dendritic arbor. Therefore, one of the key features of cluster organization is the iso-orientation of their dendrites.

9.5. VARIOUS AFFERENTS IN THE BF SHOW REGIONALLY RESTRICTED LOCALIZATION

Using a double strategy of recording the location of putative contact sites between identified axons and cholinergic profiles as well as identifying in representative cases under the electron microscope the presence of synapses, one can get a fairly good idea about the extent of potential transmitter interactions in the BF (for references, see *[4,38]*). Although the noradrenergic and dopaminergic axons contact cholinergic neurons in extensive portions of the BF, the majority of afferents (cortical, amygdaloid, striatal, peptidergic) appear to have a preferential distribution in the BF; thus a specific input can contact only a subset of neurons. Figure 7 gives examples of the distribution of restricted vs more diffuse afferents.

9.6. PROBABILITY OF CONNECTIONS

In many cases examined, labeled terminal varicosities detected in the BF were related to both cholinergic and noncholinergic postsynaptic elements. In fact, the detectability of synapses on cholinergic neurons was usually proportional to the density of terminals present in a given area. Thus, at first approximation, the probability of synapses between cholinergic neurons and various afferents may depend on the geometry of the dendritic arbor and axonal ramifications. Since a systematic study comparing the dendritic arbor of cholinergic neurons with various afferent orientations would require a substantial time, we only briefly comment on this issue here, by documenting the case of calcitonin-gene-related-peptide (CGRP)-containing axons in the internal capsule. Figure 8A shows cholinergic cells and their traced dendrites at about 1.5 mm posterior to bregma. Figure 8B is a schematic drawing from an adjacent section that was immunostained for CGRP and whose axons in the internal capsule were traced at high magnification. Comparing the polar histograms of cholinergic dendritic trees in the internal capsule (Fig. 8C) with that of CGRP axons (Fig. 8D), it is obvious that CGRP axonal ramifications have the same prevailing direction as the dendritic arbor of cholinergic neurons. Indeed, electron microscopic studies confirmed abundant presence of CGRP in axon terminals synapsing with cholinergic neurons in this region *(48)*.

The probability of synaptic connections can be calculated from the overlap of axonal and dendritic domains *(49)*. The probability of having more than one synapse between any given presynaptic axon and a postsynaptic cell is maximal in the case when the terminal axon and the receiving dendrite are running in parallel ("climbing" fiber type contact). However, only one synapse is possible if the axon runs at right angles to the dendrite ("crossing over" type of geometry). If the terminal axon is oriented obliquely to the receiving dendrite, the probability of synaptic contacts is a cosine function of the angle between the axon and the dendrite *(50)*. Since various afferents show specific localization, it is likely that cholinergic cells in various BF subdivisions can sample a unique combination of afferents. It seems that in each major subdivision of the BF along the axis of the major orientation of the cell bodies, a specific type of axon is

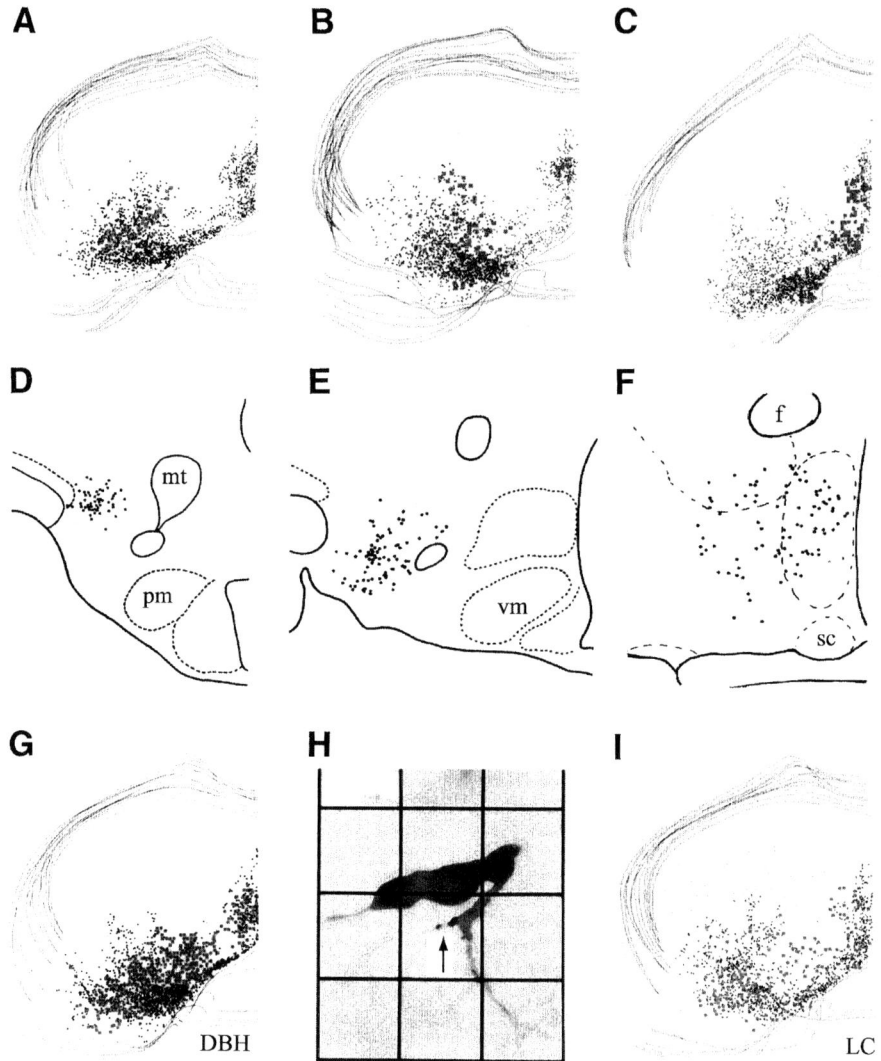

Fig. 7. Differential distribution of various afferents in the cholinergic forebrain. (**A–C,G,I**) Composite maps illustrating putative zones of contacts between afferent fibers and cholinergic neuronal elements following *Phaseolus vulgaris* leucoagglutinin (PHA-L) injections into the (**A**) far-lateral hypothalamus, (**B**) midlateral hypothalamus, (**C**) medial hypothalamus, (**I**) locus coeruleus. (**G**) Shows the distribution of putative contact sites from a material stained for dopamine-ß-hydroxylase and choline acetyltransferase. (**H**) PHA-L-labeled terminal varicosities (arrow) in close apposition to a proximal dendrite of a cholinergic neuron. The grid simulates the ocular reticle used to screen sections for high magnification (63×) light microscopic analysis. One division of the grid = 16 µm. Cholinergic neurons are represented by dots. Zones of putative contacts between cholinergic elements and terminal varicosities are depicted as solid squares (corresponding to 80 × 80 µm areas in the section). (**D–F**) Location of labeled cells at the PHA-L injection sites from cases depicted in panels **A–C**. Panels **A–C,H** are modified from Cullinan and Zaborszky *(57)* with kind permission from Wiley-Liss. Panels **G** and **I** are modified from Zaborszky et al. *(51)*, with permission from Elsevier Science. A color version of this figure is enclosed in the CD-ROM.

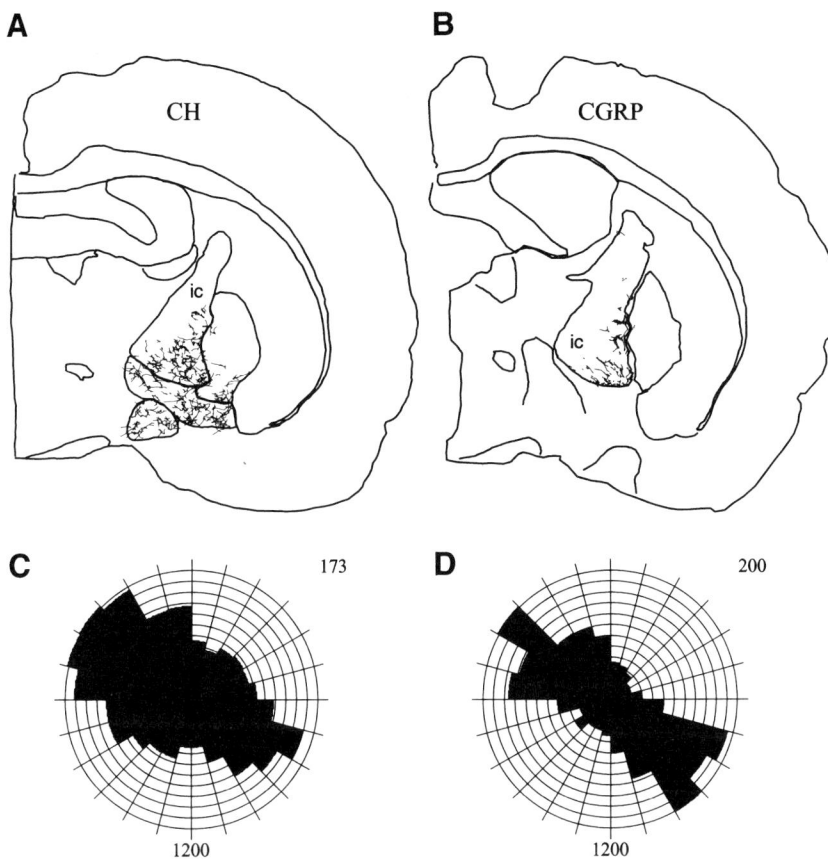

Fig. 8. Comparison of the 2D orientation of cholinergic dendritic segments (**A**) and CGRP-containing axons (**B**) in the internal capsule (ic). (**C and D**) Polar histograms of cholinergic dendrites (**C**) and CGRP axonal ramifications (**D**) from the same general area. Note that the majority of dendrites and axons occupy the same sector of the polar histograms. Upper right numbers indicate the number of segments in the analysis. The outer circle of the polar histogram correspond to a 1200-µm diameter around the origin.

maximally aligned with the preferred orientation of cholinergic dendrites of that area. Different cells, or perhaps different dendrites of the same cell, can sample the same input differently according to the spatial organization of the dendrites and corresponding axons. For example, the majority of dopamine-ß-hydroxylase positive varicosities (used to stain noradrenaline and adrenaline containing neurons) establish single synapses with cholinergic dendrites, while a small population of cholinergic neurons (at most 5%) appears to receive multiple contacts on their dendrites in the form of climbing-type arrangements. Such climbing-type synapses were most often detected in the substantia innominata (Fig. 3 in *[51]*), but were also occasionally seen in other BF regions. It is unclear whether cholinergic neurons with climbing-type inputs are different in other respects, however, one can speculate that the noradrenaline released at these climbing-type synapses must have a more powerful action on these selected neurons as compared to the single synapses at random locations.

Our earlier assumptions about the randomness of connections *(46,52)* had to be modified when we realized that prefrontal axons seem to terminate exclusively on noncholinergic cells, including parvalbumin-containing GABA-ergic cells, in spite of the fact that many of these axons arborize in the immediate vicinity of cholinergic neurons *(53)*. It is expected that the detailed reconstruction of local axon collaterals of BF neurons may add to the specificity of the connectional scheme in the BF *(38)*.

9.7. MERGING DATAFILES CONTAINING NEURONS OF DIFFERENT COMPLEXITIES

To understand how individual neurons with complete axonal and dendritic arborizations fit into the global structure of the BF as outlined in the preceding paragraphs, we took advantage afforded by the juxtacellular staining of individual neurons *(54,55)*. This technique can also be combined with extracellular recording, electroencephalogram (EEG) monitoring, and subsequent chemical identification of the filled neurons. As recorded in anesthesia, neuropeptide Y (NPY) neurons are silent during spontaneous or tail pinch-induced cortical desynchronization, but accelerate their activity during episodes of cortical delta oscillations. In contrast, the firing of cholinergic neurons increases during cortical low-voltage fast electrical activity *(55)*. Since NPY-positive neurons also contain γ-aminobutyric acid (GABA) and have been shown to contact, with their local axon collaterals, cholinergic corticopetal cells, a hypothetical scenario can be suggested of how these two cell types may be involved in modulating cortical activity *(38)*. Obviously, the proper interpretation of these electrophysiological data would require understanding of the precise input–output relationships of these and other neuronal populations. Using the Neurolucida program, such fully reconstructed neurons can be "implanted" into a larger database as the one used for the generation of Figure 4; thus individual electrophysiologically and chemically identified neurons can virtually be placed into their natural environment. In this way a functional property such as "content" can be placed into the anatomical maps as "context". Figure 9C and D display a locally arborizing NPY neuron and a cortically projecting and also locally arborizing cholinergic neuron, respectively. Although both of these neurons are located (Fig. 9A,B) in the same general BF area (horizontal limb of the diagonal band), their local axons may contact different postsynaptic target, and similarly, their dendrites should sample, at least in quantitative terms, different inputs. According to our estimations, this particular cholinergic neuron gives rise to about 1400 local axonal varicosities, and in the space defined by its axonal arbor, there are approximately 1500 cells. On the other hand, the NPY neuron presented here distributes about 2900 varicosities in a space that contains 1250 neurons. Whether or not these varicosities represent synapses and whether or not they address postsynaptic targets selectively, remains to be investigated.

9.8. CONCLUDING REMARKS

Since the seminal paper of Schwaber et al. *(56)*, who first used computer-aided data acquisition and 3D reconstruction of BF cholinergic neurons, the progress in understanding the organization of the BF has been very slow. It is likely that, in the coming years, the sophisticated use of multi-electrode recordings in awake behaving animals

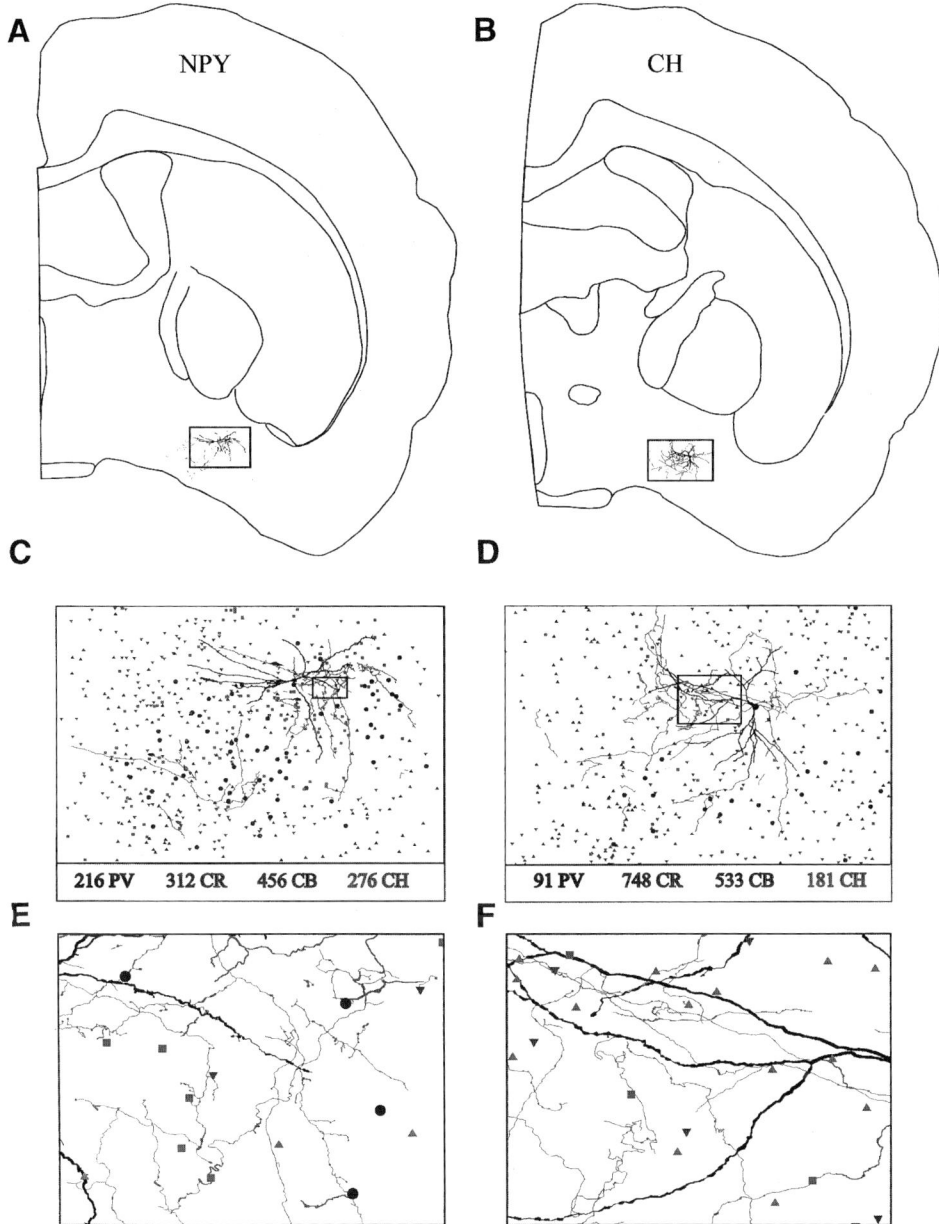

Fig. 9. Distribution of different cell types in the neighborhood of identified NPY (**A,C,E**) and cholinergic (**B,D,F**) neurons. (**A and B**) Schematic drawings illustrating the location of the electrophysiologically and chemically identified neurons. (**C and D**) Coronal view of the identified neurons embedded into the same general region of the BF derived from a different database that contains four different cell populations. Filled circles, parvalbumin; up triangles, calretinin; down triangles, calbindin; squares, cholinergic neurons. The approximate number of cell bodies from each cell population that can be found in the 3D volume of the single-cell axonal arbor is indicated below. (**E and F**) Enlarged view of the boxes from panels **C** and **D**. Thicker black indicates dendritic processes, and thinner lines indicate axonal ramifications. Note that the small varicosities along the axonal collaterals correspond to putative synaptic boutons. A color version of this figure is enclosed in the CD-ROM.

and the application of promising new computational tools will define how anatomical features constrain the extraction of information processed in the BF. For the time being, we can only speculate on how the BF, in particular the cholinergic neurons, process specific information despite the apparently diffuse organization of its elements.

The territory of the BF populated by cholinergic corticopetal cells can be viewed as a large interconnected network where a systematic directional variation of dendritic clouds and presynaptic axonal clouds permeate each other intimately. In spite of the lack of internal borders, within this large cholinergic assembly, smaller subassemblies can be delineated by differential cell densities and dendritic orientations, input–output features, and numerical relations of the constituent neuronal populations. The cholinergic cell clusters with other local or projection neurons may represent special sites (modules) where information processed in separate streams can be integrated. The location and size of these modules may temporarily vary according to the prevalence of state-related diffuse brainstem modulatory and more specific telencephalic inputs. From this latter group of afferents, the prefrontal input may function as an external threshold control, which allocates attentional resources via the BF to distributed cortical processes in a selective self-regulatory fashion.

9.9. APPENDIX

9.9.1. Animals and Tissue Processing

The reconstructions and statistical analyses presented in this paper were prepared from data obtained from adult male Sprague-Dawley rats. All animal procedures were in compliance with the National Institutes of Health (NIH) Guidelines for the Care and Use of Animals in Research and approved by the Rutgers University Institutional Review Board for the Use and Care of Animals. The anesthesia, electrophysiological recordings, perfusion of animals, and tissue processing have been described earlier *(55,57–58)*.

9.9.2. Data Acquisition

Immunostained diaminobenzidine-labeled cell bodies were digitalized in BF areas with the aid of an image-combining computerized microscope system (Zeiss Axioscope, 20× Plan-NEOFLUAR lens) using the Neurolucida software package *(59)* (MicroBrightField, Colchester, VT). Outlines of the sections, contours of structures, and fiducial markers were drawn with a 5× Plan-NEOFLUAR lens. Dendritic branches were traced from the cell body by connecting tracing points by straight lines (Plan-APOCHROMAT 40× (NA = 1.0) or Plan-NEOFLUAR 63× (NA = 1.25) oil immersion lenses). Sections containing fluorescent-tagged cell bodies were mapped by using the epifluorescent setup of the Axioscope microscope equipped with appropriate filters. Fast Blue and Fluor-Gold-labeled projection neurons (exciter/barrier filter set 365/418) and the fluorescein isothiocyanate (FITC)-labeled (FITC exciter/barrier filter set 450–490/520) cholinergic cells could be separately visualized in the same section. Labeled cells were mapped from every 8th sections at a magnification of 20×.

Although the outlines, and contours were drawn flat, disabling z input information, dendrites were followed in the depth of the sections (50 or 100 μm) by changing the focus. Curvilinear dendrites were represented in the computer as a series of short straight lines giving a close fit to the original shape and length. The Neurolucida hardware system allows a point-to-point discrimination of 0.3 µm in all axes. Neurons traced from each section were aligned to a common reference, e.g., the lowest midline point of the corpus callosum. Mapped sections were aligned using up to 99 alignment points for best-fit matching included in the Neurolucida software program. The data generated by tracing the neurons using the Neurolucida software are later referred as the Neurolucida database. The database is composed of a stack of aligned sections.

Neurons in the Neurolucida database were represented by the x, y, and z coordinates of the cell bodies. In the database, dendritic trees originating from the same neuron at different sites were represented as separate but adjacent data blocks and were encoded independently from their cell bodies. Since cell bodies were not traced as 3D objects, the origin of the primary dendrites did not necessarily match in any dimension. Due to this independence of cell bodies from their dendrites in the encoding scheme, finding the common cell body for each dendrite was not obvious. Branching points were marked. Based on the branching points, first, second, third, and higher order dendritic segments were identified as stemming from a parent node. The hierarchical encoding system (introduced by Neurolucida) allowed us to recursively represent the complexity of any dendritic tree in the database.

9.9.3. Selection of Neurons for Dendritic Tracing

In this paper, dendritic data are derived from three different datafiles. As a preliminary material, all cholinergic cell bodies with their dendritic processes were traced from seven coronal sections (50 µm). Approximately 1300 cells were traced in this material. Figures 1B and 8A are from this datafile. To create a more complete database, a second brain was cut in horizontal planes into 34 consecutive sections (100 µm thick), and every cholinergic cell was digitized. Figures 5A and 6A are generated using this horizontal dataset. Finally, a third brain was cut in the sagittal plane (100 µm thick), and the data presented in Figure 6B–F are derived from the septal region of this sagittal dataset. In this latter brain, similarly to the horizontal set, all cholinergic cell bodies were digitized. From the horizontal and sagittal dataset, about 5% of the total cholinergic neurons were selected for dendritic reconstruction based on a random sampling of the total population. In order to obtain a sample of neurons representative of the inhomogenous distribution of the entire population, we used a combination of two different density criteria, one with high and another with low resolution. For both selections, the space that incorporated all cholinergic cells was subdivided into subspaces of identical size (voxels or unit spaces), and on these voxels, based on the cell density, different selection criteria were applied. As for a high density criterion, we selected cells from voxels of 100×100 µm × section thickness (=100 µm). To resolve a larger scale inhomogeneity, cells were selected from voxels of 500×500 µm × section thickness. The larger voxel captured the density differences at a 500 µm resolution. The 100-µm sample size was applied to sample local densities at a 100-µm scale. Next, the two samples were combined. As a result, the combination of samples reflected both the global and the local distribution features of the cells. From each voxel, where the local cell counts met the density criteria, a neuron was selected on a random basis and marked for dendritic tracing. The number of cells, *s* selected for dendritic tracing from each voxel with both v' and v" sizes was proportional to the natural logarithm of the number of cells, *n* within the given space, v_{ijk}, multiplied by a constant, *c* as follows:

$$s = c \times \left[\log(n_{v'ijk}) + \log(n_{v''ijk}) \right]$$

For the interpretation of i, j, and k voxel indices, see *(41)*. The purpose of the multiplication factor c was to provide flexibility to scale up or down the number of selected cells. The value of *c* was set to 0.5 with 100 µm grids and 0.4 with 500 µm grids. The edges of the two voxel types v' and v" were 100 µm and 500 µm, respectively. Technically, the datafile of traced cell bodies was exported from Neurolucida, parsed for different objects (cell bodies, structure outlines, etc.), and the point coordinates of cell bodies were extracted. A custom written C++ program performed the partitioning, cell counting, and selection of target cells for dendritic tracing. With this sampling scheme, we marked 750 cells from the horizontal dataset (n = 15,776) and 137 from the septal sagittal dataset (n = 2266). The generated data file with the target cells marked, was inserted to the original Neurolucida datafile for subsequent dendritic tracing.

9.9.4. Analysis of the Data

For data analysis, as described here, we extracted the x, y, and z coordinates of the cell bodies from the Neurolucida database and saved them in ASCII format, each cell type in different data files. Structure outlines were stored separately. The medial, lateral, dorsal, and ventral extremes of the cholinergic cell distribution were taken as a 3D framework to incorporate the entire database. For expressing regional density changes, the 3D framework was subdivided into virtual blocks of identical size denoted as "voxels". Section thickness served as unit size for the z dimension. If cells of different types were mapped from different but adjacent sections (plots in Figs. 4 and 9C,D), then their z coordinates were collapsed into a common 2D plane (master plane) by removing the within-section depth coordinates, but preserving the x and y coordinate of the cells. Each different cell type was separately counted in each of these master plane lattices. For visualization, we used different thresholds. Differences in the voxel size and thresholding could significantly influence the obtained results. A more detailed methodological description and discussion is given in a recent publication *(41)*.

Differential density 3D scatter plot. Density differences within or between cell populations can be represented in 2D isodensity maps *(60)*. The obvious limitations of this method is the lack of the third dimension. Our method *(41)* quantifies density differences first, then plots the density descriptors in a real 3D coordinate system. The input data is provided as position of cell bodies by their locations as points in the 3D Euclidean space. In this database, each row represents a single cell given by the x, y, and z coordinates relative to a reference point as the origin. The entire database is placed into a framework, which is partitioned into boxes of identical size (voxel) as a grid system. Cells are counted within each voxel. In contrast to the parametric representation of the space, this provides a 3D volumetric dataset where the dimensions are the x, y, and z position of the voxel and the local cell count within the voxel space. Voxels of cell counts larger than a predefined density are considered and represented by a single marker randomly selected from the neurons in the corresponding voxel. The distribution of these markers highlights locations where high density neuronal clusters occur. Figures 4A,B and 5B represent this type of analysis.

Isodensity surface mapping. The spatial distribution of different cell types may be very complicated as neuronal populations interdigitate, intersect, or overlap with one another. Instead of using scatter plots, the spatial organization of density differences is better visualized by rendering a surface around large density cell groups, especially when multiple cell types are concerned. Similar to the "differential-density 3D scatter plot," a selected set of voxels are visualized. However, instead of representing them by single points, the algorithm renders a surface around voxels of larger than certain cellular density. The procedure of subdividing the 3D database into voxels (unit spaces) and calculating the voxel cell densities is identical to that of the "differential-density scatter plot." Conversion of the 3D point-coordinate database, where the entries are the cells, to a density data constructs a volumetric database. In the volumetric data, the entries are voxels defined by their 3 coordinates and the associated cell densities. Then, a density threshold is defined, and voxels characterized by larger density than the threshold are identified. The algorithm renders a 3D skeleton and determines a 2D manifold on the skeleton that is defined by interconnecting points that separate the higher density space from a lower density space. The manifold is further partitioned onto triangles and surface elements are rendered to each of these triangles. These surface elements are then smoothed, and reflectance property as well as light source are defined. For surface rendering, the C++ program and the 3D visualization toolbox of Matlab R11® (MathWorks, Inc.) were used. Figure 3B was generated by this method.

Isorelational surface rendering. Similar to the "isorelational scatter plot", the aim of this representation is to show the codistributive association between different cell types or other variables. In contrast to the "isorelational scatter plot", this plot renders a surface around the population of cells where certain density ratio is detected. Since the association of different cell types have a typically complicated spatial configuration, the scatter plot of neuronal markers does not reveal the true 3D structure. In order to reduce the complexity, voxels, where a certain density ratio of two cell types is established, are rendered with a surface. This surface separates cells where certain density ratio is higher than a critical value. The unique feature of the "isorelational surface rendering" method is the visual representation of abstract relationships, which is more important for understanding functional connections between neurons than exact locations of cell bodies. Technically, the 3D database of cell bodies is subdivided into unit spaces for both cell types. Then cell density ratios are calculated within each unit space (voxel) shared between the two cell types. The density ratios are arranged in a 3D matrix containing various ratios. Isodensity demarcation lines are calculated and rendered by a surface in such a way that cell bodies with density ratios larger than a specific number are covered by the surface. Density ratios smaller than the critical one are located outside of the surface. For visualization purposes, a range of critical density values must be applied for testing the integrity of clouds and to make sure that there are no hollow spaces covered. The algorithm of surface rendering is the same as the one described at the "isodensity surface mapping". Complex relationships between multiple components such as density relations of multiple cell types can be decomposed into pairwise relations and visualized as merged surfaces. Color coding of surfaces of different cell types helps to interpret complicated arrangements. The plots in Figure 4C,D were generated according to this method. A more detailed description of the methods described so far in this section is given elsewhere *(41)*.

Mean 3D vector of dendritic processes. Individual dendritic branches may have a principal orientation adapted to making contacts with also oriented axons, independent from the orientation of dendritic mass. To test this, the principal orientation of dendritic branches was expressed by the mean branch orientation and approximated by the average orientation and average length of the dendritic tree. The orientation and length were combined into a $V_{(P0,P1)}$ vector originating from the point $P(x_0,y_0,z_0)_i$, where the dendrite stemmed from the cell body and pointing to the $P(x_1,y_1,z_1)_i$ point, which represented the average length and orientation. In this analysis, we first calculate the dx_i, dy_i, and dz_i vectors as Euclidean distances between adjacent branch points, and the average of the dx_i, dy_i, and dz_i is used as a single vector to represent the main orientation tendency of the dendrite. Since multiple origins of dendritic processes were possible, the orientation vector was calculated separately for each main branch resulting in different vectors per cells originating from nearby points. Parsing of the Neurolucida data files and computation of vectors was all carried out by custom written C++ codes and compiled for Silicon Graphics and Pentium class computers using Irix and Linux operating systems, respectively. For visualization purposes, vectors were rescaled by a common multiplicative factor that made it easier to appreciate the main tendency of orientation. 3D aspects of the vector space were constructed by superposition of the vectors on the structure outlines (such as anterior commissure). Rotation and navigation in the database using Matlab 3D graphical user interface made it possible to gain insights of the vector orientation in the denser cell clusters. This type of analytical tool was applied to generate Figures 5A and 6A.

2D orientation of dendritic processes. The algorithm (polar histogram) supplied in the Neruolucida software package is similar to the analysis described by McMullen et al. *(47)*. The difference is that in our case, the results are collected by a computerized system, and the artifacts of that collection process need to be filtered out. The algorithm for polar histogram breaks up the dendritic processes into line segments and determines the directions that these line segments point to and that of the lengths of these segments. The sum of the lengths of the small line segments is approx the same as the length of the original tracing. The direction of the vector is calculated by projecting the line segment onto the plane of the sectioning. This is accomplished with the arc-cosine function. The histogram represents the total length by the distance from the origin and an angle (θ) that the vector makes with the x axis plotted in the radial direction. Each sector in the polar histogram is the sum of all the dendritic growth in that particular range of angle. There is a unique value of the polar histogram for each value of the angle θ. Some information has been lost because the dendrites that are traced in the sectioning plane are always going to be longer than dendrites in the plane perpendicular to the sectioning plane. Therefore, this type of analysis is useful primarily to characterize the orientation of dendrites that are roughly coplanar. Neurolucida allowed us to outline and select cell populations based on anatomical markers from single sections or a stack of sections and to construct a polar histogram of the dendrites pooled together from all the marked neurons.

Comparison of 3D location of labeled cells from different brains. After mapping labeled cell bodies, the Neurolucida files were transferred to a Silicon Graphics (Octane) workstation for further analysis and 3D visualization using the Micro3D (Oslo Research Park) program. In order to compare data from several brains with multiple retrograde tracer injections, each section was visually aligned to the corresponding map of a "master" brain with the aid of surface contours and fiducial markers, including the corpus callosum, anterior commisure, internal capsule, stria medullaris, stria terminalis, and the fornix. To create a maximum fit, an interactive procedure was used, including moving, rotation, and shrinkage corrections along the x, y, and z axes. To avoid gaps between sections in the visualizations, individual cells were randomized in a $400 \times 400 \times 50$-µm space (z-spread). Figures 2 A,B and 3A were generated according to this procedure.

Overlap analysis. The degree of overlap between two neuronal populations was estimated by subdividing individual sections into an array of 500×500 µm voxels and counting the number of digitized coordinate pairs (cell) per voxel using a custom-made program similar to the one applied by Alloway et al. *(40)*. To avoid analyzing areas with low density of cells, only voxels containing 3 or more cells were included. Voxels containing a defined number of "population 1" or "population 2" cells are dark gray (red in the color version) or light gray (blue in the color version), respectively, while voxels containing a similarly defined number or more cells of both categories (at least 3 of each) are labeled white. The number of differently labeled voxels are counted for each section and also summed across sections and used to estimate the percentage of overlapping voxels and also the percentage of a given cell population in the overlapping voxels. The charts in Figure 2C,D are from this material.

Merging files containing cells of different complexities. Figure 9 C,D were prepared merging two different datafiles: one derived from a series of sections containing four different cell populations in the BF using immunostaining for parvalbumin (PV), calretinin (CR), calbindin (CB), and choline acetyltransferase (CH) ("four marker" brain). The other file contained a single electrophysiologically and chemically identified neuron (NPY or CH) with its axonal ramifications and dendritic trees digitized from a series (n = 10–100) of 50-µm thick sections.

The cell mapping and anatomical landmarks were extracted from the four marker brain, in which four series of alternate sections (n = 48) were stained with antibodies against CH, PV, CR, and CB. The distance between two consecutive sections stained with identical markers was 300 µm. Adjacent four sections containing markers for PV, CR, CB, and CH were aligned using standard anatomical landmarks (i.e., corpus collosum, lateral ventricle, fornix, thalamus, optic chiasm) and collapsed into a single section by removing the within section depth coordinates, but preserving the x and y coordinates of the cells, resulting in a 3D series of 2D layers. The distance from bregma of each of this composed layers was calculated using the average of the original four sections. This way, we created a set of 12 layers, each containing four different cell populations with their original x and y coordinates. The same dataset was the basis for the analysis documented in Figure 4.

Reconstruction of single identified cells was achieved by routine procedures as described in the literature (for references, see [61]). Since the tissue sections contain only one stained neuron, all axonal and denritic processes can be followed through a series of adjacent sections. After determining the distance from bregma of the single reconstructed cell body, the corresponding section from the four marker brain was merged into the section that contained the single identified cell body. The axonal arbor fields of the single identified cells were outlined, and the number and cell types that were enclosed were extracted from the Neurolucida database. Data from the cell marker and fractal analysis of the axon arbor was then used to estimate the approximate numbers and types of cells that may be embedded in the axonal ramification space of the single reconstructed cell and could come into contact with it.

ACKNOWLEDGMENTS

The research summarized in this review was supported by NIH Grant No. NS23945 and IR25 GM60826 to L.Z. and A.D. We wish to thank Mr. Jack R. Glaser, President, MicroBrightField, Inc. and other stuff members of his company with whom our interaction has been excellent over many years.

REFERENCES

1. Dunnett SB, Fibiger HC. Role of forebrain cholinergic system in learning and memory: relevance to the cognitive deficits of aging and Alzheimer's dementia. Progr Brain Res 1993; **98**:413–420.
2. Everitt BJ., Robbins TW. Central cholinergic systems and cognition. Annu Rev Psychol 1997; **48**:649–684.
3. Heimer L, de Olmos J, Alheid GF, Zaborszky L. "Perestroika" in the basal forebrain; opening the border between neurology and psychiatry. Progr Brain Res 1991; **87**:109–165.
4. Zaborszky L, Pang K, Somogyi J, Nadasdy Z, Kallo I. The basal forebrain corticopetal system revisited. Ann NY Acad Sci 1999; **877**:339–367.
5. Jones BE, Muhlethaler M. Cholinergic and GABAergic neurons of the basal forebrain: role in cortical activation. In: Handbook of Behavioral State Control—Cellular and Molecular Mechanisms (Lydic R, Baghdoyan HA, eds.) CRC Press, New York, 1999, pp. 213–234.
6. de Lacalle S, Saper CB. The cholinergic system in the primate brain: Basal forebrain and pontine-tegmental cell groups. In: Handbook of Chemical Neuroanatomy. The Primate Nervous System, Part I (Bloom FE, Bjorklund A, Hokfelt T, eds.) Elsevier, New York, 1997, pp. 217–252.
7. Geula C, Mesulam MM. Cholinergic systems and related neuropathological predilection patterns in Alzheimer disease. In: Alzheimer Disease (Terry RD, Katzman R, Bick KL, eds.) Raven Press, New York, 1994, pp. 263–291.
8. Swaab DF. Neurobiology and neuropathology of the human hypothalamus. In: The Primate Nervous System, Part I. Handbook of Chemical Neuroanatomy, Vol. 13 (Bloom FE, Bjorklund A, Hokfelt T, eds.) Elsevier, New York, 1997, pp. 39-118.
9. Whitehouse PJ, Price DL, Struble RG, Clark AW, Coyle JT, Delong MR. Alzheimer's

disease and senile dementia: loss of neurons in the basal forebrain. Science 1982; **215**:1237–1239.

10. Gritti I, Mainville L, Mancia M, Jones B. GABAergic and other noncholinergic basal forebrain neurons, together with cholinergic neurons, project to the mesocortex and isocortex in the rat. J Comp Neurol 1997; **383**:163–177.

11. Zaborszky L. Afferent connections of the forebrain cholinergic projection neurons, with special reference to monoaminergic and peptidergic fibers. In: Central Cholinergic Synaptic Transmission (Frotscher M, Misgeld U, eds.) Birkhauser, Basel, 1989, pp. 12–32,

12. Heilman KH, Watson RT, Valenstein E. Neglect and related disorders. In: Clinical Neuropsychology (Heilman KM, Valenstein E, eds.) Oxford University Press, New York, 1993, pp. 279–336.

13. Mesulam MM. Attentional networks, confusional states, and neglect syndromes. In: Principles of Behavioral and Cognitive Neurology (Mesulam, M.M., ed.), Oxford University Press, New York, 2000, pp.174–293.

14. McGaughy J, Everitt BJ, Robbins TW, Sarter M. The role of cortical cholinergic afferent projections in cognition: impact of new selective immunotoxins. Behav Brain Res 2000; **115**:251–263.

15. Szentagothai J. The modular architectonic principle of neural centers. Rev Physiol Biochem Pharmacol 1983; **98**:11–61.

16. Gerfen CR. The neostriatal mosaic. I. Compartmental organization of projections from the striatum to the substantia nigra in the rat. J Comp Neurology 1985; **236**:454–476.

17. Bjaalie JG, Diggle PJ, Nikundiwe A, Karagulle T, Brodal P. Spatial segregation between populations of ponto-cerebellar neurons: statistical analysis of multivariate interactions. Anat Rec 1991; **231**:510–523.

18. Graybiel AM, Penney JB. Chemical architecture of the basal ganglia. In: The Primate Nervous System, Part III, Handbook of Chemical Neuroanatomy, Vol. 15 (Bloom FE, Bjorklund A, Hokfelt T, eds.) Elsevier, Amsterdam, 1999, pp.227–284.

19. Mountcastle VB. Perceptual Neuroscience. The Cerebral Cortex. Harvard University Press, Cambridge, MA, 1998.

20. Malach R. Dendritic sampling across processing streams in monkey striate cortex. J Comp Neurol 1992; **31**:303–312.

21. He S-Q, Dum RP, Strick PL. Topographic organization of corticospinal projections from the frontal lobe: motor areas on the lateral surface of the hemisphere. J Neurosci 1993; **13**:952–980.

22. Malach R. Cortical columns as devices for maximizing neuronal diversity. Trends Neurosci 1994; **17**:101–104.

23. Malmierca MS, Blackstad TW, Osen KK, Karagulle T, Molowny RL. The central nucleus of the inferior colliculus in rat: a Golgi and computer reconstruction study of neuronal and laminar structure. J Comp Neurol 1993; **333**:1–27.

24. Malmierca MS, Leergard TB, Bajo VM, Bjaalie JG, Merchan MA. Anatomic evidence of a three-dimensional mosaic pattern of tonotopic organization in the ventral complex of the lateral lemniscus in cat. J Neurosci 1998; **18**:10603–10618.

25. Leergaard BT, Alloway KD, Mutic JJ, Bjaalie JG. Three-dimensional topography of corticopontine projections from rat barrel cortex: correlations with corticostriatal organization. J Neurosci 2000; **20**:8474–8484.

26. Jacobs GA, Theunissen FE. Extraction of sensory parameters from a neural map by primary sensory interneurons. J Neurosci 2000; **20**:2934–2943.

27. Sofroniew MV, Eckenstein F, Thoenen H, Cuello AC. Topography of choline acetyltransferase-containing neurons in the forebrain of the rat. Neurosci Lett 1982; **33**:7–12.

28. Armstrong DM, Saper CB, Levey AI, Wainer BH, Terry RD. Distribution of cholinergic neurons in the rat brain demonstrated by immunohistochemical localization of choline acetyltransferase. J Comp Neurol 1983; **216**:53–68.
29. Mesulam MM, Mufson EJ, Wainer BH, Levey AI. Central cholinergic pathways in the rat: an overview based on an alternative nomenclature (Ch1-Ch6). Neuroscience 1983; **10**:1185–1201.
30. Rye DB, Wainer BH, Mesulam M-M, Mufson EJ, Saper CB. Cortical projections arising from the basal forebrain: a study of cholinergic and noncholinergic components combining retrograde tracing and immunohistochemical localization of choline acetyltransferase. Neuroscience 1984; **13**:627–643.
31. Amaral DG, Kurz J. An analysis of the origins of the cholinergic and noncholinergic septal projections to the hippocampal formation in the rat. J Comp Neurol 1985; **240**:37–59.
32. Carlsen J, Zaborszky L, Heimer L. Cholinergic projections from the basal forebrain to the basolateral amygdaloid complex: a combined retrograde fluorescent and immunohistochemical study. J Comp Neurol 1985; **234**:155–167.
33. Zaborszky L, Carlsen J, Brashear HR, Heimer L. Cholinergic and GABAergic afferents to the olfactory bulb in the rat with special emphasis on the projection neurons in the nucleus of the horizontal limb of the diagonal band. J Comp Neurol 1986; **243**:488–509.
34. Woolf NJ. Cholinergic system in mammalian brain and spinal cord. Progr Neurobiol 1991; **37**:475–524.
35. Wainer BH, Steininger TL, Roback JD, Burke-Watson MA, Mufson EJ, Kordower J. Ascending cholinergic pathways: functional organization and implications for disease models. Progr Brain Res 1993; 98:9–30.
36. Mesulam MM, Geula C. Nucleus basalis (Ch4) and cortical cholinergic innervation in the human brain: observations based on the distribution of acetylcholinesterase and choline acetyltransferase. J Comp Neurol 1988; **275**:216–240.
37. Baskerville KA, Chang HT, Herron P. Topography of cholinergic afferents from the nucleus basalis of Meynert to representational areas of sensorimotor cortices in the rat. J Comp Neurol 1993; **335**:552–562.
38. Zaborszky L, Duque A. Local synaptic connections of basal forebrain neurons. Behav Brain Res 2000; **15**:143–158.
39. Felleman DJ, Van Essen DC. Distributed hierarchical processing in the primate cerebral cortex. Cerebral Cortex 1991; **1**:1–47.
40. Alloway D, Crist J, Mutic JJ, Roy SA. Corticostriatal projections from rat barrel cortex have an anisotropic organization that correlates with vibrissal whisking behavior. J Neurosci 1999; **19**:10908–10922.
41. Nadasdy Z, Zaborszky L. Computational analysis of spatial organization of large scale neural networks. Anat Embryol 2001; **204**: 303–317.
42. Valverde F. Reticular formation of the pons and medulla oblongata. A Golgi study. J Comp Neurol 1961; **116**:71–99.
43. Leontovich TA, Zhukova GP. The specificity of the neuronal structure and topography of the reticular formation in the brain and spinal cord of carnivora. J Comp Neurol 1963; **121**:347–381.
44. Ramon-Moliner E, Nauta WJH. The isodendritic core of the brain stem. J Comp Neurol 1966; **126**:311–336.
45. Das GD, Kreutzberg GW. Evaluation of interstitial nerve cells in the central nervous system. Adv Anat Embryol 1968; **41**:1–58.
46. Zaborszky L. Synaptic organization of basal forebrain cholinergic projection neurons. In: Neurotransmitter Interactions and Cognitive Functions (Levin E, Decker M, Butcher L, eds.) Birkhauser, Boston, 1992, pp. 27–65.

47. McMullen NT, Goldberger B, Suter CM, Glaser EM. Neonatal deafening alters nonpyramidal dendrite orientation in auditory cortex: a computer microscope study in the rabbit. J Comp Neurol 1988; **267**:92–106.
48. Csillik B, Rakic P, Knyihar-Csillik E. Peptidergic innervation and the nicotinic acetylcholine receptor in the primate basal nucleus. Eur J Neurosci 1998; **10:**573–585.
49. Braitenberg V, Schutz A. Cortex: Statistics and Geometry of Neuronal Connectivity. Springer, Berlin, 1998.
50. Szentagothai J. "Specificity versus (quasi-) randomness" revisited. Acta Morph Hung 1990; **38**:159–167.
51. Zaborszky L, Cullinan WE, Luine VN. Catecholaminergic-cholinergic interaction in the basal forebrain. Prog Brain Res 1993; **98:**31–49.
52. Zaborszky L, Cullinan WE, Braun A. Afferents to basal forebrain cholinergic projection neurons: an update. In: Basal Forebrain: Anatomy to Function (Napier TC, Kaliwas PW, Hanin I, eds.) Plenum Press, New York, 1991, pp. 43-100.
53. Zaborszky L, Gaykema RP, Swanson DJ, Cullinan WE. Cortical input to the basal forebrain. Neuroscience 1997; **79**:1051–1078.
54. Pang K, Tepper JM, Zaborszky L. Morphological and electrophysiological characteristics of non-cholinergic basal forebrain neurons. J Comp Neurol 1998; **394**:186–204.
55. Duque A, Balatoni B, Detari L, Zaborszky L. EEG correlation of the discharge properties of identified neurons in the basal forebrain. J Neurophysiol 2000; **84**:1627–1635.
56. Schwaber JS, Rogers WT, Satoh K, Fibiger HC. Distribution and organization of cholinergic neurons in the rat forebrain demonstrated by computer-aided data acquisition and three-dimensional reconstruction. J Comp Neurol 1987; **263:**309–325.
57. Cullinan WE, Zaborszky L. Organization of ascending hypothalamic projections to the rostral forebrain with special reference to the innervation of cholinergic projection neurons. J Comp Neurol 1991; **306**:631–667.
58. Gaykema RPA, Zaborszky L. Direct catecholaminergic-cholinergic interactions in the basal forebrain: II. Substantia nigra and ventral tegmental area projections to cholinergic neurons. J Comp Neurol. 1996; **374**:555–577.
59. Glaser JR, Glaser EM. Neuron imaging with Neurolucida—a PC-based system for image combining microscopy. Comput Med Imaging Graph 1990; **14**:307–317.
60. Vassbo K, Nicotra G, Wiberg M, Bjaalie JG. Monkey somatosensory cerebrocerebellar pathways: Uneven densities of corticopontine neurons in different body representations of areas 3b, 1, and 2. J Comp Neurol 1999; **406**:109–128
61. Heimer L, Zaborszky L. (eds.) Neuroanatomical Tract-Tracing Methods 2. Recent Progress. Plenum Press, New York, 1989.

10
Architecture of Sensory Map Transformations
Axonal Tracing in Combination with 3D Reconstruction, Geometric Modeling, and Quantitative Analyses

Trygve B. Leergaard and Jan G. Bjaalie

ABSTRACT

In this chapter, we review recent investigations of sensory-related brain stem maps in the rat cerebro-cerebellar and cat auditory systems. Sensitive axonal tracing techniques were used to identify specific components of these neural systems, such as the distribution of pontine terminal fields originating in primary somatosensory cortex, and the connectivity between different auditory brain stem nuclei. Distribution of labeled axonal plexuses and cell bodies, and the outlines of brain regions and nuclei, were recorded with the use of image-combining computerized microscopy and reconstructed in three dimensions. Local coordinate systems were established to allow comparison of data from different experiments and to facilitate data sharing in neuroinformatics databases. The distribution patterns were investigated with geometric modeling of the clustered patterns of labeling, density gradient analysis, and analysis of spatial overlap of terminal fields in dual tracing experiments. Further, series of slices through the 3D reconstructions (sections at chosen angles of orientation) were used for a more detailed analysis. These computerized methods allowed us to discover new principles of organization pertaining to sensory map transformations in somatosensory cerebro-cerebellar pathways and ascending auditory pathways.

10.1. INTRODUCTION

The functions of a given brain region are, to a high degree, determined by the architecture of its afferent and efferent connections *(1–6)*. The major pathways of the mammalian brain are characterized by orderly spatial relationships. This orderly arrangement of afferent and efferent projections among components of the nervous system is commonly referred to as topographic organization, as opposed to randomized or chaotic organization *(7–11)*. In the afferent pathways, the orderly representations of the body surface, the visual field or the tonal frequencies, are generally well preserved. Thus, the mapping of the sensory periphery onto receiving regions of the central nervous system is often described as continuous, with an overall point-to-point connectivity.

From: *Computational Neuroanatomy: Principles and Methods*
Edited by: G. A. Ascoli © Humana Press Inc., Totowa, NJ

Some pathways of the brain, sensory-related and others, contain more complex interdigitating or fractured discontinuous patterns *(12–16)*.

To explore topographic patterns at the level of sensory-related brain stem nuclei, we have used sensitive axonal tracing techniques in combination with electrophysiological characterization and computerized 3D reconstruction. In this chapter, we will outline basic data acquisition procedures used for digitizing brain regions and the distribution of retrogradely labeled cells and anterogradely labeled terminal fields of axons *(17)*. Further, the methods for 3D reconstruction from digitized sections will be discussed together with new approaches for data presentation in local coordinate systems, which is important in the context of data sharing and comparison of results from different experiments and different laboratories *(18)*. Finally, we will describe various steps of analysis, such as slicing of the reconstructions, surface modeling of labeled structures, density gradient analysis, stereoimaging, and spatial overlap analysis. We exemplify the use of these computerized methods for the study of two major brain pathways: the somatosensory cerebro-cerebellar system and the ascending auditory pathways. In the cerebro-cerebellar system, we focus on transformations from cortical 2D via brain stem 3D to cerebellar 2D representations *(19,20)*. In the auditory pathways, we study changes from straightforward (point-to-point) to more complicated, interdigitating patterns of tonotopic organization *(21,22)*.

10.2. MAP TRANSFORMATIONS IN CEREBRO-CEREBELLAR AND AUDITORY SYSTEMS

A variety of anatomic, histologic, immunocytochemical, electrophysiologal, and tomographic techniques are used to map various levels of organization in the brain *(23,24)*. In brain pathways, knowledge about topographical order is essential for understanding basic structural and functional organization. A map of a sensory surface in a given cortical area may be reproduced or modified in other regions of the brain receiving inputs from this cortical area. Similarly, projections from a brain stem nucleus to other brain structures may preserve spatial relationships or introduce new neighboring relationships among components represented in the projection. Such map transformations are believed to reflect different computational properties of brain regions *(3,14,16)*.

Two examples of projection systems will be dealt with here: (*i*) the cerebro-pontine projection, which is the first link in one of the largest pathways in the brain, connecting the cerebral cortex with the cerebellum (for review, see *[25]*); and (*ii*) the pathways through the lemniscal nuclei, which is one of the least understood parts of the auditory system (for review, see *[15]*). The data shown will exemplify complicated 3D patterns of axonal labeling originating within functionally well-defined somatosensory and auditory brain maps. The labeling patterns preserve topographical order, but introduce various degrees of new neighboring relationships.

Cerebro-cerebellar projections originate in large parts of the cerebral cortex and reach almost all regions of the cerebellum (for reviews, see *[25–28]*). The projections are synaptically interrupted in the pontine nuclei. The somatosensory representations in the cerebro-cerebellar pathway have been outlined in detail in both source (cerebral cortex) and target regions (the cerebellar cortex). Both regions receive primary somatosensory information. Nevertheless, the patterns of topographic organization are different. The primary somatosensory cortex (SI) contains a relatively continuous map of the body surface *(7,9)*, (for review, see *[29]*). By contrast, the tactile responses in the cerebellar granule cell layer form a highly discontinuous, or fractured, map *(12,13,30)* (for review, see *[31]*). From physiological data, it is evident that the tactile-related cerebro-cerebellar circuit exhibits precise projection patterns *(12)*. We have studied the structural nature of the transformation from the continuous cerebral to the fractured cerebellar map *(19,20)*. The basic methods and results are reviewed below.

The nuclei of the lateral lemniscus are intercalated in the ascending auditory pathways. The dorsal division is organized in a lamellar fashion, whereas the ventral part of these nuclei has been reported to contain a complex, widespread, and patchy organization (for review, see *[15]*), analogous to the patchy mosaic organization previously reported in the pontine nuclei (for review, see *[27]*). While tonotopic order is maintained by point-to-point (frequency-specific) connections between most central nervous components of the auditory system (for review, see *[32]*), such principles have not been found in the ventral part of the lemniscal nuclei. Complex tonotopic arrangements have been demonstrated in rats *(33)*, and bats *(34,35)*. In cats, results from physiological and anatomic investigations are incongruent and have failed to demonstrate clear-cut tonotopic order *(36–42)*. Thus, the question whether there is a frequency-specific organization within this region of the brain has remained open. One possibility could be that the presence of a complex clustered organization has hampered the understanding of tonotopy in this particular auditory nucleus. Alternatively, structural order might be absent or brought about by some other modality than frequency specificity. We employed the same methodological tools as used in our investigations of the cerebro-cerebellar system to search for topographic organization in the nuclei of the lateral lemniscus (21,22).

10.3. NEURAL TRACING TECHNIQUES

The anatomic organization of connections between brain regions, and between the periphery and the central nervous system, are currently studied with increasingly sensitive tracing techniques. These techniques allow a high level of precision and detail in the mapping of topographical patterns *(24,43–46)* (for a historical review, see *[47]*). In our investigations of the cerebro-cerebellar *(19,20)* and auditory systems *(21,22)*, we have used the neural tracers wheat germ agglutinin-horsesradish peroxidase (WGA-HRP; *[48]*), *Phaseolus vulgaris*-leucoagglutinin (PHA-L; *[49]*), biotinylated dextran amine (BDA; *[50,51]*), and rhodamine conjugated dextran amine (FluoroRuby, FR; *[52]*) to identify the target regions of particular populations of projection neurons (anterograde tracing; Fig. 1). The tracers were applied either in small amounts, in injection sites with a diameter of 200 – 400 µm *(20)*, in larger amounts *(19,21,22)* (Fig. 1), or with the use of multiple repeated injections into the same functionally defined region *(19)*. The equally sensitive tracers BDA and FR were used together in a dual-tracer approach to investigate patterns of segregation and overlap in the projections to the pontine nuclei from neighboring cerebral locations *(20)*. In the auditory system we took advantage of the bidirectional axonal transport of dextran amines: BDA first labeled the cells of origin of afferent projections to the injection site in the inferior colliculus (retrograde tracing) and was then transported anterogradely from these cells

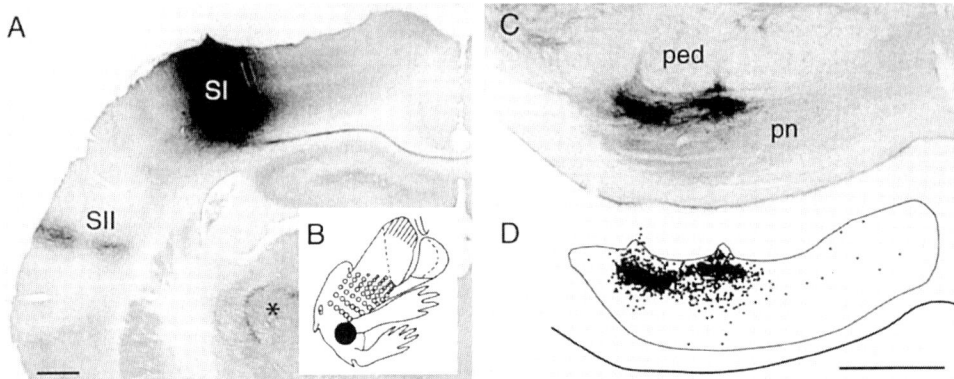

Fig. 1. Anterograde axonal tracing of pontine projections from SI in rats (modified from [19]). (A) Photomicrograph of a frontal section through the center of a BDA injection site placed in the right SI trunk representation under electrophysiological guidance. A bundle of labeled callosal fibers emerge from the injection site. Labeling is also visible in the secondary somatosensory cortex, SII, and thalamus (asterisk). (B) Line drawing of the SI somatotopic map (modified with permission from [9]). The black dot indicates the position and size of the injection site. (C) Photomicrograph of a transverse section through the midpontine level in the same animal, showing two dense plexuses of BDA-labeled fibers dorsolaterally in the right pontine nuclei, close to the descending peduncle. (D) Computerized plot of the same section as shown in panel C. The thick contour line represents the ventral surface of the pons and the thin contour lines the boundaries of the gray matter. Dots represent the distribution of labeled axons in the pontine nuclei. Bars, 500 µm. SI, primary somatosensory cortex; SII, secondary somatosensory cortex; ped, peduncle; pn, pontine nuclei.

to label collateral axons. In this way, we could label cells in the lemniscal nuclei that project to the inferior colliculus, as well as the collateral projections to the lemniscal nuclei from other sources of input to the inferior colliculus (21,22).

While these and other available techniques offer new and unique possibilities for outlining detailed topography, problems related to presentation, and efficient use of data remain to be solved. In this context, precise data acquisition, collection of data from complete series of sections, and full 3D reconstruction, represent important steps towards an improved understanding of structure–function relationships.

10.4. IMAGE-COMBINING MICROSCOPY FOR DATA ACQUISITION

Data entry for the studies reviewed here was made with an image-combining computerized microscope. With this method, data from histological sections, including labeling patterns, were coded as lines and points. The principle of image-combining computerized microscopy was first introduced by Glaser and van der Loos (53,54) and has later been used by numerous investigators (for a review of anatomical data acquisition methods, see [55]). This system mixes a computer graphical image of digitized structures with the image of the specimen (Fig. 2). Movement of the microscope stage (controlled by stepping motors via the computer) is accompanied by a translation of the graphical image. The user thus obtains direct feedback during the data entry procedure. The system is optimized for high-resolution recording of numerous x, y (or x, y, and z)

Sensory Map Transformations   203

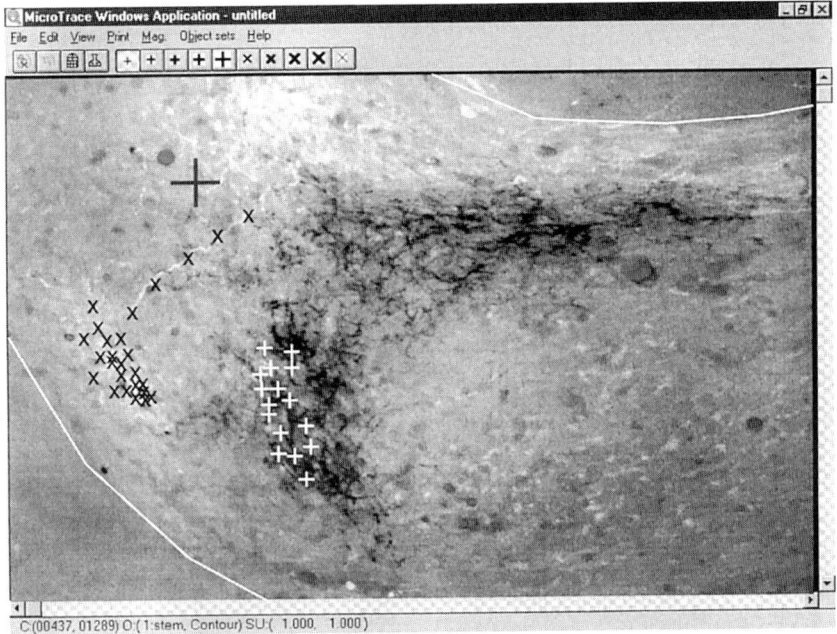

Fig. 2. The graphical user interface of the data acquisition program MicroTrace: field of view in the image-combining computer microscope. The specimen is a transverse section through the right pontine nuclei viewed with fluorescence microscopy (through the Leitz N2.1 rhodamine filter block). Axonal plexuses were anterogradely labeled after injection of rhodamine-conjugated dextran amine and biotinylated dextran amine in electrophysiologically defined individual whisker representations in SI (data from [20]). The borders of the pontine gray substance are digitized as lines, and two categories of symbols are placed above parts of the labeled regions. The large cross illustrates the computer screen cursor.

coordinates across large brain regions. The precision of the data recording depends on the quality of the microscope optics, the stepping motors, and the resolution of the graphic feedback image (54). With our technical configuration (17), recording of positions over short distances, close to the center of the field of view, were performed without any detectable error (other than those imposed by the pixel size of the screen). With a 25× objective lens, we recorded the length of a 2 mm micrometer with a precision of ±2 µm.

In our investigations of the cerebro-pontine system, we digitized contour lines for a number of structures (the ventral surface of the pons, the outlines of the pontine gray, the contours of the corticobulbar and corticospinal fiber tracts, the midline of the brain, and the outline of the fourth ventricle) as a reference for the alignment of the sections. The density and distribution of anterogradely labeled axonal plexuses within the pontine nuclei was digitized semiquantitatively as points (19,20,56). In areas with low density of labeling, point coordinates were placed at regular intervals along the length of single axons. In areas with dense labeling, a rough correspondence was sought between the density of labeling and the number of digitized points (Fig. 2). This approach allowed subsequent analyses of density gradients and overlap patterns.

A simplified data entry approach was used for the auditory system investigations *(21,22)*. Here, series of detailed camera lucida drawings were digitized using essentially the same approach as outlined above. The external lemniscal nuclear borders were digitized as lines. Labeled neurons were recorded as single points, whereas the sharply defined patches of labeling within the lemniscal nuclei were surrounded with contour lines.

10.5. 3D RECONSTRUCTION

The precision of a 3D reconstruction from serial sections depends upon the section quality, the data acquisition method, and the procedure for section alignment. It is essential to minimize section distortions and to obtain complete series of sections. In our experiments, brains were fixed by paraformaldehyde perfusion, cryoprotected with buffered sucrose, and sectioned at 50 µm on a freezing microtome. To minimize distortions, microtomy was performed carefully at stable temperatures, and cerebellar tissue *(18)* was embedded in gelatin prior to sectioning. Series of sections that were incomplete or contained damaged sections were not used for reconstruction.

The digitized sections from our investigations were imported into our Open Inventor-based application for 3D reconstruction, visualization, and analyses of brain regions and neuronal labeling patterns. This application and its predecessors have been used in more than 20 publications (recent examples include, *[18–22,56–61]*). The current version (Micro3D 2001) runs on Silicon Graphics workstations and PCs equipped with Red Hat Linux and Open Inventor for Linux (TGS Inc., San Diego, CA) and is made available through the Oslo Research Park (see also http://www.nesys.uio.no/).

With the use of our application, the digitized sections were aligned interactively on the computer screen, using multiple anatomic landmarks, and real time rotation of the reconstruction during alignment (Fig. 3; for an impression of real time rotation, see e.g., movie sequence 1 on the accompanying CD-ROM). Sections were assigned z values defined by section thickness and serial numbers, before they were maneuvered in position, using a handlebox and specific sliders for translocation and rotation of sections, available in the program Micro3D (Fig. 3A). Examples of anatomic landmarks used for alignment of sections through the pontine nuclei, include the ventral surface of the pons, the outlines of the pontine gray, the contours of the corticobulbar and corticospinal fiber tract, the midline and the floor of the fourth ventricle. To further aid the inspection of the 3D reconstructions from various angles of view, sections from control brains cut in section planes orthogonal to the sections from the experimental animals were digitized and used as templates for alignment. To ensure that the reconstructions retained natural proportions, sections that had been submitted to histologic or immuncytochemical processing were measured in the xy plane, and linear size adjustments were introduced in the final reconstruction to maintain correct in vivo proportions. Below, we exemplify some of the features of the Micro3D application in the context of our investigations of the cerebro-pontine and lemniscal nuclei systems.

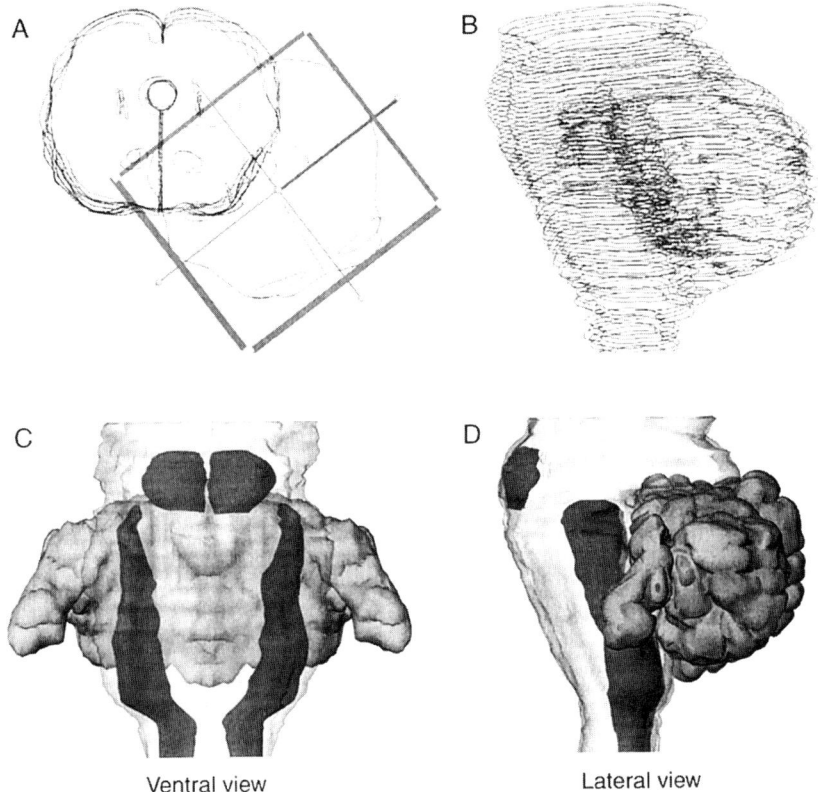

Fig. 3. Assembly and visualization of a 3D reconstruction of the rat brain stem and cerebellum (modified from *[18]*). **(A)** Series of digitized transverse brain stem sections are aligned according to multiple anatomical landmarks. **(B)** Oblique lateral view of the complete aligned 3D reconstruction. **(C and D)** Ventral and lateral views of the complete 3D reconstruction. The outer surface of the brain stem is represented as a transparent surface, and the outer boundaries of the cerebellum, pontine nuclei, and trigeminal sensory nuclei are represented as solid surfaces. The reconstruction of the brain stem is based on a series of transverse sections and is combined with a series of sagittal sections (cf. *[18]*). The left half of the cerebellar reconstruction is a mirror copy of the right half. The program Micro3D was used to visualize the individual components of the reconstruction separately or in different combinations. A movie of a 360° rotation sequence of this reconstruction is available in the companion CD-ROM (sequence 1).

10.6. LOCAL COORDINATE SYSTEMS FOR INDIVIDUAL BRAIN STEM NUCLEI

10.6.1. Comparison of Results

To make efficient use of data from individual neural tracing experiments, it is necessary to compare results from different experiments. 3D computerized reconstructions are well suited for direct comparison by superimposing data from multiple experiments into a common 3D space. On the technical side, comparisons of results are typically hampered by variation in the resolution of drawings or photomicrographs, the plane of

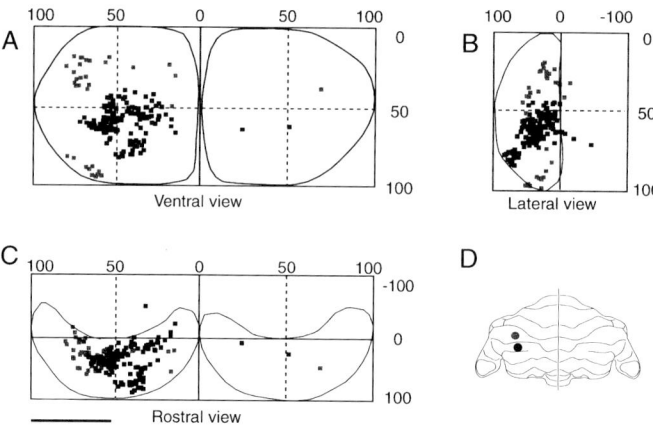
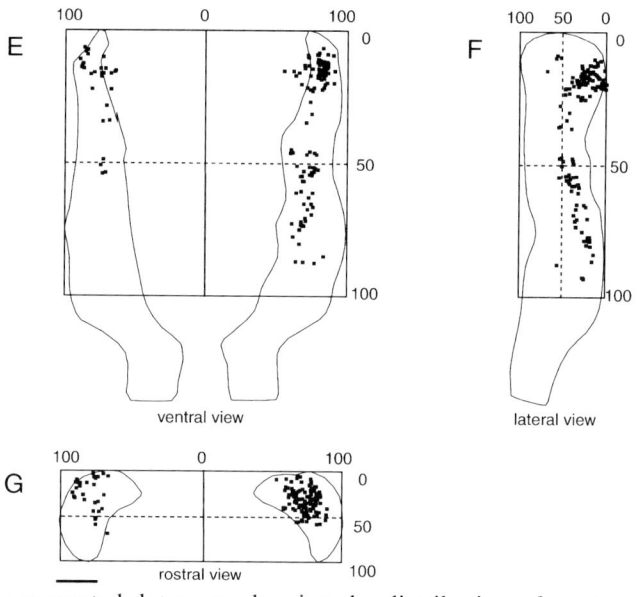

Fig. 4. Computer-generated dot maps showing the distribution of pontocerebellar (**A–C**) and trigeminocerebellar *(E–G)* projection neurons labeled after tracer implantation into crusI and IIa (**D**) of the rat cerebellum (modified from *[18]*). Each dot represents one labeled neuron. The diagrams show the internal coordinate systems for the pontine and the trigeminal nuclei from three angles of view. Coordinate systems of relative values from 0 to 100% are used. The halfway (50%) reference lines are shown as dotted lines. Curved solid lines represent nuclear boundaries. Black dots represent pontine neurons labeled by implantation of FR into the crown of crusIIa, and gray dots represent pontine neurons labeled by implantation of FE into the crown of crusI. The size and location of the implantation sites are shown in the line drawing in panel **D**. Movies showing rotation sequences of these reconstructions are available on the accompanying CD-ROM (sequences 2 and 3). Pontocerebellar projection neurons labeled by tracer implantations in the left crusIIa are located centrally in the right pontine nuclei, whereas neurons labeled by tracer implantation in crusI are located more rostrally and caudally. Trigeminocerebellar projection neurons labeled by tracer into the crown of the left crusIIa are distributed in two main clusters located dorsally in the left (ipsilateral) trigeminal sensory nuclei. No trigeminal neurons were labeled after tracer implantation in crusI. Bars, 1 mm.

sectioning, the use of different section spacing, and dissimilar techniques for data documentation. These problems are common to most anatomical brain mapping investigations *(24,47,62–66)*. For our investigations of the rat pontine nuclei, we have therefore implemented an internal coordinate system suitable for detailed analysis of experimental labeling patterns and presentation of neural tracing data in standardized diagrams (Fig. 4A–C) *(19)*. Internal coordinate systems facilitate the comparison of experimental data from different experiments and allow positional coordinates to be translated to standard atlas coordinates (e.g., the stereotaxic atlas of Paxinos and Watson *[67]*). We are currently implementing internal coordinate systems for several brain stem regions of the rat (Fig. 4E,F) *(18)*.

The origin of the pontine coordinate system was defined as the crossing of the midline and a line tangential to the rostral end of the pontine nuclei. A rectangular frame of reference was introduced, oriented perpendicular to the long axis of the brain stem. Planes, tangential to the pontine gray, defined the caudal and lateral borders of the pontine nuclei. To avoid the interference of inter-individual size variability, relative coordinates (0–100%) were introduced (for details, cf. *[18]*). Based on this coordinate system, we made diagrams showing the outlines of the pontine nuclei in standard ventral, lateral, and dorsal angles of view. These diagrams facilitated data presentation (Figs. 4A-C, 5, and 6) and comparison of results from different animals *(19,20)*.

10.6.2. Databasing

The standardized data presentation scheme described above allows efficient communication of large quantities of experimental neuroanatomic data. By sharing original data files with the neuroscience community, quality control and alternative interpretations of data are more readily achieved *(68–70)*. We have made available original data and additional illustrations from two recent publications *(19,20)* in a neuroinformatics knowledge base [(http://www.nesys.uio.no/Database/), see also (http://www.cerebellum.org/)]. The aim was to explore the possibilities for Web-based publication of data, in addition to providing full documentation of our material. The original data were published simultaneously with the peer-reviewed journal articles *(19,20)*. Standardized illustrations and more advanced rotation sequences were made available together with the original data files (ASCII coordinate files). We believe that these data presentation formats might be developed further to serve database purposes. Examples of 3D rotation sequences, supplementing ordinary illustrations, are provided on the accompanying CD-ROM.

10.7. VISUALIZATION AND QUANTITATIVE ANALYSES OF THE DISTRIBUTION OF LABELED AXONS AND CELLS

10.7.1. Slicing of 3D Reconstructions

Traditional anatomic studies are based on observations made in serial sections. The plane of sectioning may, however, lead to incomplete interpretations of topographic organization. In addition to real time rotations of 3D reconstructions and diagrams showing the total projection patterns from several angles of view (Figs. 3 and 4), dynamic subdividing of the complete reconstruction into sections (here referred to as slices) of chosen thickness and orientation add a new dimension to the analysis of brain

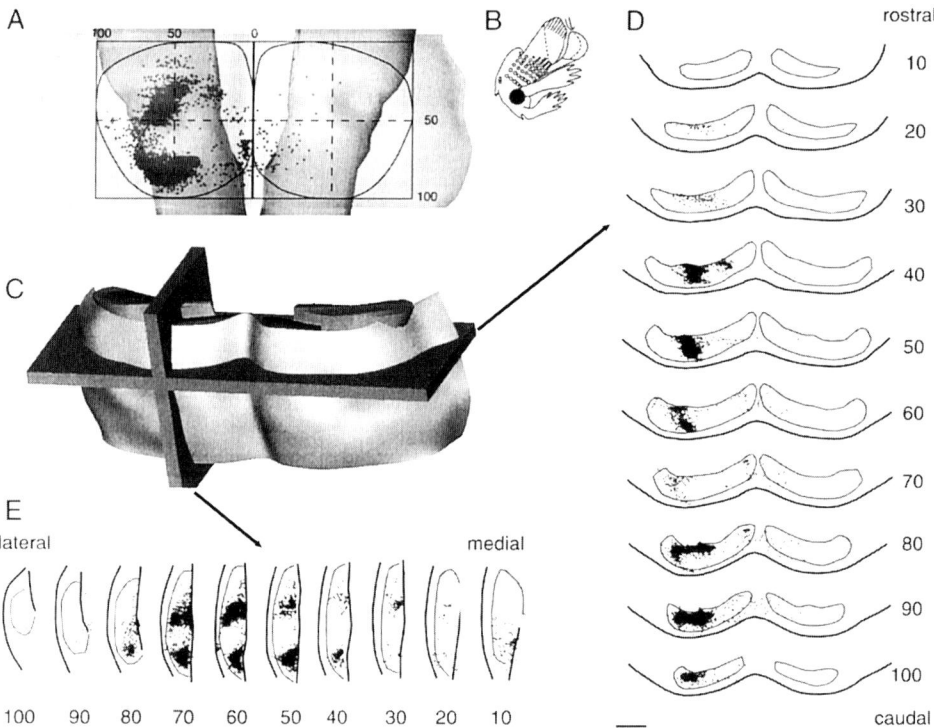

Fig. 5. Computer-generated 3D reconstruction of the rat pontine nuclei subdivided into consecutive series of sagittal and transverse slices (modified from *(19)*; see also Fig. 1), showing the distribution of pontine terminal fields after injection of BDA into the trunk representation of SI (**B**). (**A**) 3D reconstruction of the pontine nuclei, ventral view. The ventral surface of the pons is represented as a transparent surface, and the dorsally located descending peduncles are visible as solid surfaces. The dots represent the distribution of BDA-labeled fibers within the pontine nuclei. Presentation otherwise as in Figure 4A. The rectangular boxes in panel **C** illustrate 200-µm thick transverse and sagittal slices (shown in panels **D** and **E**, respectively) from the computer reconstruction. The numbers assigned to each slice refer to the internal pontine coordinate system. Depending on the chosen plane of sectioning, the shapes of the terminal fields appear different. Bar, 500 µm.

topography. Figure 5 shows two consecutive series of slices, transverse and sagittal, through a reconstruction of the pontine nuclei. In this experiment, an injection was placed into a limited part of SI. Depending on the plane of slicing, the distribution and shape of the labeling in the pontine nuclei appeared different. In the transverse section plane, which is used in most studies of the pontine nuclei *(71–73)* (for review, see *[25,27]*), the labeling appeared fairly continuous from rostral to caudal ("columnar"). By contrast, the sagittal series through the same reconstruction revealed two major separated zones of labeling, rostrally and caudally, in the pontine nuclei (compare Fig. 5D and E). Careful inspection of the reconstructions was, therefore, needed to fully understand the 3D distribution and shape of the labeling. In our investigations of the cerebro-pontine system we systematically compared slices of different orientation for

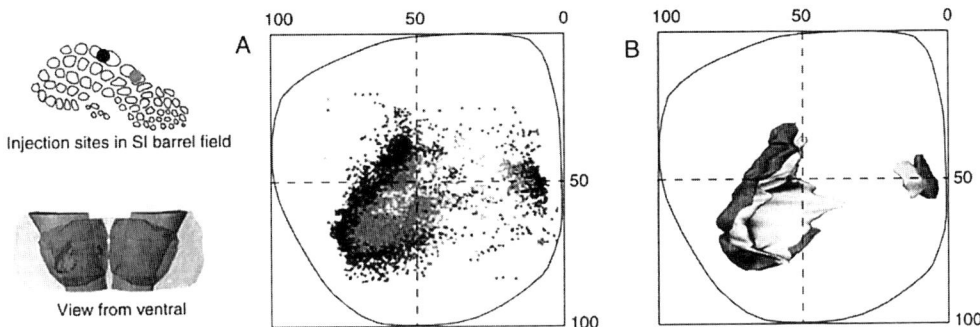

Fig. 6. 3D reconstruction showing the topography of pontine terminal fields arising in the rat SI whisker barrel field (modified from *[20]*). The anterograde tracers BDA (gray) and FR (black) were injected into electrophysiologically defined individual whisker representations in SI (shown in the upper left inset), and the distribution of labeling was computer-reconstructed in 3D (lower left inset). **(A)** Computer-generated dot map showing the distribution of BDA (gray)- and FR (black)-labeled fibers within the ipsilateral pontine nuclei. The clusters of black dots surround the clusters of gray dots externally. **(B)** The outer boundaries of labeled clusters are demonstrated by solid surfaces. The labeled clusters arising from the same row of SI barrels are located in dual lamellae that are shifted from internal to external. Movie sequences of this reconstruction are available on the accompanying CD-ROM. Sequence 4 corresponds to panel **A** (labeling rendered as dots), and sequence 5 corresponds to panel **B** (labeling shown as solid surfaces).

the detailed analyses of the distribution patterns and for inter-individual comparison *(19,20)* (see also http://www.nesys.uio.no/Database/).

10.7.2. Surface Modeling of Labeled Structures

Computerized surface modeling is an important tool for visualization of the size, shape, and extent of labeling patterns, as exemplified in our investigations using surface modeling based on series of manually digitized contour lines surrounding the zones of labeling *(19–22)*. Three different approaches for surface modeling were used. The surface modeling of the brain stem and cerebellum in Figure 3 was performed with a simple triangulation method, as outlined by Toga *(74)*. The surfaces of the brain stem nuclei in Figures 3, 5, and 6 were resynthesized from contour lines using the software library SISL (Sintef Spline Library). The latter surfaces are piece-wise polynominal (spline) approximations to the given data *(75)*. The surfaces of Figures 7 and 8 were modeled using the program Nuages (Prisme, INRIA, *[76]*).

In the rat pontine nuclei, surfaces were used to demonstrate the shape and distribution of the terminal fields (anterogradely labeled) originating in electrophysiologically defined parts of SI *(19,20)* (Fig. 6). To determine the localization of the contour lines used for surface modeling, density gradient analysis (see below) was used to exclude regions containing the lowest density levels of labeling. With this approach, we demonstrated that the terminal fields of fibers were distributed in an inside-out shell-like fashion, and that neighboring relationships in source (SI) were largely preserved in the target region (pontine nuclei) *(19)* (Fig. 6; these findings are illustrated to advantage in

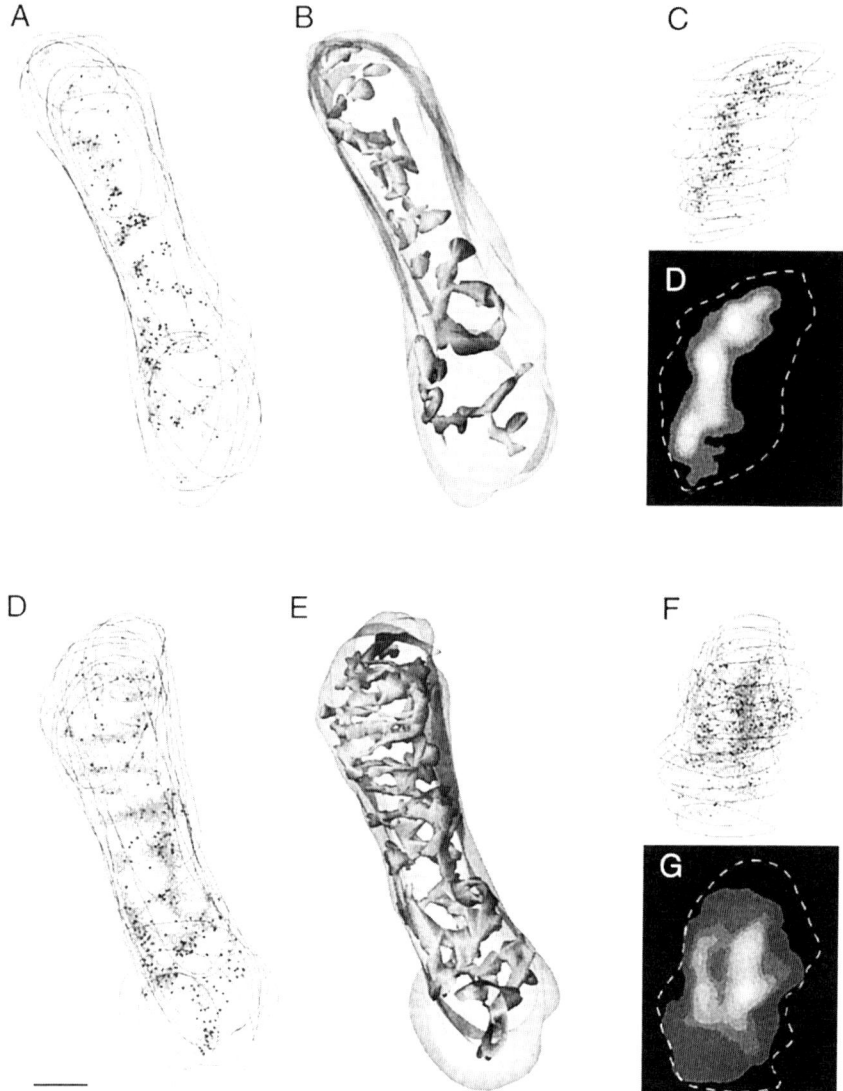

Fig. 7. Computerized 3D reconstructions of the ventral part of the lemniscal nuclei in cat, showing the tonotopic distribution of axonal clusters and neurons labeled after injections of BDA into high-frequency (**A–D**) or low-frequency (**D–G**) areas of the inferior colliculus (modified from [21]). (**A,C,D,F**) Computer-generated 3D dot maps showing the distribution of labeled fibers and cells in a view from caudal (**A** and **D**) and dorsal (**C** and **F**). Labeled fibers are shown as gray dots, retrogradely labeled cells as black (larger) dots, and the external surface of this nuclear region as contour lines. (**B,E**) The external nuclear boundary is represented as a transparent surface, and the outer boundaries of the clusters containing labeling are visualized as solid surfaces. (**D,G**) Density gradient maps of labeling projected along the long axis of the lemniscal nuclei (dorsal view). The gray scale gradient shows the highest densities in white and the lowest in dark gray. Densities lower than 5% of the maximum value are not shown. The maps reveal a tendency for a banded distribution, and the majority of high-frequency representations are located laterally, whereas the low-frequency representations are located more medially. Scale bar, 500 μm.

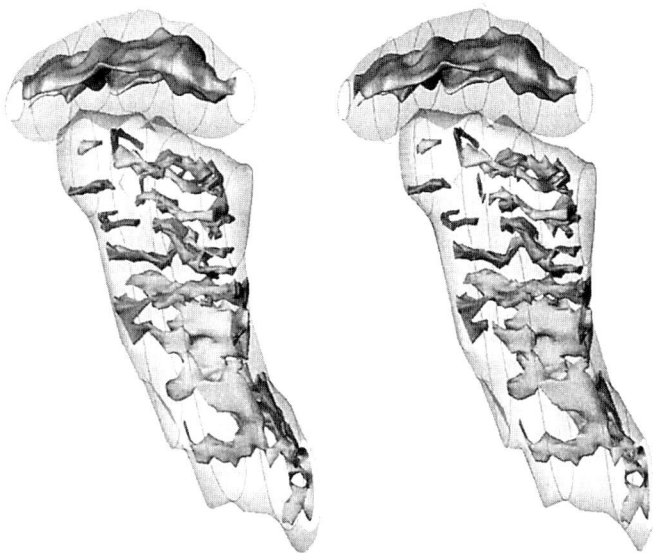

Fig. 8. Computer-generated stereo pairs showing the 3D distribution of low-frequency representations within the cat lemniscal nuclei in a view from lateral (data from *[21,22]*). The external boundaries of the nuclei are represented as transparent surfaces and labeling as solid surfaces. In the dorsal division (top), labeling is distributed in a continuous lamina. In the ventral division (bottom), labeling is distributed in multiple clusters of various shapes. Note differences in shape of clusters from dorsal to ventral (top to bottom). To see a 3D image, the viewer must cross the eye axis to let the pair of images merge.

sequence 4 and 5 on the accompanying CD-ROM). We also demonstrated spatial order at the level of projections from individual whisker barrels *(20)*. In the cat lemniscal nuclei, surfaces were used to demonstrate the shape and distribution of the regions containing terminal fields of axons (Figs. 7 and 8), following tracer injection into frequency-specific parts of the inferior colliculus *(21,22)*. In the ventral part of these nuclei, we demonstrated complicated 3D mosaic patterns, presumably representing discontinuous frequency band compartments *(21)*, as a counterpart to the continuous laminar compartments found in remaining auditory nuclei *(15,22)*. Surface modeling of tracer-labeled internal fiber plexuses was useful for showing the 3D shape of the sharply defined frequency band laminae in the inferior colliculus *(57)*. The outer boundaries of these sharply defined clusters of labeling in the inferior colliculus were determined automatically, using a procedure based on triangulation of the convex hull of the point data set (representing the labeling), followed by removal of triangles along the boundary *(75)*.

10.7.3. Density Gradient Analysis

The surface modeling approach outlined above demonstrates overall distribution and shape of individual clusters of labeling. Other approaches are required to extract more quantitative information about the labeling. Stereological methods are efficient for estimating the numbers of labeled structure within defined regions *(77,78)*. Nadasdy and Zaborszky *(79)* used density values to selectively visualize subsets of 3D reconstructed neuronal populations, either by representing isodensity regions with a marker or by rendering surfaces around isodensity volumes. A less sophisticated but useful tool is density gradient analysis as employed in our investigations of pontine nuclei and lemniscal nuclei organization *(19,21)*. This analysis is based on the 2D "collapsed" projection of a 3D point data set (representing labeling). The analysis may be repeated for different 2D projections (angles of view). A square grid was superimposed on the 2D map, and each square was assigned a gray or color level corresponding to the density of points within a user-defined radius centered on the square. Thereby, a gray scale (or color)-coded density map is constructed (Fig. 7). With the use of small grid size and short radius, it is possible to demonstrate changes in densities across short distances. An example of the use of such density gradient analysis is shown in Figure 7D,G. By selectively visualizing regions of high and low density, we were able to recognize a horizontal frequency gradient in the ventral part of the lemniscal nuclei, not previously identified *(21)*. We also used this method to demonstrate the nonuniform distribution of labeling in the complete SI projection to the pontine nuclei *(19)* and to compare the average distribution of retrogradely labeled cerebro-pontine neurons among cytoarchitectonically defined areas in SI cortex of the monkey *(61)*.

10.7.4. Stereoimaging

The traditional journal format is not suited for visualization of 3D reconstructions. Perception of depth is, however, offered by stereoimages. The stereoscopic effect is mimicked by image pairs that have approximately 8° different vertical rotation. To see a 3D image, the viewer must cross the eye axis to let the pair of images merge. Figure 8 demonstrates the 3D spatial organization of the clustered high and low-frequency representations within the cat lemniscal nuclei *(21,22)*. Stereoscopic illustrations are readily made from 3D reconstructions and have been helpful tools in several investigations *(18,20,57,80,81)*.

10.7.5. Analysis of Spatial Overlap

Single axonal tracing is useful for the mapping of large-scale topographic patterns, as exemplified in our studies of the overall projections from SI to the pontine nuclei *(19)* and our investigations of the lemniscal nuclei system *(21,22)*. High-precision investigation of segregated and overlapping projection patterns requires a dual tracing approach, with two equally efficient axonal tracers producing clearly different labeling products in the same sections. We used BDA in combination with FR to outline the detailed spatial relationships between projections to the pontine nuclei originating in different SI whisker barrels *(20)* (see also *[82,83]*). Spatial overlap and segregation of the differently labeled axonal clusters was readily seen in the sections (Fig. 3) and in the dot map representations (Fig. 6). Estimates of the amount of overlap were based on the data files (with point coordinates representing the distribution and density of terminal fields of labeling) and were performed by subdividing each section into an array of bins of a chosen size. The numbers of digitized points were counted for each bin, summed across sections, and used to estimate an index of total overlap. Overlap was defined as co-location of points of both categories in one bin. The overlap estimate was highly influenced by bin size and threshold criteria. If small bins (e.g., 5 μm^2) and high thresholds (e.g., >10 points per bin) were used, the amount of overlap is minimal, whereas by comparison, large bin and low thresholds resulted in higher overlap indexes. Similar solutions were used previously by others *(83–85)*.

Analysis of spatial overlap of terminal fields in the pontine nuclei, labeled after injection of two sensitive axonal tracers into individual whisker representations in SI, revealed a statistically significant decrease of overlap index with increasing distance separating the two injection sites *(20)* (see also *[82]*). The use of high magnification plotting of individual fibers of labeling was important for revealing overlapping projections, which may not have been detected with more automated image acquisition followed by segmentation of terminal fields of labeling.

10.8. CONCLUSIONS

We have developed customized computerized tools for reconstruction, visualization, and analysis of neuronal populations and terminal fields of fibers within defined brain regions. We have used these tools to investigate complex 3D distribution patterns and to provide new insights into map transformations in sensory-related brain stem nuclei. By combined use of neuronal tract-tracing and 3D reconstruction techniques, we have demonstrated topographic distribution gradients in the cerebro-pontine projection. Clustered terminal fields are arranged inside-out within lamellar subspaces in the pontine nuclei, and linear shifts in the location of cortical sites of origin correspond to predictable smooth shifts in the location of terminal fields. Employing a similar experimental and analytic approach, we demonstrated a novel mosaic topography in the cat VCLL. Thus, the cat VCLL contains a complex tonotopic organization with frequency-specific representations distributed with a clustered mosaic pattern.

While the essentially 2D cortical maps are possible to investigate as surface maps *(8–10,12,13)*, a 3D reconstruction approach has been a prerequisite for our anatomic investigations of 3D brain stem maps. The computerized approach introduces new flexibility in the analysis of data, and facilitates more efficient use (and reuse) of data collected with high-resolution axonal tracing technology. It is our belief that the combination of experimental anatomical methods, computerized data acquisition and analysis, and structural modeling of the ensuing results will open new avenues for studying structure–function relationships in the brain.

ACKNOWLEDGMENTS

Financial support was provided by European Community Grant Bio4 CT98-0182, the Research Council of Norway, and the Jahre Foundation.

REFERENCES

1. Leise EM. Modular construction of nervous systems: a basic principle of design for invertebrates and vertebrates. Brain Res Rev 1990; **15**:1–23.
2. Nelson ME, Bower JM. Brain maps and parallel computers. Trends Neurosci 1990; **13**:403–408.
3. Brown LL. Somatotopic organization in rat striatum: evidence for a combinational map. Proc Natl Acad Sci USA 1992; **89**:7403–7407.
4. Katz PS. Neurons, networks, and motor behavior. Neuron 1996; **16**:245–253.

5. Kaas JH. Topographic maps are fundamental to sensory processing. Brain Res Bull 1997; **44**:107–112.
6. Mountcastle VB. The columnar organization of the neocortex. Brain 1997; **120**:701–722.
7. Woolsey CN. Organization of somatic sensory and motor areas of the cerebral cortex. In: Biological and Biochemical Bases of Behavior (Harlow HF, Woolsey CN, eds.). University of Wisconsin Press, Madison, 1958, pp. 63–81.
8. Woolsey TA, Van der Loos H. The structural organization of layer IV in the somatosensory region (SI) of mouse cerebral cortex. The description of a cortical field composed of discrete cytoarchitectonic units. Brain Res 1970; **17**:205–242.
9. Welker C. Microelectrode delineation of fine grain somatotopic organization of (SmI) cerebral neocortex in albino rat. Brain Res 1971; **26**:259–275.
10. Killackey HP, Rhoades RW, Bennett-Clarke CA. The formation of a cortical somatotopic map. Trends Neurosci 1995; **18**:402–407.
11. Chklovskii DB. Optimal sizes of dendritic and axonal arbors in a topographic projection. J Neurophysiol 2000; **83**:2113–2119.
12. Bower JM, Beermann DH, Gibson JM, Shambes GM, Welker W. Principles of organization of a cerebro-cerebellar circuit. Micromapping the projections from cerebral (SI) to cerebellar (granule cell layer) tactile areas of rats. Brain Behav Evol 1981; **18**:1–18.
13. Bower JM, Kassel J. Variability in tactile projection patterns to cerebellar folia crus IIA of the Norway rat. J Comp Neurol 1990; **302**:768–778.
14. Graybiel AM, Aosaki T, Flaherty AW, Kimura M. The basal ganglia and adaptive motor control. Science 1994; **265**:1826–1831.
15. Merchán MA, Malmierca MS, Bajo VM, Bjaalie JG. The nuclei of the lateral lemniscus: old views and new perspectives. In: Acoustical Signal Processing in the Central Auditory System. (Syka J, ed.), Plenum Press, New York, 1997, pp. 211–226.
16. Bower JM. Control of sensory data acquisition. Int Rev Neurobiol 1997; **41**:489–513.
17. Leergaard TB, Bjaalie JG. Semi-automatic data acquisition for quantitative neuroanatomy. MicroTrace—computer programme for recording of the spatial distribution of neuronal populations. Neurosci Res 1995; **22**:231–243.
18. Brevik A, Leergaard TB, Svanevik M, Bjaalie JG. Three-dimensional computerised atlas of the rat brain stem precerebellar system: approaches for mapping, visualization, and comparison of spatial distribution data. Anat Embryol 2001; **204**:319–332.
19. Leergaard TB, Lyngstad KA, Thompson JH., et al. (2000) Rat somatosensory cerebropontocerebellar pathways: spatial relationships of the somatotopic map of the primary somatosensory cortex are preserved in a three-dimensional clustered pontine map. J Comp Neurol **422**:246–266.
20. Leergaard TB, Alloway KD, Mutic JJ, Bjaalie JG. Three-dimensional topography of corticopontine projections from rat barrel cortex: correlations with corticostriatal organization. J Neurosci 2000; **20**:8474–8484.
21. Malmierca MS, Leergaard TB, Bajo VM, Bjaalie JG, Merchan MA. Anatomic evidence of a three-dimensional mosaic pattern of tonotopic organization in the ventral complex of the lateral lemniscus in cat. J Neurosci 1998; **18**:10603–10618.
22. Bajo VM, Merchán MA, Malmierca MS, Nodal FR, Bjaalie JG. Topographic organization of the dorsal nucleus of the lateral lemniscus in the cat. J Comp Neurol 1999; **407**: 349–366.
23. Welker W. Mapping the brain. Historical trends in functional localization. Brain Behav Evol 1976; **13**:327–343.
24. Toga AW, Mazziotta JC. Brain Mapping. The Methods. Academic Press, San Diego, 1996.
25. Brodal P, Bjaalie JG. Salient anatomic features of the cortico-ponto-cerebellar pathway. Prog Brain Res 1997; **114**:227–249.

26. Brodal P. The cerebropontocerebellar pathway: salient features of its organization. In: The Cerebellum-New Vistas. Exp. Brain Res., Suppl. 6 (Chan-Palay V, Palay S, eds.), Springer-Verlag, Berlin and Heidelberg, 1982, pp. 108–132.
27. Brodal P, Bjaalie JG. Organization of the pontine nuclei. Neurosci Res 1992; **13**:83–118.
28. Schmahmann JD, Pandya DN. The cerebrocerebellar system. Int Rev Neurobiol 1997; **41**:31–60.
29. Chapin JK, Lin CS. The somatic sensory cortex of the rat. In: The Cerebral Cortex of the Rat (Kolb B, Tees RC, eds.), MIT Press, Cambridge, MA, 1990, pp. 341–380.
30. Shambes GM, Gibson JM, Welker W. Fractured somatotopy in granule cell tactile areas of rat cerebellar hemispheres revealed by micromapping. Brain Behav Evol 1978; **15**:94–140.
31. Welker W. Spatial organization of somatosensory projections to granule cell cerebellar cortex: functional and connectional implications of fractured somatotopy (summary of Wiscounsin studies). In: New Concepts in Cerebellar Neurobiology (King JS, ed.), A.R. Liss, New York, 1987, pp. 239-280.
32. Irvine DRF. Physiology of the auditory brainstem. In: The Mammalian Auditory Pathway: Neurophysiology (Popper A, Fay RR, eds.), Springer, New York, 1992, pp. 153–231.
33. Merchán MA, Berbel P. Anatomy of the ventral nucleus of the lateral lemniscus in rats: a nucleus with a concentric laminar organization. J Comp Neurol 1996; **372**:245–263.
34. Covey E, Casseday JH. The monaural nuclei of the lateral lemniscus in an echolocating bat: parallel pathways for analyzing temporal features of sound. J Neurosci 1991; **11**:3456–3470.
35. Covey E, Casseday JH. The lower brainstem auditory pathways. In: Hearing in Bats (Popper AN, Fay RR, eds.). Springer, New York, 1995, pp. 235-295.
36. Aitkin LM, Anderson DJ, Brugge JF. Tonotopic organization and discharge characteristics of single neurons in nuclei of the lateral lemniscus of the cat. J Neurophysiol 1970; **33**:421–440.
37. Guinan Jr. JJ, Norris BE, Guinan SS. Single auditory units in the superior olivary complex: II. Locations of unit categories and tonotopic organization. Int J Neurosci 1972; **4**:147-166.
38. Adams JC. Ascending projections to the inferior colliculus. J Comp Neurol 1979; **183**:519–538.
39. Kudo M. Projections of the nuclei of the lateral lemniscus in the cat: an autoradiographic study. Brain Res 1981; **221**:57–69.
40. Whitley JM, Henkel CK. Topographical organization of the inferior collicular projection and other connections of the ventral nucleus of the lateral lemniscus in the cat. J Comp Neurol 1984; **229**:257–270.
41. Friauf E. Struktur und funktion von motorischen und sensorischen neuronen im hirnstamm der ratte. Thesis. Der Eberhard-Karls-Universität Tübingen, 1987.
42. Glendenning KK, Hutson KA. Lack of topography in the ventral nucleus of the lateral lemniscus. Microsc Res Tech 1998; **41**:298–312.
43. Ugolini G. Specificity of rabies virus as a transneuronal tracer of motor networks: transfer from hypoglossal motoneurons to connected second-order and higher order central nervous system cell groups. J Comp Neurol 1995; **356**:457–480.
44. Tang Y, Rampin O, Giuliano F, Ugolini G. Spinal and brain circuits to motoneurons of the bulbospongiosus muscle: retrograde transneuronal tracing with rabies virus. J Comp Neurol 1999; **414**:167–192.
45. Köbbert C, Apps R, Bechmann I, Lanciego JL, Mey J, Thanos S. Current concepts in neuroanatomical tracing. Prog Neurobiol 2000; **62**:327–351.
46. Van Haeften T, Wouterlood FG. Neuroanatomical tracing at high resolution. J Neurosci Methods 2000; **103**:107–116.
47. Swanson LW. A history of neuroanatomical mapping. In: Brain Mapping. The Systems (Toga AW, Mazziotta JC, eds.). Academic Press, San Diego, 2000, pp. 77–109.

48. Mesulam M-M. Principles of horseradish peroxidase neurohistochemistry and their applications for tracing neural pathways—axonal transport, enzyme histochemistry and light microscopic analysis. In: Tracing Neural Connections with Horseradish Peroxidase (Mesulam M-M, ed.). John Wiley & Sons. Chichester, 1982, pp. 1–151.
49. Gerfen CR, Sawchenko PE. An anterograde neuroanatomical tracing method that shows the detailed morphology of neurons, their axons and terminals: immunohistochemical localization of an axonally transported plant lectin, *Phaseolus vulgaris* leucoagglutinin (PHA-L). Brain Res 1984; **290**:219–238.
50. Veenman CL, Reiner A, Honig MG. Biotinylated dextran amine as an anterograde tracer for single- and double-labeling studies. J Neurosci Methods 1992; **41**:239–254.
51. Lanciego JL, Wouterlood FG. Dual anterograde axonal tracing with *Phaseolus vulgaris*-leucoagglutinin (PHA-L) and biotinylated dextran amine (BDA). Neuroscience Protocols 1994; **94-050-06-01-13**.
52. Glover JC, Petursdottir G, Jansen JK. Fluorescent dextran-amines used as axonal tracers in the nervous system of the chicken embryo. J Neurosci Methods 1986; **18**:243–254.
53. Glaser EM, Gissler M, Van der Loos H. An interactive camera lucida computer-microscope. Soc Neurosci Abstr 1979; 5:1697.
54. Glaser EM, Tagamets M, McMullen NT., Van der Loos H. The image-combining computer microscope—an interactive instrument for morphometry of the nervous system. J Neurosci Methods 1983; **8**:17-32.
55. Bjaalie JG. Three-dimensional computer reconstructions in neuroanatomy: basic principles and methods for quantitative analysis. In: Quantitative Methods in Neuroanatomy (Steward MG, ed.). John Wiley & Sons. Chichester, 1992, pp. 249–293.
56. Leergaard TB, Lakke EA, Bjaalie JG. Topographical organization in the early postnatal corticopontine projection: a carbocyanine dye and 3-D computer reconstruction study in the rat. J Comp Neurol 1995; **361**:77–94.
57. Malmierca MS, Rees A, Le Beau FE, Bjaalie JG. Laminar organization of frequency-defined local axons within and between the inferior colliculi of the guinea pig. J Comp Neurol 1995; **357**:124–144.
58. Bjaalie JG, Sudbø J, Brodal P. Corticopontine terminal fibres form small scale clusters and large scale lamellae in the cat. Neuroreport 1997; **8**:1651–1655.
59. Berg BG, Almaas TJ, Bjaalie JG, Mustaparta H. The macroglomerular complex of the antennal lobe in the tobacco budworm moth *Heliothis virescens*: specified subdivision in four compartments according to information about biologically significant compounds. J Comp Physiol A 1998; **183**:669–682.
60. Hallem JS, Thompson JH, Gundappa-Sulur G, Hawkes R, Bjaalie JG, Bower JM. Spatial correspondence between tactile projection patterns and the distribution of the antigenic Purkinje cell markers anti-zebrin I and anti-zebrin II in the cerebellar folium crus IIA of the rat. Neuroscience 1999; **93**:1083–1094.
61. Vassbø K, Nicotra G, Wiberg M, Bjaalie JG. Monkey somatosensory cerebrocerebellar pathways: uneven densities of corticopontine neurons in different body representations of areas 3b, 1, and 2. J Comp Neurol 1999; **406**:109–128.
62. Swanson LW. Mapping the human brain: past, present, and future. Trends Neurosci. 1995; **18**:471–474.
63. Schmahmann JD, Doyon J, McDonald D, et al. Three-dimensional MRI atlas of the human cerebellum in proportional stereotaxic space. Neuroimage 1999; **10**:233–260.
64. Thompson PM, MacDonald D, Mega MS, Holmes CJ, Evans AC, Toga AW. Detection and mapping of abnormal brain structure with a probabilistic atlas of cortical surfaces. J Comput Assist Tomogr 1997; **21**:567–581.
65. Toga AW, Thompson P. An introduction to brain wharping. In: Brain Wharping (Toga AW, ed.). Academic Press, San Diego, 1999, pp. 1-26.

66. Thompson PM, Woods RP, Mega MS, Toga AW. Mathematical/computational challenges in creating deformable and probabilistic atlases of the human brain. Hum Brain Mapp 2000; **9**:81–92.
67. Paxinos G, Watson C. The Rat Brain in Stereotaxic Coordinates. Academic Press, San Diego, 1998.
68. Shepherd GM, Mirsky JS, Healy MD, et al. The Human Brain Project: neuroinformatics tools for integrating, searching and modeling multidisciplinary neuroscience data. Trends Neurosci 1998; **21**:460–468.
69. Chicurel M. Databasing the brain. Nature 2000; **406**:822–825.
70. Koslow SH. Should the neuroscience community make a paradigm shift to sharing primary data? Nat Neurosci 2000; **3**:863–865.
71. Mihailoff GA, Burne RA., Woodward DJ. Projections of sensorimotor cortex to the basilar pontine nuclei in the rat: an autoradiographic study. Brain Res 1978; **145**:347–354.
72. Mihailoff GA, Lee H, Watt CB, Yates R. Projections to the basilar pontine nuclei from face sensory and motor regions of the cerebral cortex in the rat. J Comp Neurol 1985; **237**:251–263.
73. Wiesendanger R, Wiesendanger M. The corticopontine system in the rat. II. The projection pattern. J Comp Neurol 1982; **208**:227–238.
74. Toga AW. Three-Dimensional Neuroimaging. Raven Press, New York, 1990.
75. Bjaalie JG, Daehlen M, Stensby TV. Surface modelling from biomedical data. In: Numerical Methods and Software Tools in Industrial Mathematics (Daehlen M, Tveito A, eds.). Birkhauser, Boston, 1997, pp. 9–26.
76. Geiger B. Three-Dimensional Modelling of Human Organs and Its Application to Diagnosis and Surgical Planning. Report 2105, Sophia-Antipolis, Institut National de Recherche en Informatique et en Automatique, France, 1993.
77. Pakkenberg B, Gundersen HJ. Neocortical neuron number in humans: effect of sex and age. J Comp Neurol 1997; **384**:312–320.
78. Selemon LD, Rajkowska G, Goldman-Rakic PS. Elevated neuronal density in prefrontal area 46 in brains from schizophrenic patients: application of a three-dimensional, stereologic counting method. J Comp Neurol 1998; **392**:402–412.
79. Nadasdy Z, Zaborszky L. Computational analysis of spatial organization of large scale neural networks. Anat Embryol 2001; **204**:303–317.
80. Blackstad TW, Osen KK, Mugnaini E. Pyramidal neurones of the dorsal cochlear nucleus: a Golgi and computer reconstruction study in cat. Neuroscience 1984; **13**:827–854 (Erratum in Neurosci 1985; **15**:923).
81. Blackstad TW, Karagulle T, Malmierca MS, Osen KK. Computer methods in neuroanatomy: determining mutual orientation of whole neuronal arbors. Comput Biol Med 1993; **23**:227–250.
82. Alloway KD, Crist J, Mutic JJ, Roy SA. Corticostriatal projections from rat barrel cortex have an anisotropic organization that correlates with vibrissal whisking behavior. J Neurosci 1999; **19**:10908–10922.
83. Alloway KD, Mutic JJ, Hoffer ZS, Hoover JE. Overlapping corticostriatal projections from the rodent vibrissal representations in primary and secondary somatosensory cortex. J Comp Neurol 2000; **426**:51–67.
84. He SQ, Dum RP, Strick PL. Topographic organization of corticospinal projections from the frontal lobe: motor areas on the lateral surface of the hemisphere. J Neurosci 1993; **13**:952–980.
85. Alloway KD, Mutic JJ, Hoover JE. Divergent corticostriatal projections from a single cortical column in the somatosensory cortex of rats. Brain Res 1998; **785**:341–346.

11
Competition in Neuronal Morphogenesis and the Development of Nerve Connections

Arjen van Ooyen and Jaap van Pelt

ABSTRACT

During the development of the nervous system, neurons form their characteristic morphologies and become assembled into synaptically connected networks. In many of the developmental phases that can be distinguished, e.g., axonal differentiation, neurite elongation and branching, and synapse rearrangement, competition plays an important role. Focusing on competition, we review model studies on neuronal morphogenesis and the development of nerve connections.

11.1. INTRODUCTION

During development, neurons become assembled into functional networks by growing out axons and dendrites (collectively called neurites), which connect synaptically to other neurons. A number of developmental phases can be distinguished.

Early dendritic and axonal morphogenesis. The neurons begin to grow by projecting many broad, sheet-like extensions, called lamellipodia, which subsequently condense into a number of small undifferentiated neurites of approximately equal length *(1)*. Eventually, one of the neurites (usually the longest) increases its growth rate—while at the same time the growth rate of the remaining neurites is reduced—and differentiates into an axon. The remaining neurites become later differentiated as dendrites and form characteristic branching patterns. The development of dendritic morphology proceeds by way of the dynamic behavior of growth cones, which are specialized structures at the terminal ends of outgrowing neurites and which mediate neurite elongation and branching *(2)*. Among the many intra- and extracellular mechanisms involved in growth cone behavior are intracellular calcium levels, signal transduction cascades and cytoskeletal changes *(3)*.

Axon guidance and synapse formation. The axons need to migrate to their targets, and one of the mechanisms by which this is achieved is by the diffusion of chemoattractant molecules from the target through the extracellular space *(4)*. This creates a gradient of increasing concentration, which the growth cone at the tip of a migrating axon can sense and follow *(5)*. Axons are also repelled by diffusible molecules that are secreted by tissues the axons need to grow away from. In addition to diffusible molecules, axons are also attracted and repelled by surface-bound molecules on other cells, and in the extracellular matrix *(4)*. Once the axons have arrived at their targets, they form synaptic connections by transforming their growth cones into synapses.

Synapse rearrangement. The phase of synapse formation is followed by a phase of refinement, including both the formation of new synapses and the elimination of existing synapses *(6,7)*. This process often involves withdrawal of some axons and, thus, a reduction in the number of axons innervating an individual target cell. In some cases, withdrawal of axons continues until the target is innervated by just a single axon, whereas in most other cases several innervating axons remain *(8–10)* (see also Subheading 11.4.).

Competition plays an important role in many of the above described developmental phases. In axonal differentiation, all the neurites have the potential to develop into the axon *(11,12)*. In experiments in which the axon is transected at various distances from the soma, the longest neurite remaining after transection usually becomes the axon, regardless of whether it was previously an axon or a dendrite *(12)*. Thus, axonal differentiation appears to be a competitive process in which the growth rate of the longest neurite is accelerated at the expense of all the other neurites, whose growth become inhibited *(1,13)* (see also *[14]*).

Some form of competition is also operative in the formation of dendritic trees: the branching probability of an individual growth cone (i.e., terminal segment) appears to decrease with the number of other growth cones in the tree. Such a dependence turns out to be necessary for reducing the proliferating effect of the increasing number of growth cones (see Subheading 11.2.). Competitive effects between neurites are expected to occur also in the elongation of neurites: the proteins upon which elongation depend (namely, tubulin and microtubule-associated proteins) are produced in the soma and need to be divided among all the growing neurites of a neuron *(15)*. Competition for tubulin could explain the observation that sometimes only one of the daughter growth cones propagate after branching, while the other stays dormant for a long time (see Subheading 11.3.1.).

Competition between innervating axons for target-derived neurotrophic factors is thought to be involved in the withdrawal of axons *(10,16)*. The cells that act as targets for the innervating axons release limited amounts of neurotrophic factors, which are taken up by the axons via specific receptors at their terminals and which affect the growth and branching of the axons *(17,18)*.

To gain a real understanding of nervous system development and function, experimental work needs to be complemented by theoretical analysis and computer simulation. Even for biological systems in which all the components are known, computational models are necessary to explore and understand how the components interact to make the system work and how phenomena at different levels of organization or description are linked. In this chapter, we discuss (*i*) models of the development of dendritic morphology, focusing on competitive phenomena (Subheadings 11.2. and 11.3.); and (*ii*) models of competition between innervating axons in the refinement of connections (subheading 11.4.). For a model on the role of competition in axonal differentiation, see *(14)*. For models of axon guidance and fasciculation, see *(19)*.

For all models we present, the various components of the model are relatively uncontroversial biologically, and the aim is to explore and understand quantitatively the consequences of the interactions between these components, in terms of the phenomena and data the model can generate. This provides hypotheses and predictions about these phenomena and data at a lower level of organization or description. For example, neuronal morphology is linked to the actions of growth cones (Subheading 11.2.); the phenomenon of dormant growth cones is linked to competition, at the molecular level, for tubulin (Subheading 11.3.1.); and axonal competition is linked to the actions and biochemistry of neurotrophins (Subheading 11.4.2.).

11.2. DEVELOPMENT OF DENDRITIC MORPHOLOGY: A STOCHASTIC MODEL

Dendritic branching patterns emerge from a developmental process of neurite elongation and branching. This process is mediated by growth cones, which, under the influence of intracellular and extracellular mechanisms, show highly dynamic behavior, such as advance, reorientation, splitting, shape and speed changes, retraction, and even complete disappearance. Outgrowth is, therefore, not a regular process of continued elongation and branching; nevertheless, it eventually results in dendritic branching patterns that are typical for the type of neurons under consideration.

Modeling dendritic branching patterns from a developmental point of view raises the question at which level of detail the growth process should be described. In our modeling approach, we assume that at a sufficiently coarse time scale the averaged outcome of all the underlying growth processes can be described as a sustained stochastic process of elongation and branching. The stochasticity assumption is warranted because of the multitude of mechanisms that determine the behavior of growth cones.

In our dendritic growth model (recently reviewed in [20,21]), a distinction is made between topological and metrical properties of dendritic trees. Topological properties emerge from the branching process as segments increase in number and develop a particular connectivity pattern. Metrical properties emerge from both the branching and the elongation process. The dendritic growth model has first been developed and validated for the branching process. Later, elongation was included; this has the advantage that the optimization of the metrical properties can be built upon an already optimized branching process.

Modeling the dendritic branching process. To describe the branching process, the total developmental period T is divided into a series of N time bins. In each time bin i, a terminal segment (growth cone) may branch with a probability given by:

$$p_i(n_i, \gamma) = D n_i^{-E} \times 2^{-S\gamma} C_i \qquad [\text{Eq. 1}]$$

Parameter D determines the basic branching probability, which is taken equal to $D = B/N$, with B denoting the expected number of branching events of an isolated segment during the full period T. With the term n_i^{-E}, the branching probability is made dependent on the number n_i of terminal segments (this may be thought of as representing some form of competition; see below). The strength of this dependence is determined by parameter E. With the term $2^{-S\gamma}C_i$, the branching probability is made dependent on the centrifugal order γ of the terminal segment (see Fig. 1), thus allowing a modulation

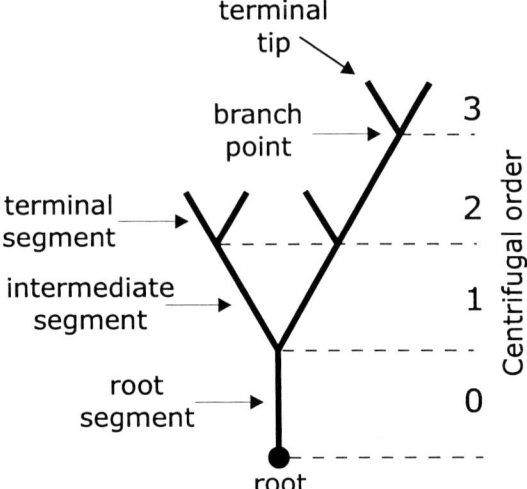

Fig. 1. Example of a rooted binary topological tree in which are distinguished a root, intermediate and terminal segments, root point, branch point, and terminal tip. Segments are labeled by a centrifugal ordering scheme.

of the branching probabilities over the different terminal segments in the dendritic tree. The strength of this modulation is determined by parameter S. The normalization constant $C_i = n_i / \sum_{j=1}^{n_i} 2^{-S\gamma_j}$ ensures that this modulation does not change the mean branching probability Dn_i^{-E}, averaged over all terminal segments in the tree. Thus, the term controls the rate of increase of the number of segments, while the topological structure is under control of the modulation $2^{-S\gamma}C_i$. The number N of time bins can be chosen arbitrarily, but such that the branching probability per time bin remains much smaller than one, thus making the probability of more than one branching event per time bin negligibly small.

During outgrowth, an increasing number of terminal segments is participating in the branching process, and this proliferation strongly determines the rate with which the number of terminal segments increases. In the model, this proliferation is kept under control by making the branching probability dependent on the total number of segments, via parameter E. Figure 2A illustrates how fast the number of terminal segments increases for the unrestricted case $E = 0$, i.e., when the branching probability per time bin remains constant (Fig. 2D). For $E = 1$, in contrast, the branching probability is inversely proportional to the total number of terminal segments in the tree (Fig. 2F), resulting in a linearly increasing number of branch points up to the value of parameter B (Fig. 2C). (Note that a binary tree with 3 branch points has 4 terminal segments.) The branching parameters B and E determine the growth rate not only of the mean number of terminal segments but also of the standard deviation of the terminal segment number distribution, as is shown in Figure 2. This is expressed also in the shape of this distribution, with smaller means and standard deviations (SDs) for increasing values of E (Fig. 3). Figure 4 shows how the mean and SD depend on the parameters B and E. For a given value of B, the mean number of terminal segments decreases with increasing

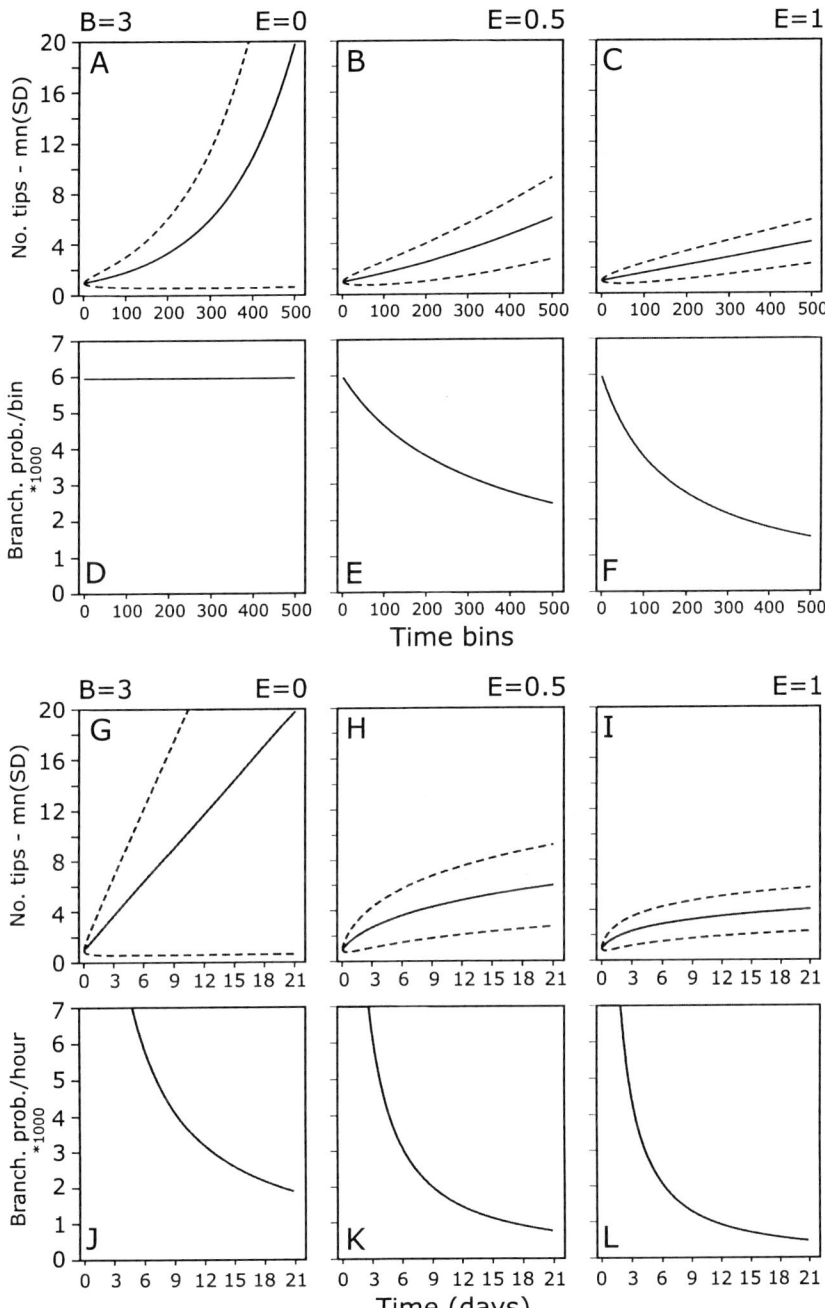

Fig. 2. (A–C) Growth curves of the mean and the SD of the number of terminal segments and **(D–F)** the course of the branching probability of a terminal segment per time bin, plotted vs a time bin scale with 500 time bins. The curves are calculated for $B = 3$, and for three values of E with $E = 0$ **(A,D)**, $E = 0.5$ **(B,E)**, and $E = 1$ **(C,F)**. The figures illustrate how parameter E influences the growth curves and the course of the branching probability. Panels **(G–L)** are obtained by a nonlinear (exponential with exponent 3) mapping of the time bin scale with 500 time bins onto a continuous time scale with an arbitrary duration of 3 wk. Panels **(J–L)** display the time course of the branching probability of a terminal segment per hour. These panels illustrate that the shape of the growth curves is changed by the mapping but that the relation between mean and SD is maintained.

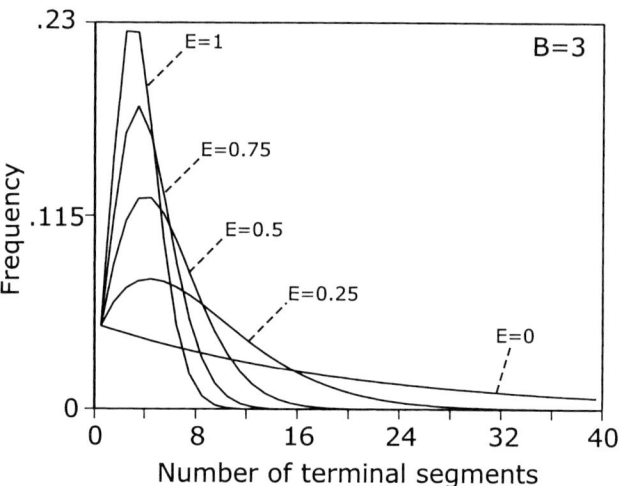

Fig. 3. Frequency distributions of the number of terminal segments per dendritic tree as produced for $B = 3$ and different values of the branching parameter E.

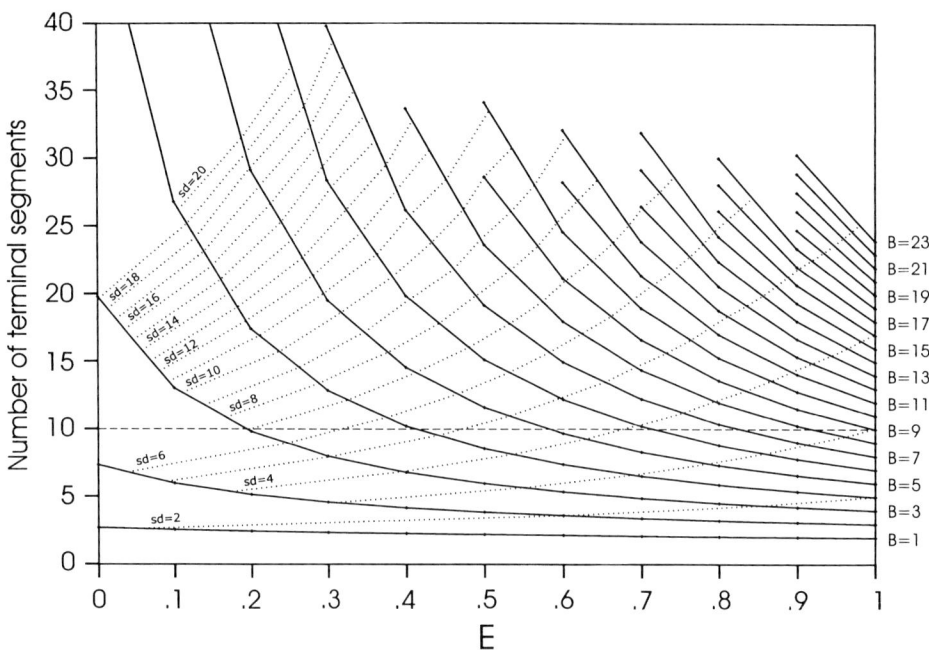

Fig. 4. Map of the number of terminal segments per dendritic tree vs the branching parameter E, for different values of the branching parameter B. The figure shows how the number of terminal segments, for a given B, decreases with increasing values of E. The figure also shows that a given number of terminal segments (say 10) can be produced by different combinations of parameters B and E, but that for these combinations the standard deviation decreases for increasing values of E. Finally, the figure shows that for $E = 1$, the mean number of branch points in a tree equals the branching parameter B. Note that for binary trees the number of terminal segments is equal to the number of branch points plus one.

Neuronal Morphogenesis and Nerve Connections

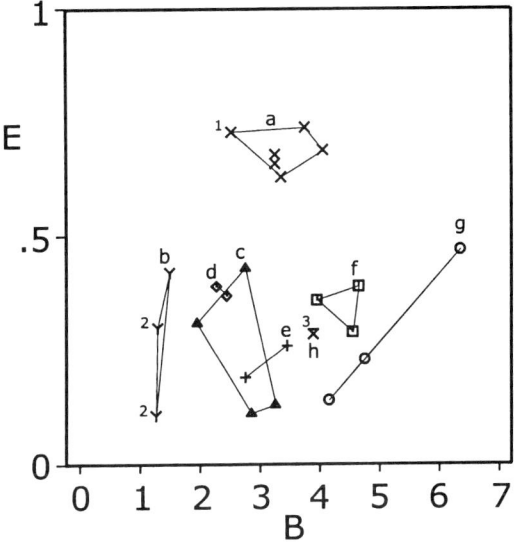

Fig. 5. Scattergram of optimized values of the branching parameters B and E to data sets of observed dendritic trees. The figure is a combination of an earlier data compilation *(22)* including the references to the data sources and later results. The numbered data points are obtained from (1) *(21)*, (2) *(25)*, and (3) *(24)*; see these references for the number of neurons each data point represent. The data points are grouped according to their cell types and refer to dendrites from (a) rat cortical pyramidal neurons, (b) rat cortical multipolar nonpyramidal neurons, (c) rat motoneurons, (d) human dentate granule cells, (e) cultured cholinergic interneurons, (f) cat motoneurons, (g) frog motoneurons, and (h) cat deep layer superior colliculus neurons. The data points show a clear clustering per cell type. The pyramidal cell group (a) differs from the other groups in their E values, while the other groups tend to differ in their B values.

values of E. This reducing effect of parameter E on the growth of the total number of terminal segments may represent some kind of competition between growth cones for branching. Trees of a given size can be produced by many combinations of the parameters B and E, where higher values of E, resulting in a lower growth rate, should be accompanied by higher values of B. However, higher E values also result in lower values of the SD, as indicated in Figure 4, where the horizontal line at, for instance, degree 10 crosses (dotted) curves of lower SD value for increasing E. Matching to observed SD values finally determines which E and B values most optimally predict the observed mean and SD values. The optimized B and E values tend to show clustering for different cell types, as was shown in *(22)* and is illustrated in Figure 5. Especially the pyramidal cell group shows significantly higher E values than the other cell groups, suggesting that these dendrites develop under stronger competitive conditions. The clustering suggests also a differentiation in B values between the other cell groups. Although the statistics are still poor, one may conclude from these findings that the branching parameters B and E indeed represent cell type-specific characteristics of dendritic branching patterns.

Modeling the dendritic branching process in continuous time. To describe the branching process in continuous time, the time bin scale needs to be mapped onto a real time scale. The equation for the branching probability per time bin then transforms into a branching probability per unit of time:

$$p_t(n_t, \gamma) = D(t) n_t^{-E} 2^{-S\gamma} C_i \qquad \text{[Eq. 2]}$$

with parameter $D(t)$ denoting the basic branching rate per unit of time. Time bins will obtain equal durations in a linear mapping, but may have different durations in a non-linear mapping. An example is given in Figure 2G–L, which illustrates how the growth curves and the probability curves change when the time bin scale (with 500 bins) is exponentially mapped onto a continuous time scale of 504 h (3 wk). Note that the relation between mean and SD is maintained, being independent of the type of mapping.

Modeling the elongation process. Once the branching process has been optimized to the observed data set, the metrical properties can be modeled. To this end, newly formed daughter segments at a branching event are given an initial length and an elongation rate for the period of time up to the moment they branch again. Both the initial lengths and the elongation rates are randomly drawn from gamma distributions, with mean and SD values of $\overline{l_{in}}, \sigma_{l_{in}}, \overline{v}$, and σ_v, respectively.

Results of the dendritic growth model. The model has been applied to dendritic data sets of a variety of cell types, including rat large layer 5 pyramidal neurons *(20)*, small layer 5 pyramidal neurons *(23)*, layer 2/3 pyramidal neurons *(21)*, guinea pig cerebellar Purkinje cells *(21)*, cat deep layer superior colliculus neurons *(24)*, and rat cortical multipolar nonpyramidal neurons *(25)*. In all these examples, the dendritic shape properties were well approximated up to the very details of their distributions.

11.3. NEURITE ELONGATION AND BRANCHING: CELL BIOLOGICAL MECHANISMS

Most models of the development of dendritic morphology describe neurite elongation and branching in a stochastic manner. Although these models are very successful at generating the observed variation in dendritic branching patterns (see Subheading 11.2.), they do not clarify how the biological mechanisms underlying neurite outgrowth are involved, namely, the dynamics of the tubulin and actin cytoskeleton. In this section, we present models that study the role of tubulin dynamics in neurite outgrowth.

11.3.1. Neurite Elongation as a Result of Tubulin Polymerization

The length of a neurite is determined by its microtubules, which are long polymers of tubulin present throughout the entire neurite. Tubulin monomers are produced in the cell body and are transported down the neurite to the growth cone. Polymerization of tubulin, which occurs mainly in the growth cone, elongates the microtubules and thus the neurite. The rates of tubulin assembly and disassembly are influenced by the actin cytoskeleton in the growth cone, by microtubule-associated proteins (MAPs), and by (activity-dependent) changes in the intracellular calcium concentration *(26–30)*.

In *(31)* and *(15)*, the consequences of the interactions between tubulin transport and tubulin (dis)assembly are explored. The model in (31) is based on the model in *(15)* and describes neurite elongation and retraction as the result of tubulin assembly and disas-

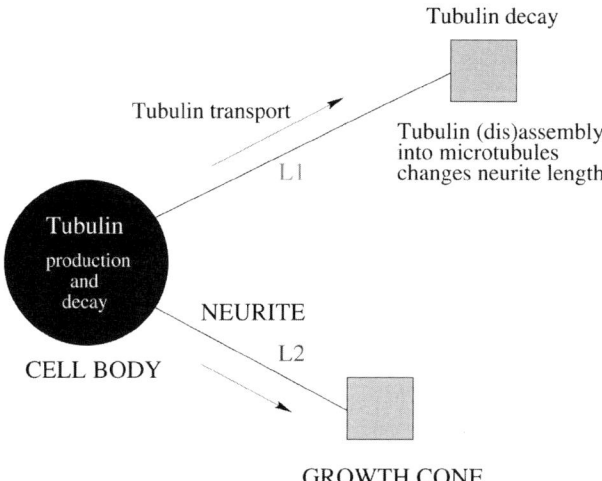

Fig. 6. The compartmental model of a single neuron with two neurites. See further section 11.3.1. From *(31)* with permission.

sembly. A simple compartmental model of a single neuron with n different neurites is considered (Fig. 6). There is one compartment for the cell body and one compartment for the growth cone of each neurite i ($i = 1,...,n$). The time-dependent changes of the neurite length L_i, the concentration C_0 of tubulin in the cell body, and the concentrations C_i of tubulin in the growth cones are modeled. Tubulin is produced in the cell body, at rate s, and is transported into the growth cones of the different neurites by diffusion and active transport, with diffusion constant D and rate constant f, respectively. At the growth cones, concentration-dependent assembly of tubulin into microtubules takes place, which elongates the trailing neurite. Disassembly of microtubules into tubulin causes the neurite to retract. The rate constants a_i and b_i for, respectively, assembly and disassembly are taken slightly different in different neurites. Differences in rate constants between neurites can arise as a result of differences between neurites in electrical activity (which affects the concentration of intracellular calcium), in the actin cytoskeleton of the growth cones, or in the state or concentration of MAPs. Finally, tubulin is also subjected to degradation, with rate constant g, both in the cell body and in the growth cone. Thus, the rates of change of L_i, C_i, and C_0 become:

$$\frac{dL_i}{dt} = a_i C_i - b_i \qquad \text{[Eq. 3]}$$

$$\frac{dC_i}{dt} = b_i - a_i C_i + \frac{D}{L_i + k}(C_0 - C_i) + fC_0 - gC_i \qquad \text{[Eq. 4]}$$

$$\frac{dC_0}{dt} = s - \sum_{i=1}^{n} \frac{D}{L_i + k}(C_0 - C_i) - \sum_{i=1}^{n} fC_0 - gC_0 \qquad \text{[Eq. 5]}$$

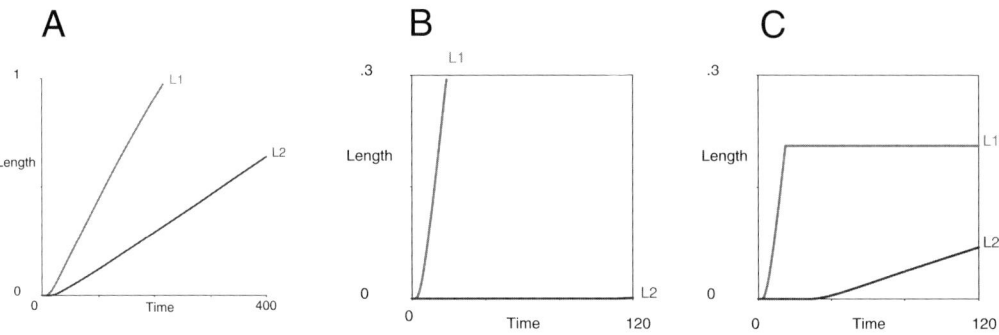

Fig. 7. Results of the compartmental model of a single neuron with two neurites. Neurite 1 has a higher rate constant for tubulin assembly. As a result, neurite 1 can slow down **(A)** or even prevent **(B)** the growth of the other neurite. Stopping the growth of neurite 1 triggers, after a time delay, the growth of the other neurite **(C)**. Parameters (all units arbitrary): $b_1 = b_2 = 0.01$, $D = 0.5$, $g = 0.1$, $s = 0.07$, $f = 0$, and $k = 1$. In **(A)**, $a_1 = 0.09$ and $a_2 = 0.06$. In **(B)** and **(C)**, $a_1 = 0.3$ and $a_2 = 0.05$. From *(31)* with permission.

where k is the distance between the centres of the cell body and growth cone compartment when $L_i = 0$. In *(15)*, there is no degradation of tubulin, which is biologically not plausible and which makes the mathematical analysis more difficult, and no active transport of tubulin.

The analysis of the model shows that small differences between neurites in their rate constants for assembly and/or disassembly (e.g., as a result of differences between neurites in intracellular calcium concentration) lead to competition between growing neurites of the same neuron (also reported in *[15]*). This competition emerges as a result of the interactions between tubulin-mediated neurite elongation and transport of tubulin. If one of the neurites has a higher rate constant for tubulin assembly and/or a lower rate of disassembly, it can slow down (Fig. 7A) or even prevent (Fig. 7B) the outgrowth of the other neurites for a considerable period of time (i.e., they are "dormant"), by using up all the tubulin produced in the soma. Only after the fastest growing neurite has reached a certain length (the longer the neurite, the smaller the amount of tubulin that is transported by diffusion per unit time) can the tubulin concentration in the growth cones of the other neurites increase, causing them to grow out. The smaller the rate of production of tubulin in the cell body, the bigger this period of dormancy.

In *(31)*, it was shown that stopping the outgrowth of the fastest growing neurite (e.g., representing the physiological situation that a neurite has reached its target) can "awaken" the dormant growth cones, which then, after a characteristic delay, start growing out (Fig. 7C). The length of the delay is determined by the time it takes for the tubulin concentration to build up to the value where the rate of assembly ($a_i C_i$) is bigger than the rate of disassembly (b_i).

Preliminary results show that the higher the relative contribution of the active component (parameter f) to the transport process, the smaller the competitive effects. In more detailed compartmental models *(32)*, in which each neurite is divided into many compartments, we found very similar results as those reported here.

The model can account for the occurrence of "dormant growth cones" *(33)*—the observation that, after branching, only one of the daughter growth cones propagates. The prediction of the model that there should be competition between growing neurites of the same neuron has recently been confirmed experimentally (G.J.A. Ramakers, unpublished results). These findings show that (*i*) when one neurite stops growing out, other neurites (after a certain delay, as in the model) start growing out; and (*ii*) when more neurites are growing out at the same time, the rate of outgrowth is smaller than when only a single neurite is growing out. To test whether this is indeed due to competition for tubulin, as our model suggests, the concentration of tubulin in growth cones should be monitored during outgrowth. The model predicts that the concentration of tubulin in growth cones that are not growing out should be below the critical value [the concentration of tubulin at which assembly ($a_i C_i$) just equals disassembly (b_i)].

11.3.2. *The Role of Microtubule-Associated Proteins in Neurite Elongation and Branching*

The tubulin dynamics is influenced by many modulators, among which the MAPs play a prominent role *(34)*. They regulate not only assembly and disassembly, but also the bundling and spacing of microtubules. The phosphorylation state of MAPs affects their function *(28,29)*. When MAPs are dephosphorylated, they promote tubulin assembly and microtubule bundling and so promote neurite elongation *(35)*. When MAPs are phosphorylated, they inhibit assembly and bundling; the spacing between microtubule bundles increases, which favors dendritic branching *(36)*. It has been proposed that the rates of elongation and branching are determined by the relative concentrations of phosphorylated and dephosphorylated MAPs *(37,38)*. This is itself dependent on the concentration of intracellular calcium, which regulate both phosphorylation and dephosphorylation through the actions of calmodulin-dependent protein kinase 2 and calcineurin *(28,29,39,40)*. Using a compartmental model for elongation and branching, Hely et al. *(38)* studied what the implications are of the interactions between the calcium dynamics (influx of calcium along, and diffusion within, the whole dendritic tree) and the effects of calcium on MAP (de)phosphorylation. In the model, the ratio of the concentrations of phosporylated and dephosporylated MAPs at the tip of a terminal segment (i.e., the growth cone) determines the branching probability and the rate of elongation. MAPs are produced in the soma and are transported to the growth cone by diffusion and active transport. One sigmoidal function is used to describe how the rate of MAP phosphorylation depends on the concentration of calcium in the growth cone; another sigmoidal function is used to describe how dephosphorylation depends on calcium. One result of the model is that the relative position of these two functions, together with the calcium dynamics, determines what dendritic structure will develop. As the tree grows, the calcium concentration in the terminal segments increases. The concentration is highest in the terminal segments and lowest in the soma (because of the higher surface-to-volume ratio in the thin terminal segments). As the tree grows, the terminal segments become farther away from the soma, which acts as a sink for calcium, so that the calcium concentration in the terminal segments increases. Depending on the relative position of the two sigmoidal functions, this increased calcium concentration leads either to a lower branching probability (producing trees in which the terminal segments are longer than the proximal segments, e.g., as in the basal dendrites of pyramidal neurons *[41]*) or to a lower branching probability (producing trees in which the terminal segments are shorter than the proximal segments, e.g., as in cultured hippocampal neurons *[42]*). Thus, given a particular branching pattern, the model predicts how the functions relating calcium with phosphorylation and dephosphorylation should be in order to produce this.

11.4. COMPETITION BETWEEN AXONS IN THE REFINEMENT OF NEURAL CIRCUITS

During development, the refinement of neural circuits involves both the formation of new connections and the elimination of existing connections *(6,7)*. Neurons, and other cell types, often are initially innervated by more axons than ultimately maintain into adulthood *(7,43)*. This initial hyperinnervation followed by elimination occurs, for example, in the development of connections between motor neurons and muscle fibers *(8,9)*, where elimination of axons continues until each muscle fiber is innervated by just a single axon, and in the formation of ocular dominance columns in the visual cortex *(44,45)*. Although there is a reduction in the number of axons that an individual target receives, the total number of synapses onto a target often increases (both in the visual and in the neuromuscular system), because of further arborization of the remaining axons *(46–48)*.

The process that reduces the number of axons innervating a postsynaptic cell is often referred to as axonal or synaptic competition. In particular, it is believed that axons compete for neurotrophic factors, survival- or growth-promoting substances, released by the postsynaptic cells upon which the axons innervate *(10,16)*. During an earlier stage of development, when initial synaptic contacts are made, these neurotrophic factors have a well-established role in the regulation of neuronal survival *(49,50)*. But many studies now indicate that neurotrophic factors may also be involved in the later stages of development, when there is further growth and elimination of innervation (for a critical review, see *[51]*). For example, neurotrophic factors have been shown to regulate the degree of arborization of axons (e.g., see *[52]*; for more references, see Subheading 11.4.2.).

Although the notion of competition is commonly used in neurobiology, the process is not well understood, and only a few formal models exist (for an extensive review, see *[48]*).

11.4.1. Competition Through Synaptic Normalization and Modified Hebbian Learning Rules

Most computational models of the development of nerve connections, especially models of the formation of ocular dominance columns, typically enforce competition rather than model its putative underlying mechanisms explicitly (for a review, see *[53]*). To see how competition between input connections can be enforced, consider n inputs, with synaptic strengths $w_i(t)$ ($i = 1,...,n$), impinging on a given postsynaptic cell at time t. Simple Hebbian rules for the change $\Delta w_i(t)$ in synaptic strength in time interval Δt state that the synaptic strength should grow in proportion to the product of the postsynaptic activity level $y(t)$ and the presynaptic activity level $x_i(t)$ of the ith input. Thus:

$$\Delta w_i(t) \propto y(t)x_i(t)\Delta t \qquad [\text{Eq. 6}]$$

If two inputs (e.g., two eyes) innervate a common target and if the activity level in both inputs is sufficient to achieve potentiation, then this rule causes both pathways to be strongly potentiated, and no segregation (ocular dominance) occurs. What is required is that when the synaptic strength of one input grows, the strengths of the other one shrinks. This can be achieved by imposing the constraint that $\sum_i^n w_i$ should be kept constant (synaptic normalization). At each time interval Δt, following a phase of Hebbian learning, in which $w_i(t + \Delta t) = w_i(t) + \Delta w_i(t)$, the new synaptic strengths are forced to satisfy the normalization constraint.

Another approach for achieving competition is to modify equation 6. With Equation 6, only increases in synaptic strength can take place; decreases in synaptic strength are brought about by enforcing synaptic normalization afterwards. Both increases in synaptic strength (long term potentiation, or LTP) and decreases in synaptic strength (long term depression, or LTD) can be obtained if we assume that the postsynaptic activity level $y(t)$ must be above some threshold θ_y to achieve LTP and otherwise yield LTD; for the presynaptic activity level $x_i(t)$, a similar possibility can be assumed *(53)*. Thus:

$$\Delta w_i(t) \propto [y(t) - \theta_y][x_i(t) - \theta_x]\Delta t \qquad [\text{Eq. 7}]$$

A stable mechanism for ensuring that when some synaptic strengths increase, others must correspondingly decrease (i.e., competition) is to make one of the thresholds variable. If the threshold θ_x^i increases sufficiently as the postsynaptic activity $y(t)$ or synaptic strength $w_i(t)$ (or both) increases, conservation of synaptic strength can be achieved *(53)*. Similarly, if the threshold θ_y increases faster than linearly with the average postsynaptic activity, then the synaptic strengths will adjust to keep the postsynaptic activity near a set point value *(54)*.

Yet another mechanism that can balance synaptic strengths is based on a form of (experimentally observed) long-term synaptic plasticity that depends on the relative timing of pre- and postsynaptic actions potentials (spike timing-dependent plasticity, or STDP) *(55)*. Presynaptic action potentials that precede postsynaptic spikes strengthen a synapse, whereas presynaptic action potentials that follow postsynaptic spikes weaken it. STDP has the effect of keeping the total synaptic input to the neuron roughly constant, independent of the presynaptic firing rates *(56)*.

11.4.2. Competition Through Dependence on Shared Target-Derived Resources

Keeping the total synaptic strength onto a postsynaptic cell constant (synaptic normalization) is a biologically unrealistic way of modeling competition. In both the neuromuscular and the visual system, the total number of synapses onto a postsynaptic cell increases during competition, as the winning axons elaborate their branches and the losing axons retract branches. Synaptic normalization is too rigid a constraint compared with the plasticity of the developing nervous system, and models based on this constraint may, therefore, become too restricted in the range of phenomena they can produce *(57,58)*. If Hebbian learning rules are modified only to enforce competition and not to represent a possible physiological mechanism, this is equally unsatisfactory. Modeling the actual mechanism of competition can give the models more flexibility and potentially a larger explanatory and predictive power. It will also be easier to interpret and extend these models, because its variables and parameters are more directly linked to biological processes and mechanisms.

If the dependence of axons on the same target-derived neurotrophic factor is modeled, competition between input connections does not have to be enforced, but comes about naturally. In most existing models of competition for target-derived neurotrophic factor, there is a fixed amount of neurotrophin that becomes partitioned among the individual synapses or axons; i.e., there is no production, decay, and consumption of the neurotrophin *(48)*. This assumption is biologically not realistic. The model by Van Ooyen and Willshaw *(59)* considers the production and consumption of neurotrophin and incorporates the dynamics of neurotrophic signaling (such as release of neurotrophin, binding kinetics of neurotrophin to receptor, and degradation processes) and the effects of neurotrophins on axonal growth and branching. The model can also incorporate the effects of electrical activity: postsynaptic activity can influence the release of neurotrophin, while presynaptic activity can influence the number of neurotrophin receptors (see the section below describing the model). The approach by Van Ooyen and Willshaw has similarities to that by Elliott and Shadbolt *(60)* and Jeanprêtre et al. *(61)*, although Elliott and Shadbolt *(60)* does not model all the processes involved in a dynamic fashion, and Jeanprêtre et al. *(61)* has to assume a priori thresholds, as well as to postulate a positive feedback rather than derive it from the underlying biological mechanisms (see further below).

The model by Van Ooyen and Willshaw (59). Important variables in the model are the total number of neurotrophin receptors that each axon has and the concentration of neurotrophin in the extracellular space. In the model, there is a positive feedback loop between the axon's number of receptors and amount of bound neurotrophin. Unlike in the work by Jeanprêtre et al. *(61)*, this positive feedback, which enables one or more axons to outcompete the others, was derived directly from underlying biological mechanisms. Following binding to receptor, neurotrophins can increase the terminal arborization of an axon *(52,62–70)* and, therefore, the axon's number of synapses. Because neurotrophin receptors are located on synapses, increasing the number of synapses means increasing the axon's total number of receptors. Thus, the more receptors an axon has, the more neurotrophin it will bind, which further increases its number of receptors, so that it can bind even more neurotrophin, at the expense of the other axons.

Instead of increasing the terminal arborization of an axon, neurotrophins might increase the axon's total number of receptors by increasing the size of synapses *(71)* or by up-regulating the density of receptors *(72)*.

Description of the model. The simplest situation in which we can study axonal competition is a single target at which there are n innervating axons, each from a different neuron. Each axon has a number of terminals, on which the neurotrophin receptors are located (see Fig. 8A). Neurotrophin is released by the target into the extracellular space, at rate σ, and is removed by degradation, with rate constant δ. In addition, at each axon i, neurotrophin is bound to receptors, with association and dissociation constants $k_{a,i}$ and $k_{d,i}$, respectively. Bound neurotrophin (the neurotrophin–receptor complex) is also degraded, with rate constant P_i. Degradation of the neurotrophin–receptor complex also removes receptor molecules; therefore, new unoccupied receptors need to be inserted (at rate ϕ_i) into the axon terminals. In addition, there is turnover of unoccupied receptors, with rate constant γ_i. Thus, the rates of change of the total number R_i of unoccupied receptors on axon i, the total number C_i of neurotrophin–receptor complexes on axon i, and the extracellular concentration L of neurotrophin are:

$$\frac{dC_i}{dt} = (k_{a,i} L R_i - k_{d,i} C_i) - \rho_i C_i \quad \text{[Eq. 8]}$$

$$\frac{dR_i}{dt} = \phi_i - \gamma_i R_i - (k_{a,i} L R_i - k_{d,i} C_i) \quad \text{[Eq. 9]}$$

$$\frac{dL}{dt} = \sigma - \delta L - \sum_{i=1}^{n} (k_{a,i} L R_i - k_{d,i} C_i)/v \quad \text{[Eq. 10]}$$

where v is the volume of the extracellular space. The term $(k_{a,i} L R_i - k_{d,i} C_i)$ represents the net amount of neurotrophin that is being bound to receptor. Equations 8 and 9 are similar to the ones used in experimental studies for analyzing the cellular binding, internalization, and degradation of polypeptide ligands such as neurotrophins (73).

The biological effects of neurotrophins (all of which, as explained above, can lead to an axon getting a higher total number of receptors) are triggered by a signaling cascade that is activated upon binding of neurotrophin to receptor (17). In order for the total number of receptors to increase in response to neurotrophin, the rate ϕ_i of insertion of receptors must be an increasing function, f_i (called growth function), of C_i. To take into account that axonal growth is relatively slow, ϕ_i lags behind $f_i(C_i)$ with a lag given by:

$$\tau \frac{d\phi_i}{dt} = f_i(C_i) - \phi_i \quad \text{[Eq. 11]}$$

where the time constant τ for growth is of the order of days. Setting immediately $\phi_i = f_i(C_i)$ does not change the main results. Different classes of growth functions were studied, all derived from the general growth function:

$$f_i(C_i) = \frac{\alpha_i C_i^m}{K_t^m + C_i^m} \quad \text{[Eq. 12]}$$

Depending on the values of m and K, the growth function is a linear function (Class I: $m = 1$ and $K_i \gg C_i$), a Michaelis-Menten function (Class II: $m = 1$ and $K_i \not\gg C_i$), or a Hill function (Class III: $m = 2$). Within each class, the specific values of the parameters α_i and K_i, as well as those of the other parameters, will typically differ between the innervating axons, e.g., as a result of differences in activity or other differences. For example, increased presynaptic electrical activity can increase the axon's total number of receptors (e.g., by up-regulation [74,75] or by stimulating axonal branching [26]) which implies that, for example, α_i is increased or γ_i is decreased.

The whole model thus consists of three differential equations for each axon i (Equations 8, 9, and 11) and one equation for the neurotrophin concentration (Equation 10). By means of numerical simulations and mathematical analysis, we can examine the outcome of the competitive process. Axons that at the end of the competitive process have no neurotrophin ($C_i = 0$, equivalent to $\phi_i = 0$) are assumed to have withdrawn or

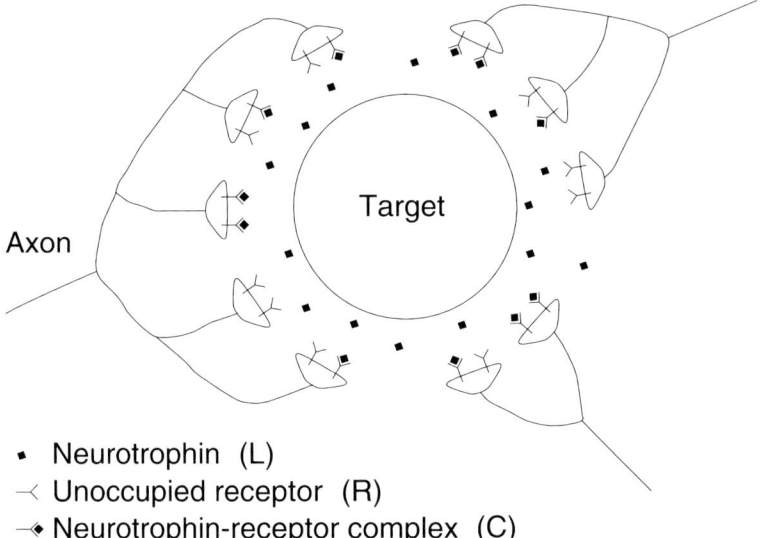

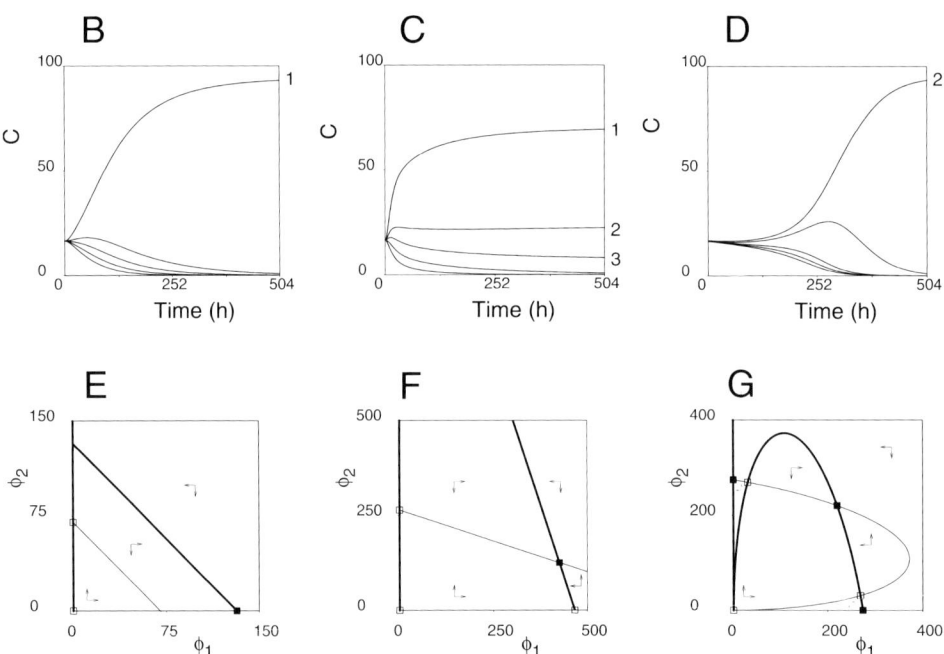

Fig. 8. The model by Van Ooyen and Willshaw *(59)* (see Subheading 11.4.). **(A)** Target cell with three innervating axons, each with a different degree of branching. The target releases neurotrophin, which binds to neurotrophin receptors at the axon terminals.

For three different classes of growth functions, **(B–D)** show the development of innervation for a system of five innervating axons, where each axon has a different competitive strength, β_i (defined in Subheading 11.4.) The values of C_i are in number of molecules. Panels **(E–G)** show died, while axons that do have neurotrophin ($C_i > 0$, equivalent to $\phi_i > 0$) are regarded as having survived.

All parameters in the model have a clear biological interpretation. For the numerical simulations, the parameter values were taken from the data available for nerve growth factor (NGF) (see also *[59]*). Mathematical analysis *(59)* shows that the results do not depend on specific choices for the parameters and are, therefore, also relevant for other neurotrophic factors.

Results of the model. For class I, starting with any number of axons, elimination of axons takes place until a single axon remains (single innervation) (Fig. 8B,E). The axon that survives is the one with the highest value of the quantity $\beta_i \equiv [k_{a,i}(\alpha_i/K_i - \rho_i)]/[\gamma_i(k_{d,i} + \rho_i)]$ which is interpreted as the axon's competitive strength. For class I, the number of surviving axons cannot be increased by increasing the rate σ of release of neurotrophin: the higher amount of neurotrophin results in further growth of the winning axon and thus more uptake of neurotrophin, so again not enough neurotrophin is left to sustain the other axons. This shows that the widely held belief that competition is a consequence of resources being produced in limited amounts is too simplistic. If the growth function is a saturating function (classes II and III), then more axons may survive if the rate σ of release of neurotrophin is increased (Fig. 8C,D,F,G). A saturating growth function means that the "size" of an axon (in terms of number of neurotrophin receptors) is bounded, so that when an axon is at its maximum, a higher amount of neurotrophin does not result in further growth and more uptake, so that other axons can profit.

For classes I and II, there is, for a given choice of the parameter values, only one stable innervation pattern (either single or multiple innervation). For class III, in contrast, stable equilibria of single and multiple innervation can coexist, and which of these will be reached in any specific situation depends on the initial conditions (Fig. 8D,G).

Fig. 8. *(continued)* the nullcline pictures for a system of two innervating axons [the variables R_i, C_i, $i = 1,2$ and L are set at quasisteady state; in (**E**) and (**F**), $\beta_1 > \beta_2$; in (**G**), $\beta_1 = \beta_2$]. In (**E–G**), the bold lines are the nullclines of ϕ_1, and the light lines are the nullclines of ϕ_2 (the x- and y-axes are also nullclines of ϕ_2 and ϕ_1, respectively). Intersection points of these lines are the equilibrium points of the system. A filled square indicates a stable equilibrium point; an open square indicates an unstable equilibrium point. Vectors indicate direction of change. (**B**) Class I. Elimination of axons takes place until a single axon remains. The axon with the highest value of the competitive strength, β_i, survives. (**C**) Class II. For the parameter settings used, several axons survive. (**D**) Class III. Dependence on initial conditions; although axon one has the highest value of the competitive strength, axon two survives because its initial value of ϕ_i is sufficiently higher than that of axon one. (**E**) Class I. The nullclines do not intersect at a point where both axons coexist. (**F**) Class II. The nullclines intersect at a point where both axons coexist. For a sufficiently lower rate of release of neurotrophin, for example, the nullclines would not intersect, and only one axon would survive. (**G**) Class III. There is a stable equilibrium point where both axons coexist, as well as stable equilibrium points where either axon is present [the stable equilibrium point at ($\phi_1 = 0$, $\phi_2 = 0$) is not indicated, because it is too close to another unstable point]. For a sufficiently higher value of K_i, for example, the stable equilibrium point where both axons coexist would disappear. From *(48)* with permission.

For all classes, axons with a high competitive strength β_i survive, and the activity dependence of β_i (e.g., via α_i) means that these are the most active ones, provided that the variation due to other factors does not predominate.

The coexistence of several stable equilibria for class III implies that an axon that is removed from a multiply innervated target may not necessarily survive ("regenerate") when replaced with a low number of neurotrophin receptors (Fig. 9A,B). To enhance the possibility that a damaged axon can return and survive on its former target, the model suggest that it is more efficient to increase the number of receptors on the regenerating axons than to increase the amount of neurotrophin (which also makes the already existing axons on the target "stronger").

Comparison with empirical data. The model can account for the development of both single and multiple innervation following a stage of hyperinnervation. Examples of single innervation are the innervation of skeletal muscle fibers *(9)*, autonomic ganglion cells with few dendrites *(76)*, and the climbing fiber innervation of cerebellar Purkinje cells *(77)*. Although undergoing a reduction in innervation, most other cell types remain multiply innervated. In agreement with the model, increasing the amount of target-derived neurotrophin delays the development of single innervation (class I) *(78)* or increases the number of surviving axons (classes II and III) (e.g., in epidermis *[79]*).

The model can also explain the coexistence of stable states of single and multiple innervation (class III) in skeletal muscle. Persistent multiple innervation is found in denervation experiments after reinnervation and recovery from prolonged nerve conduction block *(80)*. In terms of the model, conduction block changes the sizes of the basins of attraction of the equilibria (via changes in the competitive strength β_i or in the rate σ of release of neurotrophin), so that the system can go to an equilibrium of multiple innervation, while under normal conditions single innervation develops. Once the conduction block is removed, the system will remain in the basin of attraction of the multiple innervation equilibrium (Fig. 9C,D).

For competition to occur, it is not necessary that there is presynaptic or postsynaptic activity or that there is activity-dependent release of neurotrophin (cf. *[51]*). Differences in competitive strength (β_i) between axons can arise also as a result of differences in other factors than presynaptic activity, such as intrinsic differences in neurotrophic signaling (e.g., insertion or degradation of neurotrophin receptors). Thus, both presynaptic and postsynaptic activity may be influential but are not decisive *(81,82)*. This is in agreement with recent findings in the neuromuscular system, which show that activity is not necessary for competitive synapse elimination *(82)*; silent synapses can displace other silent synapses.

The model can be, and is (Ribchester, R.R., personal communication), tested experimentally. The model predicts that the axons that are being eliminated will have a small number of neurotrophin receptors. The shape of the growth function, which determines what innervation can develop, can be determined experimentally in vitro by measuring, for different concentrations of neurotrophin in the medium, the total number of terminals of an axon or, better, the axon's total number of neurotrophin receptors that it has over all its terminals. In relating axon survival to neurotrophin concentration, the

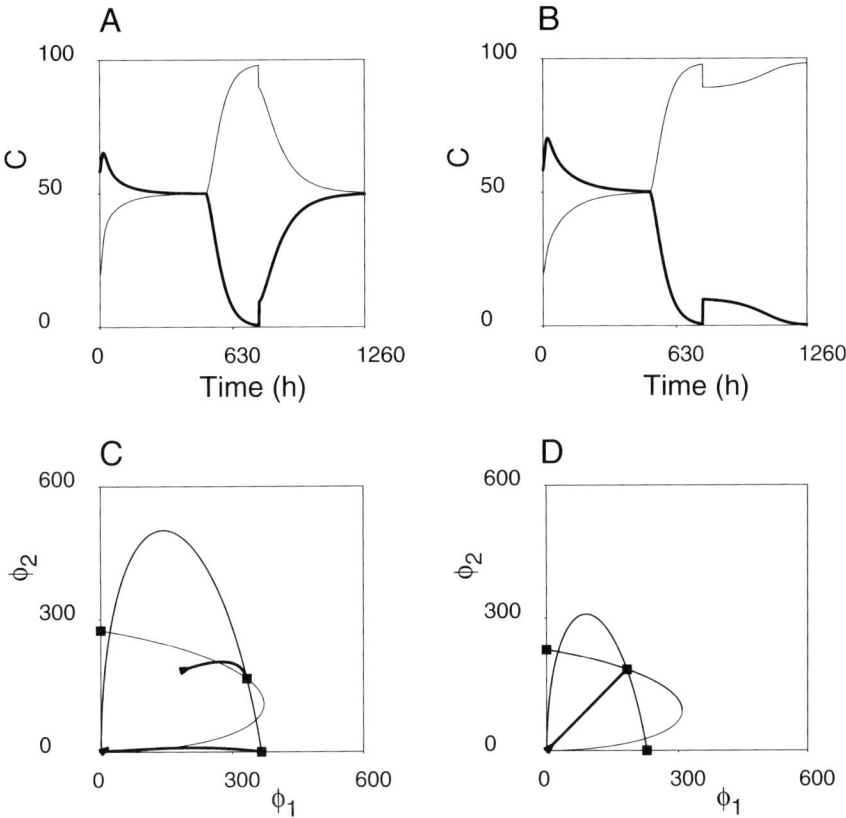

Fig. 9. The implications of the coexistence of stable states of single and multiple innervation for class III in the model by Van Ooyen and Willshaw *(59)* (see subheading 11.4.). In (**A** and **B**), removal of an axon from a multiply innervated target and subsequent replacement, for (**A**) class II and (**B**) class III. At $t = 504h$, axon 1 (bold line) is removed by setting $\alpha_1 = 0$. At $t = 756h$, axon 1 is replaced by setting α_1 back to its original value, with initial conditions $\phi_1 = 30$, $R_1 = \phi_1/\gamma$, and $C_1 = 0$. Only for class II the replaced axon can survive. For class III, in order for the replaced axon to survive, a much higher initial value of ϕ_1 would be required. From *(59)* with permission. The phase-space plots of (**C** and **D**) illustrate how, for class III, persistent multiple innervation can arise after recovery from nerve conduction block, in a system of two innervating axons. For explanation of nullclines and symbols, see Figure 8 (for clarity, the unstable equilibria are not indicated). The triangles mark the starting points of trajectories (bold lines). As shown in (**C**), under normal conditions, with electrically active axons that have a different level of activity (represented by $\alpha_1 = 400$ and $\alpha_2 = 300$; other parameter values as in Fig. 8G) and a low initial number of receptors (i.e., ϕ_i is low: $\phi_1 = \phi_2 = 0.25$), single innervation develop. When activity is blocked (values of α_i lower and the same: e.g., $\alpha_1 = 250$ and $\alpha_2 = 250$), as in (**D**), the same initial conditions lead to multiple innervation. Subsequent restoration of activity means that the nullclines are again as in (**C**), but now the starting values of ϕ_i are those reached as in (**D**), i.e., in the basin of attraction of the polyneuronal equilibrium point. The system goes to this equilibrium and will remain there forever, i.e., persistent polyneuronal innervation. Another way in which persistent multiple innervation can arise following nerve conduction block is through altering the rate of release of neurotrophin, σ, which also changes the sizes of the basins of attraction of the equilibria. From *(48)* with permission.

model predicts, for example, that the smaller the value of K_i of the growth function, the lower the concentration of neurotrophin needed to rescue more axons.

Further extensions of the model. In the model as described above, it is assumed that the concentration of neurotrophin is uniform across the extracellular space, so that all axons "sense" the same concentration. This is a good assumption if all the axons are close together on the target structure, as, for example, at the endplate on muscle fibers *(83)*. However, if the target structure is large (e.g., a large dendritic tree), the spatial dimension of the extracellular space should be taken into account. Modeling local release of neurotrophin along the target and diffusion of neurotrophin in the extracellular space, Van Ooyen and Willshaw *(84)* showed that distance between axons mitigates competition, so that if the axons are sufficiently far apart on the target, they can coexist (i.e., even under conditions, e.g., a class I growth function, where they cannot coexist with a uniform extracellular space). This can explain that (*i*) when coexisting axons are found on mature muscle cells, they are physically separated *(85–87)*; and (*ii*) a positive correlation exists between the size of the dendritic tree and the number of innervating axons surviving into adulthood *(46,76,88)*. In the ciliary ganglion of adult rabbits, for example, neurons that lack dendrites are innervated by a single axon, whereas neurons with many dendrites are innervated by the largest number of axons. In newborn animals, in contrast, all neurons are innervated by approximately the same number of axons.

In another extension of the model, Van Ooyen and Willshaw *(84)* considered a single target that releases two types of neurotrophin *(89–91)* and at which there are two types of innervating axons. Each axon type can respond to both neurotrophin types, but with different affinities (e.g., each axon type may have a different type of neurotrophin receptor, with each receptor type binding to both types of neurotrophin, but with a different affinity). The results show that different types of axons can coexist (i.e., even under conditions, e.g., a class I growth function, where they cannot coexist with a single type of neurotrophin) if they respond to the neurotrophins with sufficiently different affinities. By having axons respond with different affinities to more than one type of neurotrophin, the model can account for competitive exclusion among axons of one type while at the same time there is coexistence with axons of another type innervating the same target. This occurs, for example, on Purkinje cells *(77)*, where climbing fibers compete with each other during development until only a single one remains, which coexists with parallel fibers innervating the same Purkinje cell.

11.5. DISCUSSION

Stochastic dendritic growth models appear to be successful in describing the shapes of dendritic branching patterns, as shown in section 11.2. and by other authors *(92–94)*. The parameter values, obtained after a process of optimization, are assumed to reflect basic characteristics of the branching process. Emphasis has been given to competitive phenomena as becoming apparant by the size-dependent branching probabilities. We have shown that the "competition parameter" E significantly differentiates between different cell types. Competitive interactions were also suggested by Nowakowski et al. *(92)* as underlying a suppression of further branching immediately after a branching event.

Other successful approaches for reconstructing dendritic complexity are based on stochastic algorithms, in which segment lengths and diameters are obtained by sampling the observed distributions of shape characteristics directly *(95,95)*. These approaches do not include a phase of parameter optimization.

For a further interpretation of the results of stochastic models in terms of underlying mechanisms, one needs to model elongation and branching at more detailed levels (such as in Subheading 11.3.). In Subheading 11.3.2., we have introduced a model that explicitly includes some of the cellular mechanisms involved in elongation and branching. In the model, we have studied the consequences of the interactions between the calcium dynamics in dendritic trees and the effects of calcium on MAP (de)phosphorylation (which influences elongation and branching). With respect to producing the variability in dendritic morphologies, the model compares well with the stochastic model (Subheading 11.2.) Reproducing the data (particularly the terminal length data) using the stochastic model required separate phases of elongation–branching and elongation only, with different rates of elongation in each phase. These phases emerge automatically in the MAP model, in which both elongation and branching are generated from the same intrinsic mechanism and need not be manipulated independently.

In Subheading 11.3.1., we have shown that competition between growing neurites can emerge as a result of the interactions between the transport of tubulin and the tubulin-mediated elongation of neurites. The model can account for "dormant growth cones" and for recent experimental findings in tissue culture (G.J.A. Ramakers, unpublished results) that show that when one neurite stops growing out, other neurites, after a delay, start growing out. These results are also relevant for understanding the formation of nerve connections, because it shows that changes in the growth of a subset of a neuron's neurites (e.g., as a result of changes in electrical activity, or as a result of neurites finding their targets) can affect the growth of the neuron's other neurites (see also [97]).

At their target, axons from different neurons compete for target-derived resources. Our model of axonal competition suggests that the regulation of axonal growth by neurotrophins is crucial to the competitive process in the development, maintenance, and regeneration of nerve connections. Among the many axonal features that change during growth in response to neurotrophin (degree of arborization and, consequently, number of axon terminals; size of terminals; and density of receptors), the consequent change in the axon's total number of neurotrophin receptors, thus changing its capacity for removing neurotrophin, is what drives the competition. The form of the dose–response curve between neurotrophin and axonal arborization (or better, the total amount of neurotrophin receptors) determines what patterns of innervation can develop and what the capacity for axon regeneration will be.

REFERENCES

1. Dotti CG, Sullivan CA, Banker GA. The establishment of polarity by hippocampal neurons in culture. J Neurosci 1988; **8**:1454–1468.
2. Letourneau PC, Kater SB, Macagno ER. (eds.) The Nerve Growth Cone. Raven Press, New York, 1991.
3. Kater SB, Mattson MP, Cohan C, Connor J. Calcium regulation of the neuronal growth cone. Trends Neurosci 1988; **11**:315–321.

4. Tessier-Lavigne M, Goodman CS. The molecular biology of axon guidance. Science 1996; **274**:1123–1133.
5. Goodhill GJ. Diffusion in axon guidance. Eur J Neurosc 1997; **9**:1414–1421.
6. Wolff JR, Missler M. Synaptic reorganization in developing and adult nervous systems. Anat Anz 1992; **174**:393–403.
7. Lohof AM, Delhaye-Bouchaud N, Mariani J. Synapse elimination in the central nervous system: functional significance and cellular mechanisms. Rev Neurosci 1996; **7**:85–101.
8. Brown MC, Jansen JKS, Van Essen D. Polyneuronal innervation of skeletal muscle in newborn rats and its elimination during maturation. J Physiol (Lond) 1976; **261**:387–422.
9. Jansen JKS, Fladby T. The perinatal reorganization of the innervation of skeletal muscle in mammals. Prog Neurobiol 1990; **34**:39–90.
10. Purves D, Lichtman JW. Principles of Neural Development. Sinauer, Sunderland, MA, 1985.
11. Dotti CG, Banker GA. Experimentally induced alterations in the polarity of developing neurons. Nature 1987; **330**:254–256.
12. Goslin K, Banker GA. Experimental observations on the development of polarity by hippocampal neurons in culture. J Cell Biol 1989; **108**:1507–1516.
13. Goslin K, Banker GA. Rapid changes in the distribution of GAP-43 correlate with the expression of neuronal polarity during normal development and under experimental conditions. J Cell Biol 1990; **110**:1319–1331.
14. Samuels DC, Hentschel HGE, Fine A. The origin of neuronal polarization: a model of axon formation. Philos Trans R Soc B 1996; **351**:1147–1156.
15. Van Veen MP, Van Pelt J. Neuritic growth rate described by modeling microtubule dynamics. Bull Math Biol 1994; **56**:249–273.
16. Purves D. Body and Brain: A Trophic Theory of Neural Connections. Harvard University Press, Cambridge, MA, 1988.
17. Bothwell M. Functional interactions of neurotrophins and neurotrophin receptors. Annu Rev Neurosci 1995; **18**:223–253.
18. Lewin GR, Barde Y-A. Physiology of the neurotrophins. Annu Rev Neurosci 1996; **19**:289–317.
19. Hentschel HGE, Van Ooyen A. Models of axon guidance and bundling during development. Proc R Soc Lond B 1999; **266**:2231–2238.
20. Van Pelt J, Uylings HBM. Natural variability in the geometry of dendritic branching patterns. In: Modeling in the Neurosciences: From Ionic Channels to Neural Networks. (Poznanski, RR, ed.). Harwood Academic Publishers, Amsterdam, 1999. pp. 79–108.
21. Van Pelt J, Van Ooyen A, Uylings HBM. Modeling dendritic geometry and the development of nerve connections. In: Cannon RC (CD-ROM) Computational Neuroscience: Realistic Modeling for Experimentalist. De Schutter E (ed.), CRC Press, Boca Raton, 2001. pp. 179–208.
22. Van Pelt J, Dityatev AE., Uylings HBM. Natural variability in the number of dendritic segments: model-based inferences about branching during neurite outgrowth. J Comp Neurol 1997; **387**:325–340.
23. Van Pelt J, Uylings HBM. Modeling the natural variability in the shape of dendritic trees: application to basal dendrites of small rat cortical layer 5 pyramidal neurons. Neurocomputing 1999; **26–27**:305–311.
24. Van Pelt J, Schierwagen A, Uylings HBM. Modeling dendritic morphological complexity of deep layer cat superior colliculus neurons. Neurocomputing 2001; **38–40**:403–408.
25. Van Pelt J, Graham B, Uylings HBM. Formation of axonal and dendritic branching patterns. In: Modeling Neural Development. (van Ooyen A, ed.). MIT Press, Cambridge, MA, in press.
26. Ramakers GJA, Winter J, Hoogland TM, et al. Depolarization stimulates lamellipodia for-

mation and axonal but not dendritic branching in cultured rat cerebral cortex neurons. Dev Brain Res 1998; **108**:205–216.
27. Raper JA, Tessier-Lavigne M. Growth cones and axon pathfinding. In: Fundamental Neuroscience (Zigmond MJ, Bloom FE, Landis SC, Roberts JL, Squire LR, eds.). Academic Press, San Diego, 1999. pp. 519–546.
28. Gelfand VI, Bershadsky AD. Microtubule dynamics: mechanisms, regulation, and function. Annu Rev Cell Biol 1991; **7**:93–116.
29. Sánchez C, Díaz-Nido J, Avila J. Phosphorylation of microtubule-associated protein 2 (MAP2) and its relevance for the regulation of the neuronal cytoskeleton function. Prog Neurobiol 2000; **61**:133–168.
30. Schilstra MJ, Bayley PM, Martin SR. The effect of solution composition on microtubule dynamic instability. Biochem J 1991; **277**:839–847.
31. Van Ooyen A, Graham BP, Ramakers GJA. Competition for tubulin between growing neurites during development. Neurocomputing 2001; **38–40**:73–78.
32. Van Ooyen A, Graham BP. Compartmental models of growing neurites. Neurocomputing 2001; **38–40**:31–36.
33. Bray D. Branching patterns of individual sympathic neurons in culture. J Cell Biol 1973; **56**:702–712.
34. Maccioni R, Cambiazo V. Role of microtubule associated proteins in the control of microtubule assembly. Physiol Rev 1995; **75**:835–864.
35. Yamamoto H, Saitoh Y, Fukunaga K, Nishimura H, Miyamoto E. Dephosphorylation of microtubule proteins by brain protein phosphatases 1 and 2A, and its effect on microtubule assembly. J Neurochem 1985; **50**:1614–1623.
36. Hall G, Lee V, Kosik K. Microtubule destabilization and neurofilament phosphorylation precede dendritic sprouting after close axotomy of lamprey central neurons. Proc Natl Acad Sci USA 1991; **88**:5016–5020.
37. Audesirk G, Cabell L, Kern M. Modulation of neurite branching by protein phosphorylation in cultured rat hippocampal neurons. Dev Brain Res 1997; **102**:247–260.
38. Hely, TA, Graham BP, Van Ooyen A. A computational model of dendrite elongation and branching based on MAP2 phosphorylation. J Theor Biol 2001; **210**: 375–384.
39. Giese KP, Fedorov NB, Filipkowski RK, Silva AJ. Autophosphorylation at Thr286 of the alpha calcium-calmodulin kinase II in LTP and learning. Science 1998; **279**:870–873.
40. Klee CB, Draetta GF, Hubbard MJ. Calcineurin. Adv Enzymol Relat Areas Mol Biol 1988; **61**:149–200.
41. Larkman, A. Dendritic morphology of pyramidal neurones of the visual cortex of the rat: I. Branching patterns. J Comp Neurol 1991; **306**:307–319.
42. Diez-Guerra FJ, Avila J. MAP2 phosphorylation parallels dendrite arborization in hippocampal neurones in culture. Neuroreport 1993; **4**:412–419.
43. Purves D, Lichtman JW. Elimination of synapses in the developing nervous system. Science 1980; **210**:153–157.
44. Hubel DH, Wiesel TN, LeVay S. 1977; Plasticity of ocular dominance columns in the monkey striate cortex. Philos Trans R Soc Lond Ser B **278**:377–409.
45. Wiesel TN. Postnatal development of the visual cortex and the influence of environment. Nature 1982; **299**:583–591.
46. Purves D. Neural Activity and the Growth of the Brain. Cambridge University Press, Cambridge, UK, 1994.
47. Sanes JR, Lichtman JW. Development of the vertebrate neuromuscular junction. Annu Rev Neurosci 1999; **22**:389–442.
48. Van Ooyen A. Competition in the development of nerve connections: a review of models. Network Comput Neural Syst 2001; **12**:R1–R47.

49. Oppenheim RW. Cell death during development of the nervous system. Annu Rev Neurosci 1991; **14**:453–501.
50. Oppenheim RW. The concept of uptake and retrograde transport of neurotrophic molecules during development: history and present status. Neurochem Res 1996; **21**: 769–777.
51. Snider WD, Lichtman JW. Are neurotrophins synaptotrophins? Mol Cell Neurosci 1996; **7**:433-442.
52. Cohen-Cory, S. and Fraser, S. E. Effects of brain-derived neurotrophic factor on optic axon branching and remodelling in *in vivo*. Nature 1995; **378**:192–196.
53. Miller KD. Synaptic economics: competition and cooperation in correlation-based synaptic competition. Neuron 1996; **17**:371–374.
54. Bienenstock EL, Cooper LN, Munro PW. Theory for the development of neuron selectivity: orientation specificity and binocular interaction in visual cortex. J Neurosci 1982; **2**:32–48.
55. Zhang LI, Huizhong WT, Holt CE, Harris WA, Poo M-M. A critical window for cooperation and competition among developing retinotectal synapses. Nature 1998. **395**:37–44.
56. Song S, Miller KD, Abbott LF. Competitive Hebbian learning through spike-timing dependent synaptic plasticity. Nat Neurosci 2000; **3**:919–926.
57. Swindale NV. The development of topography in the visual cortex: a review of models. Network Comput Neural Sys 1996; **7**:161–247.
58. Elliott T, Shadbolt NR. Competition for neurotrophic factors: ocular dominance columns. J Neurosci 1998; **18**:5850–5858.
59. Van Ooyen A, Willshaw DJ. Competition for neurotrophic factor in the development of nerve connections. Proc R Soc B 1999; **266**:883–892.
60. Elliott T, Shadbolt NR. Competition for neurotrophic factors: mathematical analysis. Neural Comput 1998; **10**:1939–1981.
61. Jeanprêtre N, Clarke PGH, Gabriel J-P. Competitive exclusion between axons dependent on a single trophic substance: a mathematical analysis. Math Biosci 1996; **133**:23–54.
62. Campenot RB. Development of sympathetic neurons in compartmentalized cultures. I. Local control of neurite outgrowth by nerve growth factor. Dev Biol 1982; **93**:1–12.
63. Campenot RB. 1982; Development of sympathetic neurons in compartmentalized cultures. II. Local control of neurite survival by nerve growth factor. Dev Biol 1982; **93**:13–22.
64. Edwards RH, Rutter WJ, Hanahan D. Directed expression of NGF to pancreatic β cells in transgenic mice leads to selective hyperinnervation of the islets. Cell 1989; **58**:161–170.
65. Yasuda T, Sobue G, Ito T, Mitsuma T, Takahashi A. Nerve growth factor enhances neurite arborization of adult sensory neurons; a study in single-cell culture. Brain Res 1990; **524**:54–63.
66. Yunshao H, Zhibin Y, Yaoming G, Guobi K, Yici C. Nerve growth factor promotes collateral sprouting of cholinergic fibres in the septohippocampal cholinergic system of aged rats with fimbria transection. Brain Res 1992; **586**:27–35.
67. Diamond J, Holmes M, Coughlin M. Endogenous NGF and nerve impulses regulate the collateral sprouting of sensory axons in the skin of the adult rat. J Neurosci 1992; **12**:1454-1466.
68. Causing CG, Gloster A, Aloyz R, et al. Synaptic innervation density is regulated by neuron-derived BDNF. Neuron 1997; **18**:257–267.
69. Schnell L, Schneider R, Kolbeck R, Barde Y-A, Schwab, ME. Neurotrophin-3 enhances sprouting of corticospinal tract during development and after adult spinal cord lesion. Nature 1994; **367**:170–173.
70. Funakoshi H, Belluardo N, Arenas E, et al. Muscle-derived neurotrophin-4 as an activity-dependent trophic signal for adult motor neurons. Science 1995; **268**:1495–1499.
71. Garofalo L, Ribeiro-da-Silva A, Cuello C. Nerve growth factor-induced synaptogenesis and hypertrophy of cortical cholinergic terminals. Proc Natl Acad Sci USA 1992; **89**:2639–2643.

72. Holtzman DM, Li Y, Parada LF, et al. p140^{trk} mRNA marks NGF-responsive forebrain neurons: evidence that *trk* gene expression is induced by NGF. Neuron 1992; **9**:465–478.
73. Wiley HS, Cunningham DD. A steady state model for analyzing the cellular binding, internalization and degradation of polypeptide ligands. Cell 1981; **25**:433–440.
74. Birren, S. J., Verdi, J. M., and Anderson, D. J. Membrane depolarization induces p140^{trk} and NGF responsiveness, but not p75^{LNGFR}, in MAH cell. Science 1992; **257**:395–397.
75. Salin T, Mudo G, Jiang XH, Timmusk T, Metsis M, Belluardo N. Up-regulation of trkB mRNA expression in the rat striatum after seizures. Neurosci Lett 1995; **194**:181–184.
76. Hume RI, Purves D. Geometry of neonatal neurones and the regulation of synapse elimination. Nature 1981; **293**:469–471.
77. Crepel F. Regression of functional synapses in the immature mammalian cerebellum. Trends Neurosci 1982; **5**:266–269.
78. Nguyen QT, Parsadanian AS, Snider WD, Lichtman JW. Hyperinnervation of neuromuscular junctions caused by GDNF overexpression in muscle. Science 1998; **279**:1725–1729.
79. Albers KM, Wright DE, Davies BM. Overexpression of nerve growth factor in epidermis of transgenic mice causes hypertrophy of the peripheral nervous system. J Neurosci 1994; **14**:1422–1432.
80. Barry JA, Ribchester RR. Persistent polyneuronal innervation in partially denervated rat muscle after reinnervation and recovery from prolonged nerve conduction block. J Neurosci 1995; **15**:6327–6339.
81. Ribchester RR. Activity-dependent and -independent synaptic interactions during reinnervation of partially denervated rat muscle. J Physiol. (Lond) 1988; **401**:53–75.
82. Costanzo EM, Barry JA, Ribchester RR. Competition at silent synapses in reinnervated skeletal muscle. Nat Neurosci 2000; **3**:694–700.
83. Balice-Gordon RJ, Chua CK, Nelson CC, Lichtman JW. Gradual loss of synaptic cartels precedes axon withdrawal at developing neuromuscular junctions. Neuron 1993; **11**:801–815.
84. Van Ooyen A, Willshaw DJ. Development of nerve connections under the control of neurotrophic factors: parallels with consumer-resource systems in population biology. J Theor Biol 2000; **206**:195–210.
85. Kuffer D, Thompson W, Jansen JKS. The elimination of synapses in multiply-innervated skeletal muscle fibres of the rat: dependence on distance between end-plates. Brain Res 1977; **138**:353-358.
86. Lømo T. What controls the development of neuromuscular junctions? Trends Neurosci 1980; **3**:126–129.
87. Lo Y-J, Poo M-M. Activity-dependent synaptic competition in vitro: heterosynaptic suppression of developing synapses. Science 1991; **254**:1019–1022.
88. Purves D, Hume RI. The relation of postsynaptic geometry to the number of presynaptic axons that innervate autonomic ganglion cells. J Neurosci 1981; **1**:441–452.
89. Barde YA. Trophic factors and neuronal survival. Neuron 1989; **2**:1525–1534.
90. McManaman JL, Crawford F, Clark R, Richker J, Fuller F. Multiple neurotrophic factors from skeletal muscle: demonstration of effects of bFGF and comparisons with the 22-kdalton CAT development factor. J Neurochem 1989; **53**:1763–1771.
91. Lindsay RM, Wiegand SJ, Altar CA, DiStefano PS. Neurotrophic factors: from molecule to man. Trends Neurosci 1994; **17**:182–189.
92. Nowakowski RS, Hayes NL, Egger MD. Competitive interactions during dendritic growth: a simple stochastic growth algorithm. Brain Res 1992; **576**:152–156.
93. Uemura E, Carriquiry A, Kliemann W, Goodwin J. Mathematical modeling of dendritic growth in vitro. Brain Res. 1995; **671**:187-194.
94. Devaud JM, Quenet B, Gascuel J, Masson C. Statistical analysis and parsimonious modelling of dendrograms of *in vitro* neurones. Bull Math Biol 2000; **62**:657–674.

95. Burke RE, Marks WB, Ulfhake B. A parsimonious description of motoneuron dendritic morphology using computer simulation. J Neurosc 1992; **12**:2403–2416.
96. Ascoli G, Krichmar JL. L-Neuron: a modeling tool for the efficient generation and parsimonious description of dendritic morphology. Neurocomputing 2000; **32–33**:1003–1011.
97. Gan WB, Macagno ER. Competition among the axonal projections of an identified neuron contributes to the retraction of some of those projections. J Neurosci 1997; **17**:42293–42301.

12
Axonal Navigation Through Voxel Substrates
A Strategy for Reconstructing Brain Circuitry

Stephen L. Senft

ABSTRACT

Recent advances in obtaining and analyzing 3D volumetric scans of entire brains and in algorithmic generation of neuron-like tree structures now make possible the representation of simulated brain networks having both large-scale anatomical fidelity and submicron specification. The present work illustrates a strategy for embedding detailed compartmental models of brain circuitry within voxel-based anatomical Atlases. Groups of simulated neuronal somata are given coordinates corresponding to brain nuclei segmented from the Atlas data. Other cellular details (soma number and size and statistical dendritic form) can be derived from published studies of these nuclei. Simulated axons are then made to navigate from these sources into specified target regions, with their paths being guided by tracts detected in the 3D scans. Although the actual substrates traversed by living axons involve evanescent chemical markings that are not captured by the voxel data, enough boundaries remain to usefully constrain the axons, and a rough scaffolding of brain circuitry may be constructed. Terminal arborization and synapse with target cells is achieved via cell-biologically-motivated growth algorithms. One can activate segmental compartments in the resulting structure to emulate signals propagating in 3D through the synthetic networks. This new combination of methods should allow one to generate plausible visualizations of brain structure and activity that are obtainable by no other means and which can be improved as scans and algorithms are refined. Such networks could serve to relate visually microscopic and macroscopic brain anatomy. The simulated behavior of these circuits, while still far from precisely replicating in vivo dynamics, nevertheless, by being mapped onto realistic pathways, may suggest novel patterns of activity flow through the central nervous system (CNS).

12.1. INTRODUCTION
12.1.1. Volume Data

Volumetric anatomical data is becoming increasingly prevalent. These 3D raster formats provide plenary views of tissue organization wherein attributes assigned to every surface and interior voxel can be inspected and measured. Natural objects, including brains, are routinely converted into voxel format by a number of scientific scanning methods, such as computerized X-ray tomography (CT), magnetic resonance imaging (MRI), and confocal microscopy. The now routine exchange of scientific information in this data format has been facilitated by a growing variety of 3D rendering tools that permit sophisticated and reproducible visualization and analysis of the scanned objects (e.g., VoxelView [1], and Analyze [2]).

However, in comparison with what is desired by the research biologist, the degree of structural detail in presently available voxel representations of tissue is severely limited. Pertinent levels of comprehensive analysis ideally should extend to the cell biological level of channel densities, local ion dynamics, and subcellular regulation of genes and gene-products and should extend to all cells contained in the full span of the organs studied.

Currently, 3D subcellular detail can be achieved only for small regions (e.g., using polarizing or confocal microscopes [3–5]). But even then, such data represents the transduced signals only of a tiny fraction of the ongoing biochemistry encompassed by the scan. For certain tissues, whose organization varies little with position and whose capacity scales with mass (such as liver or lung), sampling such relatively restricted volumes may suffice for evaluating the microscopic causes of macroscopically observed changes. But this is not true of the central nervous system (CNS). The most salient (and still very imperfectly understood) feature of the nervous system is the network: spatially extended skeins of connected cell ensembles. Here, because of highly specific axonal projections, activity at long distances has profound local importance. In principle, decisive information can be transported into a remote region even by single slender fibers. Even more pervasively consequential may be the local integration of multiple subthreshold inputs from distant sources.

The present publication addresses this resolution constraint as it pertains to brain scans. It introduces a methodological strategy whereby information about axonal connectivity can be integrated with volumetric data to produce flexible representations of distributed 3D brain networks having subcellular resolution.

12.1.2. Network Data

Network features currently are captured by volumetric data in very sketchy form. It is true that the confocal microscope can digitize tissue in 3D at resolutions high enough to identify sites of synaptic contact with reasonable confidence, particularly when they are augmented with multicellular electrophysiological recordings *(5,6)*. But this achievement is restricted to composited vignettes little larger than the working distance and field-of view of a high numeric aperture objective (less than a cubic millimeter). Moreover, it requires painstaking tissue preparation. Extensive manual tracing or image analysis also is needed to extract even a single neuron's dendritic arbor, and it becomes dauntingly difficult to acquire the geometry of an extensive multicellular axonal network, at the requisite better than 1 μm resolution. Reconstructions from electron micrographs are even more constrained in the spatial extents achievable, although they do represent the gold standard for verifying connectivity *(7,8)*. At the other spatial extreme, the water content of very large (100 mm) areas can be surveyed with ease using magnetic resonance imaging (MRI). Using tensor analysis *(9–11)*, one can now begin to highlight the brain's major (50–500 µm scale) fiber pathways, perhaps even during development *(12,13)*. Even though MRI technology under some conditions *(14)* can detect individual cells, even then it does not define their branches or synapses, and thus in general, this method of data collection also does not provide sufficiently detailed representations of network connectivity.

While no practical method currently is able to document the detailed connectivity of networks throughout the brain, there may be a practical optimum in the presently achievable compromise between resolution and extent of coverage. Recently, Atlas information has become available *([15]*; and see Subheading 12.2.1) having a level of voxel resolution (approx 10 µm) intermediate between MRI and confocal. This data, carefully reconstructed from histologically-stained physical sections, contains relatively high resolution cellular and pathway information throughout the entire adult mouse brain. Similar atlases are available for the rat and other strains and ages of mouse *(16–18)*. Because they are based on optical image formation and on specific staining methods (as yet unavailable to MRI) these sources of data can provide information on the relatively smaller nuclei and tracts throughout the brain, which is important for the work presented here. However, like MRI data, these Atlases do not resolve cell-to-cell contacts.

12.1.3. Histochemical Data

There is, in addition to voxel data, a vast corpus of cell biological information that catalogs brain network components. This body of work represents the combined efforts of generations of anatomists and biochemists, using an extreme variety of approaches. As a result, highly detailed and diverse types of data are available on cells local to virtually any nameable region in the CNS and for a large variety of species (e.g., mouse, rat, rabbit, monkey, man). One can learn, for almost any given region and stage of development, the numbers and types and densities of cells, and often their biochemical attributes, such as transmitter and channel and control protein information. Because of Golgi stains, intracellular dye injections, and most recently, genetic labeling methods (e.g., *[19]*), cell arborization characteristics often are known as well, having been acquired by visual observation and by *camera lucida* tracing (increasingly in computerized and even confocal format).

Classically, this variety of cell biological information reposes, piecemeal, in the printed literature, not amenable to rapid review but obtainable only through laborious manual library search. More recently, this original literature is becoming available electronically, and for certain of these computerized data *(20,21)*, compendia are progressively being compiled. For neuroanatomy in particular, the on-line availability of arbor data *(22,23)* is particularly encouraging, although in comparison to the trillions of cells in even a single brain, the several hundred or so electronically traced cells represent a very small sample. Fortunately, this trend towards electronically-based biological information is accelerating.

12.1.4. Integrative Aims

Neuroscientists have a wish to recombine this multifaceted information into coherently functioning quantitative systems. Until that is better achieved, functional parcellation and integration can be assessed for the CNS only in a very approximate sense. For, while an astonishing number of facts are known about the brain and its components, one can at present rarely (if at all) predict accurately the consequences for local neurite activation of network properties operating at a distance (the essence of brain activity). Such predictive insights require densities of parameter measurement and levels of consistency among the resulting quantitative or statistical data that are generally unattained at present.

Ideally, we are interested in coherently representing rich networks of neurites having forms like those seen through the microscope, but without the constraints of scale or extent that are imposed by the practical limits of optical technology. However, we remain far from having any techniques that will allow us to systematically record volumetric anatomy (much less dynamic physiology) at enough resolution to view the network organization of the CNS with such accuracy. Until technological breakthroughs of large magnitude provide this (predictable at this point in time for neither MRI, nor Atlases, nor confocal methods), we must look for other means of obtaining such views. One possibility is that this can be done through computer simulation and 3D image synthesis.

As a contribution to this integrative effort, the present work describes a method for generating detailed 3D views of arborized neuronal networks, embedded in and constrained by voxel-based substrates, and informed by observations from cell biology. Construction of such models (which can explicitly contain very many consistently framed parameters) could lead towards a representation of region-wide physiology operating at the synaptic level.

12.2. METHOD AND RESULTS

Two principal components are newly linked in this work: (*i*) voxel-based Atlases of the mouse brain *(15)*; and (*ii*) a system for creating networks of compartment-based neuronal models made of branched tubules (ArborVitae, *[24]*). This work also relies on 3D visualizations and analyses provided by additional software packages (Vital Images' VoxelView and VoxelMath, the latter written in large part by the author). VoxelView allows one to view 3D voxel data sets from arbitrary vantage points (including interior views) and to highlight portions of the data based on intensity differences and on coherency of structural regions. VoxelMath adds a suite of 2D and 3D image-processing routines, which in this project were used to help to align sections and to extract brain regions on the basis of location, intensity, and texture. The Atlas data is used as a 3D matrix within which simulated CNS networks are grown.

12.2.1. Mouse Atlas

Dr. Richard Sidman and colleagues at Harvard University generated the voxel data used in this work. The data consist of two interleaved series of sections cut at 20 µm from a single mouse brain, and processed alternately with Nissl and myelin stains. These sections were among those originally published as an Atlas in 1971 *(25)*. In 1998, the original slides were rescanned in 24-bit color at 10 µm lateral resolution using a Leaf Lumina camera. They were reduced to gray scale (for ease of manipulation

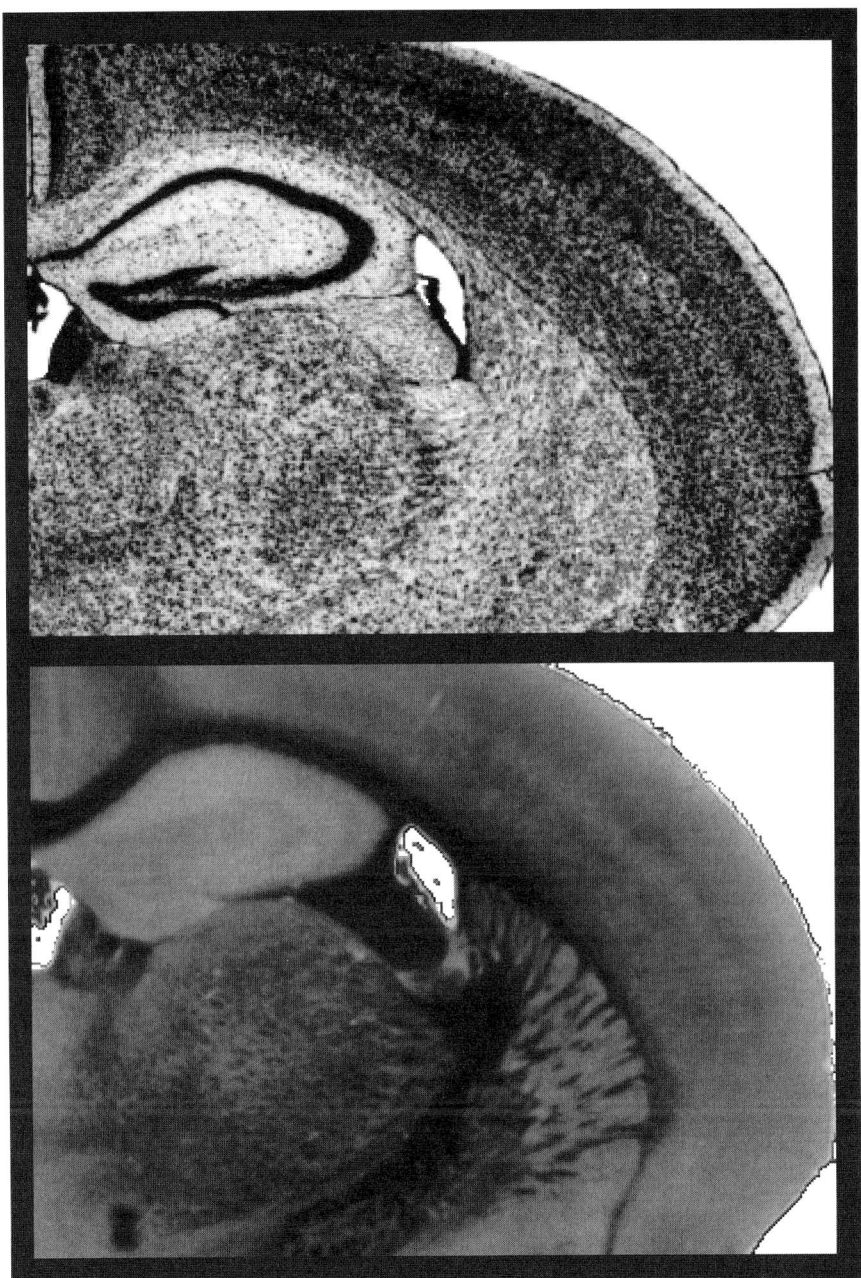

Fig. 1. Nissl and myelin slice data: frontal. These 8-bit raw cropped sections extracted from the Sidman mouse Atlas show the degree of anatomical voxel detail underlying this simulation. The thalamus, thalamic radiations, and portions of the target cortex are prominent. The dark cluster of cells in the lower middle of the upper Nissl-stained panel is the thalamus, and includes the VB on its right. The region with dark lines through it (in myelin stain, lower panel) is the thalamic radiation. Cerebral cortex is along the top (truncated) and at the far right. The hippocampus in the upper panel has its classical appearance showing the V-shaped dentate gyrus and, in the lower panel, lies under a dark myelinated band (the white matter) running along the under surface of the cortex. An electronic version of this figure is available in the companion CD-ROM.

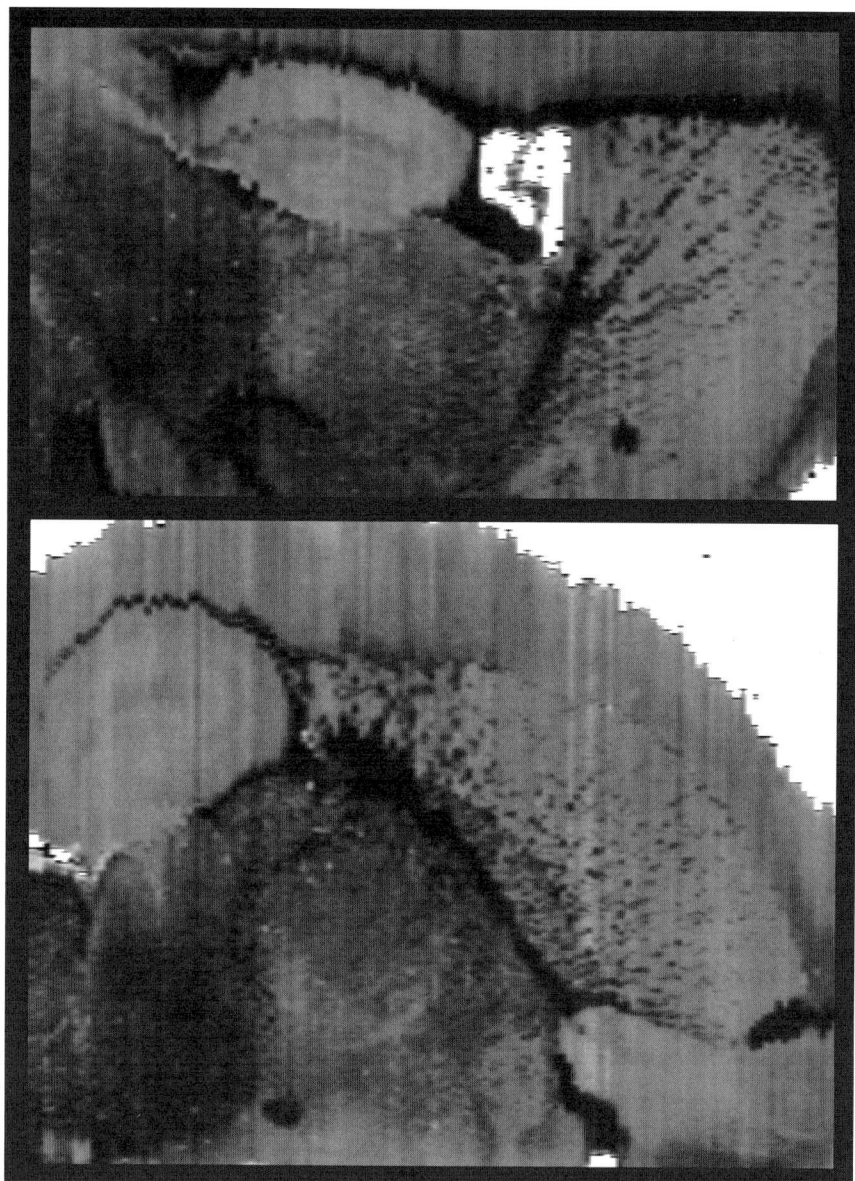

Fig. 2. Myelin slice data: sagittal and horizontal. This figure shows the myelin voxel data viewed along the two other canonical axes. The cerebral cortex runs along the top (truncated) in the upper, sagittal, view. The hippocampus is the light gray oblate structure at the upper middle. The thalamic radiations are at the right, containing dark obliquely-running myelinated line segments (which are continuous in 3D, but truncated in this plane). The thalamus itself is in the lower center. The bright region in the middle of the upper image is ventricle, bisected by a band of choroid plexus. The lower panel shows a horizontal view. Anterior of the brain is to the right. Hippocampus is at the far left, thalamic radiations are to its right, cerebral cortex is above both of these structures, and VB is the large circular region in the lower center of the picture. Vertical banding is from differential staining of the original sections, which is seen on the edge in these views. Some alignment errors remain, in part due to differential shrinkage of adjacent sections during histological processing. An electronic version of this figure is available in the companion CD-ROM.

with the then available computers and software) and assembled into two series (Nissl and myelin), each 281 sections in number.

For the present purpose it is important that voxel data be well aligned in order to minimize nonbiological boundary artifacts that might interfere with simulated axon navigation. Atlas sections were aligned approximately, by eye, during the scanning procedure, then more carefully matched using interactive and semiautomatic methods in VoxelMath. This intermediate process did not apply rotational alignment, in order to maintain image sharpness, which generally becomes degraded with multiple rotations. It also repaired a few out-of-order sequences. Next the data were aligned more precisely using both offset and rotation in Bitplane's *AutoAligner*. By comparing the final result with orthogonal views in the printed Atlas, one can observe a small residual drift in 3D down the central axis of the brain, which is not eliminated by these alignment steps. Alignment instead primarily optimized the match between adjacent images, and the accuracy in matching was limited by differential shrinkage of the sections.

The results of this work are two coherent 100 MB ($768 \times 496 \times 281 \times 8$-bit) data sets that are in approximate alignment with each other. In one can be seen clearly the cellular distributions of numerous brain nuclei and in the other the myelinated pathways interconnecting them. As one aspect of this Atlas, a number of CNS structures have been segmented into adjoining 3D regions. The raw and segmented voxel data can be paged through, in any of the three canonical orthogonal axes, or on any oblique axis, and can be very effectively perused in 3D using the large variety of rendering options in VoxelView. A selected portion of the Atlas series is shown here in Figures 1 and 2.

These Atlas visualizations do not show the cellular connectivity of the mouse brain because, even at 10 μm and with these stains, that information is not recorded in the voxels. Future plans are to improve the resolution of such Atlases to approx 2 μm (Sidman, personal communication). This would much better capture the topologies of myelinated pathways and the textural details within brain nuclei, but it still would be insufficient to follow arbor branches. Nevertheless, this intermediate resolution data can be very useful in providing a realistic 3D template to constrain a higher resolution, algorithmically reconstructed, anatomy.

12.2.2. ArborVitae

The compartmental modeling program ArborVitae uses algorithmic procedures to grow dendrites on clusters of source cells and to extend simulated axonal arbors from those cells so that they come close to, and arborize near, clusters of target cells. Axons can crawl along the target cells to synapse selectively on the somata or dendrites. The program can mingle many such sources and targets and emulates simple forms of signal propagation through the networks, which result from connecting these cell groups. The cell structures produced also can be exported to additional simulation packages, such as Neuron *(26)*. ArborVitae permits free 3D navigation through a network, with the ability to zoom into regions of interest to view features such as varicosities, spines, or synaptic contacts. In addition, the program can recapitulate, as a 3D movie, the developmental trajectory taken during the ontogeny of the network. In brief, ArborVitae enables the creation of a 3D library of neuronal cell morphology and provides a detailed anatomical framework in which to emulate localized neuronal physiology. In this framework, the internal state of any compartment can be sensitive to distant events in an extended circuit, because the program tracks both the afferent trajectories and their internal activity.

In an earlier instantiation of the program, the emulated cells and neurites grew in free space, as it were, and the distributions of cells were specified using simple statistical constraints. For many of the named brain nuclei, data exist that could permit one, in this statistical way, to establish cell locations in a common 3D coordinate system. The groups of cells thus simulated (see *[24]*) appeared as globular clusters or laminate structures (or as germinal zones, from whence cells could migrate to a final position). Without any additional constraints, the tracts interconnecting such cell groups were relatively straight, but while this may be a biological design goal (to conserve material and maximize signal transmission speed), this is only sometimes observed in the CNS. Instead, the meander of tracts brings much of the interesting morphology to the brain. However, it is difficult even to approximately define the peculiar cell group and tract geometries observed in the CNS, such as in the hippocampus *(27)*, by analytical means. Now this important geometrical aspect of simulated brain reconstruction is greatly simplified because one can use, as 3D templates, anatomical shapes segmented from voxel data. Two important improvements result: the location of cell bodies can be more accurately modeled and more natural constraints on axonal projection can be applied.

12.2.3. In Voxo Tissue Culture

To merge simulated cellular detail with voxel intensity data, solid regions containing source and target cells are extracted from the voxel data by a combination of manual and automatic segmentation procedures *(15)* and saved as data files containing those voxel locations and intensities. The data files are resampled to provide 3D coordinates for the generated source and target cells. Simulated networks then are elaborated among these cell populations by ArborVitae, which in addition applies constraints derived from cell biology. These show up as statistical parameters governing cell size and branching and in the design of algorithms emulating general mechanisms of neurite outgrowth *(24)*.

A shared memory data structure is used to provide the outgrowth algorithms with access to the 3D voxel intensities (which can be viewed and manipulated simultaneously with VoxelView and VoxelMath) in the regions being traversed by the emerging arbors. The resulting neuronal geometries can be shown as vectors or rendered tubules in 3D and can be reinserted into the underlying voxels by modifying those memory locations that contain (have the same world coordinates as) the geometry.

As a test case, a portion of the mouse somatosensory thalamocortical afferent (TCA) pathway was chosen (in part because it represents a nontrivial projection and in part because the author's PhD thesis *(28)* focused on the development of such afferents in this region of the mouse brain). First, the ventrobasal complex (VB) and somatosensory layer IV cortex were delimited (Fig. 3). To define them precisely would have involved considerable manual intervention. However, for expediency, their boundaries were roughly hewn by hand tracing and region-growing procedures (for instance, the pial surface was eroded inward to provide superior and inferior margins for layer IV, but the resulting region strictly also encompasses much of layer V). Numerous individual

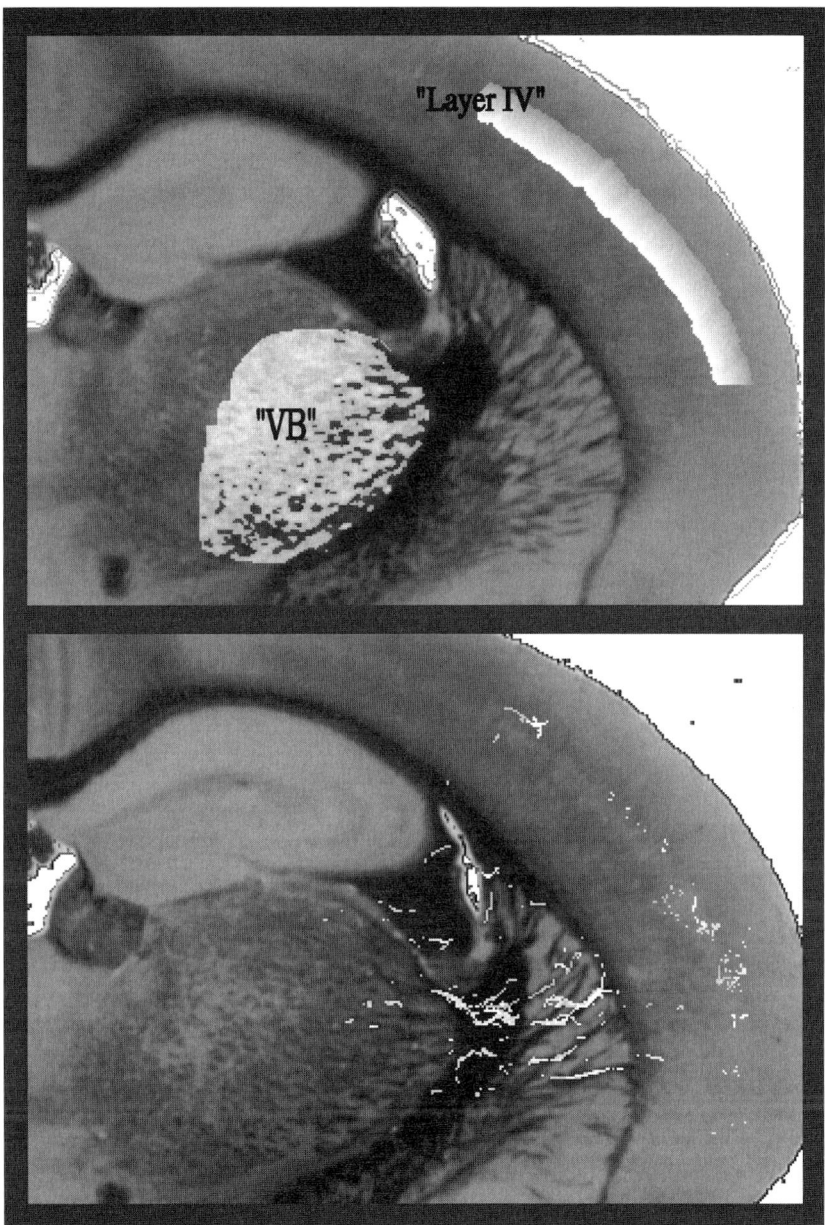

Fig. 3. Slice data: outlined regions and embedded fibers. The top panel indicates (in overlay) the approximate location of the source (VB, lower center) and target (Layer IV, upper right) regions used in these simulations. VB was carved out of the raw data by drawing a boundary in one section and propagating the contour to all sections containing the nucleus. Layer IV was defined by eroding the brain surface (in 3D) to two levels, roughly defining the top and bottom of layer IV (but also including some of layer V), and then restricting the cortical region to that sector lying directly above the radiations. The bottom panel shows portions of the paths (crossing this section) taken by simulated axons, *en route* between these two regions, as they cross the plane of this section. For the most part, axons travel along the dark channels in the radiations. Some errant fibers cross the dark fimbria of the hippocampus. Fragments of these axons' terminal arbors (intact in 3D) are visible in the cortical target zone. An electronic version of this figure is available in the companion CD-ROM.

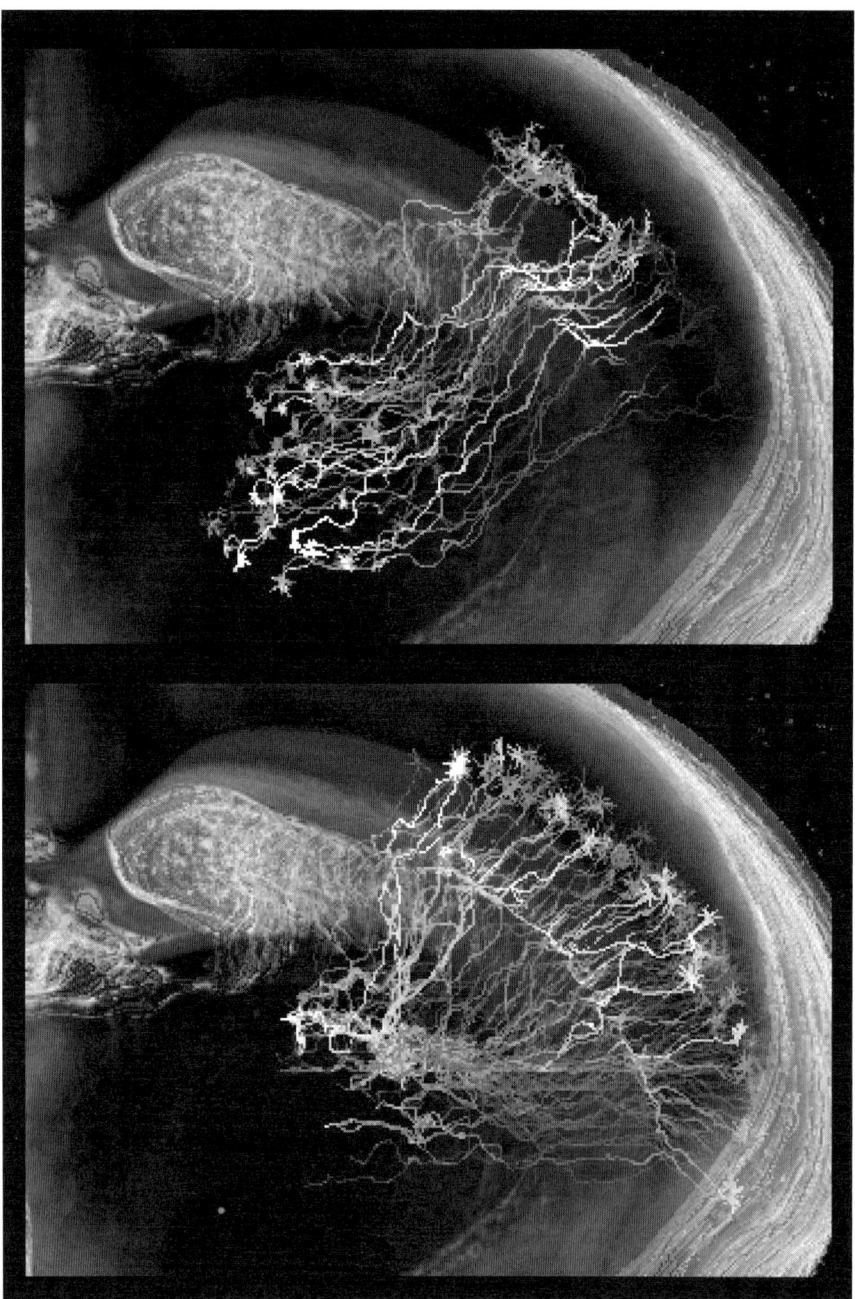

Fig. 4. Rendered data: embedded pathways. This figure shows a number of thalamic projection cells with dendritic arbors in the thalamus (see also Fig. 7), sending their axons through the thalamic radiations and into the cerebral cortex. Each cell is given a different hue. The axonal tracks run to the upper right, past the voxels containing the hippocampus and ventricular choroid plexus (bright region at top left), but in a plane tilted towards the viewer. The bottom panel shows the fiber paths taken by cells projecting from the cortex back to VB (in this simulation no attempt was made to give their dendrites pyramidal form, as in *24*). Color version of this figure is available in the companion CD-ROM.

synthetic axons then were made to navigate, starting from the dorsal thalamus, through channels which represent the thalamic radiations in the myelin-stained voxel data and into the cerebral cortex, where, in the vicinity of nominal layer IV, they arborize (Fig. 4).

12.2.4. Navigation

The elements of afferent navigation in this system include a dispersal of tropic (growth directing) factor, outward from targets, plus hill-climbing behavior for axons (up concentration gradients of tropic factor) modified by the 3D texture of the voxel terrain that they pass through. Tropic dispersal is modeled as a ($1/r^3$) drop in steady state concentration with 3D distance, integrated from all source cells (which can vary in the amount of factor that they emit, based on cell size or activity).

Using a $3 \times 3 \times 3$ cubic neighborhood, axonal "growth cones" advance towards their targets, traversing one voxel at a time. Each axon tip selects, as the "most appropriate" path, that voxel containing the highest (darkest) local myelin signal, from any of the 26 neighbor voxels that also contain an increased concentration of tropic factor. "Boundary" (high image intensity) voxels, such as found at ventricular and pial surfaces, are not crossed and can cause an axon tip to stop growing. Many classes of fiber populations can be navigating through the same region, towards differing targets if the grids are made multivalued. Each class of axon also can have a specified maximum allowed bend angle, or "stiffness", so that in general, they do not readily double back on their paths in regions having low gradients of tropism or myelin. Lastly, so that every axonal path is unique, the specific coordinates used are randomized within each voxel selected by the growing tip. Best results (i.e., more precise travel within the myelin channels) were obtained when the asymmetric ($10 \times 10 \times 40$ μm) raw voxel data was interpolated in shared memory to contain isotropic 10 μm cubic voxels.

Once within the target tissue (defined as any low resolution grid point that contains one or more factor-releasing cells), this axonal navigation method is augmented by a "homing-in" mechanism. This feature orients the growth cone towards specific nearby simulated target somata or dendrites, and mediates the process of synapse onto specifiable regions of the target cells.

These sets of constraints allow simulated axons to navigate with biologically plausible pathways through voxel data (see Figs. 4 and 5). In this test case, they converged from VB and funneled into the thalamic radiations, then splayed apart as they entered into the cortex. The simulated thalamocortical afferents generally refrained from taking wildly errant paths (e.g., caudally, down the cerebral peduncle), because of the requirement to climb up a specific tropic gradient, in this case, derived from layer IV cortex. But, as observed in real preparations, some of the simulated fibers did meander out of the predominant route and corrected their paths later, to arrive at their target nonetheless. In nature, multiple classes of projection fiber can use the same primary conduit (e.g., the medial forebrain bundle *[29]*) and sort themselves out at its destination. Similar behavior could occur here if each afferent system is made sensitive to its own tropic compound.

12.2.5. Arborization

In nature, TCAs have stereotypic paths and branching patterns en route to and within the cortex (see Discussion). They rarely branch in the thalamic radiations after they

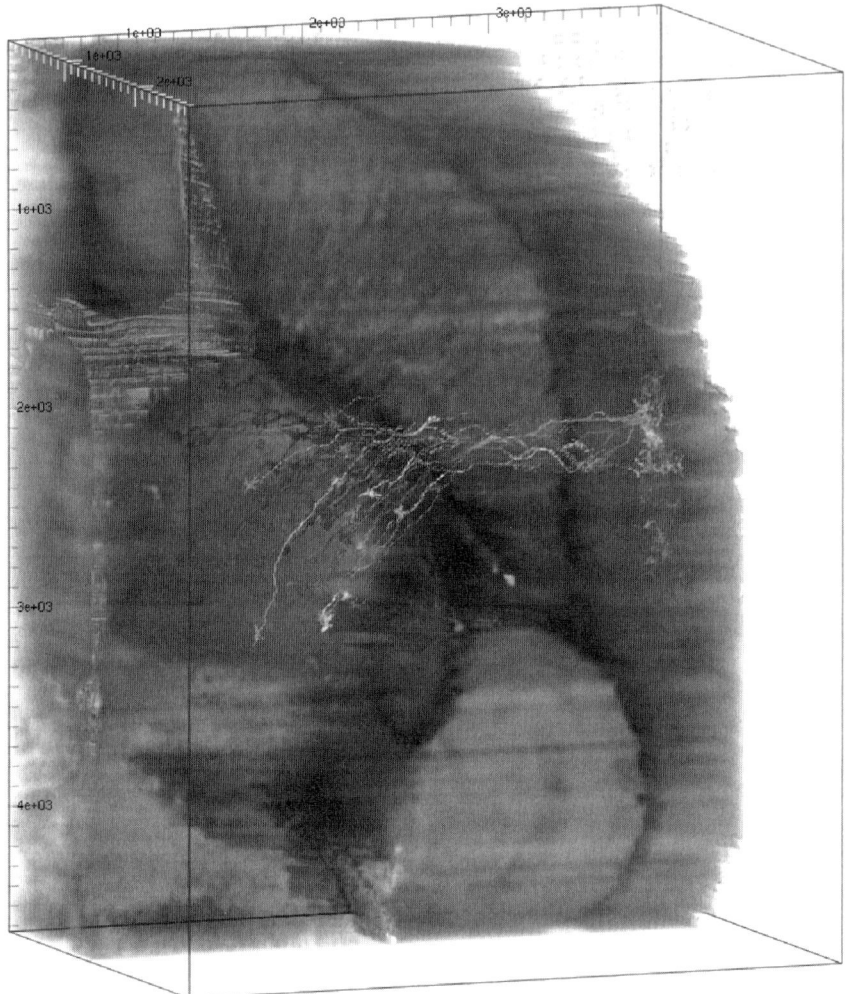

Fig. 5. Rendered data: brain context (orthogonal). This figure shows the same TCA fibers as in Fig. 4, employing more translucent and 3D settings in VoxelView. Hippocampus is below, at right, radiations are at top center, cortex is to the right, and brain midline is on the left. Fibers exit the thalamus (center left), pass through the thalamic radiations, and terminate near a band of target cells in the cerebral cortex. Color version of this figure is available in the companion CD-ROM.

pass through and synapse with the thalamic reticular nucleus, but they emit collaterals repeatedly (and usually at right angles) as they travel in the white matter beneath cortex, and they arborize in layer VI and, most profusely, in layer IV *(30)*. While the precise causes of TCA branching are not fully understood, this pathway exemplifies the phenomenon of axonal growth towards tissues that emit attractants *(31,32)*, as well as target-specific elicitation of collaterals *(33,34)*.

In ArborVitae, arborization is governed by both intrinsic and extrinsic mechanisms. Simulated axons are given fundamental branching repertoires that they would express in free space and in the absence of any target. These consists of a finite set of (concep-

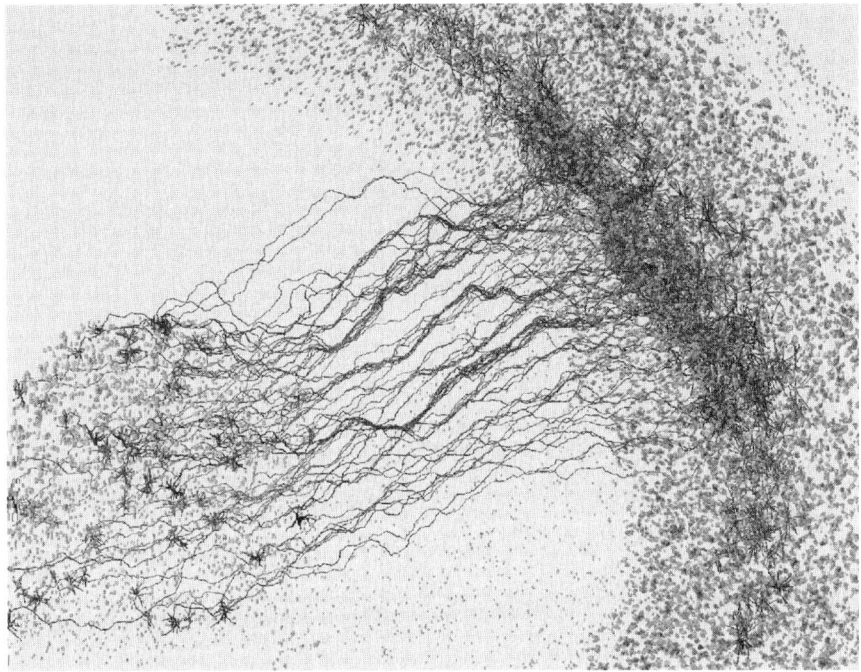

Fig. 6. Simulated data: thalamocortical pathway. This figure shows the same simulated thalamocortical pathway data, represented in ArborVitae using geometrical primitives rather than voxels. The ventrobasal complex is at the lower left. Some of the cells have elaborated dendrites and have sent axons running obliquely up to the cerebral cortex (upper right). As a simplification, the cortex consists of a reduced number of layers, one of which has target cells that likewise have elaborated some dendrites. The TCAs meet these cells and arborize, predominantly along the closest among the band of cells in the layer. Each cell is given a separate hue. Color version of this figure is available in the companion CD-ROM.

tually, genetically instanced) states. For each state the branch number, frequency, and angle are constrained by a different set of statistics (e.g., mean and standard deviation) *(35)*. Growth states can play out sequentially (free space), or they can be selected by voxel-based inhibitory or excitatory (environmental) influences. For instance, travel in fiber bundles might inhibit bifurcation, passage adjacent and into the more permissive cortex might elicit collateralization and sporadic branching, whereas entrance into a target region might trigger a "terminal" mode of arborization.

Many of these morphologically distinct growth phases appear to be conserved among differing types of axon (see also *[36]*). However, detailed experimental evidence for this state formalism is not yet available. It may prove tedious to collect (or to estimate iteratively) statistics for each candidate growth epoch. For instance, in some cases, it may be difficult to distinguish shifts among multiple endogenous states vs the effect which entry into the sphere of influence of a new target might have on a single state. Hence, measurements from nature might reflect an entanglement of state parameters with environmental modulations. (For means of deriving such statistics from measured data, see *[23]*.) Regardless of its biological validity, the state transition construct pro-

Fig. 7. Simulated data: thalamus. This is a close up of the ventrobasal complex in Figure 6, showing the (spiny) dendritic branching pattern of the projection cells, and the initial portions of their axons, decorated with varicosities. The dendritic branching patterns have not been tuned to morphometry data, which, however, is available in the literature. The pale rounded structures are VB somata that have not been given dendrites or axons, but merely indicate the extent of this brain nucleus. Color version of this figure is available in the companion CD-ROM.

vides a flexible empirical tool for recreating numerous hierarchical classes of complex arbors, and it can coexist with other mechanisms.

This formalism in ArborVitae can permit one to piecewise approximate the observed TCA behavior by quantitatively specifying statistics to match the differing angles and frequencies of axon branching in the radiations, white matter, and various layers of cortex. Axons could be configured to shift state based on detection of all of these boundary landmarks along their projection path. But, for the purposes of introducing this method, TCA branching behavior is further simplified. It omits an explicit collateralization phase in the subcortical matter and, instead, has two intrinsic states of branching (low frequency before reaching cortex and moderate frequency within cortex), plus a terminal mode triggered by entry into layer IV. These behaviors provide a basic mechanism whereby a TCA can invade and colonize a small region of cortex (Fig. 6). The omitted collaterals, while important, may primarily provide the spatial variance needed for effective arbor refinement by competition (see [37]), which also is not modeled at present. The dendritic arbors of the thalamic projecting neurons (Fig. 7) were created using algorithms presented previously in (23), but were not tuned to particular morphometry data available for these cells (see [38]).

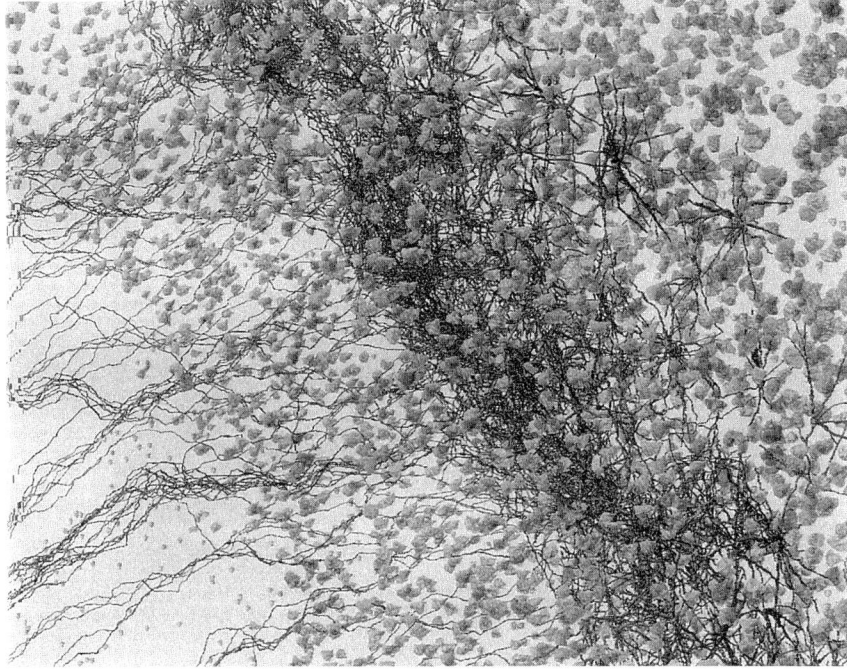

Fig. 8. Simulated data: cortex. This figure is a close-up of the cerebral cortex. Incoming TCAs, singly and in bundles, enter in the inferior margin of layer IV cortex and arborize among the target cells (various hues). The cells without dendrites (pseudo-Nissl-stained) depict additional nontarget cells. Color version of this figure is available in the companion CD-ROM.

Up to and within the target zone, three general means were used by simulated axons to regulate their growth: (*i*) the sign of the slope of local tropic factor concentration; (*ii*) state-dependent branching; and (*iii*) contact with nearby target cells.

The gradient of tropic factor is a helpful cue for leading axons up to layer IV, but, if used alone, would bar afferents from the upper part of that layer. (Assuming simple diffusion, concentration will be maximal in the middle of a target.) Procedurally, therefore, it helped to attenuate the axons' sensitivity to attractant (i.e., emulating a form of "saturation") when within the target. This made axons meander in and through layer IV more uniformly (Fig. 8), rather than building up at its center (or lower edge; depending on parameters, explosive branching can occur just as a target is reached).

A threshold of some sort is needed for the target to trigger a change of growth (branching) state. It can be modeled as a level of exterior concentration experienced only near to the target, or as a level of tropic factor internally accumulated (and dynamically degraded) *en route* to the target. Both methods appeared to work in specific instances. Thresholds based on extracellular concentration more consistently signaled the boundary of layer IV, but effective rate constants also were occasionally found when testing metabolizing models (and, while these rates might be hard for us to set, one can imagine much evolutionary time for nature's fine-tuning of them). Nonetheless, it was difficult to configure terminal arborization thresholds, of either sort, which remained adaptable over a wide range of conditions. Thus, both methods were

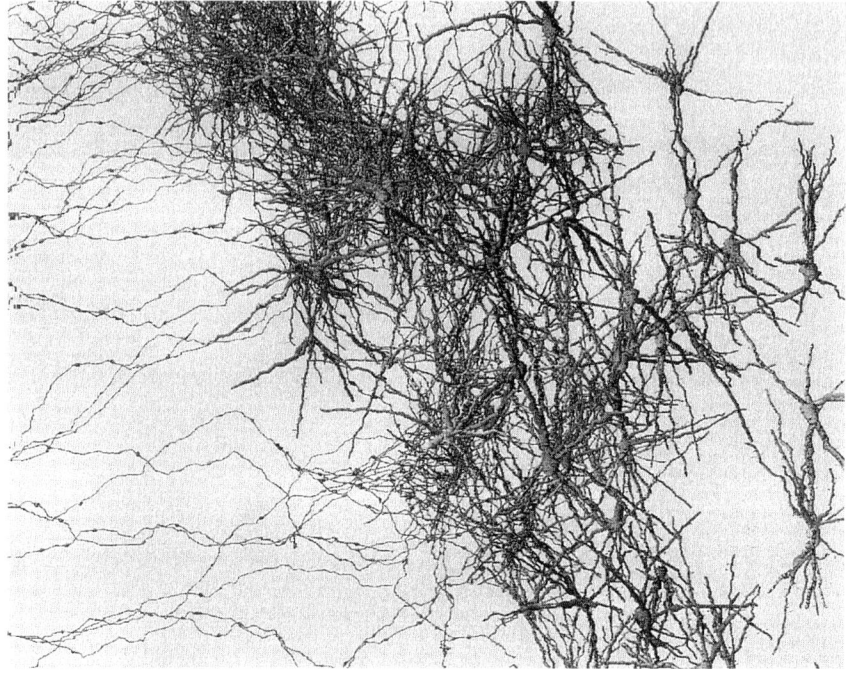

Fig. 9. Simulated data: cortex (detail). This shows an additional close-up from a region nearby (but not identical to) that of Figure 8. The Nissl-stained somata have been rendered invisible to better show the relationship of the incoming afferents to their target cells. Each afferent and each cortical cell has had all of its segments set to the same uniform hue, to help visually to disentangle the network, which, however, is better appreciated with a 3D display than in a 2D image. Note that some of the afferents branch before they reach the layer of target cells, and that some target cells (e.g., at the far right) receive fewer or no afferent inputs (see Discussion). Color version of this figure is available in the companion CD-ROM.

supplanted by a more reliable means (using global variables), which identified as "within target" those axon tips entering any grid point (see below) containing tropism-releasing cells.

Once triggered, ArborVitae's terminal arborization behavior (Fig. 9) can free run, or it can be an aspect of "homing-in" behavior. In the former mode, intrinsic state parameters govern the frequency of bifurcation and eventually will stop elongation and branching by randomly terminating the growing tips. The latter mode is more interesting, as it tends to produce a more uniformly innervated target. When homing-in, if a terminal mode growth cone fails to detect a target cell within a given radius and within a given forward-facing acceptance angle, it branches using terminal mode angle statistics. This increases the likelihood of future contact. Conversely, contact with a target cell increases the probability that that portion of the afferent will terminate. But if an afferent ending does not terminate, it disengages from (become insensitive to) its current target and resumes searching.

The number of synapses made by each afferent currently is not tightly regulated, but instead is set for the TCA population as a whole. The number per axon can be affected,

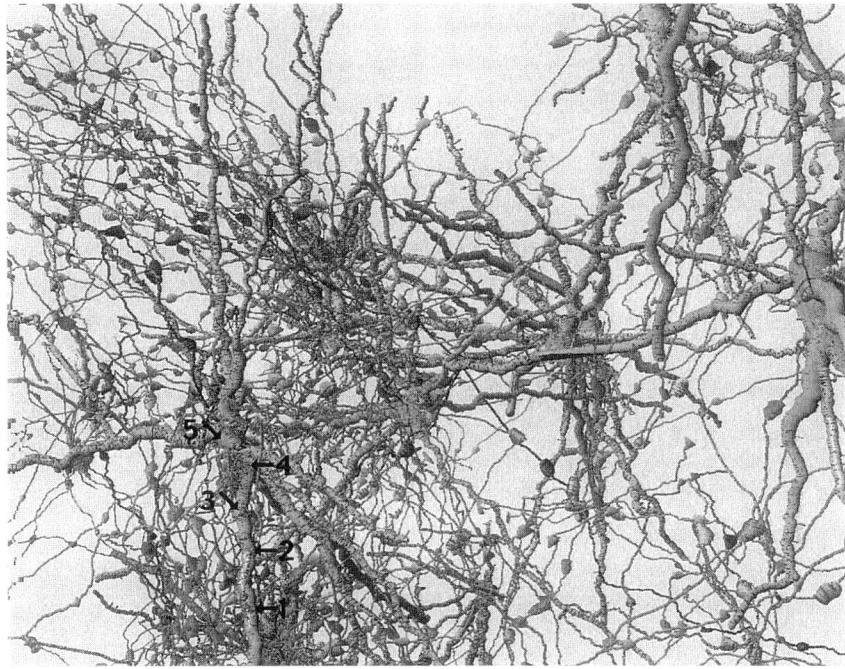

Fig. 10. Simulated data: cortex (high power). This figure is at a higher magnification than Figure 9, to show the manner in which the afferents "home-in" onto and cluster around the dendrites and somata of some target cells. The tiny spots (arrows) on the target cells' somata and dendrites are ArborVitae's representations for synapses. An axon entering the field in the lower left walks up the nearby dendrite, making contacts (1–5) on it at regular intervals, until it reaches the soma. Some postsynaptic cells have spines (visible, for instance) on the large dendrites in the upper center of the image. Color version of this figure is available in the companion CD-ROM.

however, by the size of the afferent's parent cell body, by how many targets are encountered, and by what type of morphological and physiological synapse is specified for each type of target *(39)*. Afferents can crawl to specifiable portions of the target neuron depending on whether the input is making axosomatic or axodendritic synapses (Fig. 10). Note that those cells which release long-range tropic attractant need not be the same ones preferentially synapsed upon, although in this simulation, the stellate cells of granular cortex play both roles.

This method for constructing ingrowing afferents provides a workable approximation (Figs. 4–10) to the arbors seen in nature *(40)*, although collaterals in the white matter were rarer here, as expected (since a distinct process for collateralization was not explicitly modeled). However, the simulated terminal arbors often varied among axons much more in size than anticipated, due to preponderant branching (see also *[23]*) by the earliest afferents to invade the target (this increases the numbers of tips that are candidates for subsequent branching). Such imbalances even out somewhat by scaling up the simulation by a factor of 10 to have more target cells (100–1000) and more incoming axons (25–250). However, that raised a broader issue: with additional target cells (and in the absence of any regulation of factor production or catabolism) more tropic factor is released. Thus, any behavior tied in the simulation to thresholds in the concentration or in the gradient of such a molecule (as seen naturally, e.g., with sonic hedgehog in the spinal cord *[41]*, can be altered, perhaps dramatically). Consequently, one may unintentionally get varying behaviors simply due to scaling the simulation. It is not clear whether this, in general, is a quirk only of the simulation or also a concern in nature.

12.2.6. Approximations

While having the advantage of filling space, the representational flexibility of voxels is not itself sufficient to depict intricate networks of branched neurons at currently practicable scales. One would require over a terabyte of voxel data merely to store a gray scale representation of the mouse's brain sampled to 0.5 µm. Even the thalamocortical slab simulated here (Fig. 4), would take about 50 GB at that resolution. Moreover, one would need efficient methods for acquiring, importing, and processing that information. In particular, one would need ways to link together those voxels that comprised each neuron. Consequently, the current simulation is performed using a sparser geometric method based on tubule primitives (each containing numerous physiological variables). Depending on cell number, it uses approx 50–250 MB of memory to represent the TCA pathway and runs on an SGI Indigo-2 with 384 MB. Yet, implemented in floating point, it far exceeds its underlying voxel substrate in useable resolution. To conserve computational time, no attempt was made to avoid tubule superposition or collision (although ArborVitae can be set to test for this).

Our current understanding of cell biology rules out action-at-a-distance as an explanation for interneuronal behavior. We instead posit that, were all of the pertinent biochemistry known to us, axonal outgrowth decisions would be explicable in purely local terms. Therefore, if an axon is to reach a distant target it must repeatedly sample its local environment (within a filopodial reach of approx 50 µm). In nature, the axon's milieu is largely conditioned by compounds that have diffused from distant sources. Natural diffusion takes place in parallel at very fine spatial scales and proceeds at biochemical clock speed. To mimic this effect on a computer obviously requires some approximations.

For the simulation, we restrict the growth cone to 26 possible changes of direction, each associated with information about the local environment in one neighboring voxel (here, 10 µm on each side). The decision of where to orient next (and whether to advance or retreat) is a function of that information plus historical data internal to the growth cone. Simulated axonal growth cones therefore need to know the tropic "concentration" for each of its neighboring voxels. One might emulate diffusion by finite element computation carried out at the voxel level, using a large auxiliary 3D array. Instead, an even lower resolution grid was used for calculating dispersal of tropic factors (using an inverse-distance-cubed rule), and an effective diffusion value was interpolated at the finer voxel level only when a previously unvisited voxel was encountered. This was efficient, since relatively few of the Atlas voxels were traversed by the TCA pathway. Interestingly, axonal navigational accuracy was degraded when this intermediate grid was made too detailed.

Grids and voxels, thereby, constitute effective intermediates for converting action-at-a-distance into local interaction. Of necessity, it involves some less than ideal approximations. One example, alluded to above, is in the determination of when an axon is within a target that is emitting tropic molecules. This method (based on the cell count within each grid) will work consistently in this framework for any set of axons and targets. But it leaves unspecified a cellular-based rationale for how an axon could know how near it is to a target (when still out of filopodial range). Hence, a more biologically plausible general solution for this part of the navigation problem is needed, in which the concentrations, diffusion, and metabolism of tropic factors are more precisely defined. Similarly, competition for growth factor clearly is in operation in the biological setting, and in the computational setting is potentially stabilizing. However, it involves additional metabolic parameters that have not yet been modeled.

12.3. DISCUSSION

This paper presents a general method for specifying simulated axonal pathways so that they bear a close correspondence to tracts seen in the brain, yet can be manipulated logically. The primary intent of this modeling is to use voxel data as a blueprint upon which to erect a scaffolding of plausible connections. It will be an advance if even the most prominent projection pathways can be visualized with this level of detail. It is desirable, but not essential, that the means used for wiring up the system mimic the biological. It is not claimed that these constructed pathways are the actual paths taken, but that they represent a significant step in that direction. Models created in this way have didactic utility, and, because propagation of activity can be emulated, they could play the role of a dynamic 3D breadboard for hypothesizing about brain function.

If one aims further to recapitulate pathway development, then one must bear in mind that many biological cues are not captured as voxel data, and that numerous embryological structural elements, such as radial glia, are removed from the brain itself during the course of ontogeny. Surfaces relied on by navigating axons may no longer exist, and short open spaces crossed easily in the embryo may have grown too large to traverse in the adult. Pathways, including the thalamocortical, also may be established mutually by more than one outgrowing population *(42)* or by transient markers *(43,44)*. Brain tissue also can undergo regional torsion, which may drag axons passively into new positions (e.g., medial geniculate, 7th nerve). Thus, even if there is an extent to which these constraints and algorithms eventually do model the underlying biological strategies, such earlier developmental features will not be addressed unless one obtains high resolution scans of appropriately labeled developing embryos (e.g., *[12,13]*).

A more direct relevance might be to better understand regeneration or the behavior of stem cells (e.g., *[45]*) when introduced into brain regions that have preexisting pathways. Depending on their degree of differentiation, stem cells could possess, or could acquire, sensitivity to target tropic factors and, thereby, be selective in their patterns of projection. In principle, descendents of this kind of a simulator could help to predict the behavior of implanted cells. But if so, such a program would have to address many additional biological features.

12.3.1. Biological Navigation

In nature, axons grow out, during or shortly after migration of the cell body from its site of origin, by polarizing so that one of its emerging neurites becomes the axon *(46)*. To get to and recognize their targets, axons use multiple categories of molecular cues. These include long-range diffusible chemoattractants, such as netrin *(47)*, and chemorepellants, such as semaphorins (*[48]*, but see also *[49]*), as well as contact adherents (*eph*, *[50]*) and contact repellants collapsin and cadherin (see *[51]*).

Axons often tend to travel on interfaces, or at least don't cross boundaries unless given overarching reasons (optic tract, perforant path, massa intermedia). They prefer to follow preexisting pathways formed by substrate markers *(52)*, blood vessels, or other axons, with which they selectively fasciculate and defasciculate. They can avoid nonpermissive tissue altogether *(53)*, and can be selectively sensitive to molecular boundary markers, such as roundabout, found at midline tissue *(51)*. Large families of such markers help to specify cortical areas *(54)* so that axonal access can be regulated regionally. Axons can carry and respond to molecular gradients, such as repulsive axon guidance signal (RAGS) *(55)* and *Elf1* *(56)*, and this enables afferent populations to developmentally transmit sensory maps into the CNS.

The balancing of these many molecular classes of influence is mediated, within axonal growth cones, by elaborate and sensitive regulation systems (e.g., *rac/rho*, *[48,57]*), linked both to receptors and to the subcellular machinery responsible for neurite advance and retraction (e.g., *[58]*).

12.3.2. Biological Branching

Axons have additional molecular transport machinery for depositing and removing biochemical components from their growing tips. It conveys samples of the environment centripetally to the nucleus and exports genomic instructions centrifugally to the growing arbor. An important class of sampled compounds includes the neurotrophins (brain-derived neurothrophic factor [BDNF], NT-3, NT-4/5, nerve growth factor [NGF]), whose bioavailability can regulate cell survival as well as neurite complexity *(36,59)*. Many other control points are coming to light at the nuclear level, including the *wnt* genes, which regulate neurite branching in cerebellum *(60)* and elsewhere.

Trophic effects on axon growth can extend to single collaterals and even in laminar-specific ways *(61)*. For instance, *foxb1* has been shown to regulate specific collateralization in the mammillothalamic tract *(62)*, and the pons elicits collaterals from the corticospinal tract using a diffusible signal, even after its initial phase of projection *(33)*.

Time-lapse microscopy demonstrates highly dynamic branching behaviors of navigating axons *(63–66)*, including transitions between lamellate and compact growth cone shapes at apparent choice points. When growth cones reach their target zones, there is additional and profuse branching (e.g., *[67]*), and an intricate mutual and simultaneous remodeling of axons and dendrites *(68)*.

12.3.3. Growth Algorithms

The ArborVitae program currently can emulate, and in simplistic form, only a portion of this list (migration of somata, axonal attractants, fasciculation, boundary behavior, contact termination, simple synapsis). More extensive and quantitative algorithms are needed, therefore, to match the variety and subtlety of observed axonal and dendritic outgrowth behavior.

It is dramatically clear, though (even from the features briefly alluded to above), that there is a wealth of pertinent biochemical information that could be brought to bear to inform this kind of a growth model, and to fine-tune it, as needed, for a number of specific pathways. Even for the TCA pathway, numerous factors are known to affect tract formation: chondroitin *(69)*, polysialic acid *(32)*, membrane-bound *(67)*, and diffusible *(31)* tropic compounds, and the *wnt* transcription factor *pax6 (70)*.

However, the many entities implicated thus far do not yet constitute a coherent molecular explanation of navigation through tissue. There is an increasing variety of interacting biochemical components known to affect brain wiring, yet an absence of a currently applicable comprehensive theory of biochemical control of outgrowth. It will be a challenge to keep models manageable (and computable) and still account qualitatively for the extraordinary diversity in observed axonal branching patterns. Such wide diversity makes it probable that any generally conserved mechanism controlling axonal growth and branching will be exquisitely sensitive to perturbation. But two design criteria must be balanced: one requires stable control similarities across pathways in order to simplify models of outgrowth (many thousands of axons will have to be simulated), and one wishes to be able to generate numerous variations. It is not yet clear whether this is achieved in nature by multiple genomic instructions or by high reactivity.

One potentially useful point of view is to consider that, whatever the complexity of the underlying biochemical machinery, axonal behavior can be expressed using a small "vocabulary" of final common growth states that form the neurite's geometry: elongation, shortening, change of diameter, change of angle, bifurcation, and anchoring. Filopodial evaluation of potential targets will modulate these underlying variables, but the overall sensorimotor transfer function of the growth cone (its "language") might be emulated effectively by state transition models informed by statistics (as has been helpful in other fields, such as linguistics, e.g., *[71]*). Hopefully, combinatorial differences in marking afferents and their selected targets *(72,73)* additionally will account compactly for much of the apparent diversity in behavior. It is unknown though (either biologically or heuristically) how many distinct tropic factors would be needed to establish all of the recognized major tract systems. (It seems like a 3D biochemical analogue of the 4-color mapping problem, which additionally allows for inductive changes over time). Multiple diffusing signals, attractive and repulsive, can operate within ArborVitae, but thus far they have been used only sequentially, not in parallel.

Consonant with an expectation of high sensitivity in the branching process, while a variety of reasonably approximate TCAs were made (e.g., Figs. 4,6, and 9), it proved difficult to configure an interaction of these simulated axons with their targets so that they would mimic the branching patterns seen and traced in biological tissue *(39,40,66,74,75)*. The afferent population often failed to uniformly innervate the entirety of the emulated layer IV target, and it was common to see a few axons overarborize at the expense of their companions. This is not surprising, given our limited quantitative knowledge of the control of axon branching, the ensuing large parameter space, and the lack thus far of simulating either the reticular nucleus, subplate cells

(76), layer VI, or control mechanisms internal to the growth cone, such as *rac* and *rho* *(57,77)* or competition for neurotrophin, or activity-based neurite remodeling. Many of these phenomena are becoming well enough understood that they can, and should, be added. Prototypes for some of these autoregulatory outgrowth processes exist (e.g., microtubule dynamics *[78]* and neurotrophin metabolism *[79]*).

12.3.4. Future Directions

This is a new technique, and it is not yet clear how much spatial resolution in voxels is needed, or if the high contrast given by the Loyez myelin stain is required. But similar results (not shown) were obtained after down sampling this thalamocortical voxel data by a factor of two, and any intensity difference above the noise should suffice for evaluating which neighboring voxel to select. Clearly, though, very small pathways will require higher resolution voxel data or better heuristics.

While the thalamocortical pathway has been used to illustrate this method, preliminary tests have been made using other large projection systems with encouraging results (corticothalamic, lower panel in Fig. 4, and lateral olfactory tract, not shown). It is possible, however, that structural features peculiar to the TCA pathway are fortuitously favorable for this method, or that subtle details in the navigation algorithm lead it to work well only in certain contexts. On the other hand, the quality of contrast used to inform TCA growth is widespread through the Atlas, and analogous data from a variety of sources can be expected to become equally, or more, distinct as scanning and reconstruction methodologies mature. Too, the navigation algorithm accommodates a considerable amount of tract meandering. Hence, this method should be applicable to a wide range of anatomical pathways, for which auxiliary data also are being generated and systematically compiled (e.g., *[80,81]*).

Even without recapitulating precise arborization patterns, by incorporating the above cell biological subtleties, it may be possible with this type of program (and larger computers) to relatively rapidly create an approximate yet functionally interconnected representation of the primary tracts in the mouse CNS. If only as a preliminary scaffolding, such a progressively tunable construct will have utility in organizing our collective thinking about the plainly complex process of brain wiring.

It will be a challenging longer-term goal to add interneurons (as well as glia and blood vessels), for which voxel-based tract information will of necessity be lacking (there being low coherence to their projections). However the attempt should be rewarding given a scaffolding constructed from the emulation of major pathways, based on geometry derived from Nissl-stained volumetric data and augmented with local circuit connectivity data obtained by multicellular physiological recording *(6,82,83)*. Interestingly, as the simulated networks are made more lifelike, progressively more powerful visualization and analysis techniques will be required merely to inspect the fine details of the resulting synthetic tissue.

ACKNOWLEDGMENTS

I would like to thank Dr. Richard Sidman for the use of his mouse brain Atlas and for his generous teachings of mouse neuroanatomy. I would also like to thank the anonymous reviewers for their helpful suggestions on the text. Portions of this report were supported by Human Brain Project Grant No. 5-RO1-NS39600-2 (to Giorgio Ascoli) jointly funded by the National Institute of Neurological Disorders and Stroke and the National Institute of Mental Health (National Institute of Health).

REFERENCES

1. Senft SL, Argiro VJ, Van Zandt WL. Volume microscopy of biological specimens based on non-confocal imaging techniques. IEEE Scanning 1990; **90**:424–428.
2. Robb RA. Visualization methods for analysis of multimodality images. In: Functional Neuroimaging: Technical Foundations (Thatcher RW, Hallett M, Zeffiro T, John ER, Huerta M, eds.). Academic Press, San Diego, 1994.
3. Inoue S. Whither video microscopy? Towards 4D imaging at the highest resolution of the light microscope. In: Digitized Video Microscopy (Herman B, Jacobson K, eds.). Alan R. Liss, New York, 1989.
4. Hosokawa T, Bliss TVP, Fine A. Quantitative three-dimensional confocal microscopy of synaptic structures in living brain tissue. Microsc Res Tech 1994; **29**:290–296.
5. Mainen ZF, Maletic-Savatic M, Shi SH, Hayashi Y, Malinow R, Svoboda K. Two-photon imaging in living brain slices. 1999; Methods **18**:231–239.
6. Markram H, Wang Y, Tsodyks M. Differential signaling via the same axon of neocortical pyramidal neurons. Proc Natl Acad Sci USA 1998; **95**:5323–5328.
7. White EL, Hersh SM. A quantitative study of thalamocortical and other synapses involving the apical dendrites of corticothalamic projection cells in mouse SmI cortex. J Neurocytol 1982; **11**:137–157.
8. Harris KM. Serial electron microscopy as an alternative or complement to confocal microscopy for the study of synapses and dendritic spines in the central nervous system. In: Three-Dimensional Confocal Microscopy: Volume Investigation of Biological Specimens (Stevens JK, Mills LR, Trogadis JE, eds.). Academic Press, New York, 1994, pp. 421–445.
9. Inglis BA, Yang L, Wirth III E, Plant D, Mareci TH. Diffusion anisotropy in excised normal rat spinal cord measured by NMR microscopy. Magn Reson Imaging 1997; **15**:441–450.
10. Peled S, Gudbjartsson H, Westin C, Kikinis R, Jolesz F. Magnetic resonance imaging shows orientation and asymmetry of white matter fiber tracts. Brain Res 1998; **780**:27–33.
11. Mori S. (this text) Ch. 13.
12. Smith BR, Linney E, Huff DS, Johnson GA. 1996; Magnetic resonance microscopy of embryos. Comput Med Imaging Graph **20**:483–490.
13. Jacobs RE, Ahrens ET, Dickinson ME, Laidlaw D. Towards a micro-MRI atlas of mouse development. Comput Med Imaging Graph 1999; **23**:15–24.
14. Jacobs RE, Fraser SE. Magnetic resonance microscopy of embryonic cell lineages and movements. Science 1994; **263**:681–684.
15. Sidman RL, Kosaras B, Misra BM, Senft S. Digital mouse brain atlas in 3-D at 10 μm voxel resolution. Soc Neurosci Abs 1999; **25**:1508. (*http://www.hms.harvard.edu/research/brain*)
16. Toga AW, Santori EM., Hazani R, Ambach K. A 3D digital map of rat brain. Brain Res Bull 1995; **38**:77–85.
17. Altman J, Bayer SA. Atlas of Prenatal Rat Brain Development. CRC Press, Boca Raton, 1995.
18. Williams RW. Mapping genes that modulate mouse brain development: a quantitative genetic approach. In: Mouse Brain Development (Goffinet AF, Rakic P, eds.). Springer, New York, 2000, pp. 21-49.
19. Okada A, Lansford R, Weimann JM, Fraser SE, McConnell SK. Imaging cells in the devel-

oping nervous system with retrovirus expressing modified green fluorescent protein. Exp Neurol 1999; **156**:394–406.
20. Flybase, The Biological Laboratories, Harvard University, Cambridge, MA. The FlyBase database of the *Drosophila* genome projects and community literature. The Fly Base Consortium. Nucleic Acids Res 1999; **27**:85–88. (*http://flybase.bio.indiana.edu*).
21. Marenco L, Nadkarni P, Skoufos E, Shepherd G, Miller P. Neuronal database integration: the Senselab EAV data model. Proc AMIA Symp 1999; 102–106.
22. Ascoli GA, Krichmar JJ, Nasuto SJ, Senft SL. Generation, description and storage of dendritic morphology data. Philos Trans R Acad Sci B 2001; **356**:1131–1145.
23. Ascoli GA, Krichmar JJ, Scorcioni R, Nasuto SJ, Senft SL. Computer generation and quantitative morphometric analysis of virtual neurons. Anat Embryol 2001; **204**:283–301.
24. Senft SL. A statistical framework for presenting developmental neuroanatomy. In: Neural-Network Models of Cognition, Biobehavioral Foundations (Donahoe J, Dorsel VP, eds.). Elsevier Press, New York, 1997, 37–57.
25. Sidman RL, Angevine JB Jr, Pierce ET. Atlas of the Mouse Brain and Spinal Cord. Harvard University Press, Cambridge, MA, 1971, p. 261.
26. Hines ML, Carnevale NT. The NEURON simulation environment. Neural Comput 1997; **9**:1179–1209.
27. Ascoli GA, Hunter L, Krichmar JL, Olds JL, Senft SL. Computational neuroanatomy of the hippocampus. Neurosci Abs 1998; 24, p.2015.
28. Senft SL. Development of mouse somatosensory cortex: anatomical and mathematical descriptions and analyses. Thesis, Washington University, 1989.
29. Valverde F. Studies on the Pyriform Lobe. Harvard University Press, Cambridge, MA, 1965.
30. Senft SL, Woolsey TA. Growth of thalamic afferents into mouse Barrel cortex. Cereb Cortex 1991; **1**:308–335.
31. Rennie S, Lotto RB, Price DJ. Growth-promoting interactions between the murine neocortex and thalamus in organotypic co-cultures. Neuroscience 1994; **61**:547–564.
32. Yamamoto N, Inui K, Matsuyama Y, et al. Inhibitory mechanism by polysialic acid for lamina-specific branch formation of thalamocortical axons. J Neurosci 2000; **20**:9145–9151.
33. Bastmeyer M, O'Leary DDM. Dynamics of target recognition by interstitial axon branching along developing cortical axons. J Neurosci 1996; **16**:1450–1459.
34. Castellani V, Bolz J. Membrane-associated molecules regulate the formation of layer-specific cortical circuits. Proc Natl Acad Sci USA 1997; **94**:7030–7035.
35. Senft SL, Ascoli GA. Reconstruction of brain networks by algorithmic amplification of morphometry data. Lect Notes Comp Sci 1999; **1606**(Foundations and Tools for Neural Modeling):25–33.
36. Lentz SI, Knudson CM, Korsmeyer SJ, Snider WD. Neurotrophins support the development of diverse sensory axon morphologies. J Neurosci 1999; **19**:1038–1048.
37. Senft SL, Woolsey TA. Mouse Barrel cortex viewed as Dirichlet Domains. Cereb Cortex 1991; **1**:348–363.
38. Arnold PB, Li CX, Waters RS. Thalamocortical arbors extend beyond single cortical barrels: an in vivo intracellular tracing study in rat. Exp Brain Res 2001; **136**:152–168.
39. Lübke J, Markram H, Frotscher M, Sakmann B. Frequency, number and dendritic distribution of autapses established by layer 5 pyramidal neurons in the neocortex: comparison with synaptic innervation of neighboring neurons of the same class. J Neurosci 1996; **16**:3209–3218.
40. Senft SL, Woolsey TA. Computer-aided analyses of thalamocortical afferent ingrowth. Cereb Cortex 1991; **1**:336–347.
41. Roelink HR, Porter J, Chiang C, et al. Control of cell pattern in the neural tube: induction

of floor plate and motor neuron differentiation by different concentrations of the amino terminal cleavage product of *sonic hedgehog* autoproteolysis. Cell 1995; **81**:445–455.
42. Molnar Z, Blakemore C. How do thalamic axons find their way to the cortex? Trends Neurosc 1995; **18**:389–397.
43. Emerling DE, Lander AD. Laminar specific attachment and neurite outgrowth of thalamic neurons on cultured slices of developing cerebral neocortex. Development 1994; **120**:2811–2822.
44. Tuttle R, Nakagawa Y, Johnson JE, O'Leary DDM. Defects in thalamocortical axon pathfinding correlate with altered cell domains in Mash-1-deficient mice. Development 1999; **126**:1903–1916.
45. Snyder EY, Park KI, Flax JD, et al. Potential of neural "stem-like" cells for gene therapy and repair of the degenerating central nervous system. Adv Neurol 1997; **72**:121–132.
46. Bradke F, Dotti CG. Establishment of neuronal polarity: lessons from cultured hippocampal neurons. Curr Opin Neurobiol 2000; **10**:574–581.
47. Tessier-Lavigne M, Placzek M. Target attraction: are developing axons guided by chemotropism? Trends Neurosci 1991; **15**:303–310.
48. Nakamura F, Kalb RG, Strittmatter SM. Molecular basis of semaphorin-mediated axon guidance. J Neurobiol 2000; **44**:219–229.
49. Hotary KB, Tosney KW. Cellular interactions that guide sensory and motor neurites identified in an embryo slice preparation. Dev Biol 1996; **176**:22–35.
50. Zisch AH, Pasquale EB. The *Eph* family: a multitude of receptors that mediate cell recognition signals. Cell Tissue Res 1997; **290**:217–226.
51. Tessier-Lavigne M, Goodman CS. The molecular biology of axon guidance. Science 1996; **275**:1123–1133.
52. Sheppard AM, Hamilton SK, Pearlman AL. Changes in the distribution of extracellular matrix components accompany early morphogenetic events of mammalian cortical development. J Neurosci 1991; **11**:3928–3942.
53. Tosney KW, Oakley RA. The peri-notochordal mesenchyme acts as a barrier to axon advance in the chick embryo: implications for a general mechanism of axonal guidance. Exp Neurol 1990; **109**:75–89.
54. Donoghue MJ, Rakic P. Molecular evidence for the early specification of presumptive functional domains in the embryonic primate cerebral cortex. J Neurosci 1999; **19**:5967–5979.
55. Drescher U, Kremoser C, Handwerker C, Loschinger J, Noda M, Bonhoeffer F. In vitro guidance of retinal ganglion cell axons by RAGS, a 25kDa tectal protein related to ligands of *eph* receptor tyrosine kinases. Cell 1995; **82**:359–370.
56. Nakamoto M, Cheng HJ, Friedman GC, et al. Topographically specific effects of ELF-1 on retinal axon guidance *in vitro* and retinal axon mapping *in vivo*. Cell 1996; **86**:755–766.
57. Hall A. *Rho* GTPases and the actin cytoskeleton. Science 1998; **279**:509–514.
58. Kater SB, Rehder V. The sensory-motor role of growth cone filopodia. Curr Opin Neurobiol 1995; **5**:68–74.
59. McAllister AK, Lo DC, Katz LC. Neurotrophins regulate dendritic growth in developing visual cortex. Neuron 1995; **15**:791–803.
60. Hall AC, Lucas FR, Salinas PC. Axonal remodeling and synaptic differentiation in the cerebellum is regulated by *wnt*-7a signaling. Cell 2000; **100**:525–535.
61. Bolz J, Castellani V, Mann F, Henke-Fahle S. Specification of layer-specific connections in the developing cortex. Prog Brain Res 1996; **108**:41–54.
62. Alvarez-Bolado G, Zhou X, Voss AK, Thomas T, Gruss P. Winged helix transcription factor *foxb1* is essential for access of mammillothalamic axons to the thalamus. Development 2000; **127**:1029–1038.

63. Edmondson JC, Hatten ME. Glial-guided granule neuron migration *in vitro*: a high-resolution time-lapse video microscopic study. J. Neurosci. 1987; **7**:1928–1934.
64. Godement P, Wang LC, Mason CA. Retinal axon divergence in the optic chiasm: dynamics of growth cone behavior at the midline. J Neurosci 1994; **14**:7024–7039.
65. Yamamoto N, Higashi S, Toyama K. Stop and branch behaviors of geniculocortical axons: a time-lapse study in organotypic co-cultures. J Neurosci 1997; **17**:3653–3663.
66. Skaliora I, Adams R, Blakemore C. Morphology and growth patterns of developing thalamocortical axons. J Neurosci 2000; **20**:3650–3662.
67. Gotz M, Novak N, Bastmeyer M, Bolz J. Membrane-bound molecules in rat cerebral cortex regulate thalamic innervation. Development 1992; **116**:507–519.
68. Dailey ME, Smith SJ. The dynamics of dendritic structure in developing hippocampal slices. J Neurosci 1996; **16**:2983–2994.
69. Bicknese AR, Sheppard AM, O'Leary DDM, Pearlman AL. Thalamocortical axons extend along a chondroitin sulfate proteoglycan-enriched pathway coincident with the neocortical subplate and distinct from the efferent path. J Neurosci 1994; **14**:3500–3510.
70. Pratt T, Vitalis T, Warren N, Edgar JM, Mason JO, Price DJ. A role for *pax6* in the normal development of dorsal thalamus and its cortical connections. Development 2000; **127**:5167–5178.
71. Charniak E. Statistical techniques for natural language parsing. AI Magazine 1997; **18**: 33–44.
72. Holt CE, Harris WA. Target invasion, mapping and cell choice. Curr Opin Neurobiol 1994; **8**:98–105.
73. Sestan N, Rakic P, Donoghue MJ. Independent parcellation of the embryonic visual cortex and thalamus revealed by combinatorial *Eph/ephrin* gene expression. Curr Biol 2001; **11**:39–43.
74. Erzurumlu RS, Jhaveri S. Thalamic axons confer a blueprint of the sensory periphery onto the developing rat somatosensory cortex. Brain Res Dev Brain Res 1990; **56**:229–234.
75. Catalano SM, Robertson R, Killackey HP. Individual axon morphology and thalamocortical topography in developing rat somatosensory cortex. J Comp Neurol 1996; **366**:36–53.
76. Ghosh A, Shatz CJ. A role for subplate neurons in the patterning of connections from thalamus to neocortex. J Neurosci 1992; **12**:39–55.
77. Luo L, Jan LY, Jan YN. *Rho* family of GTP-binding proteins in growth cone signaling. Curr Opin Neurobiol 1997; **7**:81–86.
78. Van Veen MP, Van Pelt J. Neuritic growth rate described by modeling microtubule dynamics. Bull Math Biol 1994; **56**:249–273.
79. Van Ooyen A, Willshaw DJ. Competition for neurotrophic factor in the development of nerve connections. Proc R Soc Lond B 1999; **266**:883–892.
80. Anderson CH, Olshausen BA, Van Essen DC. Routing networks in visual cortex. In: Handbook of Brain Theory and Neural Networks (Arbib M, ed.), MIT Press, Cambridge, MA, 1995, pp. 823–826.
81. Burns GAPC, Young MP. *Neurobase*: a neuroanatomical connection database and its use in providing a description of connections in the rat hippocampal system. Brain Res Assoc Abs 1996; **13**:85, Newcastle upon Tyne.
82. Deuchars J, Thomson AM. Innervation of burst firing spiny interneurons by pyramidal cells in deep layers of rat somatosensory cortex: paired intracellular recordings with biocytin filling. Neuroscience 1995; **69**:739–755.
83. Gupta A, Wang Y, Markram H. Organizing principles for a diversity of GABAergic interneurons and synapses in the neocortex. Science 2000; 287:273–278.

13
Principle and Applications of Diffusion Tensor Imaging
A New MRI Technique for Neuroanatomical Studies

Susumu Mori

ABSTRACT

Diffusion tensor imaging (DTI) is an emerging magnetic resonance imaging (MRI) technology. Using this technique, we can characterize the way water diffuses inside imaging objects. For example, water molecules inside a cup can diffuse freely in all directions ("free diffusion" or "isotropic diffusion"). On the other hand, water molecules inside living systems often experience numerous "obstacles", such as protein fibers, membrane, and organelles. If the water diffusion is restricted by these structures it is called "restricted diffusion." If water molecules are in an environment with highly ordered (or aligned) structure, they tend to diffuse along the structure, resulting in so-called "anisotropic diffusion." In other words, the water diffusion has "directionality". The water diffusion, thus, carries a wealth of information on the micro-architecture of the imaging object. Using the DTI, we can characterize the water diffusion process. DTI can answer questions about diffusion like, "is it free or restricted?" or "is it isotropic or anisotropic?" Using the DTI technique, the water diffusion process can be characterized on a pixel-by-pixel basis. Application of the DTI to the brain has revealed that the water diffusion in the brain white matter is highly anisotropic, which is attributed to the highly ordered axonal tracts. The characterization of the anisotropic diffusion can provide detailed information on the white matter architectures, which cannot be obtained by any other radiological tools. In this chapter, we discuss the theory and history of DTI and introduce the state-of-the-art application studies.

13.1 BACKGROUND ON DIFFUSION TENSOR IMAGING

13.1.1. Conventional Magnetic Resonance Imaging and Diffusion Tensor Imaging

It is widely accepted that magnetic resonance imaging (MRI) is one of the most versatile radiological techniques to study the human brain noninvasively. A reason for its versatility stems from its capability to create many different patterns of contrast in the brain, which depend on the data acquisition techniques employed. Each contrast

From: *Computational Neuroanatomy: Principles and Methods*
Edited by: G. A. Ascoli © Humana Press Inc., Totowa, NJ

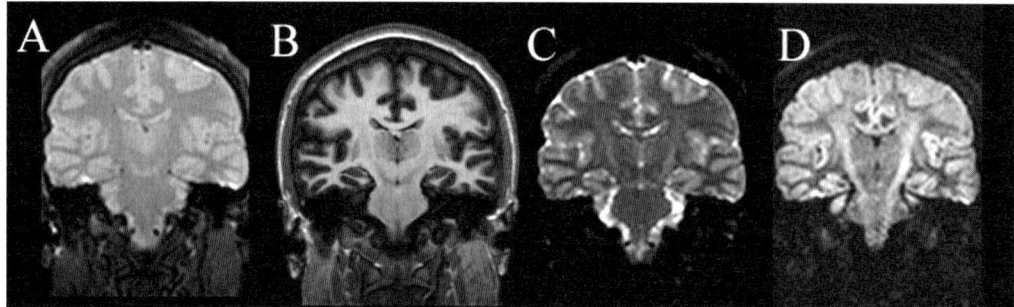

Fig. 1. Comparison of proton density (A), T_1- (B), T_2- (C), and diffusion-weighted (D) images of a human brain.

mechanism is based on different physical and chemical properties of water molecules and, thus, each pattern often reflects different physiological and/or anatomical properties of the brain. In Figure 1, four images with different contrasting (weighting) methods are compared. Image intensity of the proton density image (Fig. 1A) is proportional to the tissue water content. Image intensities in the T_1- and T_2-weighted images (Fig. 1B,C) reflect relaxation properties of water molecules inside each pixel. Differences in T_1 and T_2 relaxation in different brain regions, such as gray and white matter, is believed to reflect differences in the physical and chemical environment of water molecules, such as viscosity, susceptibility, and proton exchange with macromolecules. In late 1980s, a new weighting scheme called, "diffusion weighting" was introduced *(1,2)*. An example of a diffusion-weighted image is shown in Figure 1D.

As will be discussed later, the intensity of this image is weighted by translational motion (diffusion) of water molecules. The faster the diffusion process is, the darker the image becomes. In practice, the contrast created in diffusion-weighted images is more complicated. First of all, the absolute image intensity of diffusion-weighted images is determined not only by the extent of diffusion, but also by proton density, T_1 and T_2, depending on the data acquisition techniques (very often, so-called diffusion-weighted images are also T_2-weighted). Second, inside biological systems, translational molecular motion often has directionality *(3,4)*. In other words, whether water molecules are moving fast or slow depends on their directions. This orientation effect can be seen in Figure 2. These three images are diffusion weighted along three different orientations. It can be seen that the contrasts in these images are very different, although they were acquired with exactly the same image parameters except for the orientation of diffusion weighting.

13.1.2. Diffusion Process

Before starting to describe diffusion imaging, the translational motion of water molecules should be defined. Inside living systems, there are many factors that affect the movement of water molecules. The motion can be classified into two categories; one is coherent and the other is incoherent motion. An example of the coherent motion is blood flow, in which water molecules move unidirectionally along a certain axis. Bulk tissue motion, such as pulsation and respiration, also causes a large-scale coherent motion of water molecules (they move along with tissues). Thermal motion (Brownian

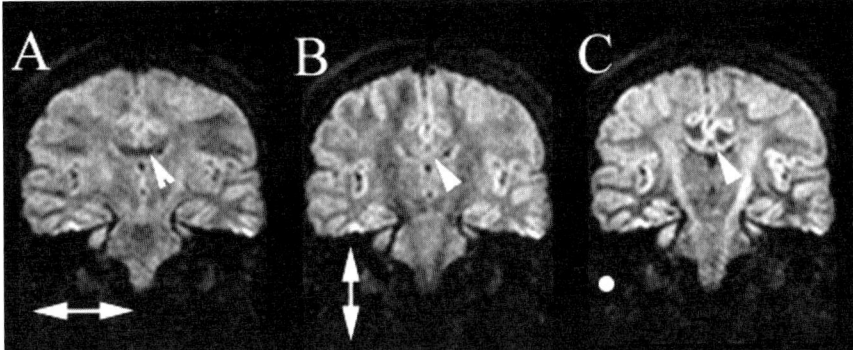

Fig. 2. Images that are diffusion-weighted along different axes. In diffusion-weighted images, brain regions where water molecules diffuse faster have lower intensities. This diffusion weighting can be applied along any desired axis (the weighting orientations are shown by arrows) and as can be seen in these three images, the image contrast depends heavily on the orientation of the diffusion weighting (panel **C** is perpendicular to the plane). For example, white arrowheads indicate the corpus callosum that seems to have high diffusion when the diffusion weighting is applied along the left-right axis (**A**), but seems to have low diffusion along the superior-inferior (**B**) or anterior-posterior axes (**C**).

motion) is an example of incoherent motion, in which the probability of water movement along any arbitrary axis is always gaussian (the probability to go to the right or left along an axis is identical). The distinction between the two may not always be clear. For example, if there are many capillary blood vessels within a pixel, water movement as a whole may look incoherent (random). Water molecules can also be moved by active transport across membranes or along protein filaments. Water motion is further complicated by the existence of barriers and obstacles such as membranes and macromolecules. The results of diffusion imaging (diffusion-weighted images) are influenced by all the factors that affect water movement. In order to understand how these factors affect the diffusion image, I would like to introduce four parameters that describe water movement; these are "shift", "size", "shape", and "orientation" as shown in Figure 3. Suppose ink is dropped on an object, and how it spreads is observed. After a while, the initially concentrated ink will form more a diluted cloud of ink due to water movement. If the center of the cloud is shifted from the initial location (shift in Fig. 3), this indicates the existence of coherent motion (or flow). If the center of the cloud does not move, the water motion is incoherent (called diffusion, hereafter). The extent of the motion is represented by the size of the cloud (size in Fig. 3). As mentioned above, the shape of the cloud, which is supposed to be a sphere for free diffusion, can be an ellipse in 2D, or "ellipsoid" in a 3D space (called diffusion ellipsoid) in living systems (shape in Fig. 3). In other words, if a sample consists of homogeneously ordered structure (e.g., actinemyocin filaments in muscle, neuronal filaments in axon), water tends to diffuse along such an ordered structure and, as a consequence, the extent of the diffusion has directionality. Unless the sample consists of multiple populations of fibers with different orientations, or the fibers have significant curvature, the way water diffuses in such an ordered environment is known to be the ellipsoid. This mode of diffusion is called anisotropic diffusion as opposed to isotropic diffusion for the spherical case. If the

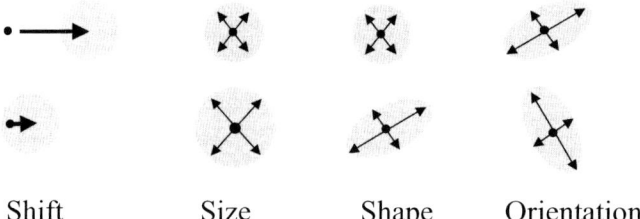

Shift   Size   Shape   Orientation

Fig. 3. Four modes of water movement; shift, size, shape, and orientation. Small black dots indicate locations where ink is dropped and shaded areas indicate how it spreads.

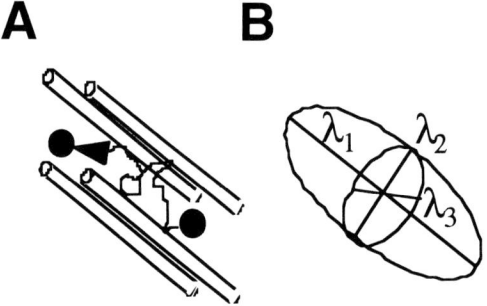

Fig. 4. Schematic view of water diffusion in an environment with strongly aligned fibers (**A**) and its expression as a diffusion ellipsoid (**B**). In (**A**), a trajectory of a solid sphere shows an example of water diffusion that is restricted by fibers depicted by bars. In this environment the diffusion properties can be expressed by an ellipsoid (**B**). Three orthogonal axes that align to the longest (λ_1), shortest (λ_2), and middle (λ_3) axes are called principal axes.

diffusion is anisotropic, the orientation of diffusion becomes an important issue (orientation in Fig. 3). One interesting fact is that the coherent motion (shift), in principle, does not have an effect on diffusion-weighted images, and, thus, size, shape, and orientation of diffusion are of central interest in the diffusion imaging.

If water molecules are in an environment where they can diffuse freely, they do not have directionality (isotropic); they diffuse in all directions with the same amount of movement. However, inside fibrous systems such as brain white matter, water diffusion tends to have directionality (Fig. 4) and the "shape" of the diffusion becomes ellipsoid (anisotropic) as mentioned above. This diffusion ellipsoid can be fully described by 6 parameters involving length and orientation (Fig. 5). There are three parameters to define the length of the longest, middle, and shortest axes (called λ_1, λ_2, and λ_3), which are orthogonal to each other, and the orientations of these three axes (unit vectors called v_1, v_2, and v_3). In other words, the λ_1, λ_2, and λ_3 define the shape, and the v_1, v_2, and v_3 define the orientation of the ellipsoid. Only one parameter is needed if the diffusion ellipsoid is spherical (1 parameter for its size [diameter] and no need for shape and orientation parameters).

13.1.3. Importance of Studying the Water Diffusion Process in the Brain

Conventional imaging contrasts such as T_1, T_2, and magnetization transfer are based on differences in chemical compositions of the brain tissue. On the other hand, contrasts obtained from the diffusion tensor imaging (DTI) technique are unique because they are based on the existence and orientation of ordered structures. Parameters we

Diffusion Tensor Imaging

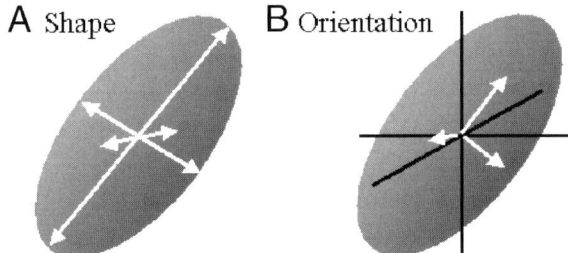

Fig. 5. Six parameters to represent an ellipsoid. Three are needed to describe the shape (**A**); the lengths of the three principal axes. Other three are needed for orientation (**B**); unit vectors that describes the orientation of the principal axes in the measurement coordinates. In our case, the measurement coordinates are defined by the orientation of the magnet as described below.

can investigate from the DTI are the size, shape, and orientation of water diffusion processes, and each parameter provides us with unique information on the physiology and anatomy of the brain.

First, the extent of water diffusion (size of the diffusion sphere–ellipsoid) is known to decrease when ischemia occurs. This is one of the most sensitive and specific markers of ischemic tissue at its early phase and has proven to be an important diagnostic tool for stroke patients (see Subheadings 13.3.1. and 13.3.2.).

Second, the anisotropy (shape) tells us where densely packed axonal fibers are located (not surprisingly, white matter has higher anisotropy than gray matter). It is expected that the anisotropy will be a good marker to study the integrity of axonal tracts in the white matter. This will be discussed in more detail in Subheading 13.3.3. Third, the orientation information of white matter tracts provides us with detailed white matter anatomy. In conventional MRI, the white matter often looks homogeneous. However, the white matter is far from homogeneous in terms of the axonal orientation. Using the orientation information, the white matter can be parcelled into different white matter tracts, which might in the future allow us to examine effects of brain diseases on individual tract systems (see Subheading 13.3.4.).

Finally, by extending the fiber orientation information into a 3D space, it has been demonstrated that 3D trajectories of white matter tracts can be reconstructed *(5–14)*. Because neuronal connectivity provides such important information to understand brain anatomy and function, and because there have been no noninvasive techniques that allow us to investigate the neuronal connectivity in humans, the development of DTI-based tract reconstruction techniques and applications to in vivo human brains are being received with much enthusiasm. This topic is covered in Subheading 13.4.

13.2. MEASUREMENT AND CALCULATION

The diffusion MRI technique can measure water diffusion along any desired direction in each pixel of the images. The diffusion is measured using the so-called "magnetic field gradient" (called "gradient" hereafter) *(15–17)*. MRI scanners are equipped with three units of the gradient devices in x, y, and z directions. By using one of these gradient axes, or by combining them, it is possible to "sensitize" or "diffusion-weight" images. Figure 6 shows an example of application of such a gradient in a simple spin-

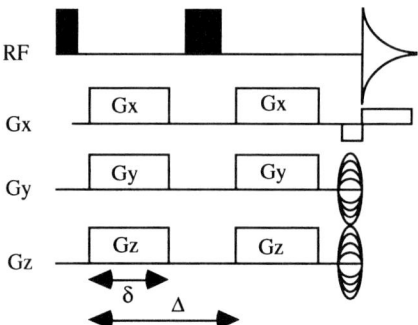

Fig. 6. An example of a diffusion MRI sequence. Diffusion weighting gradient pulses (Gx, Gy, and Gz) are applied around a 180 RF pulse. Narrow and wide solid boxes are 90 and 180 radio frequency (RF) pulses.

echo sequence. A detailed explanation of the exact mechanism of the diffusion measurement is available in our recent review *(6)*. Examples of such diffusion-weighted images using various axes of gradients are shown in Figure 7. The left most image in the top row is the so-called nondiffusion weighted image, which is acquired without the gradient (T_2-weighted image). The second image on the top row is acquired with the x-gradient (Gx in Fig. 6). Namely, the image is diffusion weighted along the x axis (indicated by arrow). By combining the gradients, such as x and y, diffusion along oblique angles can also be measured as shown in the right most image in the top row. It can be seen that signal intensities decrease by applying the diffusion weighting. The amount of this signal loss obeys Equation 1, assuming isotropic diffusion *(17)*:

$$\ln\left[\frac{S}{S_0}\right] = -\gamma^2 G^2 \delta^2 (\Delta - \delta/3) D \qquad \text{[Eq. 1]}$$

where S and S_0 are signal intensities with and without the diffusion weighting, γ is a constant (gyromagnetic ratio), G and δ are gradient strength and length, and λ is the separation between a pair of gradient pulses (see Fig. 6 for the symbol definition). Because these parameters are all known, including the amount of signal decrease (S/S_0), diffusion constants (D) at each pixel can be obtained, which is shown in the second row of Figure 7. These calculated diffusion constant maps are called "apparent diffusion constant (ADC)" maps. From the images in Figure 7, it can be seen that cerebrospinal fluid (CSF) regions where water diffuses freely have a large amount of signal loss in the diffusion-weighted images (upper row) and high values in the ADC maps (second row), which is logical.

There are two important points that should be realized from Figure 7. First, in this so-called diffusion MRI technique, a diffusion process along any desired axis, which is defined by the applied gradient axis, can be measured. Second, as can be seen in the second row of Figure 7, the ADC maps obtained using different gradient axes have markedly different contrasts. In other words, water diffusion inside the brain has directionality, and the amount of diffusion depends on the direction, which is called "diffusion anisotropy" *(3,4,18,19)*. This dependence on the direction of the diffusion measurement indicates that the water diffusion property cannot be expressed by a single

Diffusion Tensor Imaging

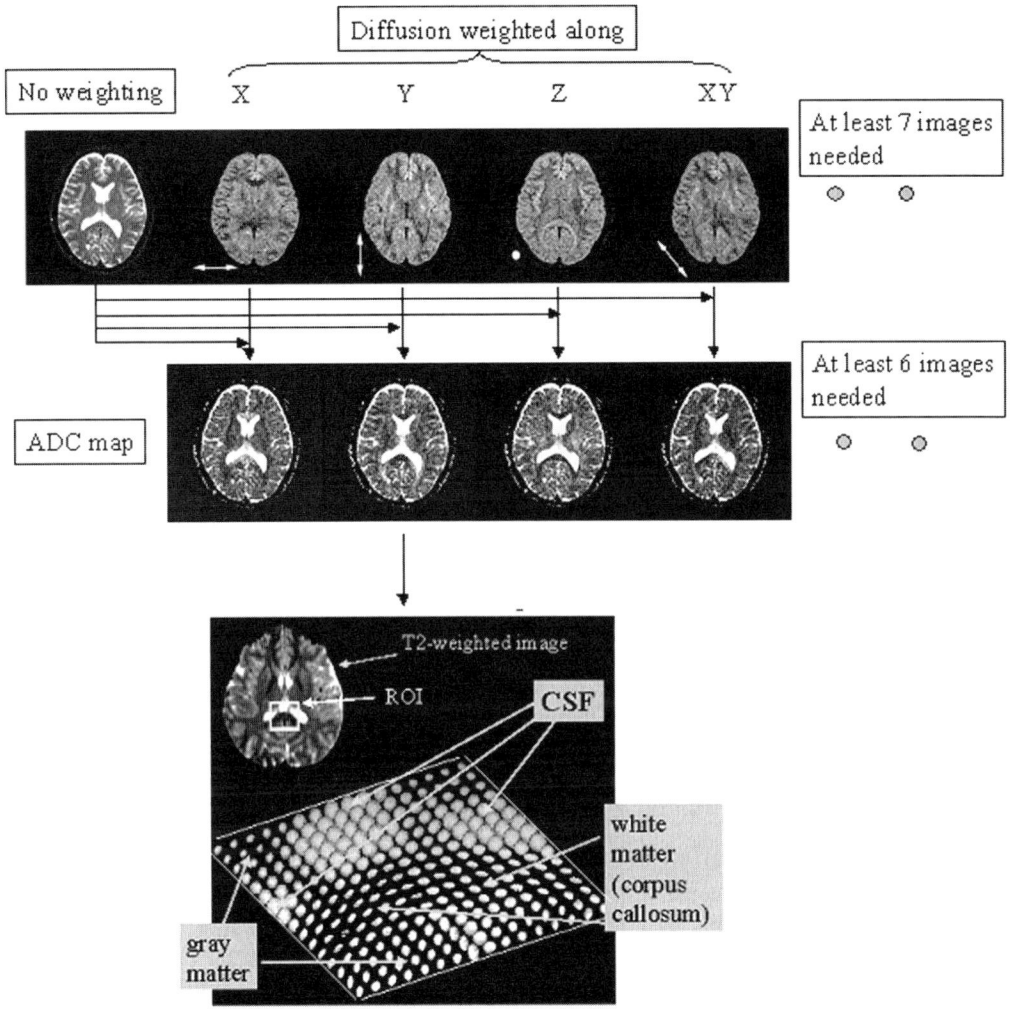

Fig. 7. Examples of diffusion-weighted images using various axes of gradients (upper row). Arrows indicate the directions of diffusion measurements, where the z axis is perpendicular to the plane. From changes in intensities between nonweighted and diffusion-weighted images, ADC at each pixel are calculated (the second row). By acquiring at least 7 diffusion-weighted images, six ADC maps can be obtained. Then, the diffusion ellipsoid at each pixel can be fully characterized as shown in the bottom row. (The image at the bottom row was reproduced from Pierpaoli et al. *[24]* with permission)

number. To fully characterize this diffusion anisotropy, the shape and orientation of the diffusion ellipsoid, introduced in Figure 5, have to be determined. It is known that such a system can be described by a 3×3 tensor (called diffusion tensor $\overline{\overline{\mathbf{D}}}$), which consists of nine elements *(19–23)*:

$$\overline{\overline{\mathbf{D}}} = \begin{bmatrix} D_{xx} & D_{xy} & D_{xz} \\ D_{yx} & D_{yy} & D_{yz} \\ D_{xz} & D_{yz} & D_{zz} \end{bmatrix}$$

[Eq. 2]

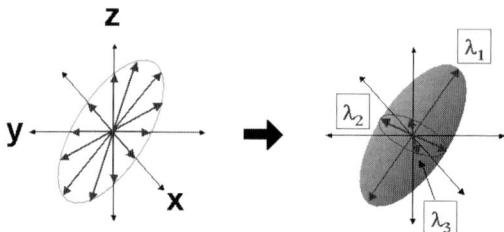

Fig. 8. From diffusion measurements along 6 independent axes, a diffusion ellipsoid can be fully characterized, which includes 3 values to describe its shape (λ_{1-3}) and 3 vectors (v_{1-3}) for its orientation.

Although this tensor $\overline{\overline{D}}$ has 9 elements, it is a symmetry tensor (i.e., $D_{xy} = D_{yx}$, $D_{xz} = D_{zx}$, and $D_{yz} = D_{zy}$) and, thus, it has 6 independent variables (3 diagonal and 3 off-diagonal terms). This makes sense because, as shown in Figure 5, 6 parameters are needed to fully characterize the diffusion ellipsoid (3 for dimension and 3 for orientation). In this way, the diffusion tensor, $\overline{\overline{D}}$, contains all the necessary information about the diffusion ellipsoid.

We have discussed how the diffusion ellipsoid can be characterized by 6 parameters that are mathematically represented by a 3 × 3 symmetric tensor (Equation 2). It has also been demonstrated that we can measure an apparent diffusion constant along any desired axis (Equation 1). In order to link these diffusion measurements (Equation 1) and the tensor representation (Equation 2), Equation 1 has to be rewritten in a more rigorous form *(19)*:

$$\ln\left[\frac{S}{S_0}\right] = -\int \gamma^2 \left[\int \overline{G}(t'') \, dt''\right] \bullet \overline{\overline{D}} \bullet \left[\int \overline{G}(t'') \, dt''\right]^T dt' \quad \text{[Eq. 3]}$$

where $\overline{G}$ is a gradient vector (orientation of diffusion measurement). To solve this equation, we need S_0 that corresponds to the image intensity without diffusion weighting (the left most image in the upper row of Fig. 7) and at least six images (S) with different $\overline{G}$ (diffusion weighted images in Fig. 7). This can be more easily understood from the visual presentation in Figure 8. The radius of a diffusion ellipsoid along a particular axis represents the extent of diffusion (ADC) along its axis. Therefore, if we measure the ADC along numerous axes, the shape of the ellipsoid can be delineated. The question here is, what is the minimum number of the measurements required to mathematically calculate the shape and orientation of the ellipsoid or, in other words, to determine 6 unknown elements of the diffusion tensor? The answer is, not surprisingly, 6 diffusion measurements as also shown in Figure 7. In practice, a total 7 images (one S_0 and six S) are needed to obtain 6 ADCs and to solve Equation 3. Once 6 elements of the diffusion tensor are obtained from Equation 3, the 6 parameters of the diffusion ellipsoid, λ_{1-3} and v_{1-3}, can be calculated by using the so-called diagonalization process. This process is repeated for each pixel. The end product of this imaging technique, called DTI, is an image with a fully characterized diffusion ellipsoid at each pixel (Fig. 7, bottom row) *(24)*. To summarize up to this point, diffusion-weighted images are "raw" images that we can obtain from scanners. From two diffusion-

Diffusion Tensor Imaging

weighted images, an ADC map can be calculated, in which an apparent diffusion constant along one predetermined axis is measured. From at least 6 ADC maps, a diffusion tensor image can be calculated.

The DTI is unique in the sense that each pixel of the image contains 6 parameters, which requires unique presentation methods. The most intuitive way to present the result is to visualize the diffusion ellipsoid at each pixel as shown in the bottom row of Figure 7 *(24)*. From this image, it can be seen that the water diffusion in the gray matter and CSF is spherical, indicating the lack of coherent fiber structures. On the other hand, the white matter has high anisotropy (the diffusion ellipsoid is elongated), as expected. However, this presentation technique produces rather complex images and it is not always easy to perceive differences in the size, shape, and orientation of the diffusion ellipsoids. To overcome these issues, many presentation and analysis techniques are postulated. In the following section, I would like to introduce 2D and 3D visualization techniques and their applications.

13.3. 2D DTI DATA ANALYSIS AND VISUALIZATION TECHNIQUES AND THEIR APPLICATION IN BRAIN STUDIES

In Figures 2 and 7, it can be seen that neither the individual diffusion-weighted image nor the ADC map is informative, because their contrast depends on the orientation of diffusion weighting or orientation of the brain with respect to the magnet. These images are called orientation-dependent images. It is highly preferable to obtain orientation-independent images, so that the images are independent of the orientation of the brain, and imaging parameters and the images can be compared between different patients and different research sites. In this section, orientation-independent images for the size and shape of the diffusion ellipsoid will be introduced.

13.3.1. Trace Image: An Orientation-Independent Visualization Technique for the Size of the Diffusion Ellipsoid

If diffusion is isotropic (spherical), the size of the diffusion (Fig. 3) can be easily represented by its diameter. However, when diffusion is anisotropic, how can the size be compared among brain regions with different shapes of diffusion ellipsoids? The most widely used orientation-independent parameter that represents the size of the ellipsoid is called "trace"*(21,25,26)*. Trace can be obtained by summing the diagonal terms of the diffusion tensor ($= D_{xx} + D_{yy} + D_{zz}$). In practice, the diagonal term D_{xx} can be directly obtained by measuring ADC using the x-gradient, D_{yy} using the y-gradient, and D_{zz} using the z-gradient. Therefore, unlike the full tensor calculation, which requires 6 ADC measurements, the trace can be obtained from just 3 ADC measurements. An example is shown in Figure 9. From this figure, it can be seen that the trace is quite homogeneous throughout the brain, indicating that regardless of various shapes and orientations of diffusion ellipsoids, their size is uniform *(21,24)*.

13.3.2 Application of Trace Image: Stroke Studies

In 1989, Moseley and coworkers found that the diffusion constant of water decreases drastically during the acute phase of stroke *(27,28)*. At present, the exact mechanism of the decrease is not completely understood, but it is believed to be related to the break-

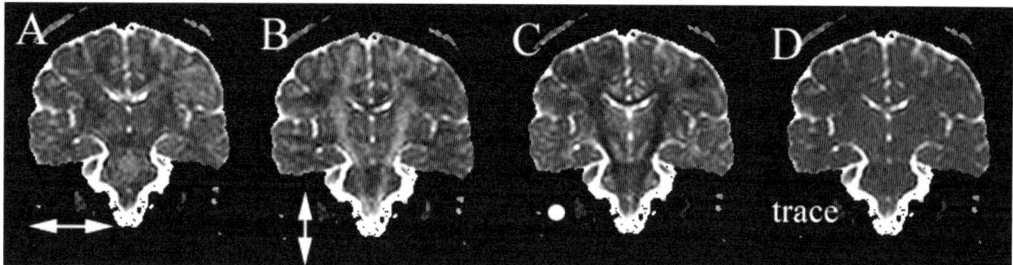

Fig. 9. Examples of ADC maps measured using x (**A**), y (**B**), and z (**C**) gradients. A trace map can be obtained by adding these three ADC maps, which is shown in (**D**). The trace image (**D**) is very uniform throughout the brain.

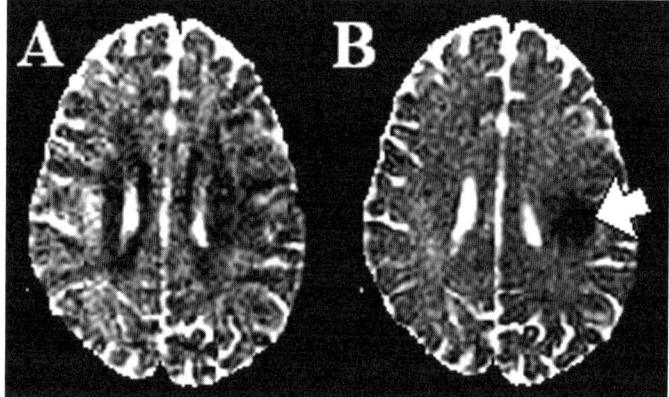

Fig. 10. An example of stroke study using diffusion imaging. An ADC map along the y (left-right) axis (**A**) and a trace image (**B**) of a stroke patient are compared. Image (**A**) contains contrast due to the orientation effect, which was removed in the trace image (**B**). The low diffusion constant caused by the stroke remains in (**B**), indicating that a reduction in the size of diffusion ellipsoids occurred. The trace image is far superior to specifically delineate the stroke affected region. Images are reproduced from Ulug et al. Stroke 1997; **28**:483, with permission.

down of membrane potential and subsequent cell swelling. This phenomenon can be described as the reduction of the size of the diffusion ellipsoids. An example is shown in Figure 10, in which an ADC map along the x axis and a trace map are compared. As can be seen from this figure, the trace image is free of unwanted contrast, which is caused by the anisotropy effect that interferes with the detection of the stroke regions. Diffusion imaging is the only radiological technique that can detect the physiological change in brain parenchyma at the acute stroke phase when the parenchyma is still alive and the lesion is possibly still reversible. There is no doubt that it is now an indispensable tool for stroke research using animal models, and it is also becoming a promising diagnostic tool in clinical situations.

13.3.3. Anisotropy Map: An Orientation-Independent Visualization Technique for Anisotropy

There are many ways to characterize the shape of diffusion ellipsoids, which are orientation-independent and are also not affected by the size. The simplest and most intuitive method is to calculate the ratio of the length of the longest and shortest axes. This method, however, has several unwanted properties. For example, the range of its value is 1 (sphere) – infinity, which is difficult to visualize, and the length of the shortest axis (thus the ratio) is very susceptible to noise. It is preferable to use a parameter that ranges 0 (isotropy) – 1 (anisotropy) for the visualization purpose. For example:

$$\frac{(\lambda_1 - (\lambda_2 + \lambda_3)/2)}{(\lambda_1 + \lambda_2 + \lambda_3)/3}$$ [Eq. 4]

has the preferable property. Namely, it becomes 0 when $\lambda_1 = \lambda_2 = \lambda_3$ (isotropic) and the maximum is λ when $\lambda_1 \gg \lambda_2, \lambda_3$ (extreme anisotropy). The most widely used parameters are *(19,22–24,29)*:

$$FA = \frac{\sqrt{(\lambda_1 - \lambda_2)^2 + (\lambda_2 - \lambda_3)^2 + (\lambda_1 - \lambda_3)^2}}{\sqrt{2}\sqrt{\lambda_1^2 + \lambda_2^2 + \lambda_3^2}}$$

$$RA = \frac{\sqrt{(\lambda_1 - \lambda_2)^2 + (\lambda_2 - \lambda_3)^2 + (\lambda_1 - \lambda_3)^2}}{\sqrt{\lambda_1 + \lambda_2 + \lambda_3}}$$ [Eq. 5]

$$VR = \frac{\lambda_1 \lambda_2 \lambda_3}{[(\lambda_1 + \lambda_2 + \lambda_3)/3]^3}$$

Here FA = fractional anisotropy, RA = relative anisotropy, and VR = volume ratio. These parameters all have 0 – 1 range. Information provided by these parameters is essentially the same. They all indicate how elongated the diffusion ellipsoid is. However, the contrasts they provide are not the same. Among these parameters probably the FA is most widely used. An example of the FA map is shown in Figure 11. It can be seen that the white matter has high FA values, which makes sense because it consists of densely packed axonal fibers. Segmentation of the white matter and gray matter can also be achieved with conventional T_1- and T_2-weighted images. T_1- and T_2-weighted images are widely used contrast mechanisms in MRI, and they are used to differentiate white and gray matter. Anisotropy, which also shows a very high contrast between white and gray matter, is based on a completely different contrasting mechanism; the directionality of water diffusion given by axonal fibers. Unlike T_1- and T_2-based contrasts, which are related to molecular tumbling rates and exchange processes, the anisotropy is a more direct indicator of highly packed axonal fibers, a hallmark of white matter. The comparison shown in Figure 11 clearly illustrates more detailed structures within the white matter when anisotropy is used.

The exact mechanism underlying the anisotropy map is not completely understood *(30)*. What is known is that the anisotropy increases during early development *(31–37)*, and its time course is different from other conventional MRI parameters, such as T_1/T_2 relaxation properties *(32–34,38–40)*. The changes during development may suggest involvement of the myelin sheath. However, anisotropy has been reported in axonal fibers without myelin sheaths *(35,38,41)*, implying that the increased anisotropy during development may be due to increased fiber density. In any case, it is apparent that the anisotropy provides a new contrasting mechanism that was formally inaccessible and, thus, it is worth pursuing its clinical possibilities as a new diagnostic tool.

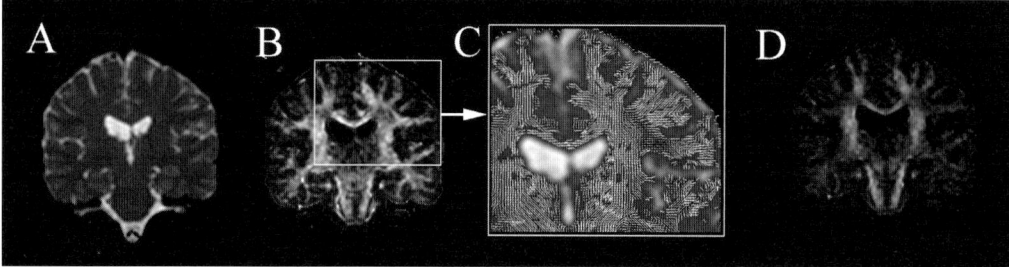

Fig. 11. Example of T_2-weighted image (**A**), FA (**B**), vector (**C**) and color map (**D**). In the vector map, a small line at each pixel represents the average orientation of fibers (lines perpendicular to the plane look like small dots). In the color map, red represents tracts running left-right, green superior-inferior, and blue anterior-posterior directions (the color figure is available in the companion CD-ROM).

13.3.4. Color Map: Visualization Technique for Orientation

The last parameter that can be obtained from the DTI is the orientation of diffusion ellipsoids. The most intuitive way to show the orientation is the vector presentation, in which small lines (vectors) indicate the orientations of the longest axis of diffusion ellipsoids. An example is shown in Figure 11C. However, unless a small region is magnified, the vector orientation is often difficult to see. To overcome this problem, a color-coded scheme was postulated (42–44) (for the color image, please refer to MORI-fig11 in the companion CD-ROM). An example is also shown in Figure 11. In the color map, three orthogonal axes (e.g., right-left, superior-inferior, and anterior-posterior) are assigned to three principal colors (red, green, and blue), with which every orientation can be represented by color. A detailed discussion about the techniques to assign colors to orientation can be found in a paper by Pajevic and Pierpaoli (44).

In Figure 12, a postmortem anatomical preparation is compared to a T_1-weighted image and a color map (for the color image, please refer to MORI-fig12 in a companion CD-ROM). The brain white matter consists of bundles of neuronal fibers connecting different parts of brain functional centers. The sizes and directions of these fiber bundles vary considerably. However, in conventional MR images, the white matter looks rather homogeneous as can be seen in Figure 12B. This is understandable, because MRI relaxation parameters, such as T_1 and T_2 are sensitive to the chemical composition of the environment, which is rather homogeneous in the brain white matter. However, it is fiber direction that makes the structure of the white matter so complex as seen in Figure 12A. The beauty of the DTI technique is that it can reveal the white matter architecture much more clearly, as seen in Figure 12C. From what has been shown so far, it is clear that one of the most important and unique functions of DTI is the parcellation of the white matter into multiple white matter tracts (44,45). White matter tracts that can be discretely identified are annotated in Figure 13 (for the color image, please refer to MORI-fig13 in the companion CD-ROM).

Clinical application of the anisotropy and color maps is still mostly in a research phase. The most extensively researched area involves white matter diseases such as multiple sclerosis (46–48), in which loss of white matter integrity is expected to lead to

Diffusion Tensor Imaging 283

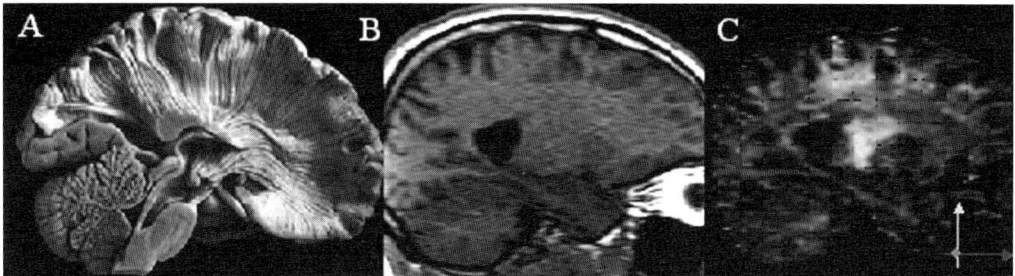

Fig. 12. Comparison of a postmortem anatomical preparation (**A**), T_1-weighted image (**B**), and the DTI-based color map (**C**). In the anatomical preparation, the gray matter was removed to reveal the structural details of the white matter. In the color map, blue pixels have fibers running horizontally (anterior-posterior), green vertically (superior-inferior), and red perpendicular to the plane (left-right) (the color figure is available in the companion CD-ROM). The fiber directions based on the DTI measurement agree very well with the anatomical preparation (**A**), which can not be appreciated in the conventional MRI (**B**). (The postmortem image was reproduced from Williams et al. *[67]* with permission).

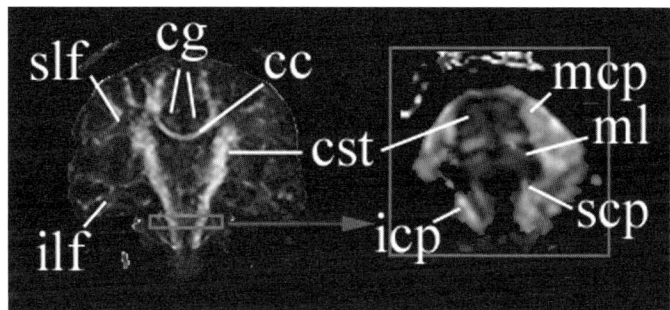

Fig. 13. Parcellation of the white matter by the DTI-based color maps (the color figure is available in the companion CD-ROM). Homogeneous-looking brain white matter can be parcellated into different tract systems depending on their orientations (colors). The left image shows a color map of a coronal slice, and the right image shows an axial slice at the location shown by a pink box. Abbreviations are: slf, superior longitudinal fasciculus; ilf, inferior longitudinal fasciculus; cg, cingulum; cc, corpus callosum; cst, corticospinal tract; scp, superior cerebellar peduncle; mcp, middle cerebellar peduncle; icp, inferior cerebellar peduncle; and ml, medial lemniscus.

decrease in anisotropy. Other reports include amyotrophic lateral sclerosis (ALS) *(49,50)*, stroke *(51)*, schizophrenia *(52)*, and reading disability *(53)*.

13.4. 3D-BASED DTI TECHNIQUES AND THEIR APPLICATIONS

Knowledge of neuronal connections by axonal projections is of critical importance for understanding brain function and its abnormalities. In conventional white matter tract-tracing methods, axonal projections have been traced in experimental animals. With these methods, it is possible to observe neuronal degeneration following carefully placed experimental brain lesions or by injecting and subsequently localizing radioisotopes or other chemicals that are taken up by nerve cells and actively trans-

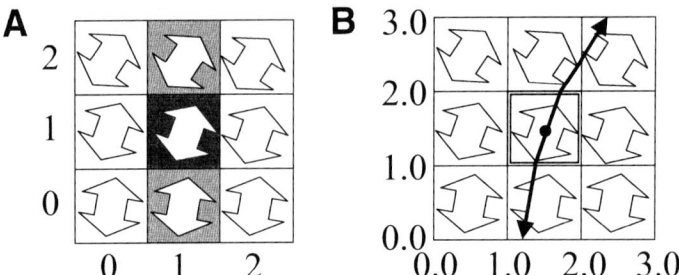

Fig. 14. Examples of fiber tracking. Double head arrows indicate the orientation of tracts at each pixel, which are obtained from the DTI measurement. Since the DTI study, which is based on water movement, and because one cannot judge effluent and affluent, tracking has to be made in both orthograde and retrograde directions.

ported along their axons. Comparable human data are much more limited because they can only be obtained from postmortem examinations of patients with naturally occurring lesions such as injuries or infarcts. One cannot control the size, the position, or the timing of these lesions. As has been demonstrated, DTI can provide the orientation of axonal fibers at each pixel. By extending this information to 3D space, it is possible to reconstruct a trajectory of a tract of interest *(5,7–12,54)*.

An example of a 2D vector map is shown in Figure 11. It is straightforward to extend the DTI measurement to 3D and obtain a 3D vector field, from which information on 3D tract trajectories can be extracted. In this step, there are two important issues to be considered.

First, the raw data, which is a vector field, is discrete information from which we have to reconstruct continuous trajectory coordinates. Second, each vector carries linear information, while trajectories of interest have curvature. This situation is shown in Figure 14A using a 2D example. Suppose a tracking is initiated from a pixel coordinate (1,1). The most intuitive way to reconstruct a trajectory is to connect pixels from the initiation pixel. However, the first problem we encounter in this approach is to judge which pixel to connect, because the vectors usually are not pointing to the center of neighboring pixels. In other words, there are only 8 surrounding pixels (26 for 3D), and by selecting one of them, the information of the vector direction cannot be fully obtained. In this case, pixels above and below the initiating pixel are chosen (shaded pixels), but they do not correctly reflect the vector angles of the involved pixels.

In order to solve this problem, which stems from the discreteness of vector information, tracking must be made in a continuous number field. An example of this is shown in Figure 14B. In this case, tracking is initiated from the center of a pixel (1, 1 in the discrete coordinate) at the coordinate (1.5, 1.5) in the continuous coordinate. Then a line is propagated in the continuous coordinate along the direction of the vector of the pixel. Now the line exits the initiating pixel at the coordinate of, e.g., (1.8, 2.0), and enters the next pixel (2, 2 in the discrete coordinate), in which the tracking starts to observe the vector direction of the new pixel. We called this approach FACT (fiber assignment using continuous fiber tracking), in which vector information is propagated linearly within a pixel *(6,8)*. The tracking line can be smoothed out to obtain a curva-

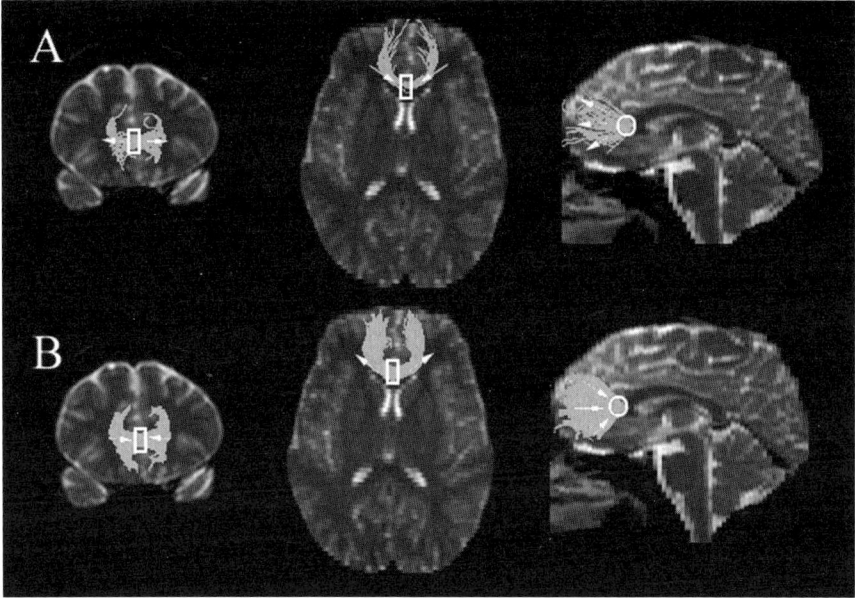

Fig. 15. Comparison between two approaches for the fiber tracking. In this example, an ROI delineated the genu of the corpus callosum, which included 21 pixels. In (**A**), tracking was initiated from the 21 pixels, resulting in 21 lines to reveal the callosal connections between the frontal lobes. In (**B**), tracking was initiated from all pixels in the brain, and tracking results that penetrated the ROI were searched. This approach resulted in identifying 1880 pixels that were connected to the ROI, thus, revealing more comprehensive structure of the callosal connections. A color version of this figure is included in the CD-ROM.

ture by using an interpolation technique, such as distant-weighted vector averaging or so-called Runge-Kutta methods *(6,9,10)*. This can reduce an error that may accumulate in the simple linear propagation technique, especially when there are large angle transitions from a pixel to a pixel.

One of the pitfalls of these types of "propagation techniques" is the accumulation of noise errors along the tract. In order to minimize this problem, techniques that are based on energy minimization are also proposed *(11)*. The diffusion tensor theory introduced in an earlier section assumes that there is only one population of tracts that have the same orientation, which is not always the case due to partial volume effects or interdigitating (or crossing) tracts in certain brain areas. It should be remembered that the tracking technique might have systematic errors in such regions. Recently, a new DTI technique called the "diffusion spectrum" method was introduced to analyze pixels with more than one population of tracts, in which water diffusion is measured along many axes *(54)*. This type of new pixel-by-pixel analyzing technique, combined with a tract reconstruction technique, may improve the quality of the reconstruction in the future.

In the line propagation technique, the process starts by identifying an anatomical landmark and drawing a region of interest (ROI). An example is shown in Figure 15. In this example, the genu of the corpus callosum was delineated, which included 21 pixels. Then tracking was performed from each pixel, resulting in 21 lines propagated

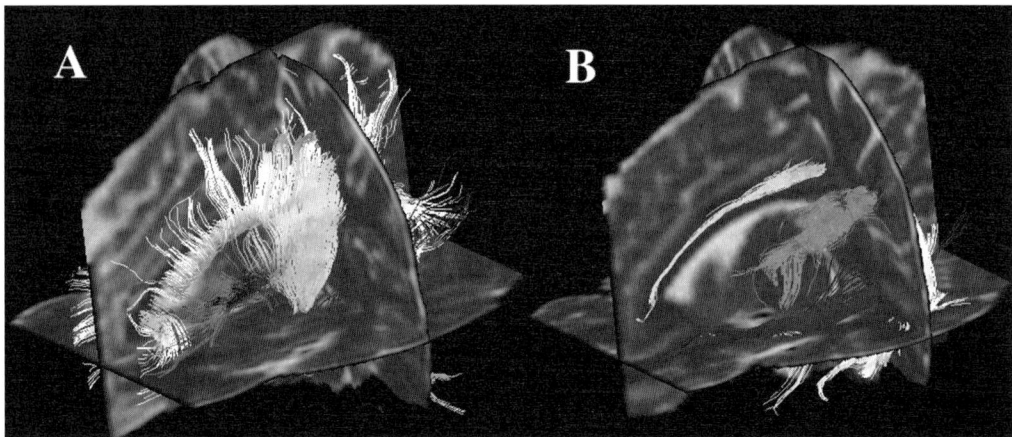

Fig. 16. Examples of 3D brain fiber reconstruction (color figures are available in the companion CD-ROM). In **(A)** white matter tracts that form corona radiata are shown: corpus callosum (yellow), anterior thalamic radiation (red), corticobalbar–corticospinal tract (green), optic radiation (blue). In **(B)**, association fibers and tracts in the limbic system are shown: cingulum (green), fimbria (red), superior (pink) and inferior (yellow) longitudinal fasciculus, uncinate fasciculus (light blue), and inferior fronto-occipital fasciculus (blue). The images were produced by Meiyappan Soleiyappan, Johns Hopkins University School of Medicine, Department of Radiology.

from the ROI (Fig. 15A). A problem that may be encountered in this approach is the branching. Whenever the tracking meets a bifurcation, it can delineate only one of them. In other words, this technique cannot delineate more than 21 tracts. It turns out that this problem can be addressed rather easily. Instead of initiating tracking from pixels in the ROI, tracking can be initiated from all pixels in the brain and tracking results that penetrate the ROI are searched (Fig. 15B) *(9)*. This approach leads to tracking results initiated from 1880 pixels in the frontal lobe that penetrate the ROI, thus revealing more comprehensive branching patterns of the tract.

Examples of tract reconstruction of prominent white matter tracts in the cerebral hemispheres are shown in \h Figure 16. In this figure, results of multiple tracking results are assigned different colors and superimposed on co-registered 3D anatomical data (for the color image, please refer to MORI-fig16 in a companion CD-ROM). For example, the corpus callosum was identified, and an ROI was defined at the midsagittal level. For the anterior thalamic radiation, the anterior limb of the internal capsule was defined as an ROI. The overall structures of these tracts agree well with what has been established in anatomical studies.

Currently, 3D tract reconstruction techniques are in a phase of validation and reliability examination. While these two issues, validation and reliability, are of critical importance to establish this technique as a research and/or clinical tool, it is also important to have a clear idea about what can be done and what cannot be done with this technique. First of all, the DTI-based tracking technique, which is based on water movement, cannot distinguish affluent and effluent directions of axonal tracts. Second, the image resolution of a typical human DTI study is on the order of 1 – 5 mm. Once an axon of interest enters into a pixel of this size and is mixed with other axons with different destinations, information about cellular level connectivity degenerates. Therefore, cellular level connectivity cannot be addressed with this technique.

So what can we study with the DTI-based tract reconstruction techniques? We believe that DTI can reveal macroscopic architectures of the white matter, such as those that can be identified in Figure 12A. In experimental animal models, cellular level connectivity can be directly studied using an invasive technique. However, study of entire brain tract structure by such cellular level techniques is practically impossible, simply because of the vast number of neurons inside the brain. Therefore, the DTI technique is a complementary technique that can characterize entire brain tract structures rapidly in a 3D electronic format. In human studies, it is a great advantage that macroscopic white matter anatomy can be obtained noninvasively. In the future, we expect that this technique will play an important role in identifying the involvement of specific tract systems in various neurological diseases such as neurodegenerative diseases, tumors, and developmental defects.

13.5. FUTURE DIRECTIONS AND SUMMARY

In this article, I introduced the concept of diffusion tensor imaging and its applications. The study of the white matter architecture using the color map and 3D reconstruction techniques is especially exciting because there have not been noninvasive techniques that can provide equivalent information. Comparison studies between the DTI-based white matter anatomical studies and histology-based classical anatomical knowledge have been showing encouraging correlations *(5,7–11,24,44,45,55)*. While the imaging resolution that the DTI technique can achieve is far inferior to that of histology, its noninvasive nature and capability of 3D data analyses give it a distinctive advantage for macroscopic characterization of white matter organization of living humans.

At present, the DTI technique is still very new and has not been established as a diagnostic tool for particular brain diseases. This is partly due to the fact that we do not have tools to analyze the vast amount of information that the DTI can provide. For example, the color map can visualize that a part of the white matter consists of several tracts with different orientations. However, how we can quantify this information? How we can detect abnormality? White matter tracts are 3D entities. How we can parameterize it and compare between normals and patients? Another important factor that limits application studies of the DTI is its poor resolution and long scanning time. Because the DTI requires at least 7 images for the tensor calculation, it is not only a slow imaging technique, but also prone to errors during image co-registration. The technique is inherently sensitive to motion artifacts, which further reduces it practicality. However, promising new hardware and data acquisition schemes are being developed, and these limitations are quickly diminishing. For example, signal-to-noise ratio can be improved by higher field magnets and stronger gradients. It has been shown that partially parallel acquisition schemes such as SENSE and SMASH drastically reduce the scanning time *(56–59)*. Various data sampling schemes have been postulated that are less motion sensitive *(60–66)*. 3D data analysis techniques are also being actively studied as mentioned in Subheading 13.4. In the near future, the DTI technique will most likely be a powerful investigational and diagnostic tool for studying brain anatomy and diseases.

REFERENCES

1. Le Bihan D, Breton E, Lallemand D, Grenier P, Cabanis E, Laval-Jeantet M. MR imaging of intravoxel incoherent motions: application to diffusion and perfusion in neurologic disorders. Radiology 1986; **161**:401–407.
2. Turner R, LeBihan D, Maier J, Vavrek R, Hedges LK, Pekar J. Echo-planar imaging of intravoxel incoherent motions. Radiology 1990; **177**:407–414.
3. Stejskal E. Use of spin echoes in a pulsed magnetic-field gradient to study restricted diffusion and flow. J Chem Phys 1965; **43**:3597–3603.
4. Cleaveland GG, Chang DC, Hazlewood CF, Rorschach HE. Nuclear magnetic resonance measurement of skeletal muscle: anisotropy of the diffusion coefficient of the intracellular water. Biophys J 1976; **16**:1043–1053.
5. Mori S, Crain BJ, Chacko VP, van Zijl PCM. Three dimensional tracking of axonal projections in the brain by magnetic resonance imaging. Ann Neurol 1999; **45**:265–269.
6. Mori S, Barker P. Diffusion magnetic resonance imaging: its principle and applications. Anat. Rec (New Anatomist) 1999; **257**:102–109.
7. Mori S, Kaufmann WK, Pearlson GD, et al. *In vivo* visualization of human neural pathways by MRI Ann Neurol 2000; **47**:412–414.
8. Xue R, van Zijl PCM, Crain BJ, Solaiyappan M, Mori S. In vivo three-dimensional reconstruction of rat brain axonal projections by diffusion tensor imaging. Magn Reson Med 1999; **42**:1123–1127.
9. Conturo TE, Lori NF, Cull TS, et al. Tracking neuronal fiber pathways in the living human brain. Proc Natl Acad Sci USA 1999; **96**:10422–10427.
10. Basser PJ, Pajevic S, Pierpaoli C, Duda J, Aldroubi A. In vitro fiber tractography using DT-MRI data. Magn Reson Med 2000; **44**:625–632.
11. Poupon C., Clark C. A., Frouin V., et al. (2000) Regularization of diffusion-based direction maps for the tracking of brain white matter fascicules. Neuroimage **12**:184–195.
12. Jones DK, Simmons A, Williams SC, Horsfield MA. Non-invasive assessment of axonal fiber connectivity in the human brain via diffusion tensor MRI. Magn Reson Med 1999; **42**:37–41.
13. Parker GJ. Tracing fiber tracts using fast marching. In: Proceedings, International Society of Magnetic Resonance. Denver, CO, 2000, p. 85.
14. Lazar M, Weinstein D, Hasan K, Alexander AL. Axon tractography with tensorlines. In: Proceedings, International Society of Magnetic Resonance in Medicine. Denver, CO. 2000, p. 482.
15. Carr H, Purcell E. Effect of diffusion on free precession in nuclear magnetic resonance experiments. Phys Rev 1954; **94**:630–638.
16. Torrey H. Bloch equation with diffusion terms. Phys Rev 1956; **104**:563–565.
17. Stejskal EO, Tanner JE. Spin diffusion measurement: spin echoes in the presence of a time-dependent field gradient. J Chem Phys 1965; **42**:288.
18. Moseley ME, Cohen Y, Kucharczyk J, et al. Diffusion-weighted MR imaging of anisotropic water diffusion in cat central nervous system. Radiology 1990; **176**:439–445.
19. Basser PJMattiello J, LeBihan D. Estimation of the effective self-diffusion tensor from the NMR spin echo. J Magn Reson B 1994; **103**:247–254.
20. Crank J. The Mathematics of Diffusion. Oxford Press, Oxford, England, 1975.
21. van Gelderen P, de Vleeschouwer MH, DesPres D, Pekar J, van Zijl PCM, Moonen CTW. Water diffusion and acute stroke. Magn Reson Med 1994; **31**:154–163.
22. Basser PJ, Pierpaoli C. Microstructural features measured using diffusion tensor imaging. J Magn Reson B 1996; **111**:209–219.
23. Pierpaoli C, Basser PJ. Toward a quantitative assessment of diffusion anisotropy. Magn Reson Med 1996; **36**:893–906.

24. Pierpaoli C, Jezzard P, Basser PJ, Barnett A, Di Chiro G. Diffusion tensor MR imaging of human brain. Radiology 1996; **201:**637–648.
25. Basser PJ, Mattiello J, Le Bihan D. MR diffusion tensor spectroscopy and imaging. Biophys J 1994; **66:**259–267.
26. Mori S, van Zijl PC M. Diffusion weighting by the trace of the diffusion tensor within a single scan. Magn Reson Med 1995; **33:**41–52.
27. Moseley ME, Kucharczyk J, Mintorovitch J, et al. Diffusion-weighted MR imaging of acute stroke: correlation with T2-weighted and magnetic susceptibility-enhanced MR imaging in cats. Am J NeuroRad 1990; **11:**423–429.
28. Moseley ME, Cohen Y, Mintorovitch J, et al. Early detection of regional cerebral ischemia in cats: comparison of diffusion- and T2-weighted MRI and spectroscopy. Magn Reson Med 1990; **14:**330–346.
29. Ulug A, van Zijl PCM. Orientation-independent diffusion imaging without tensor diagonalization: anisotropy definitions based on physical attributes of the diffusion ellipsoid. J Magn Reson Imaging 1999; **9:**804–813.
30. Henkelman R, Stanisz G, Kim J, Bronskill M. Anisotropy of NMR properties of tissues. Magn Reson Med 1994; **32:**592–601.
31. Sakuma H, Nomura Y, Takeda K, et al. Adult and neonatal human brain: diffusional anisotropy and myelination with diffusion-weighted MR imaging. Radiology 1991; **180:** 229–233.
32. Takeda K, Nomura Y, Sakuma H, Tagami T, Okuda Y, Nakagawa T. MR assessment of normal brain development in neonates and infants: comparative study of T1- and diffusion-weighted images. J Comput Assist Tomogr 1997; **21:**1–7.
33. Neil J, Shiran S, McKinstry R, et al. Normal brain in human newborns: apparent diffusion coefficient and diffusion anisotropy measured by using diffusion tensor MR imaging. Radiology 1998; **209:**57–66.
34. Morriss M, Zimmerman R, Bilaniuk L, Hunter J, Haselgrove J. Changes in brain water diffusion during childhood. Neuroradiology 1997; **41:**929–934.
35. Huppi P, Maier S, Peled S, et al. Microstructural development of human newborn cerebral white matter assessed in vivo by difusion tensor magnetic resonsnce imaging. Pediatr Res 1998; **44:**584–590.
36. Baratti C, Barnett A, Pierpaoli C. Comparative MR imaging study of brain maturation in kittens with t1, t2, and the trace of the diffusion tensor. Radiology 1999; **21:**13–142.
37. Klingberg T, Vaidya C, Gabrieli J, Moseley M, Hedehus M. Myelination and organization of the frontal white matter in children: a diffusion tensor MRI study. Neuroreport 1999; **10:**2817–2821.
38. Wimberger D, Roberts T, Barkovich A, Prayer L, Moseley M, Kucharczyk Z. Identification of "premyelination" by diffusion-weighted magnetic resonance imaging. J Comput Assist Tomogr 1995; **19:**28–33.
39. Battin M., Maalouf E., Counsell S., et al. Magnetic resonance imaging of the brain in very preterm infants: visualization of the germinal matrix, early myelination, and cortical folding. Pediatrics 1998; **101:**957–962.
40. Garel C, Briees H, Sebag G, Elmaleh M, Oury J-F, Hassan M. Magnetic resonance imaging of the fetus. Pediatr Radiol 1998; **28:**201–211.
41. Beaulieu C, Allen PS. Determinants of anisotropic water diffusion in nerves. Magn Reson Med 1994; **31:**394–400.
42. Douek P, TR, Pekar J, Patronas N, Le Bihan D. MR color mapping of myelin fiber orientation. J Comput Assist Tomogr 1991; **15:**923–929.
43. Nakada T, Matsuzawa H. Three-dimensional anisotropy contrast magnetic resonance imaging of the rat nervous system: MR axonography. Neurosci Res 1995; **22:**389–398.
44. Pajevic S, Pierpaoli C. Color schemes to represent the orientation of anisotropic tissues

from diffusion tensor data: application to white matter fiber tract mapping in the human brain. Magn Reson Med 1999; **42**:526–540.
45. Makris N, Worth AJ, Sorensen AG, et al. Morphometry of in vivo human white matter association pathways with diffusion weighted magnetic resonance imaging. Ann Neurol 1997; **42**:951–962.
46. Nusbaum AO, Lu D, Tang CY, Atlas SW. Quantitative diffusion measurements in focal multiple sclerosis lesions: correlations with appearance on TI-weighted MR images. AJR Am J Roentgenol 2000; **175**:821–825.
47. Tievsky AL., Ptak T, Farkas J. Investigation of apparent diffusion coefficient and diffusion tensor anisotrophy in acute and chronic multiple sclerosis lesions. AJNR Am J Neuroradiol 1999; **20**:1491–1499.
48. Werring DJ, Clark CA, Barker GJ, Thompson AJ, Miller DH. Diffusion tensor imaging of lesions and normal-appearing white matter in multiple sclerosis. Neurology 1999; **52**:1626–1632.
49. Ahrens ET, Laidlaw DH, Readhead C, Brosnan CF, Fraser SE, Jacobs RE. MR microscopy of transgenic mice that spontaneously acquire experimental allergic encephalomyelitis. Magn Reson Med 1998; **40**:119–132.
50. Ellis C, Simmons A, Jones D, et al. Diffusion tensor MRI assesses corticospinal tract damages in ALS. Neurology 1999; **22**:1051–1058.
51. Mukherjee P, Bahn M, McKinstry R, et al. Difference between gray matter and white matter water diffusion in stroke: diffusion tensor MR imaing in 12 patients. Radiology 2000; **215**:211–220.
52. Lim KO, Hedehus M, Moseley M, de Crespigny A, Sullivan EV, Pfefferbaum A. Compromised white matter tract integrity in schizophrenia inferred from diffusion tensor imaging. Arch Gen Psychiatry 1999; **56**:367–374.
53. Klingberg T, Hedehus M, Temple E, et al. 2000; Microstructure of temporo-parietal white matter as a basis for reading ability: evidence from diffusion tensor magnetic resonance imaging. Neuron **25**:493–500.
54. Wiegell M, Larsson H, Wedeen V. Fiber crossing in human brain depicted with diffusion tensor MR imaging. Radiology 2000; **217**:897–903.
55. Holodny AI, Ollenschleger MD, Liu WC, Schulder M, Kalnin AJ. Identification of the corticospinal tracts achieved using blood-oxygen-level-dependent and diffusion functional MR imaging in patients with brain tumors. AJNR Am J Neuroradiol 2001; **22**:83–88.
56. Pruessmann KP, Weiger M, Scheidegger MB, Boesiger P. SENSE: sensitivity encoding for fast MRI. Magn Reson Med 1999; **42**:952–962.
57. Griswold MA, Jakob PM, Chen Q, et al. Resolution enhancement in single-shot imaging using simultaneous acquisition of spatial harmonics (SMASH). Magn Reson Med 1999; **41**:1236–1245.
58. Heidemann R, Friswold M, Porter D, et al. Minimizing distortions and blurring in diffusion weighted single shot EPI using high performance gradients in combination with parallel imaging. In: Proceedings, International Society of Magnetic Resonance in Medicine. Glasgow, 2001, p. 169.
59. Bammer R, Keeling SL, Auer M, et al. Diffusion tensor imaging using SENSE-single-shot EPI. In: Proceedings, International Society of Magnetic Resonance in Medicine. Glasgow, 2001, p. 160.
60. Butts K, Pauly J, de Crespingy A, Moseley M. Isotropic diffusion-weighted and spiral-navigated interleaved EPI for routine imaging of acute stroke. Magn Reson Med 1997; **38**:741–749.
61. Anderson AW, Gore J. Analysis and correction of motion artifacts in diffusion weighted imaging. Magn Reson Med 1994; **32**:379–387.

62. Ordidge RJ, Helpern JA, Qing ZX, Knight RA, Nagesh V. Correction of motional artifacts in diffusion-weighted NMR images using navigator echoes. Magn Reson Imaging 1994; **12**:455–460.
63. Seifert MH, Jakob PM, Jellus V, Haase A, Hillenbrand C. 2000 High-resolution diffusion imaging using a radial turbo-spin-echo sequence: implementation, eddy current compensation, and self- navigation. J Magn Reson **144**:243–254.
64. Trouard TP, Theilmann RJ, Altbach MI, Gmitro AF. High-resolution diffusion imaging with DIFRAD-FSE (diffusion-weighted radial acquisition of data with fast spin-echo) MRI. Magn Reson Med 1999; **42**:11–18.
65. Pipe JG. Multishot diffusion weighted FSE with PROPELLAR. In: Proceedings, International Society of Magnetic Resonance in Medicine. Glasgow, 2001, p. 166.
66. Finsterbusch J, Frahm J. Diffusion tensor mapping of the human brain using single-shot line scan imaging. J Magn Reson Imaging 2000; **12**:388–394.
67. Williams TH, Gluhbegovic N, Jew JY. The Human Brain: Dissections of the Real Brain. Virtual Hospital, University of Iowa (1997; http://www.vh.org/Providers/Textbooks/BrainAnatomy).

Part III

14
Computational Methods for the Analysis of Brain Connectivity

Claus C. Hilgetag, Rolf Kötter, Klaas E. Stephan, and Olaf Sporns

ABSTRACT

The body of knowledge about the connectivity of brain networks on different structural scales is growing rapidly. This information is considered highly valuable for determining the neural organization underlying brain function, yet connectivity data are too extensive and too complex to be understood intuitively. Computational analysis is required to evaluate them. Here we review mathematical, statistical, and computational methods that have been used by ourselves and other investigators to assess the organization of brain connectivity networks.

Many available analysis approaches are based on a description of connectivity networks as simple or directed graphs. Given adjustments for specific neural properties, this description can unify analysis techniques across many dimensions of brain connectivity. It also makes available a great arsenal of analytical tools that have been developed previously for the graph theoretical evaluation of networks.

Generally, computational approaches to connectivity analysis may be grouped into two categories. On the one hand, statistical data exploration reveals local as well as extreme or average network properties and allows visualization of the global topological organization of the investigated networks. Useful routines for the exploration of networks by, for instance, similarity or cluster analyses, can be found in many general statistical packages. On the other hand, specialized computational techniques have been developed recently that allow testing of specific hypotheses about the organization of neural connectivity. These approaches have been based on optimization techniques that are employing cost measures to assess the structural or functional connectivity of networks.

We illustrate different methods of connectivity analyses with the well-known example of neuroanatomical connectivity of the primate cortical visual system and indicate how the identification of structural organization may shed light on functional aspects of brain networks. The methods reviewed here may be general enough to also prove useful for unraveling the structure of other large-scale and complex networks, such as metabolic, traffic, communication, or social networks.

From: *Computational Neuroanatomy: Principles and Methods*
Edited by: G. A. Ascoli © Humana Press Inc., Totowa, NJ

14.1. INTRODUCTION

Connectivity defines the role of individual nerve cells, or of distinct neuronal systems, within the global context of neural networks and the brain. Afferents determine the input into a cell or system, and output projections relay the processed information onto selected targets. Perhaps the most intuitive image of the nervous system is that of a network, and many models of information processing in the brain are based upon selected features of neural connectivity, e.g., *(1-3)*. However, the connectivity networks of mammalian brains are stunningly and intriguingly complex biological objects. Even at their simplest recognizable structural level, they resemble a quilt of many dozens of specialized elements (areas of the cortex and subcortical nuclei), which are interwoven with an intricate network of hundreds or thousands of fibers. Specialized regions in the cat cortex, for instance, possess on average 56 afferent and efferent connections with other cortical or thalamic structures. Considering only the connections within one cortical hemisphere of the cat brain, a given area connects, on average, with 32 other areas *(4)*.

How are these networks organized? Are connections distributed randomly between the different regions, or do they follow specific patterns? Does the distribution correlate with spatial or functional subdivisions of the brain? Does the organization of the neural networks hold clues about developmental factors or on structural features that might influence information processing in the nervous system? The intricacy of networks with such a high number of connections per node makes it very difficult to give reliable answers to these questions just by unaided intuition—computational analysis is needed.

An increasing range of measures is now becoming available that describe aspects of functional correlations between neural structures, allowing conclusions about functional and effective connectivity, e.g., *(5)*. Here, we concentrate on what, to many, appears to be a simpler problem, the understanding of structural neural connectivity. However, we indicate functional implications where they suggest themselves from our explorations. In this review, we mainly consider the analysis of large-scale neural connectivity at the systems level, an area where several theoretical studies have been carried out previously, and where fairly extensive sets of formalized data are already available. Without restricting generality, many of the approaches described here may also be applicable on other dimensional scales of organization.

Throughout this review, we use the example of one widely studied set of established neural systems connectivity data, corticocortical anatomical connectivity for the visual system of the macaque monkey as compiled by Felleman and Van Essen *(6)* and analyzed by themselves, Young *(3)*, Young et al. *(7)*, Hilgetag et al. *(8–10)*, Jouve et al. *(11)*, Sporns et al. *(12,13)*, and Kötter et al. (Ch. 16, this volume) among others.

In the following, we first summarize how neural connectivity is established experimentally and how it can be represented in more formal computational or mathematical terms. Then we review in greater detail one established avenue of network analysis, the graph theoretical analysis of connectivity. Following, we survey a number of readily available statistical methods that can be used to explore the hodology, that is, the essential global organization, of neural networks. We subsequently present some recent computational approaches that have been specifically developed to test hypotheses about the organization of neural connectivity. Finally, we give our conclusions and review how the different presented analysis techniques contribute to an understanding of the connectional organization of the example data for primate visual cortical connectivity.

14.2. DESCRIPTION OF NEURAL CONNECTIVITY

14.2.1. Experimental Identification of Connectivity

The traditional way of experimentally identifying structural connectivity is by tract-tracing. Typically, a tracer substance (e.g., a dye, marker particle or virus) is injected into the living nervous system, taken up by neurons, and actively transported along neural fibers by metabolic processing or other transport mechanisms of the living cell. After a well-timed interval that allows complete labeling of the injected cells, or of an intended sequence of cells down the postsynaptic chain, the brain is sectioned and chemically treated in order to reveal the extent of the labeling. One generally distinguishes between anterograde labeling (i.e., the tracer is taken up by the cell body and its dendrites and is transported towards the axonal terminals) and retrograde labeling (in which case the tracer is transported toward the cell body); specific experimental techniques allow either one or both of these approaches to be implemented. For comprehensive reviews of frequently used experimental methods, see *(14–16)*. Due to the nature of the utilized biological mechanisms, these approaches have only been employed in nonhuman experimental animals, and the species-specific connectivity of the human brain remains very much a mystery *(17)*. Recent developments in magnetic resonance imaging techniques may, however, soon lead to some improvements of this situation *(18,19)*. Naturally, the intended complete and unambiguous identification of all afferent and efferent projections of a specific neural system is an ideal state that, in the practical neuroanatomical experiment, may be hampered by underlabeling, overlabeling, mislabeling, and so on. These problems are compounded by the difficulties of precisely delineating distinct brain regions, e.g., *(20)*, or by the absence of unified parcellation schemes across different experiments. While there remains a large scope for subjective judgments and errors, neuroinformatic approaches have been developed recently that take such difficulties into account *(21–23)*. Further problems arise in determining the anatomical strength of the projections, in addition to identifying the connectivity patterns. For practical reasons, much of the available neuroanatomical literature describes connection strength only in qualitative categories, such as weak, intermediate, or strong. Recently however, some groups have undertaken the arduous task of determining connection strengths quantitatively, e.g., *(24–27)*, and these efforts will likely lead to an improved understanding of factors shaping connectivity *(28)* and allow the design of more reliable quantitative models of the brain *(29)*.

14.2.2. Computational Treatment of Experimental Data

Neuroanatomical tract-tracing experiments produce a large amount of intricate 2D or 3D imaging data that form the basis of conclusions about the existence or absence of connections between different cells, circuits, or systems. To date, only a small fraction of the original data has been selectively made available as images or verbal descriptions in journal publications, while the majority of the data is buried in histological archives in many different laboratories. There is, however, a rapidly growing number of approaches that attempt to make neural connectivity information more widely available in electronic format, both at the level of the original experimental data, for instance in form of computational atlases, and at more abstract levels of representation. The computational description and storage generally takes the form of databases, employing various commercially available and purpose-designed systems. A list of neural databases and models that also includes structural connectivity compilations in different species and on different scales of organization is maintained at (www.hirnforschung.net/cneuro/). As a more specific example, a comprehensive and well-documented database with the goal of collating all systems level structural connectivity for the brain of the macaque monkey is described at (www.cocomac.org) *(22)*.

Many of these approaches face substantial difficulties in trying to formalize experimental connectivity data in such a way that the information can be reliably compared across different studies *(30)*. Some of these issues, which are outside the scope of this chapter, are addressed in *(23)*. For our purposes, it is important that databases of connectivity data can be used to derive a condensed and abstract representation of neural networks, often in the shape of a connectivity matrix, that provides the starting point for connectivity analyses proper. We now turn to a more detailed description of connectivity representation at this level.

14.2.3. Formal Description

One way to conceptualize neuronal networks is by use of graphs, in particular one class of graphs referred to as directed graphs or digraphs. Graph theory, which is introduced in several textbooks and monographs, e.g., *(31)*, provides a number of connectivity measures and a wealth of mathematically grounded insights into local and global network attributes that can be utilized for the purpose of characterizing neuroanatomical connectivity patterns. Graph theory has been successfully applied in a variety of fields that deal with networks of different kinds, including social networks, economic networks, chemical and biochemical networks, as well as the Internet. Although graph theoretical methods present a very general tool set, their relevance for the specific networks under study may differ depending upon the specific structural and functional constraints that come into play. Here, we focus on some aspects of graph theory that might be of particular interest to the computational analysis of brain networks.

The distinction between nondirected graphs and digraphs is significant. Social networks (e.g., networks of acquaintances) are usually characterized as nondirected graphs, while synaptic connections linking neurons and brain areas are generally polarized, that is, directed. At the elementary structural level, any neural network may therefore be described as a digraph. In graph theoretical terms, a digraph G_{nk} is composed of n vertices (nodes, units) and k edges (connections), with k ranging between 0 (null graph) and n^2-n (complete or fully connected graph; self-connections are excluded). The graph's adjacency matrix $A(G)$ is composed of binary entries a_{ij}, with $a_{ij} = 1$ indicating the presence, and $a_{ij} = 0$ the absence of a connection between vertex j (source) and vertex i (target). The diagonal of $A(G)$ is zero. Connection weights c_{ij} may be assigned to nonzero entries of $A(G)$ for studies that incorporate information on the relative strength or density of connections or functional interactions (see Subheading 14.5.2.).

In order to approach the analysis of brain connectivity from a graph theoretical perspective, we need to define characteristics of such networks that are motivated and based in neurobiology. Concepts from graph theory, including adjacency matrices, have been used previously as tools to describe neuroanatomical patterns, e.g., *(11,32)*. Depending on the scale, nodes might represent single-cell bodies, cell assemblies such as circuits, columns or layers, or neural systems, that is, whole brain regions that are distinguished from others in terms of unique combinations of architecture and functions *(20)*. The higher levels of abstraction naturally ignore that projections are formed by nerve cells that represent both the connection and parts of the two connected systems. Notwithstanding, several investigators have used connection matrices (essentially equivalent to adjacency matrices of graphs) to display comprehensive anatomical datasets. Examples include the pathways linking visual cortical areas (*[6]* and see Fig. 1), as well as other cortical stations *(33)* in the macaque monkey, interconnections of cortical areas *(34)*, and thalamic nuclei *(4)* in the cat, as well as connection matrices of the rodent hippocampus *(35,36)*. Often, these matrices have included (qualitative) information on connection densities of pathways or their patterns of origin or termination. Additionally, information may be included that allows one to distinguish between previously unexplored and explicitly absent projections. The latter information is also worthwhile for informing on the hodology of connectivity networks, as demonstrated in our discussion of connectivity cluster analyses in Section 14.5. of this chapter. In addition to being a descriptive tool, graph theoretical analyses have also been used to generate predictions about the presence or absence of previously unobserved connection pathways, e.g., *(11)*. This aspect is linked to the problem of representing less reliable data in connectivity matrices and in their corresponding graphs. One possible approach, which is to represent such data as special weights and treat them specifically during the analyses, is also mentioned in Subheading 14.5.

The outlined aspects of complex neural connectivity data are exemplified by the dataset originally presented by Felleman and Van Essen *(6)*, Figure 1, which we use throughout this review to illustrate the introduced analyses. The dataset demonstrates that neural connectivity data are frequently incomplete, partly unreliable, partly contradictory (as for the differing indications of hierarchical relation in some pairs of reciprocal connections), and are, in any case, difficult to assess by intuition alone.

14.3. GRAPH THEORETICAL ANALYSIS

In what follows, we introduce a number of graph theoretical measures of increasing complexity. We briefly discuss potential neural correlates for each of these measures and indicate how their evaluation may aid in characterizing patterns of anatomical connectivity.

14.3.1. Average Degree of Connectivity

A rather crude estimate of the connectivity of a digraph is its average degree of connectivity, that is, the total number of connections present, divided by the total number of connections possible among the nodes. For a digraph containing n vertices (and

	TARGETS																													
SOURCES	V1	V2	VP	V3	PIP	V3A	V4	PO	V5	V4t	DP	VOT	LIP	VIP	MSTd	MSTl	PITd	PITv	7a	STPp	CITd	CITv	STPa	AITv	FEF	TF	46	FST	TH	AITd
V1		<	Ø	<	<	<	<	<	<	Ø	Ø	Ø	Ø	Ø	Ø	<?	Ø	Ø	Ø	Ø	Ø	Ø	Ø	Ø	Ø	Ø	Ø	Ø	Ø	Ø
V2	>		<	<	x	<	<	<	<	<	Ø	x	Ø	x	<	<	Ø	Ø	Ø	Ø	Ø	Ø	Ø	Ø	<?	Ø	Ø	<?	Ø	Ø
VP	Ø	>		x	<	<	<	<	<	Ø		<		<	<	Ø	Ø	Ø	Ø	Ø	Ø	Ø	Ø	Ø	<?	<	Ø	<	Ø	Ø
V3	>	>			<	<	<	<	<	<		Ø	<	<	<	Ø	Ø	Ø	Ø	Ø	Ø	Ø	Ø	Ø	<?	<	Ø	<	Ø	Ø
PIP	x		x	x			x	<	x		<?						Ø	Ø	x	Ø	Ø	Ø	Ø	Ø						Ø
V3A	>	>	x	x			<	x	<		<?				<?	<?	Ø	Ø	Ø	Ø	Ø	Ø	Ø	Ø	<?	Ø	Ø	<	Ø	Ø
V4	>?	>	>	>	>	>		Ø	≤?	≤?	x		<	Ø	Ø	Ø	<	<	Ø	Ø	<	<	Ø	<	<?	<	<	<	<	Ø
PO	x						Ø		x		<?		<	x	<	<	Ø	Ø	<?	Ø	Ø	Ø	Ø	Ø	<?	Ø		Ø	Ø	Ø
V5	>	>	>	>	x	≥?	≥?	≥?		=	Ø		x	<	<	<	Ø	Ø	Ø	Ø	Ø	Ø	Ø	Ø	<	Ø	<?	<	Ø	Ø
V4t	>		Ø	x			x	x	x						Ø	<?									<?			<		Ø
DP	Ø	Ø	Ø	Ø	>	>	>	≥?	Ø				≤?		<?	<?	Ø	Ø	<	Ø	Ø	Ø	Ø	Ø		Ø	<	x	Ø	Ø
VOT	Ø	x	>	Ø			>										<	<												
LIP	Ø		>	>	Ø	>	>	≤?	>		≥?			x	≤?	x	Ø	=?	x		Ø	Ø		Ø	<	≤?	<	x	Ø	Ø
VIP	Ø			>				x	x				x		x	x	Ø	Ø	x		Ø	Ø		Ø	x			x		Ø
MSTd	Ø	>	>	>		>	Ø	>	>	>?	=		=?	=			<?	<?	=	<	Ø	Ø		Ø	<			≥?		Ø
MSTl	Ø	>	Ø	Ø	Ø	>	Ø	>	>	Ø	Ø	Ø		≥?			Ø	Ø	Ø	<	Ø	Ø		Ø	<			≤?		Ø
PITd	Ø	Ø	Ø	Ø	Ø	Ø	>	Ø	Ø				Ø	Ø	x	Ø			Ø			<?		<?	<		<	x		<?
PITv	Ø	Ø	Ø	Ø	Ø	Ø	>	Ø	Ø					Ø	x	Ø			Ø			<?		<?	<		<	x		
7a	Ø	Ø	Ø	Ø		Ø	Ø	>?	Ø		>		>		>	x	Ø	Ø		x	Ø	Ø	x		≥?	<	≤?	Ø	<	<
STPp	Ø	Ø	Ø	Ø	Ø	Ø	Ø	Ø	Ø				Ø	Ø	>	>?			>?		=	=	<		>?	x	≤?	>	x	Ø
CITd	Ø	Ø	Ø	Ø	Ø	Ø	>	Ø	Ø		Ø		Ø	Ø	Ø	Ø		>?	Ø					<?	<		<			<?
CITv	Ø	Ø	Ø	Ø	Ø	Ø	>	Ø	Ø		Ø		Ø	Ø	Ø	Ø		>?	Ø					<?	<		<			<?
STPa	Ø	Ø	Ø	Ø	Ø	Ø	Ø	Ø	Ø				Ø	Ø					>?	>				Ø		x	<		x	x
AITv	Ø	Ø	Ø	Ø	Ø	Ø	x	Ø	Ø				Ø	Ø	Ø	Ø		>?				>?	Ø			<	Ø	Ø	≤?	
FEF	Ø							x	>?		x		>	>	>	>			x	>?	<?	<?					≤?	>		<
TF	Ø	Ø	>	>	Ø	Ø	>	Ø	Ø		Ø			Ø	>			>?	>?	>		>	>	>			x	<?		x
46	Ø	Ø	Ø	Ø		Ø					x				Ø				>?	x	x	x	>		x	≥?		Ø	=?	<
FST	Ø	Ø	Ø	>	Ø	>	>	Ø	>	>			≥?	≥?	≤?	=	≥?	≥?	x	<	Ø	Ø		Ø	x	=	Ø			Ø
TH	Ø	Ø	Ø	Ø	Ø	Ø	>	Ø	Ø				Ø	Ø	Ø	Ø		>	>	>	>	>	>	>			>?			x
AITd	Ø	Ø	Ø	Ø	Ø	Ø	Ø	Ø	Ø				Ø	Ø	Ø	Ø		>?	x			>?			>?		>?	Ø		

Fig. 1. Connectivity data for the cortical visual system of the macaque monkey, rearranged from Felleman and Van Essen (1991) *(6)*. Different colors and symbols in this table represent different types of structural links that connect these 30 cortical areas with each other. The table's information is meant to be read: source area in the left column has (...) connectional relation with target area in the top row. Ø signs stand for links that have been explored experimentally, but were found to be absent; x signs and light grey shading indicate existing connections for which no further qualifying information was available; < and > signs stand for feedforward and feedback connections, respectively; = denote lateral connections; ≤ and ≥ signs show "mixed" feedforward/lateral or feedback/lateral links, respectively. For easier recognition, connections of a feedforward type are additionally indicated by blue shading, feedback connections by yellow, and lateral by green shading. This classification of types relates to patterns of connections between different cortical layers as explained in more detail in Subheading 14.5.2. Question marks and lighter colors point out information that has been classified as less reliable by Felleman and Van Essen *(6)*; finally, empty slots indicate experimental information not yet available in this compilation. Areas MIP and MDP have been excluded, because of the limited information available for them. For area abbreviations refer to *(6)*. Reproduced with permission from *(91)*. A color version of this figure is included in the CD-ROM.

excluding self-connections) there are n^2-n possible connections. The average degree of connectivity provides a measure of the overall sparsity of the graph; values near 1 indicate graphs that are nearly completely connected, while values close to 0 indicate graphs with very sparse connections.

The average degree of connectivity for most brain networks appears to be rather low. For the human cerebral cortex as a whole, Murre and Sturdy *(37)* estimated the total number of neurons to be 8.3×10^9, with an estimated 6.6×10^{13} synapses between them. Thus, the degree of connectivity (or "connectivity factor") is exceedingly small, approximately 9.6×10^{-7}. However, this extremely low value is somewhat misleading, as it only holds for the entire cortex viewed as a homogeneous random network. Due to the fact that throughout the brain most neural connections are made with other neurons that are located nearby, average degrees of connectivity are much higher in a given local neighborhood. A physiological study of rat visual cortex revealed that two cortical neurons located within a distance of 300 µm are directly connected with a probability of around 0.09 (38). Another estimate (39) puts the fraction of neurons (within a square millimeter of cortex) that a given pyramidal cell is connected to at 3%. These largely consistent estimates translate into an average connection density of around 0.1 to 0.03 for pyramidal neurons within local neighborhoods of approximately columnar size. While these connection densities may still be considered relatively sparse, they do allow effective communication between neurons within columns and are likely contributing to similarities in receptive field properties within columns as well as to locally coherent neuronal activity. The highest levels of connection density are found at the level of cortical areas and the pathways interconnecting them. At this level of scale, connection matrices are found to have connection densities of around 0.36 (for the data from the macaque monkey visual cortex shown in Fig. 1; 30 areas, 315 pathways) or 0.27 (cat cortex; 65 areas, 1136 pathways, after [12]; for more detailed numbers see [10]). It should be noted that these values are derived from binary adjacency matrices and do not take into account relative densities of pathways.

14.3.2. Local Connectivity Indices

At a very basic level, local coefficients can be employed that describe properties of the individual network nodes such as their degree of connectedness with the rest of the network or ratios of the number of afferent and efferent connections (the in-degree and out-degree, in graph theoretical terminology). Similar ratios can be calculated for feedforward and feedback connections as defined by laminar origin and termination patterns. Nicolelis et al. (32), for instance, suggested relative afferent and efferent indices (normalized by the total number of interconnections between the given structures) for brain regions involved in the control of cardiovascular functions. This study also described an index adding the former two parameters as well as a "power index", which was formed by the product of the former two indices with each other and with the total number of connections in the network. Based on these measures, the study inferred the existence of "pathway attractors", that is, a small number of heavily connected structures.

Even such simple measures may have a bearing on understanding the functional roles that individual nodes play within the global network. See, for instance, Young et al. (40) for a detailed discussion of how the varying degree of connectedness of neural structures affects the functional impact of network lesions and how this simple property influences an observer's ability to infer the function of individual structures from lesion-induced performance changes. Other local measures, such as input–output ratios, are also open to functional interpretation. If a neural structure only sends projections, it may exert influence over other structures, but is not influenced by the others in turn. An

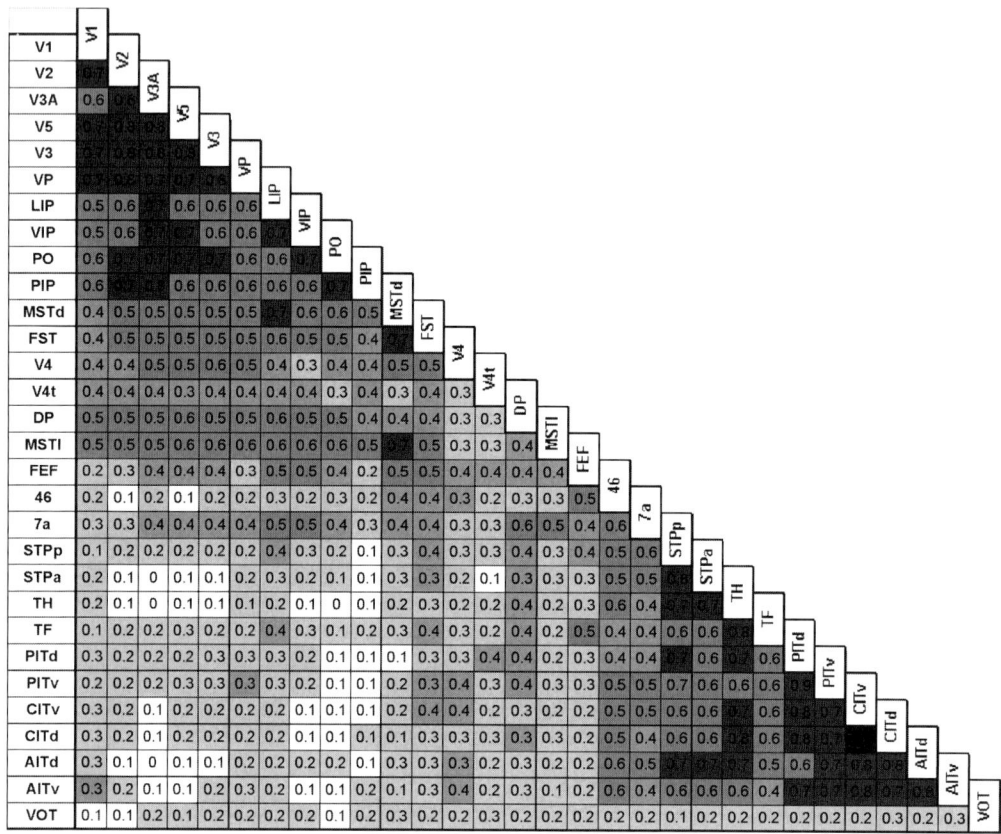

Fig. 2. Pattern similarity indices for primate visual areas based on their afferent and efferent connectivity patterns as given in Figure 1. The entries indicate the relative similarity (in grey scale-coded patterns from 0 to 1) of the connection patterns for each pair of cortical areas. As the index accounts for both afferent and efferent area connections, the entries of this triangular matrix denote symmetrical similarity relations between the areas. This matrix was reordered to facilitate the recognition of groups of areas with similar connectivity patterns. Apparent are two main distinct groups that share a high similarity in their connectional patterns, mainly peripheral visual areas (e.g., V1, V2, V3, V5, VP) on the one hand and polysensory regions of the superior temporal sulcus (STPp, STPa) and inferior temporal visual areas (PIT, CIT, AIT) on the other.

area that only receives projections, but does not have any output, on the other hand, would present the opposite extreme. In terms of control theory, such structures may be likened to controllers or receivers, respectively, and the general function carried out by them can be thought of as "transmission". For an application of this concept to prefrontal cortical connectivity, see *(41)*. For the visual system connectivity data shown in Figure 1, the average ratio of the areas' afferents to efferents is close to 1, with a standard error of 0.4. This confirms that, at least in the primate visual system, no brain region is just a recipient or originator of signals *(42)*. Such symmetry may characterize a cooperative ("give-and-take") mode of information processing, whereas asymmetry would imply information relay from the set of afferent areas to the set of target areas.

Further global graph theoretical indices, such as the cluster index discussed below, can also be expressed as properties of individual network nodes. These local indices are of particular interest in the context of neural systems networks that are made up from heterogeneous network components (e.g., nuclei or cortical areas and columns). In this case, one can compare the extrinsic, or connectional, characteristics with intrinsic (e.g., cytoarchitectonic) features, e.g., *(43,44)*.

Expanding local indices to pairwise comparisons, a matching index can be defined that describes the proportion of identical efferent and/or afferent connections of two different nodes i, j; normalized by the total of connections belonging to the two nodes *(45)*. Figure 2 shows the index applied to the data from Figure 1. This measure numerically describes the relative connectional similarity of the nodes, and according to the concepts underlying systems neuroscience, one might expect two very similar nodes to also share functional properties. For example, primate visual area FST has been only poorly characterized electrophysiologically in the past, but the entries in Figure 2 indicate that its connectivity overlaps with that of area MSTd by 71%. Thus, one might expect that the visual stimulus preferences of cells in the two areas are rather similar. This is indeed what one of the few available electrophysiological studies (46) has suggested.

14.3.3. Paths and Cycles

Within a digraph, a path is defined as any ordered sequence of distinct vertices and edges, linking a source vertex j to a target vertex i (see Fig. 3A,B). The length of a path is equal to the number of edges it contains. Importantly, no edges or vertices can be visited twice along a given path. The only exception occurs if $i = j$, in which case the resulting path is called a cycle. The shortest possible cycle (a path of length 2) consists of two vertices that are reciprocally linked by two edges. Clearly, if no path exists from j to i, j cannot act on i by means of structural connectivity. If no path exists from j to i and no path exists from i to j, i and j cannot interact. Thus, the presence or absence of paths places hard constraints on the functioning of a given network. In general, we must assume that signals sent from j to i along a given path have higher functional impact on i, in terms of information processing, the shorter the path. Similarly, the dynamic self-reinforcing tendency of excitatory cycles can be assumed to be a decreasing function of the cycle length, with short cycles producing the strongest effects. The fraction of reciprocally linked pairs of vertices may provide a good estimate for the prevalence of dynamic coupling within a network. Most cortical connection matrices exhibit abundant reciprocal coupling (a fraction on the order of 0.7 to 0.8, compared to 0.3 for equivalent random graphs). For the dataset shown in Figure 1, this fraction is 0.77. Although these reciprocal pathways may not link cells in a one-to-one fashion between areas, they have been invoked in generating reverberating, persistent, or synchronized activity *(47,48)*. The measure of cycle probability $p_{cyc}(q)$, introduced by Sporns et al. *(12)*, generalizes the concept of cyclic paths. The index, which can be computed on the basis of all non-zero shortest path lengths that connect network nodes back onto themselves, describes the likelihood of cycles of length q occurring in a given network. For $q = 2$, the index is equivalent to the fraction of reciprocal pathways. For the connection matrix of the macaque visual cortex, cycle probabilities for cycles

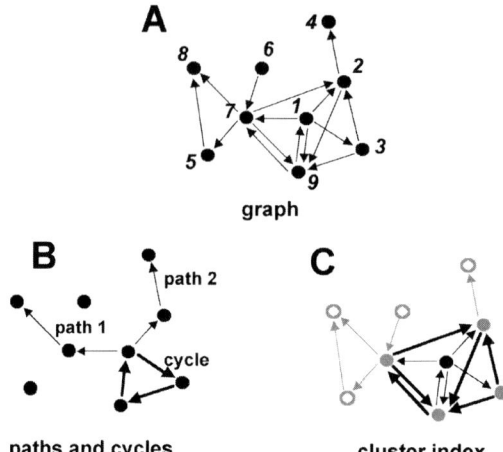

Fig. 3. An example of a directed graph and some graph theoretical measures. (**A**) A graph with $n = 9$ vertices and $k = 16$ edges is shown, vertices are numbered, edges are indicated as arrows. (**B**) Two examples of paths and one example of a cycle are shown. Paths link vertex 1 to 8 and 1 to 4, respectively. Both paths are of length 2. The cycle connecting vertices 1, 3, and 9 is of length 3. (**C**) An example for the calculation of the cluster index f_{clust} (modified after *[51]*) for vertex 1, marked as a black dot. Vertices that are direct neighbors of vertex 1 are shown as gray dots (vertices 2, 3, 7, 9), and other vertices are indicated as gray circles (4, 5, 6, 8). Edges linking the neighbors of vertex 1 are indicated by thick black arrows. The cluster index of vertex 1 is computed as the ratio of actually existing connections between the neighbors of vertex 1 and the maximal number of such connections possible. Given that vertex 1 has 4 neighbors, maximally 12 connections are possible between them. Thus, $f_{clust}(1) = 6/12 = 0.5$. The cluster index of the graph is the mean of the cluster indices for each individual vertex.

of length 2 (reciprocal connections), 3, 4, and 5 [$p_{cyc}(2) = 0.77$; $p_{cyc}(3) = 0.50$; $p_{cyc}(4) = 0.43$; and $p_{cyc}(5) = 0.41$] are significantly above values expected for an equivalent random network [random $p_{cyc}(2,3,4,5) \approx 0.31$]. High cycle probabilities for short cycles are indicative of a clustered architecture (see also cluster index below).

14.3.4. Reachability Matrix and Connectedness

If at least one path (of arbitrary length) exists between every ordered pair of vertices in the graph, the graph is connected. This condition is more easily satisfied for nondirected than for directed graphs, as in the latter, each path needs to be composed of correctly oriented edges; and the situation can arise that paths exist from nodes i to j in a digraph, while the path of opposite direction, j to i, is absent. The binary entries r_{ij} of the reachability matrix $R(G)$ record the presence or absence of all such paths, i.e., $r_{ij} = 1$ if vertex i is reachable from vertex j ($r_{ij} = 0$ otherwise). If all entries r_{ij} are ones, the graph consists of only one component and is strongly connected. A graph consists of multiple components if there are nonoverlapping subsets of vertices with no paths between them.

As discussed above, the absence of paths implies that no functional interaction can take place between components of a graph. Obviously, the existence of separate components in any biological network, including the brain, indicates their complete functional isolation. It is, at present, unknown whether all neurons of the brain, or at least the cerebral cortex, form a single giant component, in the sense that every neuron can influence every other neuron through at least one finite path. At the level of cortical systems, connection matrices that have been studied in detail are composed of a single giant component. For example, there are paths linking all cortical areas of the primate visual cortex as well as the cat cortex *(12)*. Although essential data is still missing, it seems likely that extended cortical systems (apart from lesioned or split brains) form single giant components in which interaction may occur between any pair of vertices (that is, areas).

14.3.5. Distance Matrix and Diameter

The entries of the distance matrix $D(G)$ give the distance from vertex j to vertex i *(31,49)*. The distance is defined as the shortest (directed) path between the two vertices (if no path exists between two vertices, their distance is infinite). Shortest paths can be determined in a straightforward way using Floyd's algorithm *(50)*. A review of different serial and parallel algorithms for computing shortest pathlengths can be found at (www.mcs.anl.gov/dbpp/text/node35.html). The diameter of a digraph is the global maximum of the distance matrix. The average of all the entries of the distance matrix has been called the "characteristic path length" (*[51]*, and see below).

Clearly, the distance between two neuronal units or areas can provide information about the degree or the strength of their functional relationship. As discussed above, two areas that are connected by a short path (perhaps as short as a single edge) may be exchanging signals in a more direct or efficient manner than areas that are connected by a long path (with several intervening waystations). Note that while the distance matrix has exactly one entry for each pair of vertices, multiple distinct paths of the same length may exist. The distance matrix for the visual system connectivity data of Fig. 1. is shown in Fig. 4.

In graphs, distance does not refer to any kind of metric space within which the vertices and edges are embedded. It only refers to the minimum number of distinct edges linking vertex i and j. Obviously, the physical location of neuronal units and the length of their connections play a crucial role in constraining possible connection patterns, through conservation of wiring volume, e.g., *(52)*, or issues related to conduction velocity *(53,96)*. Clearly, a purely graph-based (nonmetric) analysis of neuronal connectivity matrices would be incomplete without an appropriate consideration of physical constraints imposed by evolution, development, and adult function *(54,55)*.

14.3.6. Disjoint Paths, Edge, and Vertex Connectivity

Given two vertices i and j, two directed paths from j to i are edge-disjoint, if they have no edges in common. Similarly, two such paths are vertex-disjoint, if they have no vertices (other than i and j) in common. One of the most important theorems of graph theory (usually quoted as it applies to undirected graphs), Menger's theorem, states that the minimum number of vertices that need to be removed in order to disconnect two vertices i and j equals the maximum number of vertex-disjoint paths between them, e.g., *(49)*. For directed graphs, if three vertex-disjoint paths exist linking node j to node i, then at least three nodes need to be removed in order to cut all paths from j to i.

TARGETS

SOURCES

	V1	V2	VP	V3	PIP	V3A	V4	PO	V5	V4t	DP	VOT	LIP	VIP	MSTd	MSTl	PITd	PITv	7a	STPp	CITd	CITv	STPa	AITv	FEF	TF	46	FST	TH	AITd
V1		1	2	1	1	1	1	1	1	2	2	2	2	2	2	1	2	2	2	2	2	2	3	2	2	2	2	2	2	3
V2	1		1	1	1	1	1	1	1	1	2	1	2	1	1	1	2	2	2	2	2	2	3	2	1	2	2	1	2	2
VP	2	1		1	1	1	1	1	1	2	2	1	2	1	1	2	2	2	2	2	2	2	2	2	1	1	2	1	2	2
V3	1	1	2		1	1	1	1	1	1	2	2	1	1	1	2	2	2	2	2	2	2	2	2	1	1	2	1	2	2
PIP	1	2	1	1		2	1	1	1	2	1	2	2	2	2	2	2	2	1	2	2	2	2	2	2	2	2	2	2	2
V3A	1	1	1	1	2		1	1	1	2	1	2	2	2	1	1	2	2	2	2	2	2	3	2	1	2	2	1	2	2
V4	1	1	1	1	1	1		2	1	1	1	2	1	2	2	2	1	1	2	2	1	1	2	1	1	1	1	1	1	2
PO	1	2	2	2	2	2	2		1	2	1	3	1	1	1	1	2	2	1	2	2	2	2	3	1	2	2	2	2	2
V5	1	1	1	1	1	1	1	1		1	2	2	1	1	1	1	2	2	2	2	2	2	2	2	1	2	1	1	2	2
V4t	1	2	2	1	2	2	1	1	1		2	3	2	2	2	1	2	2	2	2	2	2	3	2	1	2	2	1	2	2
DP	2	2	2	2	1	1	1	1	2	2		3	1	2	1	1	2	2	1	2	2	2	2	2	2	2	1	1	2	2
VOT	2	1	1	2	2	2	1	2	2	2	2		2	2	2	2	1	1	3	3	2	2	3	2	2	2	2	2	2	2
LIP	2	2	1	1	2	1	1	1	1	2	1	2		1	1	1	2	1	1	2	2	2	2	2	1	1	1	1	2	2
VIP	2	2	2	1	2	2	2	1	1	2	2	3	1		1	1	2	2	1	2	2	2	2	3	1	2	2	1	2	2
MSTd	2	1	1	1	2	1	2	1	1	1	1	2	1	1		2	1	1	1	1	2	2	2	2	1	2	2	1	2	2
MSTl	2	1	2	2	2	1	2	1	1	2	2	2	2	1	2		2	2	2	1	2	2	2	3	1	2	2	1	2	2
PITd	2	2	2	2	2	2	1	2	2	2	2	3	2	2	1	2		2	2	2	2	1	2	1	1	2	1	1	2	1
PITv	2	2	2	2	2	2	1	2	2	2	2	3	2	2	1	2	2		2	2	2	1	2	1	1	2	1	1	2	2
7a	2	2	2	2	2	2	2	1	2	2	1	3	1	2	1	1	2	2		1	2	2	1	2	1	1	1	2	1	1
STPp	3	2	2	2	3	2	2	2	2	2	2	3	2	2	1	1	2	2	1		1	1	1	2	1	1	1	1	1	2
CITd	2	2	2	2	2	2	1	2	2	2	2	3	2	2	2	2	2	1	2	2		2	2	1	1	2	1	2	2	1
CITv	2	2	2	2	2	2	1	2	2	2	2	3	2	2	2	2	2	1	2	2	2		2	1	1	2	1	2	2	1
STPa	3	3	2	2	3	3	2	2	3	3	2	3	2	3	2	2	3	2	1	1	2	2		2	2	1	1	2	1	1
AITv	2	2	2	2	2	2	1	3	2	2	2	3	2	3	2	3	2	1	2	2	2	1	2		2	1	2	2	1	2
FEF	2	2	2	2	2	2	2	1	1	2	1	3	1	1	1	1	2	2	1	1	1	1	2	2		2	1	1	2	1
TF	2	2	1	1	2	2	1	2	2	2	2	2	2	2	1	2	2	1	1	1	2	1	1	1	2		1	1	2	1
46	3	3	2	2	2	2	2	2	2	3	1	3	2	2	2	2	3	2	1	1	1	1	1	2	1	1		2	1	1
FST	2	2	2	1	2	1	1	2	1	1	2	3	1	1	1	1	1	1	1	1	2	2	2	2	1	1	2		2	2
TH	2	2	2	2	2	2	1	2	2	2	2	3	2	3	2	2	2	1	1	1	1	1	1	1	2	2	1	2		1
AITd	3	3	3	3	3	3	2	2	2	3	2	4	2	2	2	2	3	1	1	2	2	1	2	2	1	2	1	2	2	

Fig. 4. Distance matrix for the example data set shown in Figure 1. Darker shading indicates shorter pathlengths. The diameter of this data set is 4 (equal to the singular longest path in the set, from area AITd to area VOT in the bottom row), the characteristic pathlength (average of all pairwise shortest nonzero paths in the set) is 1.64. This means that most areas in the visual cortex can communicate with each other either directly or via only one intermediate area, and this communication is not symmetrical.

In the brain, the removal of vertices or edges (in graph theoretical terms) is equivalent to making a lesion. Lesions may result in the functional disconnection of brain areas that are themselves not part of the lesion. In a graph, disconnection may only involve the elimination of direct (short) paths between vertices, or may result in the formation of separate components (see above). Graph theoretical methods allow the identification of vertices or sets of vertices (or edges or sets of edges) whose removal produces (functional) disconnection. For each graph, there is a minimal number κ_v of vertices (or κ_e of edges) whose removal results in a disconnected graph, that is, a graph that contains multiple components; this number is also called the graph's vertex or edge connectivity. For a computational application of this concept to the connectivity of large-scale networks see *(56)*.

14.3.7. Random Graphs

A large and important body of classical results in graph theory has been obtained by studying nondirected graphs with random adjacency matrices, e.g., *(57)*. Random graphs are generated by choosing k edges among a set of n vertices, such that all possible choices of edges are equiprobable (and no edge is chosen twice). A major characteristic of random graphs is their degree of connectivity, and they are usually taken to be of large (infinite) size. Interesting and perhaps unexpected phenomena emerge as the degree of connectivity of nondirected random graphs is varied in a systematic manner ("graph evolution", *[58,59]*). Consider a very large (potentially infinite) graph $G_{n,k}$ with $k = n^\alpha$ evolving from $\alpha = 0$ to $\alpha = 2$. As edges are added to $G_{n,k}$ up to $\alpha = 1/2$, $G_{n,k}$ contains mostly isolated edges. As α gets closer and closer to 1, more and more graphs appear that are connected but acyclic, called trees. Suddenly, as α passes through the value 1, the probability of cycles of all lengths in $G_{n,k}$ jumps from 0 to 1. Thus, the evolution of random graphs proceeds through a series of stages that can be characterized by threshold functions defining the sudden appearance of specific structural motifs.

Although brain networks are not random, and conceptualizing them as random graphs would be, for the most part, a futile and misleading exercise, the investigation of random graph evolution illustrates an important concept. Adding graph components such as edges, for example in the course of a developmental process, can produce sudden and unexpected changes, similar to phase transitions, in local or global network attributes. Threshold functions like those found in random graphs have been implied in the emergence of specific dynamical properties of chemical networks in the course of prebiotic evolution, e.g., *(60)*. One might envision a similar role in shifting between dynamical regimes during evolutionary or developmental processes involving neuronal networks *(61)*.

14.3.8. Small-World Attributes: Characteristic Path Length and Cluster Index

In the 1960s, studies of social networks produced a remarkable and counter-intuitive finding *(62)*. In many cases, any two individuals belonging to very large social networks are likely to be connected through a short sequence of intermediate acquaintances. This so-called small-world phenomenon has given rise to the popular notion of "six degrees of separation", the linking of individuals through surprisingly short sequences of intermediates that bridge significant geographical or social boundaries. Analyses by Watts and Strogatz *(51)* as well as Watts and Duncan *(63)* have demonstrated that the small-world phenomenon is not limited to social networks, but is pervasive in other kinds of networks as well. In a paradigmatic example, they studied networks that varied between perfect (local or lattice-like) order (i.e., networks with connections between nearest neighbors only) and randomness (i.e., networks in which all connections are assigned at random). As the study pointed out, most networks of real interest have intermediate structural characteristics. Such networks contain local (nearest neighbor) connections with a few additional connections added between randomly selected vertices. Watts and Strogatz defined two measures, the characteristic path length l_{path} (equivalent to the global average of the distance matrix; see Subheading 14.3.5.) and the cluster index f_{clust}. The cluster index for a given vertex in the graph measures how many connections exist between the vertex's neighbors (i.e., all the vertices that can be reached within one step), out of all possible such connections (Fig. 3C). In a sense, the cluster index expresses the "cliquishness" of a network, i.e., the tendency of each vertex's neighbors to "talk among themselves". This measure is closely related to the index introduced by *(11)*, which assesses the proportion of 'indirect connections', that is the ratio of all paths of length two, (i,k) and (k,j), connecting i and j via any intermediate vertex k. Figure 5 shows examples of graphs generated using a paradigm similar to Watts and Strogatz *(51)*. In Figure 5A, graphs are system-

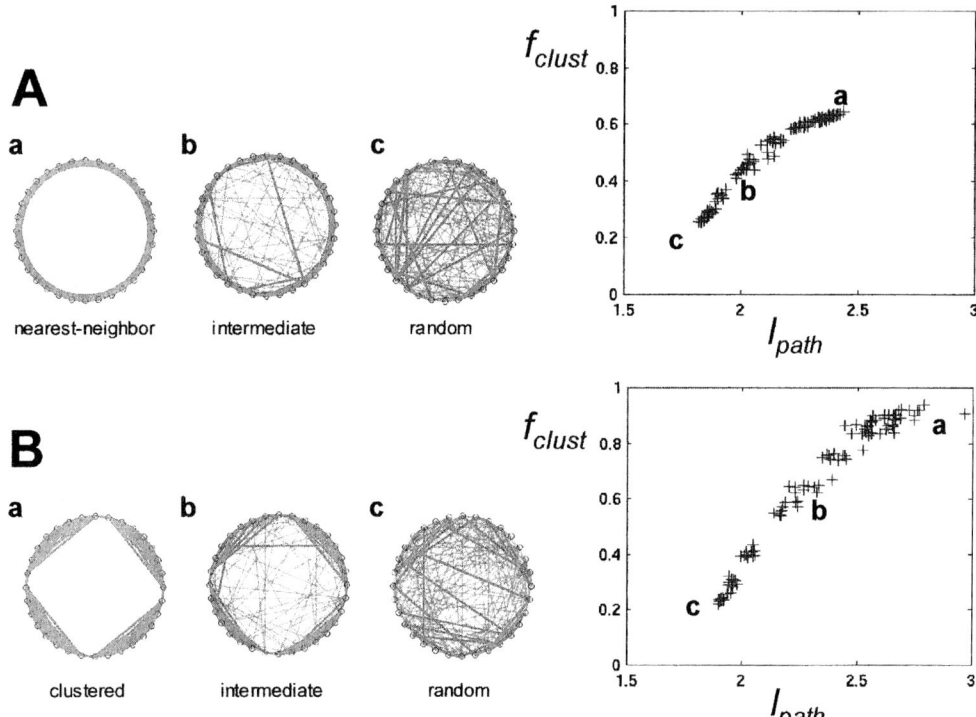

Fig. 5. Small world digraphs. (**A**) Analogous to the examples presented by Watts and Strogatz, 1998 *(51)*, graphs are varied between totally regular (case a, nearest neighbor) through intermediate cases (case b) to totally random (case c). A total of 256 connections are made between 32 vertices. Vertices are arranged in a circle (small blue circles) with red lines indicating bidirectional connections (edges) and green lines indicating unidirectional edges. A scatter plot of characteristic path length (l_{path}) and cluster index (f_{clust}) is shown on the right. Numerous graphs exist for which the value of f_{clust} is high, while the value of l_{path} is low. (**B**) Here, graphs are varied between clustered (case a) and totally random (case c). Compared to panel **A**, higher cluster index values result for case a, as well as numerous intermediate cases. Note that, in both panels **A** and **B**, characteristic path lengths are relatively short for almost all cases. This is due to the relatively high degree of connectivity (0.26) chosen to approximate that of cortical connection matrices. In Watts and Strogatz' original work *(51)*, the degree of connectivity for typical small world examples was set to about 0.04 ($n = 1000$, $k = 10$ per vertex). A color version of this figure is included in the CD-ROM.

atically varied between "nearest-neighbor" (i.e., completely ordered) and "random". The cluster index is high for graphs that contain mostly short-range connections, while the characteristic path length is relatively low. This effect is even more pronounced for graphs that are varied between "clustered" and "random" (Fig. 5B).

Analysis of cortical connection matrices in terms of these measures reveals a distinctive "small-world architecture" *(10,12)*. For example, the connection matrix of the macaque visual cortex shown in Figure 1 has a small characteristic path length $l_{path} = 1.64$ (average characteristic pathlengths of randomized networks is 1.60, standard deviation [SD] = 0.01, $n_{random} = 20$) and a high value for the cluster index $f_{clust} = 0.57$, much larger than expected for a random network of equivalent n and k (f_{clust} near average degree of connectivity, i.e., 0.36, SD = 0.01, n_{random} = 20). Similarly, the analysis of functional interactions between areas of the entire cerebral cortex also demonstrated small-world characteristics (l_{path} = 2.17, f_{clust} = 0.38) not found in equivalent random networks (l_{random}: mean = 2.15, SD = 0.02; f_{random}: mean = 0.16, SD = 0.01; n_{random} = 20) *(64)*. The high values for the cluster index indicate that local connectivity among neighboring (i.e., directly connected) areas is approximately twice as high as would be expected if connections were made at random. This confirms and quantifies the existence of densely clustered sets of areas. Small-world measures alone, however, do not allow the unbiased determination of the size, number, and membership of these clusters. Purpose-designed clustering techniques have been used to identify such clusters from anatomical matrices *(10)*. Other techniques can be brought to bear on functional connectivity matrices obtained after "running" anatomical matrices as dynamical systems *(12,64a)*. These approaches are discussed in more detail in Section 14.5 of this review.

14.3.9. Scale-Free Attributes

For a number of large self-organizing networks, including the World Wide Web, as well as scientific citation networks, it has been shown that the probability $P(k)$ that a vertex in the network interacts with k other vertices decays following a power law, i.e., $P(k) \sim k^{-\gamma}$ *(65)*. This scale-free property [in a double logarithmic plot the probability $P(k)$ scales linearly across all dimensions of the number k of connected vertices] is a result of the growth characteristics of such networks. These networks expand continuously by adding more and more vertices (i.e., Web pages or authors), while new vertices have a tendency to attach themselves to other vertices that are already well connected. Recently, scale-free attributes have been claimed to exist in biological networks as well, specifically in the network of metabolic pathways *(66)*. Are brain networks also scale-free? This property is more difficult to test reliably for neural connectivity, which spans only few levels of scale, than for networks such as the World Wide Web, and preliminary analysis of cortical connection matrices at the level of areas and pathways does not reveal the presence of a $P(k)$ power-law distribution (Sporns, unpublished observations). This may not be too surprising as there are stringent constraints on the development of corticocortical connectivity, including limits on the number of pathways a given area can maintain *(67)*. While scale-free attributes may be absent at the level of cortical systems, their existence at the level of individual cortical neurons or circuits has not yet been investigated and cannot be ruled out.

14.3.10. Conclusions and Perspectives

All of the above measures and analysis tools may be brought to bear on neuroanatomical datasets. For most of the measures, real connection matrices (at a given level of anatomical organization) will take on characteristic values. For example, it appears that cortical systems from a variety of species exhibit "small-world" attributes. It is very likely that other classes of connection patterns exist in other regions of the brain, or at other levels of organization (i.e., within local cortical circuits, or within subregions of the hippocampus). We expect a major area of interest to focus on how anatomical patterns give rise to patterns of functional connectivity, the time-varying pattern of temporal correlations between neuronal units. Even more importantly, it will be a challenge to clarify how, in a given system, anatomical connectivity constrains the temporal evolution of effective connectivity, that is, how the connectional organization constrains the causal functional influence dynamically exerted by one system element upon another of the system. As the interest in characterizing functional and effective connectivity patterns increases, due to mounting evidence for their involvement in generating distinct perceptual and cognitive states, it becomes imperative to quantitatively characterize underlying anatomical patterns as well.

14.4. STATISTICAL EXPLORATION OF CONNECTIVITY

14.4.1. General Considerations

Despite the power of graph theoretical approaches to network analysis, additional computational tools are needed to provide an intuitive overview over the global structure of networks under investigation. Such a perspective can be gained with the help of multivariate statistical approaches such as cluster or similarity analyses. Most of the statistical methods reviewed in this section are exploratory tools designed to visually detect consistent patterns or systematic relationships between variables in extensive and complex data sets. The techniques often employ dimensional reduction, to make the organization of the data accessible in 2D or 3D space. Generally, most of these procedures follow the spirit of "data mining" and are better suited to preliminarily analysis, suggesting potential organizational patterns in the data than to actually providing tests for confirming hodological concepts. Therefore, these data explorations should be followed by subsequent stages of rigorous data analyses designed to test the preliminary findings. Some analyses that might be used for this purpose are discussed in more detail in Section 14.5.

A common denominator of many exploratory statistical techniques is that they help to identify the structure of similarities between different items (here neural units or systems), using a variety of distance measures to evaluate relative similarity or dissimilarity. Naturally, similarity detection is a rather general approach not specifically adapted to connectivity data, and so difficulties may arise in interpreting the analysis results in terms of network topology. In the case of connectivity analyses, the similarity between neural units is assessed on the basis of the units' connectivity patterns, which are often only available as binary data (existing or absent connections) or patterns defined in a number of qualitative classes (such as weak or dense connections). When selecting analysis techniques and options, it is important to bear in mind that these data represent nominal or ordinal levels of measurement, rather than metric (interval or ratio) measurements. This commonly narrows the range, and in some cases also the power, of suitable data analyses techniques.

In the following, we briefly review nonmetric multidimensional scaling (NMDS), factor analysis/principal components analysis (PCA), multiple correspondence analysis (MCA), and different types of cluster analyses, which are provided, often with a large number of parameter and methods options, in general purpose statistics packages such as SYSTAT and SPSS (SPSS, Inc.), SAS (The SAS Institute, Inc.), or STATISTICA (StatSoft, Inc.). Further statistical approaches suitable for network analysis are related to graph layout techniques (e.g., De Leeuw and Michailidis [citeseer.nj.nec.com/316591.html] and linked articles). We begin with the description of one very general technique, NMDS.

14.4.2. Nonmetric Multidimensional Scaling (NMDS)

Method. Multidimensional scaling (MDS) comprises several related methods for estimating the coordinates of a set of objects in a chosen low-dimensional space from data measuring the high-dimensional distances (or dissimilarities and similarities) between pairs of objects, e.g., *(68,69)*. MDS provides routines for rearranging objects in, typically, 2D or 3D, so that the resultant configuration best matches the original high-dimensional distances between the objects. Most MDS algorithms attempt to reproduce the general rank-ordering of the original distances between the objects, rather than the distances' actual proportions; hence the nonmetric nature of the technique. This strategy makes the technique suitable for the analysis of metric as well as ordinal data, such a connectivity patterns. NMDS routines use a function minimization algorithm, which evaluates different low-dimensional configurations with the goal of maximizing the goodness-of-fit to the high-dimensional configuration. A commonly used measure for evaluating how well a particular configuration reproduces the ranks or proportions of the original distance matrix is the stress measure. The raw stress value, Φ, of a scaled configuration is defined by:

$$\Phi = \sum [d_{ij} - f(\delta_{ij})]^2$$

In this formula, d_{ij} stands for the distances reproduced in low-dimensional space, and δ_{ij} stands for the distances present in the original data. The expression $f(\delta_{ij})$ indicates a nonmetric, monotonic transformation of the observed input distances (STATISTICA manual). There are several further related transformation measures that are commonly used; however, most of them also amount to the computation of the sum of squared deviations of observed distances (or some monotonic transformation of those distances) from the reproduced distances. Thus, the smaller the stress value, the better the fit of the reproduced distance matrix to the observed distance matrix.

In the case of connectivity analyses, distances between the neural structures may be specified directly through their connectivity patterns. The strength of a connection between two structures can be taken to reflect the structures' "similarity" (strongly connected structures are more similar than unconnected structures), and distance between structures can be expressed by their relative dissimilarity. Intuitively, structures are closer to each other if they are more strongly connected and hence highly similar, and structures are more widely separated if they are unconnected. Alternatively, secondary distance and distance-like measures can be derived from the connectivity patterns, such as correlation measures or the similarity index described in Subheading 14.3.2. Although a common application of NMDS analyses is for symmetrical square distance matrices, the technique can also be applied to rectangular matrices, for instance, if there are more target than originating structures. As a rule, the fit of the data structure in lower dimensions improves with the number of different categories that the data contain (e.g., connectivity patterns typically contain only two data categories: existing and absent links); and if good fit is to be achieved, it is advisable to transform the raw data in such a way as to characterize each unit with a large number of different categories. This can be realized through a variety of conditioning methods, which may be custom-designed *(7)* or can be found as standard options in statistical packages. The MDS procedure in SPSS, for example, contains a variety of options to derive (dis)similarity distances from binary raw data, quite similar to the pattern similarity index described above. In the case of metric data, one simple way of transforming them into similarities is through Pearson's correlation. Recently, Tenenbaum and colleagues presented a powerful variant of MDS called Isomap that generalizes such data conditioning approaches by expressing all input data as local neighborhood measures *(70)*; a similar technique termed locally linear embedding (LLE) was introduced by Roweis and Saul *(71)*. A drawback of any data conditioning is, however, that the data structure apparent in the MDS output is determined through secondary measures and is, hence, less directly related to the raw connectivity data. As a consequence, careful consideration needs to be given to the organizational features actually represented, and straightforward interpretation of the analysis results may be made more difficult (cf. discussion in *[44]*).

In addition to the various measures expressing distances and similarities between the connected structures, there exist a number of models that can be used to evaluate the correspondence between the high-dimensional distances and the resulting low-dimensional configuration of the scaled data. Among the large number of parameter and method combinations for performing NMDS, no single best way of performing a scaling analysis exists, and analysis results may vary for different parameter settings. One approach to this problem is to derive solutions in different low dimensions under all available parameter settings and to evaluate their consistent features by subsequent stages of analysis, such as through cluster analysis *(36)*. This approach does, however, result in a loss of detail and interpretability. More detailed discussion of method options and data treatment in the application of MDS to neural connectivity is provided in *(7,72–74)*. These references also evaluate the potential limitations of NMDS in the analysis of neural connectivity data.

Interpretation. Relative proximity in a NMDS diagram conveys strong interconnections or high relative similarity in the high-dimensional connectional features of the linked structures. The configurations resulting from NMDS can be interpreted in terms of alignment along distinctive axes, sequences, clusters, and so on, apparent in the data arrangement. The NMDS result shown in Figure 6, for instance, provides examples for alignment along preferred axes (mainly from the bottom to the top of the diagram), as well as sequential and clustered organization. As the proximity in the diagram represents the relative similarity between structures, the actual orientation of axes in the MDS solution is arbitrary and can be rotated or reflected to facilitate interpretation. In addition to meaningful data dimensions, the scaling solution may show further peculiar configurations, such as circles, manifolds, and so on. For a detailed discussion of how to interpret the outcome of scaling analyses see *(75,76)*.

The agreement of the resultant low-dimensional configuration with the represented original high-dimensional data can be assessed by plotting the reproduced distances for a particular number of dimensions against the original (observed) input distances. This scatterplot is known as a Shephard diagram, and an example is provided in Figure 7. In an ideal low-dimensional representation of the original data structure, the rank-ordering of the original data would be perfectly reproduced by the scaled distances. Conversely, overlap in the reproduced distances indicates lack of fit. Another way of assessing how well the low-dimensional configuration approximates the original topology of the data is by plotting the stress value against different numbers of spatial dimensions on the x-axis. The idea behind this so-called scree plot is to find the value on the x-axis for which the smooth decrease of stress values appears to level off (indicating the minimal number of dimensions that should be used in order to represent the high-dimensional topology with a tolerable amount of stress) *(77)*.

Example. An example NMDS analysis was performed with the connectivity data shown in Figure 1. The table was transformed into a matrix only containing information about existing connections ("1") or absent and unknown pathways ("0"), and then reflected across the leading diagonal, so that entries in the resulting symmetrical matrix reflected reciprocal ("2") and unilateral connections ("1") or absent and unknown connections ("0"). This matrix, which can be interpreted as a similarity matrix, was transformed into a dissimilarity or distance matrix by inversion of its entries ("2"→"0"; "1"→"1"; "0" →"2") and submitted to the MDS (ALSCAL) routine in SPSS, defining the data as ordinal distances and specifying a Euclidean distance metric and the "tied" approach for the scaling algorithm. The resulting 2D configuration, with existing connections drawn between the areas, is shown in Figure 6. The solution clearly indicates the separation of visual cortical connectivity into at least two different groups of cortical areas that are more closely interconnected with each other than with the rest of the visual cortex. Also apparent are the segregation of occipital and more "peripheral" visual cortical areas (e.g., V1, V2, V3, which are synaptically closer to the retina), from parietal and more "internal" visual stations; as well as the sequential organization of connectivity between visual cortical areas. This configuration supports the commonly assumed separation of the primate visual system into two broad processing "streams" *(3,78)*. For an in-depth exploration of primate visual cortical topology on the basis of MDS analyses see *(3,7)*.

The stress and squared correlation of distances (RSQ) values of the low-dimensional arrangement suggest a moderate goodness of fit. This view is supported by the Shepard plot shown in Figure 7, which demonstrates considerable overlap between the distances in two dimensions representing the original connectional categories. The lack of fit is likely due to the low number (only three) of different categories in the analyzed data. A better fit could be achieved by data preconditioning or by representation of the scaled configuration in a higher number of dimensions. A representation in five dimensions, for instance, increases the RSQ from 0.48 for the shown arrangement to 0.63, and reduces the stress value of the configuration from 0.31 to 0.15.

14.4.3. Factor Analysis/Principal Components Analysis

Factor analysis, and one of its major variants, PCA, is a frequently used tool in the exploration of high-dimensional data structures. This type of analysis attempts to identify the factors underlying the correlation patterns of a set of variables. In a complementary way, it is also used to detect a small set of features that explains most of the variability in a larger set of variables, a strategy that can be employed for reducing the

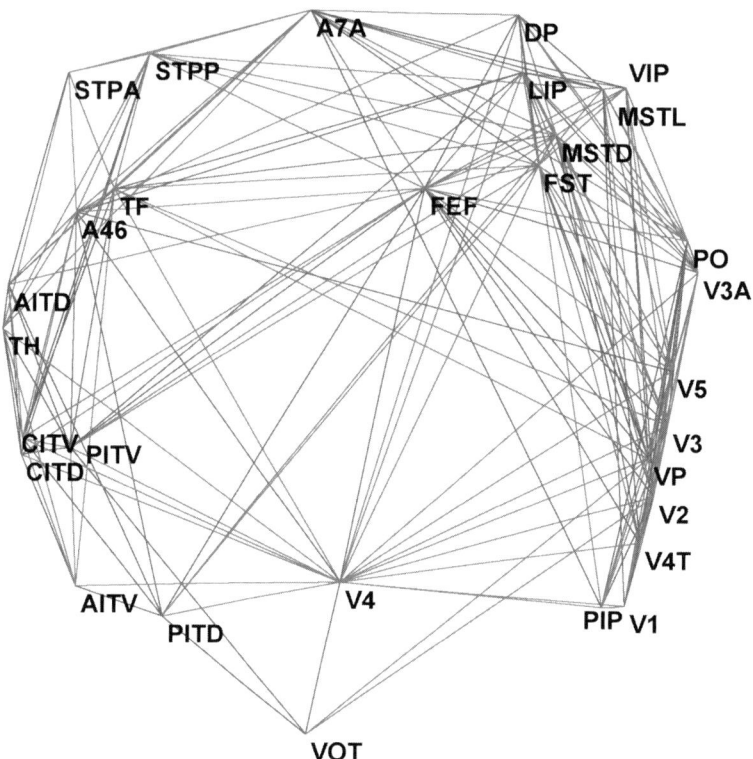

Fig. 6. NMDS representation of the dataset from Figure 1. Green lines indicate unidirectional connections between the areas, and red lines indicate reciprocal connections between the areas. For details of the analysis see main text. Starting with V1, the diagram shows the separation of the primate visual system into two main streams that are more connected internally than between each other. The ventral stream of mainly temporal areas is situated towards the left side, and the dorsal stream of occipital and occipitoparietal areas is situated towards the right side The arrangement hints that the streams might reconverge in prefrontal (e.g., A46) and polymodal (STP) cortical regions of the diagram. Also apparent is the intermediate position held by area V4, based on its interconnections with both the peripheral dorsal stream and the ventral stream, which also has been born out by another topological analysis of this system *(10)*. Additionally, the bottom to top arrangement of visual areas in this diagram broadly agrees with the arrangement from peripheral to more internal areas in the brain. For a detailed interpretation of NMDS analyses of primate visual connectivity, refer to *(3,7)*. The stress value for the shown diagram is 0.31, and the variance of the high-dimensional topology explained by the low-dimensional configuration (as represented by the configuration's squared correlation in distances) is 0.48, indicating a moderate fit. A color version of this figure is included in the CD-ROM.

number of variables for subsequent stages of analysis such as clustering or regression approaches.

The output of factor analysis methods appears superficially similar to that of NMDS, and PCA has been employed in a similar way as NMDS in the analysis of neural connectivity *(11)*. Depending on the implementation, the relationship between MDS and PCA may be more than superficial. For example, there are variants of MDS that are

Analyses of Brain Connectivity

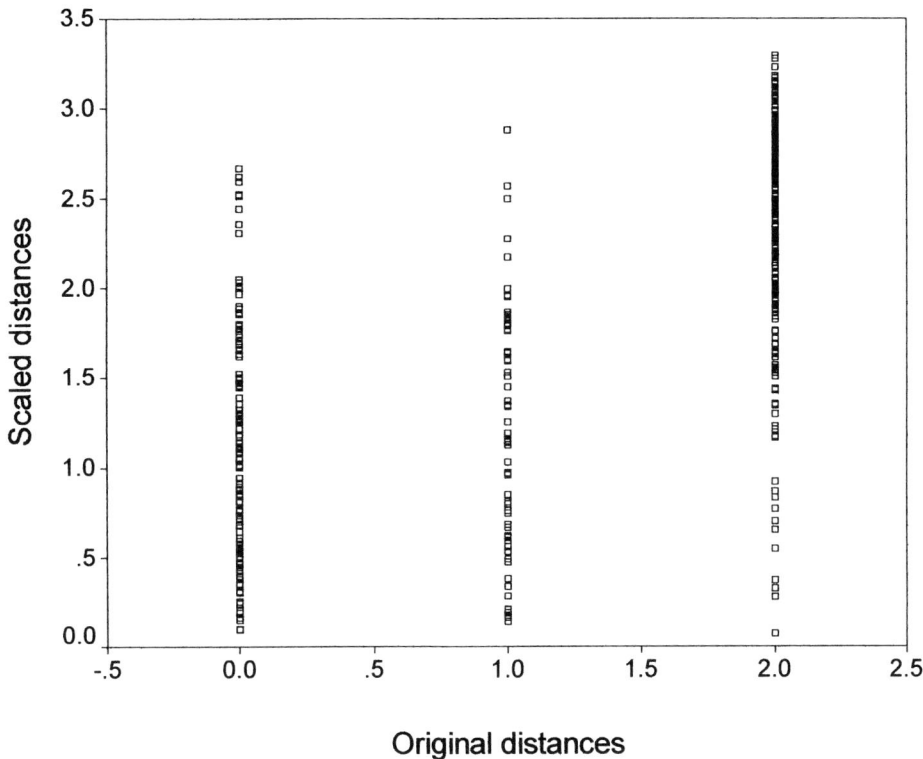

Fig. 7. Shepard scatterplot corresponding to Figure 6, for the scaled distances in 2D space (on the y-axis) versus the original dissimilarity distances (on x-axis). While there is a trend for smaller original distances to be represented by smaller reproduced distances, the scaled distances that reproduce the original three data categories overlap widely, indicating an only moderate fit between the original and the scaled configuration.

based on a decomposition of the covariance matrix of the original data into eigenvectors, and these eigenvectors are equivalent to the principal axes of the original data set as obtained by PCA (79,80).

It should, however, be kept in mind that most factor analysis techniques are based on multivariate methods for detecting linear or near-linear relationships among sets of quantitative variables. This means that these techniques are suitable only for the analysis of metric connectivity data, or for secondary metric measures, such as correlations, derived from the raw connectivity patterns. Additionally, the assumption that there are linear relationships among the connectivity patterns might be overly specific and, hence, not generally meaningful. On the other hand, specific methods among the various factor analysis approaches, such as nonlinear PCA are also capable of accepting ordinal input data and investigating nonlinear relationships between them. Unless, however, the specific goal of the analysis is data reduction, a more general type of similarity analysis such as NMDS appears more useful, since it imposes none of the restrictions applying to factor analysis approaches and is capable of detecting a large variety of possible relationships among variables. Moreover, in terms of resulting data representation, factor analysis tends to extract more factors or dimensions than NMDS; and the smaller number of factors suggested by NMDS may be more easily and readily interpretable.

14.4.4. Multiple Correspondence Analysis

Correspondence analysis is an exploratory technique designed to analyze two-way and multiway tables containing some measure of association between the rows and the columns. Correspondence analysis of more than two variables/cases is called MCA or homogeneity analysis *(81)*. Interestingly, the number of rows and columns in a correspondence table does not have to be identical. The analysis results provide information, which is similar in nature to that produced by factor analysis techniques. In contrast to factor analyses, however, correspondence analysis allows input data to be at nominal level of measurement, and it describes the relationships between the variables (for instance, neural structures represented in the columns of a connectivity matrix) as well as between the cases of each variable (the same or a different set of structures represented in the matrix rows). Therefore, correspondence analysis may be a valuable technique for analyzing connectivity between two different sets of structures or of connection matrices, in which the origins and targets of connections are described in different parcellation schemes.

MCA is also capable of analyzing data sets when linear relationships between the variables may not hold. Moreover, output interpretation is more straightforward in MCA than in other categorical techniques, such as cross-tabulation tables and log-linear modeling. The goal of the analysis is to express the data in terms of the distances between individual rows and/or columns in a low-dimensional space. To this end, each row and column are represented as a point in Euclidean space, standing for the inertia (that is, chi-square value divided by total number of observations) of the variables and cases, as determined from cell frequencies. In the graphical representation of this analysis the axes are orthogonal, and variables that are similar to each other appear close together in the graph, in an analogous way to NMDS configurations. Additionally, the distances between category points belonging to each variable in an MCA plot reflect the relationships between the categories, with similar cases plotted close to each other. While there appear to exist no worked examples of MCA in the analysis of neural connectivity, one previous application of this procedure in the structural investigation of cerebral cortex was the demonstration of similarities between laminar patterns of cells during development *(82)*.

14.4.5. Cluster Analysis

Method. Cluster analysis produces a grouping of objects based on data that represent the distances between the objects. In its basic strategy, this approach is similar to the one of MDS, though it is more specifically geared towards detecting significant groupings of objects. By varying the analysis parameters, one can gradually increase or decrease the size of the resultant clusters, and control their composition, from just one all-inclusive cluster at the one extreme to separate individual objects at the other. A hierarchical tree diagram can be used to show the linking up of separate components to larger clusters. This kind of taxonomy informs on the relative similarity among different individual objects and the clusters formed by them. The horizontal axis of the cluster tree (see example shown in Fig. 8) denotes the linkage distance. If the data contain a clear organization in terms of clusters of objects that are similar to each other, then this structure should be reflected in the hierarchical cluster tree as distinct branches with a large internode distance.

Cluster analyses that are available in general statistics packages encompass a variety of different classification algorithms. Various measures for calculating distances or (dis)similarities between individual data points or clusters are possible, and there exist many different linkage methods for combining these measures. Standard options are to compute the distances between data items by normal Euclidean metric or higher order Minkowski metrics, city-block distances, percent disagreements, or 1-Pearson's or alternative correlation measures. Attention needs to be paid to whether the routines expect coordinates or distances as standard input. In the first case, the software will provide method measures and metrics to compute the distances between the object coordinates; in the second case, the distances are directly supplied by the user, affording even greater flexibility in the treatment of the data. The latter option is, for instance, provided in clustering procedures in SAS and STATISTICA.

In addition to the different distance measures, most packages offer a variety of methods for joining separate clusters or linking individual objects into clusters. Such linkage rules might, for instance, consider the distance of clusters elements to their nearest neighbor outside of the cluster ("single linkage"), compute distance between different clusters by the average of all pairwise distances between the constituent cluster members ("average linkage"), or determine distances between clusters based on the clusters' weighted or unweighted mass centers ("centroid linkage"). These different parameter settings shape number and composition of the resulting clusters as well as the shape of the linked cluster tree. Although some of these settings are specifically adapted to the data's scale of measurement (e.g., matching similarities distances are mainly useful for ordinal data and centroid linkage may only be meaningful for metric distances), no definite general guidance can be given on the best choice of method options and parameters. Different combinations of settings may prove useful for different kinds of datasets and for the intended interpretation of the resulting cluster structure.

It is worth bearing in mind that most standard clustering methods are biased toward finding clusters possessing certain characteristics with regard to size (that is, number of members), shape, or dispersion. Many clustering methods, for instance, tend to produce compact, roughly hyperspherical clusters and are incapable of detecting clusters with highly elongated or irregular shapes. On the other hand, cluster analyses employing nonparametric density estimation, which for instance are available through the MODECLUS procedure in the SAS software, can produce clusters of unequal size and dispersion or irregular shapes, since they do not need to make specific assumptions about the form of the true density function for the given variables. Nonparametric methods may obtain good results for compact clusters of equal size and dispersion as well, but they tend to require larger sample sizes for a good recovery of such clusters, than will clustering methods that are already biased toward finding "typical" clusters.

Another noteworthy point for this type of analysis approach is that cluster analyses are really only meaningful if the investigated data do indeed possess a clustered organization. Cluster analysis will always suggest some kind of grouping, even if the data are dispersed homogeneously or are structured in another nonclustered way. Moreover, most clustering procedures do not allow testing the significance of the resulting cluster arrangement (limited significance testing is available in the SAS MODECLUS procedure). It is, therefore, a good idea to verify the clustered nature of the network at the outset of the analyses, for instance by computing the cluster index f_{clust} (51). Alternatively, NMDS can be initially performed to assess the general organization of the data. In the case of cluster analyses of neural connectivity, the distances between neural structures may again be specified directly as the similarity indicated by their pairwise connections (with existing connections indicating high similarity and small distances) or through secondary distance or similarity measures computed from the raw connectivity patterns.

Interpretation. Membership in clusters and cluster linkages inform on the family relationships between the investigated variables. The global tree structure of the clusters may also be meaningful in their own right; repeated joining of individual elements to an existing cluster, for instance, might indicate a more sequential arrangement of objects. However, the appearance of such configurations can also be reinforced or created by the chosen analysis options. Some linkage rules, for instance, may produce more stringy cluster trees than others. Generally, internode distance is a measure for the relative similarity of clusters or cluster constituents. Two clusters are more dissimilar, the longer the distance is until branches of two different clusters are joined in the cluster tree diagram.

Example. Cluster analysis was applied to the connectivity data of Figure 1, using the joining (tree clustering) cluster analysis procedure of STATISTICA. The connectivity data were analyzed as dissimilarity distances on the basis of the reflected connectivity matrix, as outlined for the NMDS analysis above. Clusters were linked by unweighted pair-group average, that is, by the average distance for all pairs of areas in two different clusters. This option produced compact cluster shapes, and the results of the analysis can be seen in the hierarchical cluster tree depicted in Figure 8.

At the lowest level (left-hand side of the diagram) clusters are formed by directly and reciprocally linked areas. These smaller clusters join up to form larger groups towards the right-hand side of the diagram. Two main clusters are apparent, consisting of occipitoparietal areas (top of the diagram) and inferior temporal areas (bottom), respectively, as well as two "outliers", areas VOT and PITd (see Fig. 8). The former cluster further separates into occipital visual areas on the one hand and parietal areas on the other (which agrees with the groupings apparent in the NMDS configuration of Fig. 6), whereas the latter cluster segregates into groups of inferotemporal and parahippocampal–polymodal areas. An equal rank of the principal clusters is indicated by their separation from the stem of the cluster tree at approximately the same point.

14.4.6. Combined Approaches

Several combinations of the outlined statistical analyses are possible. For instance, connectivity patterns may be translated into metric similarity distances by pattern matching algorithms or canonical correlations; and NMDS can be used to transform high-dimensional connectional patterns into low-dimensional (2D to 5D) Euclidean

Analyses of Brain Connectivity 319

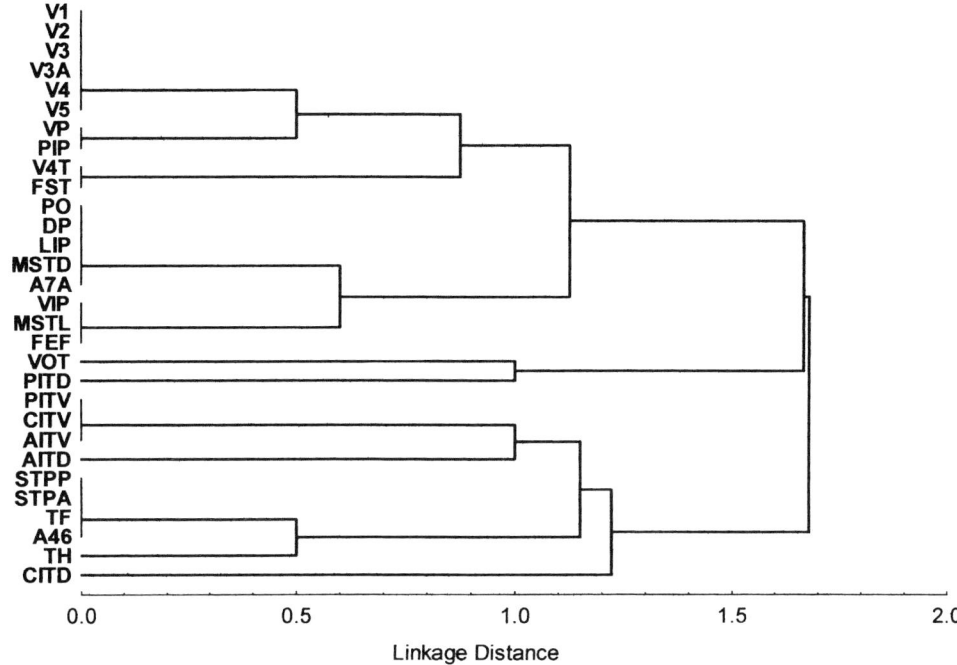

Fig. 8. Hierarchical cluster tree for direct connectivity between primate visual cortical areas. Pairwise connections between primate visual cortical areas as shown in Figure 1 were interpreted as dissimilarity distances and grouped by the average distance between all pairs of elements in different clusters. The results of this analysis demonstrate groupings of visual areas broadly similar to those apparent in the NMDS diagram of Figure 6. Most importantly, the cluster tree also indicates a separation of the primate visual system into two main groups (or streams) of areas. While both diagrams essentially indicate the same composition of the global data structure, there are minor differences in the grouped structure at the detailed level. One such incidence is the isolated position of area CITd in this cluster tree compared to the area's more integrated position in the NMDS diagram.

coordinates, whose grouping may then be further investigated by cluster analysis. See Burns and Young *(36)* for an example of the latter approach. A simple example for a combined analysis is given in the following.

For the data set depicted in Figure 1, the pattern similarity matrix (shown in Fig. 2) was computed, interpreted as a dissimilarity or distance matrix [by transforming the entries as d_{ij}(dissim): $= 1-d_{ij}$(sim)], and subsequently used in a hierarchical cluster analysis. The metric dissimilarity matrix was submitted as a distance matrix to STATISTICA's joining (tree clustering) procedure, and linkage was set to unweighted pair-group average, as for the example analysis in Subheading 14.4.5. The resulting cluster tree is shown in Figure 9. It is interesting to compare this diagram, which derives a family tree of cortical visual areas by the similarities in their global connectivity patterns, with Figure 8, which groups areas on the basis of their direct connections. In line with the principal connectional clusters expressed in Figure 8, Figure 9 also indicates two main groups of areas, which share similar connectivity patterns. This similarity is likely based on the numerous interconnections of areas within their respective

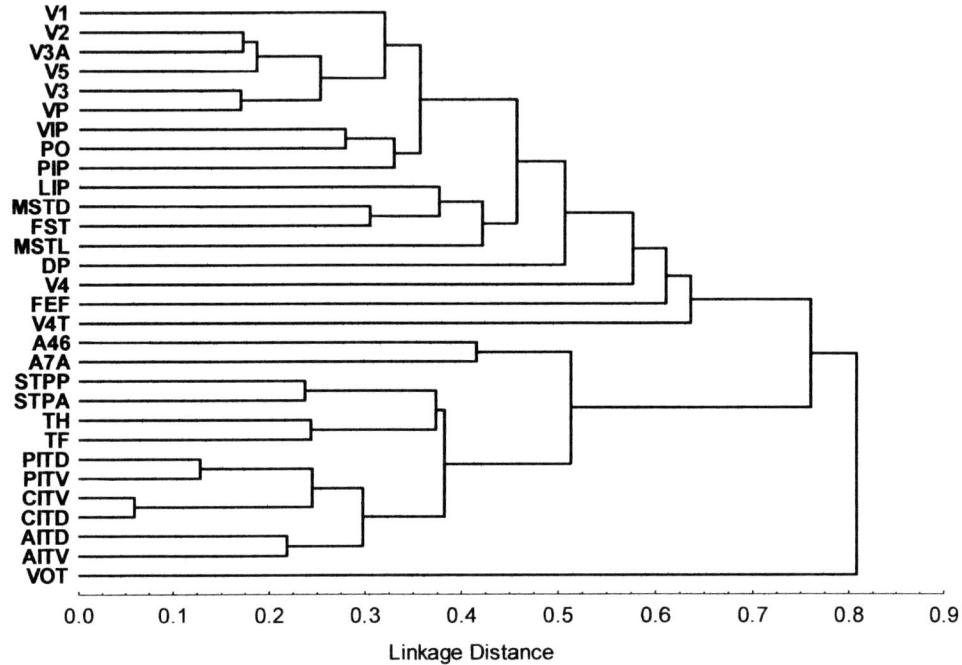

Fig. 9. Hierarchical cluster tree for global connectivity patterns of primate visual cortical areas. The binary connectivity matrix shown in Figure 1 was transformed into a metric distance matrix indicating the relative dissimilarity of the areas' global connectivity patterns; see main text for details. This distance matrix was analyzed in an analogous way to the cluster analysis for Figure 8, using pair-group averaging as linkage option to group clusters. The resulting cluster tree of connectional similarities shows broad similarities, but also some differences, to the clusters expressed in Figure 8; see main text for interpretation.

clusters, producing similar intracluster connectivity patterns. Moreover, both analyses point out area VOT as an outlier. There are, however, some differences between the diagrams on a more detailed level. For instance, Figure 9 does not replicate the separation of occipitoparietal areas into more peripheral and more internal stations that is apparent in Figure 8; instead it shows that the connectivity patterns for this group of areas form an inhomogeneous cluster, in which smaller groups of areas gradually link up to the main branch. This may indicate that areas in the occipitoparietal (or "dorsal") cluster, despite belonging to two distinct and strongly intraconnected subcomponents, generally follow a gradual shift in their global connectional patterns. This might be a sign of a partly sequential topology and would be in line with the connotation of a "stream" of visual areas. The other principal cluster of mainly temporal areas, by comparison, is relatively compact in both cluster diagrams, even though there are also several differences between the areas' patterns of direct interconnections and their global connectivity patterns.

14.5. CONCEPTUAL HYPOTHESIS TESTING

Purpose-designed computational approaches have been developed recently to test specific hypotheses about the organization of brain connectivity. Such approaches have allowed to frame hypotheses about connectional topology of neural networks more precisely than using the general-purpose statistical analyses discussed above. A shared property of many of these investigations is that they use optimization techniques to evaluate the fit of hypotheses about the hodology of networks against the background of randomly wired networks with the same number of nodes and connections. Concepts that have been tested include rules for the prediction of connections existing between neural structures, the potential minimization of wiring length between neural stations, the complex organization of neural connectivity as an adaptation to a complex environment, as well as the clustered and hierarchical organization of neural system networks. These ideas are briefly reviewed in the following, together with methods suggested to test these concepts.

14.5.1. Wiring Principles

Many researchers have been intrigued by the complex yet orderly arrangement of neural fibers in the brain and have attempted to identify general principles that shape the organization of these intricate neural networks. The understanding of such general rules might also allow predicting the existence or absence of currently unknown connectivity. One intensely debated issue is the suggestion that brain connectivity may be arranged in such a way as to minimize wiring length between neural structures *(52,83–85)*, Cherniak et al. (Chapter 4 of this text). There is indeed evidence for a correlation between topographic neighborhood relationships in the brain and a higher likelihood of connectivity among the neighboring stations compared to more spatially distant structures. For instance, Cherniak *(85)* suggested, on the basis of an exhaustive permutation analysis, that the adjacency placement of ganglions in the nematode *Caenorhabditis elegans* was an optimal arrangement for the minimization of connections between them. Similarly, Young *(3)* and Scannell *(86)* tested modeled connectivity networks with different types of neighborhood wiring (neighbors being defined by the sharing of areal borders or by being contained in the same gyrus) against the known anatomical connectivity data. They found partial agreement between the real connectivity and modeled networks, in which connections only existed between immediate neighbors, or next-neighbors-but-one, or among areas within the same gyri. Such correlations may be due to biomechanical constraints on the developing neural structures and their interconnections *(55)*. On the other hand, developmental and metabolic benefits derived from wiring optimization have to be seen in the context of the large number of simultaneously acting evolutionary, developmental, and thermodynamical constraints *(54)*, and they probably do not represent the overriding goal of connectional organization. In any case, topographic neighborhood relations may allow some prediction of connectivity within specific brain regions (e.g., "if cortical areas are bordering neighbors, then they are likely to be connected" *[54]*). Such wiring predictions have also been derived from other structural aspects of cortical organization. For instance, Barbas *(87)* suggested that cortical areas with similar structural architecture are more likely to be connected, and Jouve et al. *(11)* used graph extrapolation to predict connections between primate visual cortical areas on the basis of existing indirect connections between these areas.

14.5.2. Optimization Analyses

Optimal selection of networks using statistical measures of network dynamics. As described above, networks or graphs can be analyzed for their structural organization using graph theoretical and other tools. However, what really matters, from a neurobiological perspective, is the functional dynamics exhibited by the systems, in other words what the networks "do" while their components are active and exchanging signals along their connections. The term functional connectivity has been defined as the set of temporal correlations or deviations from statistical independence between the nodes of a network, e.g., *(5)*. An important area of investigation is concerned with the relationship between classes of structural patterns and patterns of functional connectivity that result due to the network's functional dynamics.

Naturally, for almost any realistic system, it is impossible to map this relationship exhaustively by generating all possible connectivities and testing them as dynamical systems. However, the problem becomes tractable if only certain "classes" of structural or functional connectivity are considered. In a series of studies *(88–90)*, a set of global statistical measures (based on statistical information theory) was developed that captures particular aspects of a given pattern of functional connectivity. Entropy, for example, captures the overall degree of statistical independence that exists within a network. Integration, on the other hand, captures the extent to which the network as a whole exhibits activity that is characterized by statistical dependence between its units. Complexity expresses how much a system is both functionally integrated (statistically dependent) at larger scales and functionally segregated (statistically independent) at smaller scales. At the system level, cortical networks seem to incorporate both of these features. They are composed of functionally segregated areas that are globally interconnected and functionally integrated in perception, cognition, and action. Complexity captures these dual tendencies and, therefore, can be expected to be high for modes of organization resembling the cerebral cortex.

Sporns et al. *(12,13)* have used an evolutionary search algorithm to look for connectivity patterns that produce functional connectivity characterized by high entropy, integration, or complexity (see Fig. 10). For each of these global measures and over wide ranges of structural and dynamic parameters, specific connection patterns emerge. Most interestingly, connection patterns that produce activity of high complexity closely resemble cortical connection matrices in their principal structural features. These networks with highly complex functional connectivity patterns are characterized by a clustered architecture with distinct sets of nodes that correspond to structural and functional subdivisions of the network. After MDS is applied to a complex network's covariance matrix, these sets of nodes form distinct clusters linked by relatively few connections ("bridges") (Fig. 10C). Other measures than the ones used here may be devised and used in similar search algorithms. In general, these search techniques allow the systematic exploration of small subregions of graph space under the guidance of a specific functional hypothesis (incorporated as a cost function based on functional connectivity or covariance matrices).

The CANTOR network processor. The preceding section already demonstrates the value of optimization approaches to the analysis of complex connectivity. In the fol-

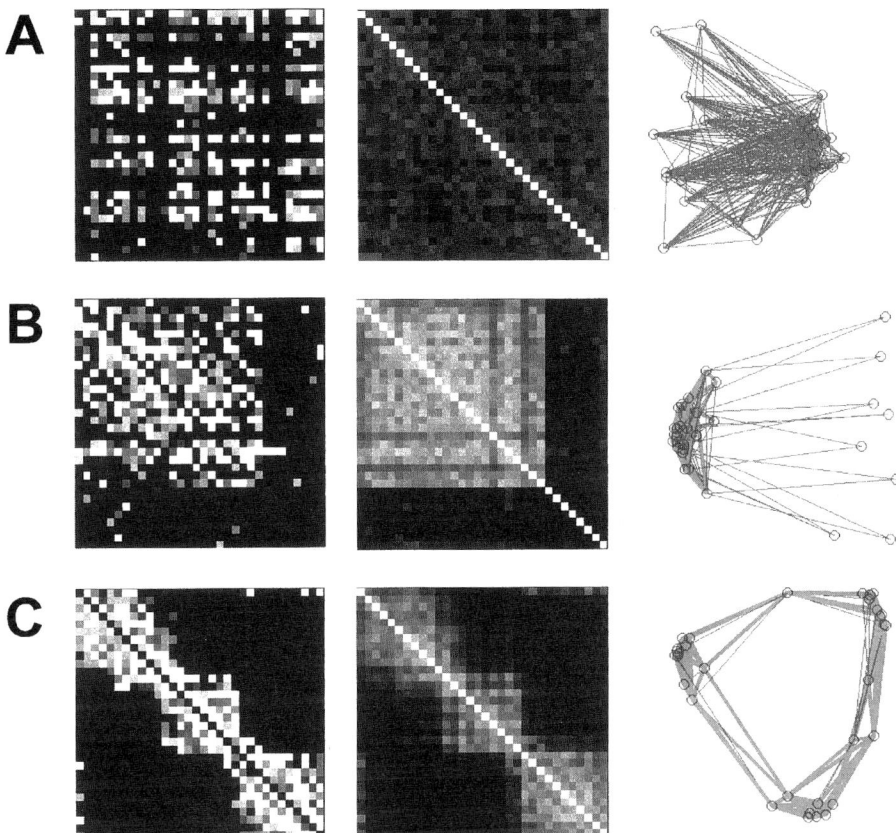

Fig. 10. Examples of graphs obtained after selection for entropy (**A**), integration (**B**), and complexity (**C**), using a variant of the graph selection algorithm of *(12)*. Connection matrices are shown on the left, and covariance matrices are shown in the middle. Displays on the right show the graphs after MDS, with vertices indicated as blue circles, and red and green lines indicating bidirectional and unidirectional edges, respectively. See text for more details. A color version of this figure is included in the CD-ROM.

lowing, we describe another computational approach based on the CANTOR system, which has been specifically designed for representing, rearranging and analyzing complex connectional data. We outline two example applications of this system in testing clustered and hierarchical organization of neural connectivity.

CANTOR, a software processor for the evaluation of complex sets of related objects, is built around a computational environment that supports stable, distributed, and persistent processing, as is required for performing computationally demanding tasks *(91)*. The system itself provides a number of analysis tools, such as an evolutionary optimization algorithm that can be combined with various multiparameter cost functions. CANTOR employs cost functions with categorical (nominal or ordinal) or metric components to evaluate the current organization of a network and to explore alternative rearrangements. For instance, the strength of interactions between nodes of a neural network may be described either metrically (e.g., for quantitatively determined structural connectivity or functional correlations between brain regions) or categorically (e.g., to qualitatively describe different classes of anatomical interconnections). Both ways can be represented and analyzed in CANTOR. The cluster analyses described previously *(10)* give examples for either type of cost function.

The values derived in cost functions can be used by CANTOR's optimization engine to initiate rearrangements of the data set. Because of the potential complexities of the data, the engine uses stochastic optimization based on an evolutionary algorithm. Briefly, the optimization starts with one or several randomly chosen initial network configurations, which are gradually improved in evolutionary epochs, according to the given cost function. Rearrangement of the network proceeds through "step mutations", which create minimal structural changes in the network's configuration. The efficiency of the optimization is increased by looking ahead two subsequent generations of candidate networks, and local minima are avoided by allowing intermediate generations with a slightly higher cost than that of their parent generation. The algorithm attempts to collect all optimal solutions within an optimal cost range, which is defined through the cost of the best solution found in all previous epochs. Therefore, all candidate network solutions have to be compared to the cost and structure of already existing solutions. This operation presents the limiting step of the algorithm for highly degenerate optimization problems, cf. *(9)*. The ability of the CANTOR system to collect a set of optimal or near-optimal solutions represents an advantage over most optimization techniques implemented in commercially available statistic packages (as described in Section 14.4.), which only deliver singular optimal solutions.

Combinations of different costs can be optimized, by default as the sum of the individual costs, although various methods of cost combination are possible. Attributes of the analyzed networks, such as connection weights, as well as the different cost components can be weighted selectively, incorporating, for instance subjective measures of reliability *(9,30)*. A list of CANTOR functions, parameter settings, as well as more detailed manual pages are available at (www.pups.org.uk).

Structural cluster analyses. Cluster analyses of connectivity have already been discussed in an earlier section of this chapter. With the help of general statistical tools such as hierarchical cluster analysis, connectivity patterns are interpreted as (dis)similarity distances, normally of a metric nature. These distances are inspected and linked into clusters, thus providing an indirect assessment of the connectivity cluster organization. Here, we return to the subject of cluster analysis, using a more direct approach. In a straightforward way, the task of finding connectivity clusters translates into identifying the groups of neural structures that share more anatomical connections or functional relationships among each other than with other neural structures. This simple concept can be expressed as a cost function in the CANTOR system and can so be used to test the cluster organization of neural networks. Previously published analyses *(10,41,64,92)* have referred to this type of cluster analysis within the CANTOR framework as optimal set analysis (OSA).

The suggested cluster cost function closely follows an intuitive concept for defining connectivity clusters *(7)*. Considering that a global optimal cluster grouping of a whole network would have as few as possible links between the different clusters, as well as a minimal number of absent connections within the clusters, a cluster cost is defined as the sum of two components: (*i*) the number of all connections between all clusters; and (*ii*) the number of absent connections with all clusters. The total combined cost has to be minimized.

Where the available data explicitly distinguish between absent and unknown data (as in the example set shown in Fig. 1), only the connections explicitly known to be absent are considered in the analysis. For less detailed datasets, both absent and unknown connections have to be treated as one category. In some cases, connectivity data compilations also contain additional information about the strength of existing connections. This information can be used to weight the strength of the connections in the cost function, using different approaches depending on the level of measurement at which the density information is available. In the metric approach, the connection density is used as a direct multiplication factor for the respective connections. In the nonmetric approach, on the other hand, ordinal density categories (weak, moderate, etc.) are used to determine the cluster organization of connectivity networks in a staged way, first taking into account all densest connections, then intermediate ones, and so on.

In the default approach, optimal cluster groupings are evolved in epochs starting from 50 different randomly chosen cluster arrangements of the structures in the connectivity network. Each epoch, by default, can yield 50 solutions within 1% of the cost of the best solution; and it was verified with the help of simple test cases that these default settings produce an exhaustive round-up of optimal configurations.

For a situation with multiple optimal solutions, a scheme can be devised that represents a summary of all solutions, by plotting the relative frequency with which two structures appear in the same cluster throughout all solutions. Figure 11 presents such a summary for the three optimal cluster arrangements found when analyzing the example data set shown in Figure 1.

The cluster structure of primate visual cortex shown supports the organization outlined by Figures 6 and 8. Once more, it indicates that visual cortical connectivity is arranged into two principal clusters of mainly occipitoparietal areas (top cluster in Fig. 11) and temporal areas (bottom cluster in Fig. 11). It also shows the dissimilarity of primary visual cortex (V1) and of area VOT, which has been previously indicated as an outlier (cf. Figs. 6 and 8). As the cluster arrangement of these data is significantly more pronounced than that of randomly arranged networks with the same size and degree of connectivity, there evidently exists a strong connectional dichotomy in the primate visual system. For a more detailed discussion of these results see Hilgetag et al. *(10)*.

One common problem in traditional cluster analyses is to determine the number of clusters to be detected in a given dataset. This number and, consequently, also the composition of clusters often depends indirectly on the parameter settings of the analysis, such as parameters for cluster diameters or specific neighborhood relationships. Frequently, no good strategy exists for choosing these values, and sometimes they cannot even be specified through user input. A typical approach pursued in standard cluster analyses, hence, is to vary these parameters systematically and represent the resulting grouping of objects as a hierarchical cluster tree. The cluster analysis presented here circumvents this problem, as CANTOR detects optimal cluster arrangements across all possible cluster sizes. In other words, the evolutionary algorithm determines the optimal number and optimal composition of clusters simultaneously, and the resultant structure depends directly on the explicit cluster cost function

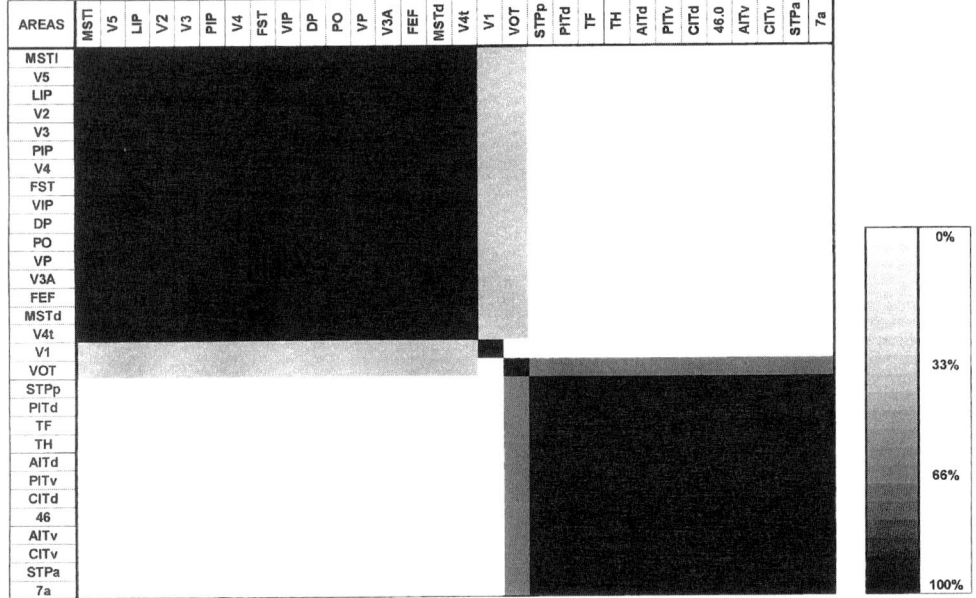

Fig. 11. Summary plot of 3 optimal cluster arrangements for the connectivity data shown in Figure 1. The shading indicates the relative frequency by which the cortical areas appeared together in the same cluster. For instance, area VOT grouped with the cluster at the bottom in 2/3 of all solutions and with the one at the top in the remaining 1/3. The analyses used the simple cluster cost function described in the text, without density- or reliability-weighting of individual connections. The shown solutions were obtained by assuming an equal weighting of the two cost components, which equates to giving equal importance to existing and absent connections between the areas. Solutions were also computed for systematic variations in the weighting of the two cost components, producing optimal cluster groupings varying from just one cluster to many small irreducible dense clusters. In the shown arrangements, the number of connections between clusters ranged between 73 and 89, while the number of absent links within the clusters varied between 58 and 42, respectively, leading to a total cost of 131. This cost is significantly smaller than the average cost for randomly organized networks with the same number of vertices and edges. See Hilgetag et al. *(10)* for a detailed description of clusters detected in macaque monkey and cat cortical connectivity. Reproduced with permission from *(91)*.

described above. Nevertheless, cluster sizes in the configuration can be controlled via selective weighting of cost components *(10)*. This feature, and the possibility of combining various metric or categorical cost components, ensure the flexibility of CANTOR's cluster algorithm and potentially make it suitable for the analysis of many different types of tangled networks, for instance traffic networks, relationship, and similarity networks (as in the evaluation of nucleic acid motifs).

Functional cluster analyses. An alternative clustering method, termed functional clustering, has been devised by Tononi et al. *(90)*, which utilizes the functional connectivity (covariance matrix) of a neural system. Based on information–theoretical considerations, these authors defined a so-called functional cluster index, which for each given set of units or areas expresses the ratio between total statistical dependence (inte-

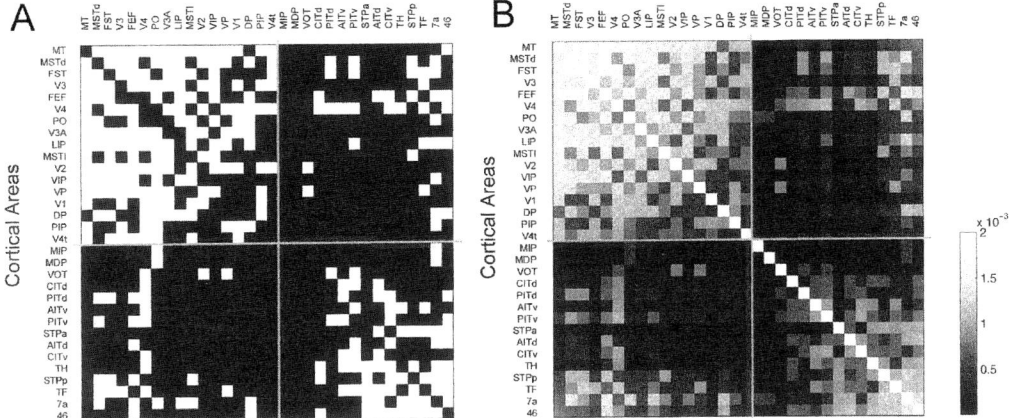

Fig. 12. Result of functional cluster analysis applied to the connectivity data set shown in Figure 1. Panel **A** shows the connection matrix and panel **B** shows the correlation (covariance) matrix of the macaque monkey visual cortex, which are displayed after reordering of rows and columns using the ranking of components of a functional cluster of high statistical significance. The cluster boundary is marked by a gray line and segregates two sets of areas largely homologous to the dorsal or occipitoparietal (upper left) and ventral or occipitotemporal (lower right) streams. Note that areas MIP and MDP are included in this analysis, and, after clustering, are positioned right between the two dominant clusters. Modified and reproduced with permission from *(13)*. A color version of this figure is included in the CD-ROM.

gration) within the set and the amount of mutual information between the set and the rest of the system. The cluster index is high if the components of a set are highly interactive and, at the same time, do not interact much with other components of the system that are not part of the set. While this measure is entirely based on the functional dynamics of the system (expressed as the system's covariance matrix), it has a certain degree of similarity with the set of two-component cost function described above, in that the measure compares the amount of within-set interaction (e.g., due to within-set connectivity) to the amount of between-set interaction (e.g., due to between-set connectivity). Sporns et al. *(12)* developed an evolutionary algorithm to efficiently evaluate large sets of functional clusters of varying sizes and derive estimates for their statistical significance (compared to a null hypothesis of a homogeneous system). Essentially, the algorithm searched for sets of units (of all sizes) that would show maximal functional clustering. When applied to the macaque monkey visual cortex (Fig. 1), a set of maximally significant clusters of different sizes is obtained *(12)*. One of the most significant cluster configurations is displayed in Figure 12, both as an optimally reordered functional connectivity matrix (Fig. 12B), as well as the corresponding anatomical connectivity matrix (Fig. 12A). The resulting cluster structure is strikingly similar to that shown in Figure 11, obtained using a clustering procedure based on anatomical, not functional, connectivity, suggesting that functional connectivity is strongly determined by underlying anatomical connectivity.

Hierarchical analyses. An established approach in the analysis of cortical connectivity is hierarchical analysis, e.g., *(6)*. Anatomical criteria have been suggested for the classification of corticocortical projections as feedforward, feedback, or lateral, on the basis of the projections' patterns of origin and termination across the cortical layers. The matrix in Figure 1 shows a large number of long-range cortical projections within the macaque monkey cortical visual system, which were classified as hierarchical relationships based on the data presented in Felleman and Van Essen *(6)*, Table 5 (also in *[6]*), and using the Felleman and Van Essen classification scheme. These hierarchical rules can be used to arrange cortical areas into global hierarchies that violate as few as possible of the individual rule constraints, e.g., *(6,9,93)*.

Hierarchical analysis can be automated with the help of the CANTOR system. For this purpose, a categorical cost function is defined that counts the number of hierarchical constraints violated by any given candidate hierarchical arrangement. When this cost was applied to optimize the data in Figure 1 hierarchically, solutions were obtained that violated 11 of the individual hierarchical constraints. This number is significantly smaller than the violation cost for optimal arrangements of the randomly redistributed data, indicating strict hierarchical regularities in the data. However, the solution space for this optimization problem proved to be highly degenerate, in the sense that a very large number of optimal solutions with the same low cost were found for the used categorical cost function. Within 1000 randomly started epochs, each of which maximally collected 30 solutions, the algorithm computed 6828 optimal hierarchies for the shown constraints, and the sustained high rate of solutions found per epoch suggested that this did not yet constitute an exhaustive set of optimal solutions for these constraints. For a slightly different, frequently referred to set of constraints (see Table 7 in *[6]*), where the number of optimal hierarchies found was even larger, with the algorithm yielding almost 200,000 visual hierarchies that all possessed a lower cost than solutions derived in previous manual analyses *(8,9)*. The degeneracy of the solution space is caused by features of the system's organization of the primate visual cortex *(9)*, as well as by the use of a simple categorical cost function in traditional hierarchical analyses. Ultimately, approaches will have to be employed that use quantitative constraints for determining sequential connectivity patterns in the brain *(9,27,94)*.

The hierarchical solutions found by CANTOR also showed several organizational features not revealed by previous manually derived schemes of the same system, such as a larger number of hierarchical stages (between 12 and 18 levels) or a different composition of constraint violations. Under this general approach, it is also possible to weight reliable and less reliable constraints (as indicated in Fig. 1) differently *(9)*. This strategy did not, however, significantly alter the resulting hierarchical arrangements.

In addition, using the CANTOR system in hierarchical analysis permits to computationally test hypotheses about the basis of the few remaining constraint violations in global visual hierarchies. Simulations of this kind suggested that one of the cortical visual areas, FST, might possess distinct subcomponents, and the approach also predicted some of the subcomponents' connectivity with other cortical structures (www.psychology.ncl.ac.uk/predictions.html) *(9)*.

14.6. CONCLUSIONS

We start by summarizing the knowledge that the presented connectivity analysis techniques revealed about the example data set of Figure 1. Subsequently, we present our general conclusions.

Simple statistical evaluation indicated that individual nodes of the primate visual cortical network possess different degrees of connectedness with the rest of the network. This picture, however, may also be due to the varying extent of knowledge gathered about the different cortical stations, cf. *(6)*. A current database project (www.cocomac.org) *(22)* is making efforts to improve this situation. Most visual cortical nodes also seem to maintain a roughly symmetrical balance between the number of input and output connections, bearing in mind that the analyzed patterns do not contain information about the strength of connections. However, there is some imbalance for ratios of nodes receiving and sending feedforward or feedback projections (as defined by laminar origin or termination patterns). Primary visual area V1, for instance, is sending a large number of connections of the feedforward type, while itself only receiving feedforward input from the thalamus. A better understanding of this imbalance will depend on gaining a better idea about the functional correlate of laminar cortical projection patterns.

The pattern similarity index introduced in Subheading 14.3.2. indicated the existence of two large groups of visual areas with similar connectivity patterns (see Figs. 2 and 9). The concepts of systems neuroscience suggest that similarly connected structures fulfill similar functional roles, which would imply that the primate cortical system is organized in such as way as to carry out at least two distinct principal tasks. The characteristic pathlength, that is, the average shortest path, in the cortical visual system was found to be not much longer than in a randomly wired network with the same number of nodes and connections. The latter type of network structure normally possesses the shortest characteristic pathlength, cf. *(51)*. Maximum distance within the example set was 4 intermediate connections, and most cortical stations can communicate with each other either directly or via one intermediate area. These short distances imply a highly efficient mode of cross-communication in the visual system.

Such efficiency is also related to the clustered, yet integrated, connectional organization of the system suggested by several different analysis approaches presented here. The existence of clusters was first established through the local cluster index f_{clust}, as well as through the local measure of cycle probability. Subsequent global exploration of the example network by NMDS and cluster analyses supported this view and helped to identify the components of the main clusters. The cluster composition was in agreement with the familiar two-stream picture of the primate visual system, e.g., *(3,78,95)*. The concept of clustered connectivity in the visual system was then tested and confirmed more explicitly using the OSA approach provided by the CANTOR optimization software.

The term streams also suggests another aspect of the organization of connections between visual cortical areas, which is a sequential arrangement of visual areas transmitting information from the sensory periphery to more internal cortical stations *(3)*. While this concept was not explored in great detail here, such sequences became apparent in the NMDS analysis of the example data in section 14.4. as well as in an analysis approach combining the computation of a connectivity pattern similarity index with hierarchical clustering (Subheading 14.4.6.). It might be a worthwhile task for the future to investigate this aspect of connectivity more closely, in a way akin to the presented analysis approaches. Similar to the network indices defined previously *(11,12,51)*, one could also define an index to measure the degree of sequential connectivity with a network. A suitable measure might be the proportion of shortest paths between the neighbors i and j of a network node k that pass through k, since this proportion is the larger the more linear and sequential a network is. The actual order of structures in the sequences could be identified in two stages. 1D NMDS, or correspondence analysis, would allow a straightforward visualization of candidate sequences, while a specific seriation cost function could be used in the CANTOR network processor to obtain all optimal arrangements of sequentially connected structures.

Finally, a specialized analysis approach investigating the hierarchical organization of the example data suggested a striking regularity of laminar connectivity patterns in the primate visual system. This result is also born out by recent quantitative approaches *(27,43)*. However, an understanding of the potential functional correlates for these intriguing regularities has only just begun *(96–99)*.

All presented methods provided interesting facets of information about the analyzed system. Several independent approaches indicated the clustered connectivity structure of the network (e.g., local cluster indices, NMDS, cluster analysis, OSA), while others provide alternative perspectives (shortest paths informing on efficiency of signal transmission, NMDS suggesting sequential organization) or focused on particular aspects of the system's organization (hierarchical analysis using CANTOR, verification of complex dynamics using graph selection). The outcome of the example analyses suggests that there is no single best method for analyzing neural connectivity. All available statistical methods help to explore the organization of connectivity data, while new computational methods are being designed to assess specific aspects of the neural connectivity, such as similarity of patterns, clustered organization, and sequences of interconnected systems. Methods specifically designed for the analysis of connectivity, however, allow a more straightforward interpretation of the results.

The increasing availability of connectivity data in electronic format will make it more feasible for the individual researcher to consider new neuroantomical data in the context of global connectivity networks, and future structural connectivity studies might address issues such as network stability and the structural and functional tolerance of different neural networks to lesions *(56,100)*. Moreover, the growing power of approaches that link the structural connectivity of neural systems to their functional properties predicts a significant expansion of this field in the future. This will be particularly important in functional imaging approaches for computing effective connectivity between neural structures from time series data (e.g., by structural equation modeling *[101–103]*) and for models describing the relation between neural activity and the blood oxygen level dependent (BOLD) signal in functional magnetic resonance imaging (fMRI), e.g., *(104)*. Finally, techniques for connectivity analysis represent valuable transferable knowledge in an age that is characterized by rapidly growing worldwide webs of information, traffic, and disease *(105)*.

ACKNOWLEDGMENTS

C.C. Hilgetag gratefully acknowledges support by the Wellcome Trust. Many thanks to R.J. Rushmore for helpful comments on this manuscript.

REFERENCES

1. Zeki S, Shipp S. The functional logic of cortical connections. Nature 1988; **335**:311–317.
2. Van Essen DC, Anderson CH, Felleman DJ. Information processing in the primate visual system—an integrated systems perspective. Science 1992; **255**:419–423.
3. Young MP. Objective analysis of the topological organization of the primate cortical visual system. Nature 1992; **358**:152–155.
4. Scannell JW, Burns GA, Hilgetag CC, O'Neil MA, Young MP. The connectional organization of the cortico-thalamic system of the cat. Cereb Cortex 1999; **9**:277–299.
5. Friston KJ. Functional and effective connectivity in neuroimaging: a synthesis. Hum Brain Mapp 1994; **2**:56–78.
6. Felleman DJ, Van Essen DC. Distributed hierarchical processing in the primate cerebral cortex. Cereb Cortex 1991; **1**:1–47.
7. Young MP, Scannell JW, O'Neill MA, Hilgetag CC, Burns G, Blakemore C. Non-metric multidimensional scaling in the analysis of neuroanatomical connection data and the organization of the primate cortical visual system. Philos Trans R Soc Lond B Biol Sci 1995; **348**:281–308.
8. Hilgetag CC, O'Neill MA, Young MP. Indeterminate organization of the visual system. Science 1996; **271**:776–777.
9. Hilgetag CC, O'Neill MA, Young MP. Hierarchical organization of macaque and cat cortical sensory systems explored with a novel network processor. Philos Trans R Soc Lond B Biol Sci 2000; **355**:71–89.
10. Hilgetag CC, Burns GA, O'Neill MA, Scannell JW, Young MP. Anatomical connectivity defines the organization of clusters of cortical areas in the macaque monkey and the cat. Philos Trans R Soc Lond B Biol Sci 2000; **355**:91–110.
11. Jouve B, Rosenstiehl P, Imbert M. A mathematical approach to the connectivity between the cortical visual areas of the macaque monkey. Cereb Cortex 1998; **8**:28–39.
12. Sporns O, Tononi G, Edelman GM. Theoretical neuroanatomy: relating anatomical and functional connectivity in graphs and cortical connection matrices. Cereb Cortex 2000; **10**:127–141.
13. Sporns O, Tononi G, Edelman GM. Connectivity and complexity: the relationship between neuroanatomy and brain dynamics. Neural Netw 2000; **13**:909–922.
14. Heimer L, Robards MJ. Neuroanatomical Tract Tracing Methods. Plenum Press, New York, 1981.
15. Heimer L, Zaborszky L. Neuroanatomical Tract-Tracing Methods 2, Recent Progress. Plenum Press, New York, 1989.
16. Köbbert C, Apps R, Bechman I, Lanciego JL, Mey J, Thanos S. Current concepts in neuroanatomical tracing. Prog Neurobiol 2000; **62**:327–351.
17. Crick F, Jones E. Backwardness of human neuroanatomy. Nature 1993; **361**:109–110.
18. Mori S, Barker PB. Diffusion magnetic resonance imaging: Its principle and applications. Anat Rec (New Anat.) 1999; **257**:102–109.
19. Conturo TE, Lori NF, Cull TS, et al. Tracking neuronal fiber pathways in the living human brain. Proc Natl Acad Sci USA 1999; **96**:10422–10427.
20. Van Essen DC. Functional organization of primate visual cortex. In: Cerebral Cortex, Vol. 3 (Peters A, Rockland K, eds.). Plenum Press, New York, 1985, pp. 259–328.
21. Stephan KE, Zilles K, Kötter R. Coordinate-independent mapping of structural and functional data by objective relational transformation (ORT). Philos Trans R Soc Lond B Biol Sci 2000; **355**:37–54.
22. Stephan KE, Kamper L, Bozkurt A, Burns GA, Young MP, Kötter R. Advanced data base methodology for the Collation of Connectivity data on the Macaque brain (CoCoMac). Philos Trans R Soc Lond B Biol Sci 2001; **356**:1159–1186.

23. Burns GA. Knowledge management of the neuroscientific literature: the data model and underlying strategy of the NeuroScholar system. Philos Trans R Soc Lond B Biol Sci 2001; **356**:1187–1208.
24. Musil SY, Olson CR. Cortical areas in the medial frontal lobe of the cat delineated by quantitative analysis of thalamic afferents. J Comp Neurol 1991; **308**:457–466.
25. Barone P, Dehay C, Berland M, Bullier J, Kennedy H. Developmental remodeling of primate visual cortical pathways. Cereb Cortex 1995; **5**:22–38.
26. MacNeil MA, Lomber SG, Payne BR. Thalamic and cortical projections to middle suprasylvian cortex of cats: constancy and variation. Exp Brain Res 1997; **114**:24–32.
27. Barone P, Batardiere A, Knoblauch K, Kennedy H. Laminar distribution of neurons in extrastriate areas projecting to visual areas V1 and V4 correlates with the hierarchical rank and indicates the operation of a distance rule. J Neurosci 2000; **20**:3263–3281.
28. Scannell JW, Grant S, Payne BR, Baddeley R. On variability in the density of cortico-cortical and thalamo-cortical connections. Philos Trans R Soc Lond B Biol Sci 2000; **355**:21–35.
29. Lennie P. Single units and visual cortex organization. Perception 1998; **27**:889–935.
30. Burns GAPC, Hilgetag CC. The computational representation and analysis of neuroanatomical knowledge: limiting the problem of information overload in neuroscience. FASEB J 2000; **14**:A544.
31. Harary F. Graph Theory. Addison-Wesley, Reading, MA, 1969.
32. Nicolelis MAL, Yu C-H, Baccala LA. Structural characterization of the neural circuit responsible for control of cardiovascular functions in higher vertebrates. Comput Bio Med 1990; **20**:379–400.
33. Young MP. The organization of neural systems in the primate cerebral cortex. Proc R Soc Lond B Biol Sci 1993; **252**:13–18.
34. Scannell JW, Blakemore C, Young MP. Analysis of connectivity in the cat cerebral cortex. J Neurosci 1995; **15**:1463–1483.
35. Patton PE, McNaughton B. Connection matrix of the hippocampal formation: I. The dentate gyrus. Hippocampus 1995; **5**:245–286.
36. Burns GAPC, Young MP. Analysis of the connectional organisation of neural systems associated with the hippocampus in rats. Philos Trans R Soc Lond B Biol Sci 2000; **355**:55–70.
37. Murre JM, Sturdy DP. The connectivity of the brain: multi-level quantitative analysis. Biol Cybern 1995; **73**:529–545.
38. Nicoll A, Blakemore C. Patterns of local connectivity in the neocortex. Neural Comput 1993; **5**:665–680.
39. Stevens CF. How cortical interconnectedness varies with network size. Neural Comput 1989; **1**:473–479.
40. Young MP, Hilgetag CC, Scannell JW. On imputing function to structure from the behavioural effects of brain lesions. Philos Trans R Soc Lond B Biol Sci 2000; **355**:147–161.
41. Kötter R, Hilgetag CC, Stephan KE. Connectional characteristics of areas in Walker's map of primate prefrontal cortex. NeuroComputing 2001; **38–40**: 741–746.
42. Zeki S. A Vision of the Brain. Blackwell, Oxford, 1993.
43. Barbas H, Rempel-Clower N. Cortical structure predicts the pattern of corticocortical connections. Cereb Cortex 1997; **7**:635–646.
44. Kötter R, Stephan KE, Palomero-Gallagher N, Geyer S, Schleicher A, Zilles K. Multimodal characterisation of cortical areas by multivariate analyses of receptor binding and connectivity data. Anat Embryol 2001; **204**:333–350.
45. Hilgetag CC. Mathematical approaches to the analysis of neural connectivity in the mammalian brain, PhD thesis, Faculty of Medicine, University of Newcastle upon Tyne, 1999.

46. Lagae L, Xiao DK, Raiquel S, Maes H, Orban GA. Position invariance of optic flow component selectivity differentiates monkey MST and FST cells from MT cells. Invest Ophthamol Vis Sci 1991; **32**:823.
47. Sporns O, Gally JA, Reeke GN, Jr, Edelman GM. Reentrant signaling among simulated neuronal groups leads to coherency in their oscillatory activity. Proc Natl Acad Sci USA 1989; **86**:7265–7269.
48. Sporns O, Tononi G, Edelman GM. Modeling perceptual grouping and figure-ground segregation by means of active reentrant connections. Proc Natl Acad Sci USA 1991; **88**: 129–133.
49. Buckley F, Harary F. Distance in Graphs. Addison-Wesley, Redwood City, CA, 1990.
50. Floyd R. Algorithm 97: shortest path. Commun ACM 1962; **5**:345.
51. Watts DJ, Strogatz SH. Collective dynamics of 'small-world' networks. Nature 1998; **393**:440–442.
52. Mitchison G. Neuronal branching patterns and the economy of cortical wiring. Proc R Soc Lond B Biol Sci 1991; **245**:151–158.
53. Ringo JL, Doty RW, Demeter S, Simard PY. Time is of the essence: a conjecture that hemispheric specialization arises from interhemispheric conduction delays. Cereb Cortex 1994; **4**:331–343.
54. Young MP, Scannell JW. Component-placement optimization in the brain. Trends Neurosci 1996; **19**:413–415.
55. Van Essen DC. A tension-based theory of morphogenesis and compact wiring in the central nervous system. Nature 1997; **385**:313–318.
56. Albert R, Jeong H, Barabasi AL. Error and attack tolerance of complex networks. Nature 2000; **406**:378–382.
57. Bollobas B. Random Graphs. Academic Press, London, 1985.
58. Erdös P, Rényi A. On the evolution of random graphs. Publ Math Inst Hung Acad Sci 1960; **5**:17–61.
59. Cohen JE. Threshold phenomena in random structures. Discr Appl Math 1988; **19**: 113–128.
60. Kauffman SA. The Origins of Order. Oxford University Press, 1993.
61. Rose G, Siebler M. Cooperative effects of neuronal ensembles. Exp Brain Res 1995; **106**:106–110.
62. Milgram S. The small world problem. Psychology Today 1967; **1**:61.
63. Watts DJ, Duncan J. Small Worlds. Princeton University Press, Princeton, NJ, 1999.
64. Stephan KE, Hilgetag CC, Burns GA., O'Neill MA, Young MP, Kötter R. Computational analysis of functional connectivity between areas of primate cerebral cortex. Philos Trans R Soc Lond B Biol Sci 2000; **355**:111–126.
64a. Kötter R, Sommer FT, Global relationship between anatomical connectivity and activity propagation in the cerebral cortex. Philos Trans R Soc Lond B Biol Sci 2000; **355**:127–134.
65. Barabasi AL, Albert R. Emergence of scaling in random networks. Science 1999; **286**: 509–512.
66. Jeong H, Tombor B, Albert R, Oltvai ZN, Barabasi AL. The large-scale organization of metabolic networks. Nature 2000; **407**:651–654.
67. Hilgetag CC, Grant S. Uniformity, specificity and variability of corticocortical connectivity. Philos Trans R Soc Lond B Biol Sci 2000; **355**:7–20.
68. Kruskal JB. Nonmetric multidimensional scaling: a numerical method. Psychometrika 1964; **29**:115–129.
69. Kruskal JB. Multidimesional scaling by optimizing goodness of fit to a nonmetric hypothesis. Psychometrika 1964; **29**:1–27.

70. Tenenbaum JB, de Silva V, Langford JC. A global geometric framework for nonlinear dimensionality reduction. Science 2000; **290**:2319–2323.
71. Roweis ST, Saul LK. Nonlinear dimensionality reduction by locally linear embedding. Science 2000; **290**:2323–2326.
72. Simmen MW., Goodhill GJ, Willshaw DJ. Scaling and brain connectivity. Nature 1994; **369**:448–449.
73. Young MP, Scannell JW, Burns GAPC, Blakemore C. Scaling and brain connectivity—reply. Nature 1994; **369**:449–450.
74. Goodhill GJ, Simmen MW, Willshaw DJ. An evaluation of the use of multidimensional scaling for understanding brain connectivity. Philos Trans R Soc Lond B Biol Sci 1995; **348**:265–280.
75. Borg I, Lingoes J. Multidimensional Similarity Structure Analysis. Springer, New York, 1987.
76. Guttmann L. A general nonmetric technique for finding the smallest coordinate space for a configuration of points. Pyrometrical 1968; **33**:469–506.
77. Cattell RB. The scree test for the number of factors. Multivariate Beh Res 1966; **1**:245–276.
78. Ungerleider LG, Mishkin M. Two cortical visual systems. In: Analysis of Visual Behaviour (Ingle DG, Goodale MA, Mansfield RJQ, eds.). MIT Press, Cambridge, MA, 1982, pp. 549–586.
79. Friston K. Characterising distributed functional systems. In: Human Brain Function (Frackowiak R, Friston K, Frith C, Dolan R, Mazziotta J, eds.). Academic Press, San Diego, 1997, pp. 107–126.
80. Healy M. Matrices for Statistics. Clarendon Press, Oxford, 2000, pp. 93–96.
81. Greenacre M. Theory and Applications of Correspondence Analysis. Academic Press, London, 1984.
82. Shankle WR, Romney AK, Landing BH, Hara J. Developmental patterns in the cytoarchitecture of the human cerebral cortex from birth to 6 years examined by correspondence analysis. Proc Natl Acad Sci USA 1998; **95**:4023–4028.
83. Ringo JL. Neuronal interconnection as a function of brain size. Brain Behav Evol 1991; **38**:1–6.
84. Cherniak C. Local optimization of neuron arbors. Biol Cybernetics 1992; **66**:503–510.
85. Cherniak C. Component placement optimization in the brain. J Neurosci 1994; **14**:2418–2427.
86. Scannell JW. Determining cortical landscapes. Nature 1997; **386**:452.
87. Barbas H. Anatomic basis of cognitive-emotional interactions in the primate prefrontal cortex. Neurosci Biobehav Rev 1995; **19**:499–510.
88. Tononi G, Sporns O, Edelman GM. A measure for brain complexity: relating functional segregation and integration in the nervous system. Proc Natl Acad Sci USA 1994; **91**:5033–5037.
89. Tononi G, Sporns O, Edelman GM. A complexity measure for selective matching of signals by the brain. Proc Natl Acad Sci USA 1996; **93**:3422–3427.
90. Tononi G, McIntosh AR, Russell DP, Edelman GM. Functional clustering: identifying strongly interactive brain regions in neuroimaging data. Neuroimage 1998; **7**:133–149.
91. O'Neill MA, Hilgetag CC. The portable UNIX programming system (PUPS) and CANTOR: a computational environment for dynamical representation and analysis of complex neurobiological data. Philos Trans R Soc Lond B Biol Sci 2001; **356**:1259–1276.
92. Hilgetag CC, Burns GAPC, O'Neill MA, Young MP. Cluster structure of cortical systems in mammalian brains. In: Computational Neuroscience: Trends in Research, 1998 (Bower JM, ed.). Plenum Press, New York, 1998, pp. 41–46.
93. Maunsell JH, Van Essen DC. The connections of the middle temporal visual area (MT) and their relationship to a cortical hierarchy in the macaque monkey. J Neurosci 1983; **3**:2563–2586.

94. Hilgetag CC, Grant S. Uniformity and specificity of long-range corticocortical connections in the visual cortex of the Cat. Neurocomputing 2001; **38–40**:667–673.
95. Milner AD, Goodale MA. The Visual Brain in Action. Oxford University Press, New York, 1996.
96. Petroni F, Panzeri S, Hilgetag CC, Kötter R, Young MP. Simultaneity of responses in a hierarchical visual network. Neuroreport 2001; **12**:2753–2759.
97. Grossberg S. How does the cerebral cortex work? Learning, attention, and grouping by the laminar circuits of visual cortex. Spat Vis 1999; **12**:163–185.
98. Grossberg S, Raizada RD. Contrast-sensitive perceptual grouping and object-based attention in the laminar circuits of primary visual cortex. Vision Res 2000; **40**:1413–1432.
99. Grossberg S, Williamson JR. A neural model of how horizontal and interlaminar connections of visual cortex develop into adult circuits that carry out perceptual grouping and learning. Cereb Cortex 2001; **11**:37–58.
100. Aharonov R, Meilijson I, Ruppin E. Measuring significance of neural elements: A quantitative approach. NeuroComputing 2002; in press.
101. McIntosh AR, Grady CL, Ungerleider LG, Haxby JV, Rapoport SI, Horwitz B. Network analysis of cortical visual pathways mapped with PET. J Neurosci **14**:655–666.
102. McIntosh A, Grady CL, Haxby JV, Ungerleider LG, and Horwitz B. Changes in limbic and prefrontal functional interactions in a working memory task for faces. Cereb Cortex 1996; **6**:571–584.
103. Büchel C, Friston KJ. Modulation of connectivity in visual pathways by attention: cortical interactions evaluated with structural equation modelling and fMRI. Cereb Cortex 1997; **7**:768–778.
104. Tagamets M-A, Horwitz B. Integrating electrophysiological and anatomical experimental data to create a large-scale model that simulates a delayed match-to-sample human brain imaging study. Cereb Cortex 1998; **8**:310–320.
105. Strogatz SH. Exploring complex networks. Nature 2001; **410**:268–276.

15
Development of Columnar Structures in Visual Cortex

Miguel Á. Carreira-Perpiñán and Geoffrey J. Goodhill

ABSTRACT

Many features of visual scenes are represented in the visual cortex in the form of maps. The best studied of these are the maps of features such as ocular dominance and orientation in primary visual cortex (V1). The beautifully regular structure of these maps and their dependence on patterns of neural activity have inspired several different computational models. In this chapter, we focus on what can be explained by models based on the idea of optimizing a trade-off between coverage and continuity, in particular, the elastic net (EN).

15.1. INTRODUCTION

Visual cortical map development is a well-studied area of computational neuroanatomy, characterized by an abundance of both experimental data and successful models. Starting with the classic experiments of Hubel and Wiesel in the 1970s, the characterization of both the adult structure of the primary visual cortex (V1) and how that structure arises developmentally have become continuously more precise and detailed as experimental techniques have improved. Current methodologies include single- and multielectrode physiology, 2-deoxyglucose and cytochrome oxidase staining, optical imaging based on both intrinsic signals and voltage-sensitive dyes, and anatomical tract-tracing techniques. In parallel, several different types of theoretical models have developed, some of which are very well understood analytically. To comprehensively review the full range of both experimental data and theoretical models relevant to map formation in visual cortex would require far more space than is available in this chapter. We, therefore, restrict ourselves to only briefly reviewing recent experimental data regarding map structure and map development and, then, focusing on one particular class of low-dimensional models, which attempt to optimize a trade-off between coverage and continuity. We further restrict ourselves to issues regarding maps overall, rather than details of the receptive fields of the individual neurons from which these maps are made. Broader reviews of models can be found in *(1,2)*.

From: *Computational Neuroanatomy: Principles and Methods*
Edited by: G. A. Ascoli © Humana Press Inc., Totowa, NJ

15.2. STRUCTURE OF ADULT MAPS

Receptive fields of V1 neurons are highly selective along a number of feature dimensions of the stimulus. These feature dimensions include position in the visual field, eye of origin (ocular dominance), orientation, direction of movement, spatial frequency of a grating, and disparity. Neurons lying along a line or column orthogonal to the surface of V1 respond in approximately the same way to visual stimuli. However, responses vary in an organized way in the tangential direction, parallel to the surface. As is common in this field, we will, therefore, discuss only the 2D structure of the visual cortex. Such 2D organization of preferred responses to a particular stimulus feature is termed a map (e.g., ocular dominance map, orientation map, and so on), and several such maps coexist on the same neural substrate. The map of preferred location in the visual field is topographic on a large scale (i.e., moving systematically across the visual field roughly corresponds to moving systematically across the cortex), though more convoluted on a fine scale *(3)*. The ocular dominance map consists of alternating stripes or blobs with a regular periodicity, with neurons in each stripe/blob responding preferentially to stimuli in one eye *(4)*. The orientation map is also striped with an overall periodicity *(5)*, but is characterized by point singularities or pinwheels, around which a circular path meets once all orientations from 0° to 180° *(6)*. The structure of the spatial frequency map is somewhat controversial *(7)*, with competing claims for both a binary *(8,9)* and a more continuous representation *(10)*.

Different maps are not independent from each other: the stripes of ocular dominance and orientation tend to run locally orthogonal to each other and orientation singularities tend to lie in the center of ocular dominance stripes *(9,11,12)*. Besides this local structure, some global structure is also apparent: in monkeys ocular dominance columns tend to run parallel to the shorter axis of V1, orthogonal to V1 boundaries, and are more irregular in the foveal region *(13,14)*. Maps of two individual animals of the same species are qualitatively similar, but maps of two individual animals of different species differ in the amount of columnar segregation and its type (stripes, blobs), periodicity, pinwheel density, and other structural characteristics. A problem which several theoretical researchers have become engaged in is to find effective ways to quantitatively characterize map structure (e.g., *[15–18]*).

15.3. MAP DEVELOPMENT: ROLE OF ACTIVITY

The view of development of V1 (layer IV) that has been universally accepted until recently can be summarized as follows (see e.g., *[19–21]*): (*i*) during early stages of circuit development, genetically specified molecular signals guide axonal outgrowth and targeting. These early connections are typically diffuse and imprecise; and (*ii*) these connections are then refined, and some are eliminated in response to visual activity, giving rise to the adult pattern of connectivity.

In particular, ocular dominance column formation requires a prolonged activity-dependent segregation process (lasting several weeks in cats and ferrets). This starts from an initial state in which lateral geniculate nucleus (LGN) afferents representing both eyes overlap extensively, and ends in a mature state in which eye-specific afferents occupy stripes. The hypothesis that column formation is activity-dependent is based on a large number of experiments dating back to the 1960s. When tetrodotoxin (TTX) is injected into both eyes of the cat during the critical period (the period when visual experience affects column formation), thus blocking all retinal activity, ocular dominance columns are not seen either anatomically or physiologically *(22)*. When one eye is occluded or sewn shut during the critical period, thus causing an imbalance in the amount of activity from the two eyes, it is found that (*i*) a higher proportion of cortical cells than normal are completely monocular; (*ii*) substantially more of the cells in layer IV can be driven by the normal eye as compared to the deprived eye; and (*iii*) ocular dominance stripes are now of different thicknesses for the two eyes. The stripes receiving input from the normal eye expand at the expense of the stripes from the deprived eye *(19,23,24)*. When an occluded eye competes with an eye injected with TTX, there is a shift towards the occluded eye *(25)*. The cortical imbalance in the representation of the two eyes is greater when an occluded eye competes with a normal eye than when a TTX-injected eye competes with a normal eye *(26)*. When strabismus is induced during the critical period, thus preventing image registration in the two retinae and decreasing the strength of between-eye correlations, all cells become entirely monocular, and the pattern of stripes becomes correspondingly sharper *(27)*. It was originally thought that overall column periodicity was left unchanged; recent data has suggested otherwise in the cat *(28)* though this is controversial *(29)*.

The data regarding the effect of activity on orientation map development is more complex and controversial; see section 15.6. and Swindale *(2)* for review.

15.4. THEORETICAL MODELS: COVERAGE AND CONTINUITY

The role of activity in shaping cortical maps has usually been modeled via Hebbian learning rules. Such rules can often be interpreted as implementing gradient ascent/descent in some objective function, so that the effect of the developmental process is to optimize (at least to some extent) that function. A particularly useful class of objective functions implements a trade-off between two competing tendencies, coverage uniformity (or completeness) and continuity (or similarity). However, even though several mathematical definitions of coverage uniformity and continuity have been given, the principles of coverage uniformity and continuity remain conceptually vague. For example, coverage uniformity and completeness are strictly different: the former means that each combination of stimuli values (e.g., any orientation in any visual field location of either eye) has equal representation in the cortex, while the latter means that any combination of stimuli values is represented somewhere in cortex. Thus, coverage uniformity implies completeness (disregarding the trivial case of a cortex uniformly nonresponsive to stimuli), but not vice versa, since it is possible to have over- and underrepresented stimuli values. Besides, it is not possible to represent all values of a continuous, higher-dimensional stimulus space with a 2D cortex[1]. A practically useful middle ground is to consider that the set of stimulus values represented by the cortex be roughly uniformly scattered in stimulus space—whatever that set is. Continuity is even

---

[1] We should say *practically* not possible, since from set theory we know that the cardinal of $\Re^D$ is equal to the cardinal of $\Re \; \forall D \geq 1$: there exists a continuous one-to-one mapping from $\Re^D$ to $\Re$.

less well defined than coverage uniformity. Loosely, we can say continuity means that neurons which are physically close in cortex tend to have similar stimulus preferences or (nonequivalently) that similar features are represented nearby in the cortex. This can be motivated in terms of economy of cortical wiring *(30)*.

The striped structure of several of the maps can be understood to represent a compromise between coverage and continuity. An early idea based on these principles is the icecube model of Hubel and Wiesel *(4)*, in which stripes of ocular dominance run orthogonally to stripes of orientation and all combinations of eye and orientation preference are represented within a cortical region smaller than a cortical point image (the collection of neurons whose receptive fields contain a given visual field location).

These general optimization principles of coverage and continuity do not in themselves support any specific development rule, since there may be different ways in which they can be optimized. However, heuristic rules that obtain local optima are rather more plausible than a global search, where the best of all possible configurations is found. It is appealing from both biological and computational perspectives to consider that visual cortical structure is the result of small developmental changes driven by neural activity. It is then plausible to think that such principles (abstractions based on physical and biological constraints and on adaptation to the environment) have a strong influence on the cortical structure, but with the following caveats: that they probably are not the only principles at work; and that they are only partly optimized in real organisms.

Particularly successful examples of such heuristic rules are the elastic net (EN) *(30–32)* and the self-organising map (SOM) *(33,34)*. In these models, the competition can be explained in a dimension reduction framework, where a 2D cortical sheet twists in a higher-dimensional stimulus space to cover it as uniformly as possible while minimizing some measure of discontinuity. These models differ in their explicit mathematical definitions, but produce maps that are similar and display a quantitatively good match to the observed phenomenology of cortical maps *(1,35)*. This includes: (*i*) the striped structure of ocular dominance and orientation columns with appropriately related periodicities and, for orientation, the existence of singularities (pinwheels); (*ii*) the interrelations between different maps, such as the tendency of orientation and ocular dominance stripes to be locally orthogonal and of the pinwheels to lie on the center of ocular dominance stripes; and (*iii*) the effect on the maps of various abnormal conditions during development, such as strabismus or monocular deprivation. Shortly, we discuss specific details of the EN and its application to cortical mapping problems. First however, we discuss coverage and continuity from a more general mathematical perspective, which helps to illustrate how the EN fits into the broader picture.

15.4.1. Mathematical Formulation of Coverage Uniformity and Continuity

Given a representation **M** of a cortical map, a mathematically convenient way of writing the trade-off between the goals of attaining uniform coverage and respecting the constraints of cortical wiring is to assume that cortical maps maximize an objective function:

$$\mathcal{F}(\mathbf{M}) \stackrel{\text{def}}{=} C(\mathbf{M}) + \lambda \mathcal{R}(\mathbf{M}) \qquad [\text{Eq. 1}]$$

where C is a measure of the uniformity of coverage, $\mathcal{R}$ is a measure of the continuity, and $\lambda > 0$ specifies the relative weight of $\mathcal{R}$ with respect to C. This formulation is formally akin to regularization theory *(36)*. We assume that maximizing either C or $\mathcal{R}$ separately does not lead to a maximum of $\mathcal{F}$ and, therefore, that maxima of $\mathcal{F}$ imply compromise values of C and C. By quantitatively defining C and $\mathcal{R}$ in terms of the map representation **M**, it is in principle possible to perform a numerical optimization of $\mathcal{F}$ to generate a map. We examine several possibilities next.

Model-based formulation. For a model, the map representation **M** is the set of model parameters, such as synapse strengths, tuning widths, mapping parameters, receptive field centers, etc. Such parameters uniquely determine the values of orientation, etc., at any point in the model cortical sheet. For the EN (which is discussed in more detail in Subheading 15.5.), the parameters are the locations in stimulus space of the reference vectors and the width k. From Equation 9, we can define:

$$C(\{\mathbf{y}_i\}_{i=1}^N, k) \stackrel{\text{def}}{=} k \sum_i \log \sum_j \Phi(\| x_i - \mathbf{y}_j \|, k) \quad \mathcal{R}(\{\mathbf{y}_i\}_{i=1}^N, k) \stackrel{\text{def}}{=} - \sum_j \| \mathbf{y}_{j+1} - \mathbf{y}_j \|^2 \quad \text{[Eq. 2]}$$

where $\{x_i\}_{i=1}^N$ is the sample of stimulus values, Φ is defined in Equation 8, and $\lambda \stackrel{\text{def}}{=} \frac{\beta}{2\alpha}$. Note that the standard EN algorithm minimizes $-\alpha \mathcal{F}$ instead of maximizing $\mathcal{F}$ and also anneals k rather than optimizing over it, but the expectation-maximization (EM) algorithm version does optimize over k and the reference vectors jointly. In the probabilistic interpretation of the EN, the C function is simply the log-likelihood of the parameters $\{y_i\}_{i=1}^N$ and k for the data sample $\{x_i\}_{i=1}^N$ i.e., the probability density that an EN density model with those parameter values generated the sample. The $\mathcal{R}$ function is a negative "length" of the net (it would be exactly the Euclidean length of the elastic net from $\mathbf{y}_1$ to $\mathbf{y}_N$ if the terms were not squared). It attains its maximal value at 0 for a point-like net where $\mathbf{y}_1 = \ldots = \mathbf{y}_N$: the more stretched the net is, the less "continuous" it is. This has been motivated in terms of economy of cortical wiring [the cost of setting up a given neural connectivity pattern; (30)].

Unlike the EN, the SOM learning rule cannot be integrated to give an objective function *(37)*. However, slight variations of its rule, leading to basically the same behavior, can be integrated and give an objective function very similar to that of the EN. For example, in the probabilistic variation of SOMs of Utsugi *(38)*:

$$C(\{y_i\}_{i=1}^N, \sigma) \stackrel{\text{def}}{=} \sum_i \log \sum_j e^{-\frac{1}{2} \left\| \frac{x_i - y_i}{\sigma} \right\|^2}$$

$$\mathcal{R}(\{y_i\}_{i=1}^N, \sigma) \stackrel{\text{def}}{=} -\frac{1}{2} \sum_j \| (\mathbf{D}y)_j \|^2 \quad \mathcal{F} \stackrel{\text{def}}{=} C + \lambda \mathcal{R} \quad \text{[Eq. 3]}$$

where **D** is a discretized differential operator, e.g., first order gives $(\mathbf{D}y)_j = \mathbf{y}_{j+1} - \mathbf{y}_j$, second order $\mathbf{y}_{j+1} - 2\mathbf{y}_j + \mathbf{y}_{j-1}$, and so on. Similarly generalized regularization terms could also be used in the EN *(39)*.

Model-free formulation. A model-free resolution-dependent representation of a map can be defined as a 2D array (not necessarily rectangular) of vector values of the stimulus variables of interest. Each position (i,j) in the array represents an ideal cortical cell. Let us call C the set of all such cortical positions (e.g., $C = \{1, \ldots, M\} \times \{1, \ldots, N\}$ represents an $M \times N$ rectangular array). There is a vector of stimulus values μ_{ij} associated to each cortical position (i,j). Stimulus variables of interest include the retinotopic position (or receptive field center in the visual field) (x,y) in degrees, the preferred orientation $\theta \in [0°, 180°])$, the ocular dominance n (–1, left eye; +1, right eye), and the spatial frequency $m \in \{-1, 1\}$. Therefore, $\mu_{ij} \stackrel{\text{def}}{=} (n_{ij}, m_{ij}, \theta_{ij}, x_{ij}, y_{ij})$ for $(i,j) \in C$ can be considered a generalized receptive field center; a receptive field could then be defined by a function sitting on the receptive field center and monotonically decreasing away from it. The collection $\mathbf{M} \stackrel{\text{def}}{=} \{\mu_{ij}\}_{(i,j) \in C}$ of such receptive field centers, together with the 2D ordering of cortical positions in C, defines the cortical map. This representation is applicable to maps measured empirically, for example with optical recording techniques.

Swindale *(40)* (see also *[41]*) introduced the following mathematical definition of coverage, which is applicable to this representation independently of any model. Given an arbitrary stimulus $\mathbf{v}$, the total amount of cortical activity that it produces is defined as:

$$A(\mathbf{v}) \stackrel{\text{def}}{=} \sum_{(i,j) \in C} f(\mathbf{v} - \mu_{ij}) \qquad \text{[Eq. 4]}$$

where f is the (generalized) receptive field of cortical location (i,j), assumed translationally invariant (so it depends only on the difference of stimulus and generalized receptive field center values); f is taken as a product of functions: Gaussian for orientation and retinotopic position (with widths derived from biological estimates of tuning curves) and delta for ocular dominance and spatial frequency. A is calculated for a regular grid in stimulus space, which is assumed to be a representative set of stimulus values. The measure of coverage uniformity is finally obtained as:

$$c' \stackrel{\text{def}}{=} \frac{\text{stdev}\{A\}}{\text{mean}\{A\}} \qquad \text{[Eq. 5]}$$

that is, the magnitude of the (normalized) dispersion of the total activity A in the stimulus space. Intuitively, c' will be large when A takes different values for different stimuli and zero if A has a positive value independent of the stimulus. Thus, it is a measure of lack of coverage uniformity, and we could define $C \stackrel{\text{def}}{=} -c'$.

Note that the function A can be seen as a kernel density estimate *(42)* for the sample $\{\mu_{ij}\}_{(ij) \in C}$ with smoothing parameter given by the width of the kernel function f. In fact, its spirit is the same as that of the probabilistic interpretation of the EN and SOM. This is because the latter are vector quantization methods, where the dimension reduction mapping is implicitly defined by the reference vectors in stimulus space. If the mapping was defined explicitly via parameters, as e.g., in the generative topographic mapping (GTM) model *(43)*, the resulting C and $\mathcal{R}$ functions would be quite different.

More difficult is to define $\mathcal{R}$ because the cortical wiring constraints are largely unknown and possibly result from the combined effect of several factors. At an abstract level, we can define $\mathcal{R}$ as a similarity measure where the preferred stimuli of nearby cortical neurons are similar, as in the expression *(44)*:

$$\sum_{(i,j),(k,l)\in C} F[(i,j),(k,l)]G(\boldsymbol{\mu}_{ij},\boldsymbol{\mu}_{kl})$$

In fact, many models of cortical development implicitly or explicitly implement such a definition of continuity for specific choices of the similarity measures F and G, like the EN, often with a flavor of wire length or local similarity of neural responses (e.g., *[45]*; see Goodhill *[46]* for discussion). However, this implies introducing model assumptions. At present, continuity is too vaguely defined to be quantitatively characterized for a model-free map representation.

15.5. THE ELASTIC NET ALGORITHM

The elastic net algorithm *(31)* was originally developed as an approximate method for the Traveling Salesman Problem (TSP), a well-known NP-complete combinatorial optimization problem. Here, the objective is to find the shortest distance a salesman can travel to visit a set of N cities in a plane and return to where he started. The key idea is that this problem is analogous to the problem of forming topographic maps in the nervous system, where cities represent input or feature points, and the tour represents the ordering of these points onto the target structure—though generally in the nervous system there is no "return to where you started" constraint. The TSP problem has been extensively discussed in the combinatorial optimization literature, being both easy to state and hard to solve. For N cities there are $\frac{(N-1)!}{2}$ possible routes: for large N, it is impossible to search them all to find the optimal tour. Therefore, many heuristic algorithms have been investigated, which aim to provide good solutions in reasonable time (for review, see Lawler et al. *[47]*). The set of valid tours for a TSP of size N can be represented as the vertices of an N-dimensional hypercube, and most techniques aim to provide good ways of stepping from one vertex to another to gradually improve the quality of the solution. However, an alternative method used by the EN is for the search to proceed through the continuous space inside the body of the hypercube, only converging to a valid solution in the final state.

The dynamics of the EN algorithm are closely related to those of the Kohonen algorithm. Both algorithms grew out of earlier models of retinotectal map formation (Kohonen: Willshaw and von der Malsburg *[48]*; EN: Willshaw and von der Malsburg *[49]*). The basic framework is an array of cortical cells that receive weighted connections from points in an input space. In the so-called low-dimensional version of both algorithms, input points represent features—such as a line segment of a particular orientation located at a particular point in the image—rather than individual image pixels. Generally, each feature is represented by an orthogonal dimension: one for x position, one for y position, one for ocularity (degree of left or right eye dominance), one for orientation, one for direction, and so on. The weight vectors of cortical cells can be represented as points in the input space. The input space is densely populated with input points, so that all appropriate feature combinations are represented. In the Kohonen algorithm, inputs are presented in turn. The initial activity of each cortical cell is the sum of pixel values times weights. This activation rule is equivalent to calculating the distance between the input vector and each weight vector (see e.g., Kohonen

[34] for mathematical details). In Willshaw and von der Malsburg (48), there was an explicit pattern of lateral cortical connections, consisting of short-range excitation and longer-range inhibition, and the activity of cortical cells was iterated until the cortical activity pattern stabilized. The most strongly responding cortical cells then had their weights updated according to a Hebbian learning rule (see Willshaw and von der Malsburg [48] for mathematical details). Kohonen's insight was that the end result of this process is usually (though not always) a blob of activity centered on the cortical cell that initially received the largest input. He therefore proposed an algorithmic shortcut, whereby it is assumed from the outset that the only cortical cells that should have their weights updated are those close to the unit that initially responded most strongly, which is the unit whose weight vector was closest to the input vector. The EN works in a similar way to this, with two important differences. Firstly, whereas Kohonen uses hard competition between cortical units, the EN uses soft competition. This means that in the EN, all cortical units are updated in proportion to how strongly they respond to each input pattern, rather than in the Kohonen algorithm, where just the most strongly responding are updated. Secondly, the EN algorithm usually operates in batch mode: all input points are considered simultaneously for updating usually cortical units, rather than presenting them one at a time as in Kohonen's algorithm. The biologically appealing interpretation that input points are being seen one at a time and that the cortex is responding to each in turn is now lost; however, this modification makes little difference from a mathematical perspective and allows a useful statistical interpretation as described below.

Refer to the positions of points in the input (feature) space as x_i and the positions of the weight vectors of cortical cells in the input space as $\mathbf{y}_j$. The change in the position $\Delta \mathbf{y}_j$ of each cortical unit at each time step is given by:

$$\Delta \mathbf{y}_j = \alpha \sum_i w_{ij}(\mathbf{x}_i - \mathbf{y}_j) + \beta k \sum_{j' \in \mathcal{N}(j)} (\mathbf{y}_{j'} - \mathbf{y}_j) \qquad [\text{Eq. 6}]$$

The first term is a matching term that represents the "pull" of feature points for cortical units, which is traded off with ratio $\alpha/\beta k$ against a regularization term representing a "tension" in the cortical sheet, i.e., a desire for neighboring cortical cells to represent neighboring points in the feature space (short-range excitation between cortical cells). The sum over all input points i indicates that the algorithm operates in batch mode. $\mathcal{N}(j)$ refers to the set of cells in the cortical sheet that are neighboring to j. The w_{ij}s (rather confusingly termed "weights" by Durbin and Willshaw [31]) say how much cortical cell j is activated by input i as a function of the difference between $\mathbf{x}_i$ and $\mathbf{y}_j$:

$$w_{ij} = \frac{\Phi(\|\mathbf{x}_i - \mathbf{y}_j\|, k)}{\sum_p \Phi(\|\mathbf{x}_i - \mathbf{y}_p\|, k)} \qquad [\text{Eq. 7}]$$

where:

$$\Phi(\|\mathbf{x}_i - \mathbf{y}_j\|, k) = \exp\left(\frac{-\|\mathbf{x}_i - \mathbf{y}_j\|^2}{2k^2}\right) \qquad [\text{Eq. 8}]$$

The sum over all cortical cells in the denominator of Equation 7 is a normalization term that says that each input produces the same overall amount of activity in the cortical

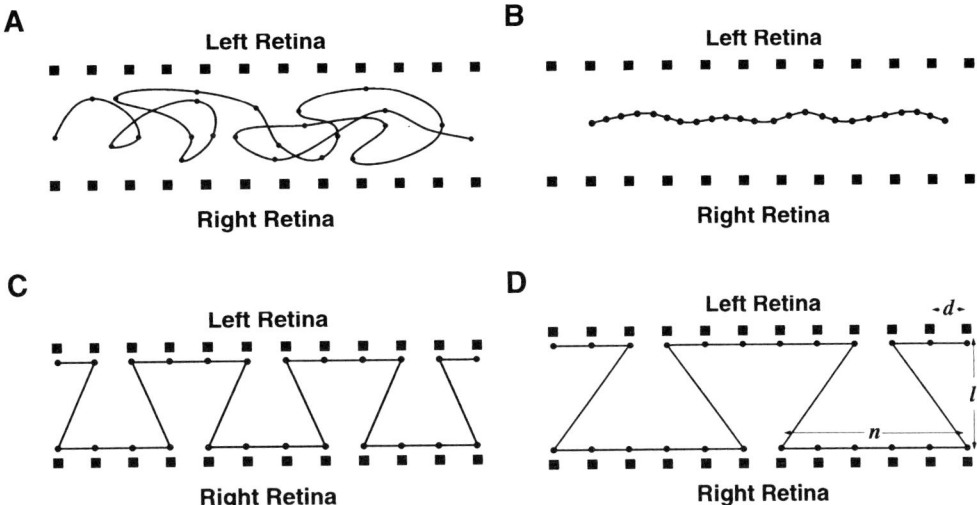

Fig. 1. Stages of development of the EN for the 1D ocular dominance problem *(32)*. Each retina is represented by a row of points in a feature space where distances are taken to represent correlations. The mapping to the cortex is represented by an elastic rope, where points on the rope represent cells in the cortex. The position of each cortical cell in this abstract space is updated iteratively, so as to simultaneously match cortical cells to retinal locations and keep the distance between neighboring cortical cells as small as possible (i.e., maximize the degree to which neighboring cortical cells receive highly correlated inputs). **(A)** Cortical cells are initially positioned randomly in the feature space, except for a crude initial topographic bias. **(B)** As development proceeds, the cortical mapping assumes an ordered topography. Cortical cells lie roughly equidistant between the two retinae, signifying a completely binocular mapping. **(C)** Eventually, cortical cells become committed to particular retinal locations, and a periodic pattern emerges (at this stage, cortical points lie on top of retinal points: for clarity in the picture, their positions have been slightly offset vertically). **(D)** Moving the two rows of retinal points further apart (increasing *l* in the terminology of the diagram) causes wider columns to be formed. For a striped solution, it is straightforward to calculate that the optimal width is $n = 2l/d$. For simplicity, the same number of cortical cells as retinal points have been drawn.

sheet: the w_{ij}s give the distribution of that activity. k is a scale parameter that determines the spread of this overall activity: if k is large, then many cortical cells are roughly equally activated by an input, whereas if k is small, then only those cortical cells whose weight vectors are closest to the input vector are significantly updated. Over the course of a simulation, the scale parameter k is gradually reduced, so that the matching term comes to dominate the regularization term. Equation 6 implements Hebbian learning in the sense that the degree to which a cortical cell is updated by a particular input depends on the degree of similarity between the input pattern and the weight vector of that cortical cell, i.e., the extent to which the activity of a cortical cell is correlated with that pattern of input activity.

Equation 6 can be integrated to produce an energy function E, which is such that

$$\Delta \mathbf{y}_j = -k \frac{\partial E}{\partial \mathbf{y}_j}$$

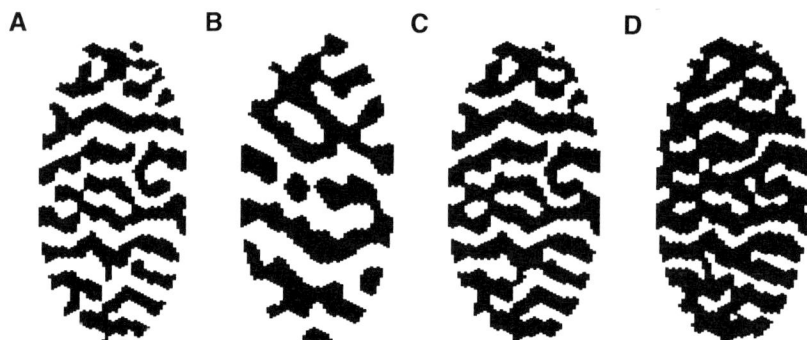

Fig. 2. Ocular dominance maps produced by the EN in two dimensions. Cortical points are colored white or black depending on the eye to which they are committed. The two retinal sheets were hexagonal grids with a circular boundary, and the cortical sheet has a hexagonal grid with an elliptical boundary. (**A**) Normal development. (**B**) Strabismic development (reduced correlation between the eyes). (**C** and **D**) Effects of monocular deprivation. (**C**) Activity in the deprived eye reduced by 25%. (**D**) Activity in the deprived eye reduced by 50%. The same initial conditions were used in each case.

$$E = -\alpha k \sum_i \log \sum_j \Phi(\| \mathbf{x}_i - \mathbf{y}_j \|, k) + \frac{\beta}{2} \sum_j \| \mathbf{y}_{j+1} - \mathbf{y}_j \|^2 \qquad \text{[Eq. 9]}$$

As explained in section 15.4.1., this function realizes the competition between coverage uniformity and continuity. The first term (coverage uniformity) simply tries to produce a one-to-one match between input features and cortical cells. The second term (continuity) says that only the similarity between neighboring cortical cells is considered and that dissimilarity in the input space is given by squared distance (the distances are squared for computational convenience). Carefully chosen versions can be interpreted as corresponding to particular patterns of lateral connections in the cortex *(39)*. This rather abstract version of Hebbian learning has the advantage of an elegant statistical interpretation *(50)*. Consider each unit on the elastic sheet as a Gaussian generator of data in the feature space. Each Gaussian has the same variance, determined by k. At each value of k, the optimization of the first term in the energy function corresponds to finding the positions of points on the sheet where they are most likely to have generated the data: a maximum likelihood model. The second term acts as a prior on the model that favours solutions with short distances between Gaussian centers: alternative forms for this term, therefore, correspond to different priors. The algorithm implements a form of graduated nonconvexity or deterministic annealing *(51)*. At large values of k, the energy function has a unique minimum. As k is reduced, the energy function bifurcates to produce several local minima, and the algorithm tracks one of these. This process continues as k is further reduced. In common with all heuristic optimization methods, the algorithm is not guaranteed to find the global minimum. Durbin et al. (50) analyzed the application of this algorithm to the TSP, where cortical units form a 1D loop of points on the tour, and feature points are cities in a plane. By calculating the first and second derivatives of the energy function with respect to the positions of cortical points, they showed that the center of gravity of the feature points (i.e., the configuration where all cortical receptive fields are coincident at this point) is always an extremum of the energy function. They calculated the value of k for which the center ceases to be a stable minimum, i.e., the energy function bifurcates, to be $k \approx \sqrt{\lambda}$, where λ is the principal eigenvalue of the covariance matrix of the feature points. At this stage, the cortical cells form a line (or sheet) along the principal axis of the covariance matrix of the feature points. By varying the structure of the covariance matrix of the feature points, it is, therefore, possible to vary the order in which different maps (e.g., orientation, ocular dominance) develop. We return to this issue later.

Ocular dominance. The behavior of the EN for the ocular dominance column mapping in one dimension is shown in Figure 1, and equivalent results in two dimensions are illustrated in Figure 2 *(32,52)*. In both figures, it can be seen that column periodicity depends on the degree of correlation between the eyes. This can be understood theoretically as follows. The EN tries to find a mapping that maximizes the degree to which neighboring cells in the cortex receive inputs that are highly correlated. In the abstract representation shown in Figure 1, this is equivalent to minimizing the length of the path that joins all input points (in fact, for reasons of computational efficiency, the elastic net minimizes the sum of squares of distances). Refer to the distance (in correlation space) between neighboring points in the same eye as d, and the distance between corresponding points in the two eyes as l (see Fig. 1D). It is easy to show that the optimal width n of a striped solution[2] is $n = 2l/d$ *(32,52)*. As l increases, the correlation between the two eyes decreases *(54)*, and the stripe width increases. An experimental prediction following from this result is that kittens raised with divergent strabismus should have wider ocular dominance columns than normal kittens *(52,55)*. Evidence in favor of this prediction was found by Löwel *(28)* (for further discussion, see Goodhill and Löwel *[56]*), though this is controversial *(29)*. The effect of reduced activity in one eye *(23,24)* can be modeled with the EN by reducing the effective pull that points in one eye exert on cortical cells relative to points in the other eye *(57)*. This corresponds to reducing the parameter α. Illustrative results are shown in Figure 2C and D.

In the macaque, besides local structure, ocular dominance stripes are also characterized by an overall orientation that varies with position in V1: ocular dominance stripes are less parallel in the foveal region, tend to be orthogonal to the borders of the neighboring visual cortical area, and decrease in width from the fovea to the periphery. Goodhill et al. *(58)* attempted to model this global structure using the EN. They identified three factors with an influence on the global structure of cortical maps and implemented them with the EN as summarized in Table 1. Thus, they concluded that (*i*) the widening of the columns in the foveal region results from stronger correlations (decreased spacing of retinal points); and (*ii*) the increased disorder in the foveal representation results from two competing effects: an elliptical cortex, which causes column

---

[2]For $l > d$, the solution that traverses the complete length of the left eye followed by the complete length of the right eye is, in fact, more optimal than any striped solution. In this case, $n = 2l/d$ is a local, rather than a global, minimum. However, in practice, the EN always finds a striped solution, since the dynamics of the algorithm always establish initial topography with the two ends of the cortex of opposite ends of the two retinae for sufficiently large values of the annealing parameter. Biologically, a biased initial topography is established by activity-independent molecular cues: see e.g., Goodhill and Richards *(53)*.

Table 1
Factors Influencing Global Structure of Cortical Maps in the Model of Goodhill et al. (58)

Factor	Reason Proposed	Elastic Net Implementation	Elastic Net Result
Spatially nonuniform correlational structure of activity in the retina.	The increase in retinal ganglion cell density proceeding from peripheral to central retina.	Foveal region of increased density of retinal points (smaller spacing), representing stronger correlations between neighboring retinal ganglion cells in the fovea.	Columns are wider and more disordered in the cortical representation of the fovea.
Spatially anisotropic correlational structure of activity in the retina.	The asymmetric way in which the retina develops.	By squashing the two retinal sheets (in the training set) so that the spacing between points is less in one direction than the other.	Columns tend to line up orthogonal to the direction of stronger correlations.
Elongated cortical shape.		Circular retina and elliptical cortex.	Columns line up parallel to the short axis of the cortical sheet.

alignment parallel to the short axis, and anisotropic input correlations, which cause column alignment parallel to the long axis. This model predicts that in a strongly anisotropic visual environment (e.g., kittens raised with cylindrical lenses), columns should tend to line up parallel to the direction of the weaker correlations.

Orientation. The EN was first applied to the formation of orientation maps by Durbin and Mitchison *(30)*. A standard trick in these types of models to capture the intrinsic periodicity of the orientation dimension is to wrap it around into a cylinder, so that orientation is now represented by a circular manifold in two dimensions rather than as just a line in one dimension. An equivalent periodicity analysis to that described earlier for ocular dominance can be performed to yield an expected periodicity of orientation columns of $2\pi r/d$ (see Fig. 3), which is in good agreement with simulation results *(59)*. A quantitative comparison of several different models of column formation *(35)* found that the EN and Kohonen algorithms actually produced the best match to real orientation and joint orientation–ocular dominance maps.

Two aspects of the EN applied to the joint formation of ocular dominance and orientation columns were examined by Goodhill and Cimponeriu *(59)*, namely, which model parameters control the order of development of the two sets of columns and how this ordering affects the final patterns of columns produced. Applying the bifurcation analysis of Durbin et al. *(50)* discussed earlier, Goodhill and Cimponeriu *(59)* derived

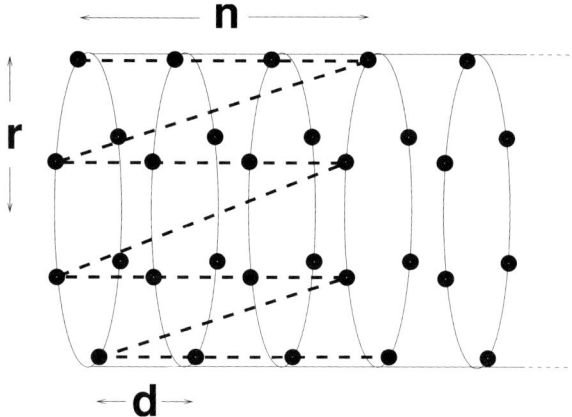

Fig. 3. One possible type of mapping between a cylindrical feature space representing orientation and one spatial dimension, and a 1D EN (solid circles, feature points; dashed line, EN; solid lines to aid visualization). Here, the net forms iso-orientation domains of smoothly varying orientation, each of which traverses n units in the spatial dimension.

expressions for the critical values of k for expansion along each of the visual space, orientation, and ocular dominance dimensions in terms of the parameters d, l, and r of the feature space (see also [60,61]). Combining these expressions with those quoted earlier for the expected ocular dominance and orientation periodicities, it is possible to relate the ratio of critical k values to the ratio of final periodicities without any reference to the parameters d, l, and r. Approximately, if the orientation wavelength is greater than the ocular dominance wavelength, then orientation columns developed first, otherwise ocular dominance columns developed first. The model, then, predicts that in normal macaque monkeys ocular dominance develops first, in normal cats orientation develops first, but in strabismic cats ocular dominance develops first. Whether this is true has not yet been experimentally determined.

An intriguing property of a particular version of the EN was discovered by Wolf and Geisel (62). They carried out a symmetry-based theoretical analysis of the dynamics of pinwheels during visual development, assuming only a very general class of activity-dependent learning rules. They showed that these symmetry properties, together with some general assumptions of Gaussian statistics and homogeneous correlations, imply that, if starting from a random unselective initial state, a minimal expected spatial density of pinwheels ρ emerges when orientation selectivity is first established: $\rho < \frac{\pi}{\Lambda^2}$ where Λ is the typical spacing of the orientation stripes. Therefore, if pinwheels are found in adult animals at a density $\rho < \frac{\pi}{\Lambda^2}$, then there must have been motion and annihilation of pinwheel pairs of opposite sign (as is also observed in the physics of defects). This result is robust to variations in the particular details of the developmental rule employed and depends only on its intrinsic symmetries. The EN falls within this general class and displays this tendency for pinwheels to move and annihilate during development, if it is run in the nonannealed regime. In this version, k is kept fixed during a simulation at a value below all relevant critical values, so that expansion occurs along all dimensions simultaneously. Pinwheel annihilation leads to growing regions of stripe-like iso-orientation domains, which locally resemble a plane wave. Using different sets of stimuli tuned to attain a variable degree of ocular dominance, the presence of ocular dominance columns slows down or even stops pinwheel annihilation, in proportion to the degree of ocular dominance segregation; thus, pronounced ocular dominance is associated to high-scaled density of pinwheels. As yet, no theoretical justification has been given for this. In the annealed version, the orientation map eventually stabilizes; whether annihilation occurs to any significant extent depends on how slow the rate of annealing is.

15.6. DISCUSSION

As we have seen above, the EN (and related models such as the SOM) can accurately reproduce a large amount of the observed phenomenology of ocular dominance and orientation maps. In addition, these models make several interesting and often surprising predictions about the outcomes of certain experiments. These include predictions of an increase in overall ocular dominance column periodicity with strabismus, a difference in the order of development between normal and strabismic cats, and pinwheel annihilation during development. While these claims are experimentally somewhat controversial, the generation of such clear and testable predictions has helped ferment an active and productive dialog between theoretical and experimental researchers. However, it is also appropriate to focus on some issues raised by recent data that these models do not account for.

15.6.1. Retinotopy Distortions

The retinotopic map of V1, given by the receptive field centers in visual space of every cortical neuron, is roughly uniform and has received far less attention than the ocular dominance and orientation maps. However, it contains both local and global distortions. Global distortions result from the fact that the magnification factor and receptive field size change monotonically from the fovea to the periphery. The retinotopic map has been approximately described by a complex logarithm *(63)*. It is not hard to produce global distortions with the EN and SOM, simply by systematically changing the density of feature points across the input space (e.g., *[64]*). More troubling are the local distortions, different from the scatter[3] of receptive fields, which result from discontinuities of the retinotopic map and are matched with the discontinuities (or pinwheels) of the orientation map *(3)*. That is, along a tangential penetration the receptive field center varies relatively smoothly from neuron to neuron, and neighboring receptive fields overlap considerably; but receptive fields of cells lying on opposite sides of a pinwheel do not overlap. Das and Gilbert *(3)* also found that the rates of variation of receptive field center (normalized by receptive field size) and of orientation preference were positively correlated. These two findings pose problems, so far unsolved, for cortical development models. On the one hand, some types of models, such as correlational models *(45)*, have relied so far on a perfect retinotopy and have not addressed topography distortions. On the other hand, dimension reduc-

---

[3]Quantitative estimates of this scatter vary, but recent results place its standard deviation between 0.1 and 0.5 times the receptive field size (e.g., *[3,65]*).

tion models, such as the EN and SOM, do not assume a perfect retinotopy, but simulations result precisely in anticorrelations rather than positive correlations: when one stimulus variable varies, the others tend to remain constant *(30)*. Why this is so is unknown: it is possible to have a rich behavior in the joint variation of stimulus variables (e.g., all variables varying at the same cortical location), while still satisfying coverage and continuity. Thus, the likely reason must lie in the specific formulation of the objective function or the training algorithm, that tends to single out solutions with anticorrelations. One factor could be the discrete character of the EN and SOM, both being vector quantization methods, which limits the different gradient values that may arise in a local optimum.

The results obtained by Das and Gilbert *(3)* used a coarse sample, since neighboring cells were recorded at approximately intervals of 50 μm for tangential penetrations and of 400 μm for a 2D grid. Ideally, the correlations and discontinuities should be obtained from a 2D sample on a wide area of the cortex at a much finer cortical separation in order to obtain meaningful gradients as a function of the cortical location. This would also show the distribution over cortex of the effects of discontinuity matching and positive correlation, which may not be uniform in view of the global distortions mentioned above. Unfortunately, such area-wide measurements, which have advanced our knowledge of the orientation and ocular dominance maps considerably, are not currently possible for the retinotopic map. Das and Gilbert *(3)* claim that their results are robust against scatter, because the latter is very small. However, the scatter could affect their results in a different way: while the retinotopic map obtained from measurements of a collection of cells at roughly the same cortical depth is noisy due to the inherent scatter of cell receptive fields, a more homogeneous map might be obtained as the aggregate receptive field resulting from cells in the same column. However, our inability to make spatial and volumetric measurements of this kind currently prevents experimental investigation of this issue.

15.6.2. Activity-Independent Mechanisms in Column Development

Earlier, we briefly reviewed the large body of evidence for a role for neural activity (both visually-evoked and spontaneous) in visual cortical map formation and plasticity. However, in the past few years, a number of pieces of experimental data have appeared, which challenge, to varying extents, the hypothesis that initial column development is activity-dependent. One reason for the rise of interest in activity-independent explanations for map formation in the cortex is the dramatic increase in our understanding of activity-independent mechanisms of axonal targeting. Since 1994, several large families of molecules, many previously unknown, have been identified to play crucial roles in the development of neuronal connections. These include the netrins, semaphorins, slits, and ephrins (reviewed in *[66,67]*). Molecules which may be particularly relevant to understanding map development in visual cortex are the ephrins, signaling through receptors of the Eph family. Low anterior to high posterior gradients of ephrins exist in the optic tectum and its mammalian homologue, the superior colliculus, while low nasal to high temporal gradients of Eph receptors exist in the retina (*[68,69]*; reviewed in *[53]*). Extensive evidence suggests that these gradients play a crucial role in guiding retinal ganglion cell axons to their targets (e.g., *[70,71]*;

other data reviewed in *[72]*). However, although ephrins have also been found in the LGN *(73)* and in somatosensory cortex *(74)*, as yet, there is no direct evidence that they play a role in guiding axons to appropriate targets in primary visual cortex. In particular, in order to control column formation, one would expect to find ephrins in an initially patchy distribution in visual cortex, and this has not so far been observed.

Ocular dominance maps. To reexamine the issue of whether the retinae or retinal activity are required for the establishment of ocular dominance columns, Crowley and Katz *(75)* enucleated ferrets[4] very early in life and let them develop. To visualize patterns of LGN axons in maturity, to determine whether ocular dominance segregation occurs, they used anterograde LGN injections (i.e., injection of a tracer into eye-specific cells in the LGN) and retrograde cortical injections. They found that, with or without information derived from the retina, geniculocortical axons organized into discrete ocular dominance stripes. Crowley and Katz *(76)* further found that columns are not present at birth, but appear as early as 16 d later (equivalent to a wk before birth in cats). Therefore, Crowley and Katz *(75)* removed the eyes *before* columns form. In addition, when they removed just one eye at an age when LGN axons have innervated the cortex, but before columns have formed, normal-looking ocular dominance columns still resulted (with the same periodicity: neither shrunk nor expanded). These results suggest that the establishment of ocular dominance columns and the plasticity of ocular dominance columns are two temporally different phases of visual cortex development.

Orientation maps. Using optical imaging of intrinsic signals and single-unit microelectrode penetrations in both eyes at postnatal d 15 (P15) or younger, Crair et al. *(77)* found that the orientation map forms before P14 in both normal and binocularly-deprived (BD) (by bilateral lid suture) cats. The similarity between the orientation maps of each eye for normal and BD cats varied with age. From P0 to around P21: the similarity increases monotonically in the same way for both normal and BD cats. From around P21: maps remain identical for normal cats, but progressively dissimilar for BD cats. They thus concluded that patterned visual experience is required for the maintenance of orientation selectivity rather than for the initial development of the orientation map, and the deterioration of maps coincides with the critical period. Orientation columns emerge independently of patterned visual experience during the second postnatal week, and patches of ipsilateral eye responses appear early in the third week. Experience then makes responses to become stronger, more selective, and nearly equal for both eyes by the beginning of the fourth wk; with continued BD, ipsilateral eye responses never become very strong or selective.

In another experiment, Gödecke and Bonhoeffer *(78)* investigated the influence of activity on the fact that orientation maps are precisely matched for both eyes (at least for binocularly-driven neurons), which is essential for disparity detection and also for stereoscopic vision. They raised kittens with reverse suturing, so that both eyes were never able to see at the same time. They found that the orientation maps in area 18 for

---

[4]Ferrets are ideal for experiments for two reasons: (*i*) they have robust ocular dominance columns and well-defined critical periods (like cats); and (*ii*) their nervous systems at birth are not yet developed (e.g., 3 wk less developed than those in cats, in which ocular dominance columns from before birth). This allows the detection of earlier developmental events.

both eyes were identical, as in normal kittens (except for minor differences such as a slight shift of some pinwheels that they imputed to technical limitations). They argued that this effect cannot be due to spontaneous retinal activity, since this would have to be synchronized between both eyes. They thus deduced that correlated visual input is not required for the alignment of orientation maps.

However, Wolf et al. *(79)* showed that it is possible to replicate the results of Gödecke and Bonhoeffer *(78)* in simulations with SOMs by using a cortex with a specific shape: a narrow elongated cortex results in matched orientation maps, while a square cortex results in unmatched orientation maps. Wolf et al. *(79)* explain this in terms of symmetry-breaking and pattern formation in physical systems, in which the qualitative behavior depends generally on the ratio between system size and characteristic wavelength of the emerging structure.

In another experiment, Weliky and Katz *(80)* examined the effect of perturbed patterns of neural activity on orientation maps in ferret V1. They implanted a stimulating cuff around the optic nerve from one eye (the other eye was removed) and stimulated it for about 2 s every 20 s from around P16 to P42. While they found effects on the receptive field structure of individual neurons, the overall structure of the orientation map looked apparently normal. Although this could be interpreted as evidence that overall orientation map structure is not activity-dependent, it is important to note that the aberrant stimulation occurred for only a small proportion of the total time (see Goodhill *[81]* for further discussion).

These studies and others suggest that the role of activity in columnar development is less determining than originally thought. However, it is important to note that even though the eyes are deprived or removed, spontaneous activity is still likely to be present in both LGN and cortex. In the LGN, spontaneous activity emerges from multiple mechanisms, including endogenous network oscillations and feedback connections *(82)*. Thus, the LGN does not simply relay patterns of retinal activity to the cortex, but rather this activity is reshaped and transformed by corticothalamic interactions.

Models Incorporating Neuronal Activity, Molecular Guidance Cues, and Gene Expression. Models such as the EN and the SOM have been very successful at accounting for much of the observed phenomenology of visual cortical maps. Although they are conventionally thought of as implementing activity-dependent learning rules, it is important to remember that the EN was originally derived from a mapping model based entirely on activity-independent molecular mechanisms *(49)*. Nevertheless, it seems reasonable to conclude that the early stages of cortical map formation are driven by a combination of molecular guidance cues and patterned gene expression, in addition to neural activity, and that a different type of model explicitly including all three factors may be necessary to account for map formation. One such kind of models are gene networks *(83)*. A gene network is a cluster of genes in which (*i*) the expression of the genes in the cluster is affected by specific stimuli, such as exposure to a hormone or neurotransmitter; and (*ii*) the protein products of some members of the cluster act as transcription factors that regulate, positively or negatively, the expression of other members. The total array of genetic regulation in neurons and other cells is a gene network with a large number (of the order of thousands) of interactions. Current models of gene networks describe the rates of change of the concentrations of gene products (mRNAs and proteins) with ordinary differential equations, as a function of the levels of transcription factors or other effector molecules and can incorporate stochastic fluctuations in molecule numbers. Logical networks, where the expression of each gene in the network is assumed to be either ON or OFF, have also been proposed. These models give rise to a rich variety of qualitative nonlinear behaviors, including multistability and oscillations. However, gene network models present some difficulties. Modeling differential equations requires short time steps and so a high computational cost, which may make them impractical for large gene networks or lengthy processes, such as development of tissues or organisms. Models of specific gene networks need to be based, insofar as possible, on values of biochemical parameters measured in vivo, which is difficult. Hence, new methods for gathering detailed data are necessary. Finally, so far, gene network models mostly do not take into account the spatial organization of gene expression, which is essential to explain biological pattern formation and, in particular, cortical map formation. Models for the *Drosophila* segmentation problem, such as those of Sharp and Reinitz *(84)* and von Dassow et al. *(85)*, are a promising step in this direction.

REFERENCES

1. Swindale NV. The development of topography in the visual cortex: a review of models. Network 1996; **7**:161–247.
2. Swindale NV. Development of ocular dominance stripes, orientation selectivity and orientation columns. In: Modelling Neural Development Ch. 12. (van Ooyen A, ed.). MIT Press, Cambridge, MA, 2001.
3. Das A, Gilbert CD. Distortions of visuotopic map match orientation singularities in primary visual cortex. Nature 1997; **387**:594–598.
4. Hubel DH, Wiesel TN. Functional architecture of the macaque monkey visual cortex. Proc R Soc Lond B Biol Sci 1977; **198**:1–59.
5. Blasdel GG, Salama G. Voltage-sensitive dyes reveal a modular organization in monkey striate cortex. Nature 1986; **321**:579–585.
6. Bonhoeffer T, Grinvald A. Iso-orientation domains in cat visual cortex are arranged in pinwheel-like patterns. Nature 1991; **353**:429–431.
7. Issa NP, Trepel C, Stryker MP. Spatial frequency maps in cat visual cortex. J Neurosci 2000; **20**:8504–8514.
8. Shoham D, Hübener M, Schulze S, Grinvald A, Bonhoeffer T. Spatio-temporal frequency domains and their relation to cytochrome oxidase staining in cat visual cortex. Nature 1997; **385**:529–533.
9. Hübener M, Shoham D, Grinvald A, Bonhoeffer T. Spatial relationships among three columnar systems in cat area 17. J Neurosci 1997; **17**:9270–9284.
10. Everson RM, Prashanth AK, Gabbay M, Knight BW, Sirovich L, Kaplan E. Representation of spatial frequency and orientation in the visual cortex. Proc Natl Acad Sci USA 1998; **95**:8334–8338.
11. Bartfeld E, Grinvald, A. Relationships between orientation-preference pinwheels, cytochrome oxidase blobs, and ocular-dominance columns in primate striate cortex. Proc Natl Acad Sci USA 1992; **89**:11905–11909.
12. Obermayer K, Blasdel GG. Geometry of orientation and ocular dominance columns in monkey striate cortex. J Neurosci 1993; **13**:4114–4129.
13. LeVay S, Connolly M, Houde J, Van Essen DC. The complete pattern of ocular dominance stripes in the striate cortex and visual field of the macaque monkey. J Neurosci 1985; **5**:486–501.

14. Horton JC, Hocking DR. Intrinsic variability of ocular dominance column periodicity in normal macaque monkeys. J Neurosci 1996; **16**:7228–7339.
15. Obermayer K, Blasdel GG. Singularities in primate orientation maps. Neural Comput 1997; **9**:555–575.
16. Löwel S, Schmidt KE, Kim D-S, et al. The layout of orientation and ocular dominance domains in area 17 of strabismic cats. Eur J Neurosci 1998; **10**:2629–2643.
17. Müller T, Stetter M, Hübener M, et al. An analysis of orientation and ocular dominance patterns in the visual cortex of cats and ferrets. Neural Comput 2000; **12**:2573–2595.
18. Kaschube M, Wolf F, Geisel T, Löwel, S. Quantifying the variability of patterns of orientation domains in the visual cortex of cats. Neurocomputing 2000; **32–33**:415–423.
19. Hubel DH, Wiesel TN, LeVay S. Plasticity of ocular dominance columns in monkey striate cortex. Philos Trans R Soc Lond B Biol Sci 1977; **278**:377–409.
20. Shatz CJ. Impulse activity and the patterning of connections during CNS development. Neuron 1990; **5**:745–756.
21. Katz LC, Shatz CJ. Synaptic activity and the construction of cortical circuits. Science 1996; **274**:1133–1138.
22. Stryker MP, Harris WA. Binocular impulse blockade prevents the formation of ocular dominance columns in cat visual cortex. J Neurosci 1986; **6**:2117–2133.
23. Shatz CJ, Stryker MP. Ocular dominance in layer IV of the cat's visual cortex and the effects of monocular deprivation. J Physiol 1978; **281**:267–283.
24. LeVay, S, Wiesel TN, Hubel DH. The development of ocular dominance columns in normal and visually deprived monkeys. J Comp Neurol 1980; **191**:1–51.
25. Chapman B, Jacobson MD, Reiter HO, Stryker MP. Ocular dominance shift in kitten visual cortex caused by imbalance in retinal electrical activity. Nature 1986; **324**:154–156.
26. Rittenhouse CD, Shouval HZ, Paradiso MA, Bear MF. Monocular deprivation induces homosynaptic long-term depression in visual cortex. Nature 1999; **397**:347–350.
27. Hubel DH, Wiesel TN. Binocular interaction in striate cortex of kittens reared with artificial squint. J Neurophysiol 1965; **28**:1041–1059.
28. Löwel S. Ocular dominance column development: Strabismus changes the spacing of adjacent columns in cat visual cortex. J Neurosci 1994; **14**:7451–7468.
29. Sengpiel F, Gödecke I, Stawinski P, Hübener M, Löwel S, Bonhoeffer T. Intrinsic and environmental factors in the development of functional maps in cat visual cortex. Neuropharmacology 1998; **37**:607–621.
30. Durbin R, Mitchison G. A dimension reduction framework for understanding cortical maps. Nature 1990; **343**:644–647.
31. Durbin R, Willshaw D. An analog approach to the traveling salesman problem using an elastic net method. Nature 1987; **326**:689–691.
32. Goodhill GJ, Willshaw DJ. Application of the elastic net algorithm to the formation of ocular dominance stripes. Network 1990; **1**:41–59.
33. Kohonen TK. Self-organized formation of topologically correct feature maps. Biol Cybern 1982; **43**:59–59.
34. Kohonen TK. Self-Organizing Maps. Springer-Verlag, Berlin, 1995.
35. Erwin E, Obermayer K, Schulten K. Models of orientation and ocular dominance columns in the visual cortex: A critical comparison. Neural Comput 1995; **7**:425–468.
36. Tikhonov AN, Arsenin, VY. Solutions of Ill-Posed Problems. Scripta Series in Mathematics. (Translation editor: Fritz John)John Wiley & Sons, New York, 1977.
37. Erwin E, Obermayer K, Schulten K. Self-organizing maps: Ordering, convergence properties and energy functions. Biol Cybern 1992; **67**: 47–55.
38. Utsugi A. Hyperparameter selection for self-organizing maps. Neural Comput 1997; **9**: 623–635.

39. Dayan P. Arbitrary elastic topologies and ocular dominance. Neural Comput 1993; **5**: 392–401.
40. Swindale NV. Coverage and the design of striate cortex. Biol Cybern. 1991; **65**:415–424.
41. Swindale NV, Shoham D, Grinvald A, Bonhoeffer T, and Hübener M. Visual cortex maps are optimised for uniform coverage. Nat Neurosci 2000; **3**:822–826.
42. Silverman BW. Density Estimation for Statistics and Data Analysis. Chapman & Hall, London, 1986.
43. Bishop CM, Svensén M, Williams CKI. GTM: the generative topographic mapping. Neural Comput 1998; **10**:215–234.
44. Goodhill GJ, Sejnowski TJ. A unifying objective function for topographic mappings. Neural Comput 1997; **9**;1291–1303.
45. Miller KD, Keller JB, Stryker MP. Ocular dominance column development: analysis and simulation. Science 1989; **245**:605–615.
46. Goodhill GJ. The influence of neural activity and intracortical connectivity on the periodicity of ocular dominance stripes. Network 1998; **9**:419–432.
47. Lawler EL, Lenstra JK, Rinnooy Kan AHG, Shmoys DB. The Travelling Salesman Problem. John Wiley & Sons, Chichester, England, 1986.
48. Willshaw DJ, von der Malsburg C. How patterned neural connections can be set up by self-organization. Proc R Soc Lond B Biol Sci 1976; **194**:431–445.
49. Willshaw DJ, von der Malsburg C. A marker induction mechanism for the establishment of ordered neural mappings: its application to the retinotectal problem. Philos Trans R Soc Lond B Biol Sci 1979; **287**:203–243.
50. Durbin R, Szeliski R, Yuille A. An analysis of the elastic net approach to the traveling salesman problem. Neural Comput 1989; **1**:348–358.
51. Rose K. Deterministic annealing for clustering, compression, classification, regression, and related optimization problems. Proc of the IEEE 1998; **86**:2210–2239.
52. Goodhill GJ. Correlations, competition, and optimality: modelling the development of topography and ocular dominance. Cognitive Science Research Paper CSRP 226, Sussex University, 1992.
53. Goodhill GJ, Richards LJ. Retinotectal maps: molecules, models and misplaced data. Trends Neurosci 1999; **22**:529–534.
54. Yuille AL, Kolodny JA, Lee CW. Dimension reduction, generalized deformable models and the development of ocularity and orientation. Neural Networks 1996; **9**:309–319.
55. Goodhill GJ. Topography and ocular dominance: a model exploring positive correlations. Biol Cybern 1993; **69**:109–118.
56. Goodhill GJ, Löwel S. Theory meets experiment: correlated neural activity helps determine ocular dominance column periodicity. Trends Neurosci 1995; **18**:437–439.
57. Goodhill GJ, Willshaw DJ. Elastic net model of ocular dominance: overall stripe pattern and monocular deprivation. Neural Comput 1994; **6**:615–621.
58. Goodhill GJ, Bates KR, Montague PR. Influences on the global structure of cortical maps. Proc R Soc Lond B Biol Sci 1997; **264**:649–655.
59. Goodhill GJ, Cimponeriu A. Analysis of the elastic net applied to the formation of ocular dominance and orientation columns. Network 2000; **11**:153–168.
60. Hoffsümmer F, Wolf F, Geisel T, Löwel S, Schmidt K. Sequential bifurcation of orientation- and ocular dominance maps. In: Proc of the Fifth Int Conf on Artificial Neural Networks (ICANN95) Vol. 1. (Fogelman-Soulie F, Gallinari R, eds.). EC2 & Cie, Paris, France, 1995, pp. 535–540.
61. Hoffsümmer F, Wolf F, Geisel T, Löwel S, Schmidt K. Sequential bifurcation and dynamic rearrangement of columnar patterns during cortical development. In: Computational Neuroscience: Trends in Research 1995. (Bower JM, ed.). Academic Press, New York, 1996, pp. 197–202.

62. Wolf F, Geisel T. Spontaneous pinwheel annihilation during visual development. Nature 1998; **395**:73–78.
63. Schwartz EL. Computational studies of the spatial architecture of primate visual cortex: columns, maps, and protomaps. In: Primary Visual Cortex in Primates Vol. 10 of Cerebral Cortex, Ch. 9. (Peters A, Rockland KS, eds.). Plenum Press, New York, 1994, 359–411.
64. Wolf F, Bauer H-U, Geisel T. Formation of field discontinuities and islands in visual cortical maps. Biol. Cybern 1994; **70**:525–531.
65. Hetherington PA, Swindale NV. Receptive field and orientation scatter studied by tetrode recordings in cat area 17. Vis Neurosci 1999; **16**:637–652.
66. Tessier-Lavigne M., Goodman CS. The molecular biology of axon guidance. Science 1996; **274**:1123–1133.
67. Mueller BK. Growth cone guidance: first steps towards a deeper understanding. Annu Rev Neurosci 1999; **22**:351–388.
68. Cheng HJ, Nakamoto M, Bergemann AD, Flanagan JG. Complementary gradients in expression and binding of ELF-1 and Mek4 in development of the topographic retinotectal projection map. Cell 1995; **82**:371–381.
69. Drescher U, Kremoser C, Handwerker C, Loschinger J, Noda M, Bonhoeffer F. In vitro guidance of retinal ganglion cell axons by RAGS, a 25 KDa tectal protein related to ligands for Eph receptor tyrosine kinases. Cell 1995; **82**:359–370.
70. Feldheim DA, Kim Y-I, Bergemann AD, Frisén J, Barbacid M, Flanagan JG. Genetic analysis of Ephrin-A2 and Ephrin-A5 shows their requirement in multiple aspects of retinocollicular mapping. Neuron 2000; **25**:563–574.
71. Goodhill GJ. Dating behavior of the retinal ganglion cell. Neuron 2000; **25**:501–503.
72. Flanagan JG, Vanderhaeghen P. The ephrins and Eph receptors in neural development. Annu Rev Neurosci 1998; **21**:309–345.
73. Feldheim DA, Vanderhaeghen P, Hansen MJ, et al. Topographic guidance labels in a sensory projection to the forebrain. Neuron 1998; **21**:1303–1313.
74. Vanderhaeghen P, Lu Q, Prakash N, et al. A mapping label required for normal scale of body representation in the cortex. Nat Neurosci 2000; **3**:358–365.
75. Crowley JC, Katz LC. Development of ocular dominance columns in the absence of retinal input. Nat Neurosci 1999; **2**:1125–1130.
76. Crowley JC, Katz LC. Early development of ocular dominance columns. Science 2000; **290**:1321–1324.
77. Crair, MC, Gillespie DC, Stryker MP. The role of visual experience in the development of columns in cat visual cortex. Science 1998; **279**:566–570.
78. Gödecke I, Bonhoeffer T. Development of identical orientation maps for two eyes without common visual experience. Nature 1996; **379**:251–254.
79. Wolf F, Bauer H-U, Pawelzik K, Geisel T. Organization of the visual cortex. Nature 1996; **382**:306–307.
80. Weliky M, Katz LC. Disruption of orientation tuning in visual cortex by artificially correlated neuronal activity. Nature 1997; **386**:680–685.
81. Goodhill GJ. Stimulating issues in cortical map development. Trends Neurosci 1997; **20**:375–376.
82. Weliky M, Katz LC. Correlational structure of spontaneous neuronal activity in the developing lateral geniculate nucleus in vivo. Science 1999; **285**:599–604.
83. Smolen P, Baxter DA, Byrne JH. Mathematical modeling of gene networks. Neuron 2000; **26**:567–580.
84. Sharp DH, Reinitz J. Prediction of mutant expression patterns using gene circuits. 1998; Biosystems **47**:79–90.
85. von Dassow G, Meir E, Munro EM, Odell GM. The segment polarity network is a robust developmental module. Nature 2000; **406**:188–192.

16
Multi-Level Neuron and Network Modeling in Computational Neuroanatomy

Rolf Kötter, Pernille Nielsen, Jonas Dyhrfjeld-Johnsen,
Friedrich T. Sommer, and Georg Northoff

ABSTRACT

While quantitative neuroanatomy produces increasingly detailed descriptions of the nervous system at all levels of organization, it remains a major challenge to integrate the information from cell, population, and systems levels in a mutually informative way. Computer simulation is a powerful tool for exploring links between structure and function of cells, tissues, and organs. It is based on mathematical models that capture the essence of anatomical, physiological, and behavioral observations, and rely on accurate quantitative descriptions. Previous computer models, including much of our own work, have focused on relating structural and functional data at nearby descriptive levels. Here, we discuss concepts in linking cell, population, and systems level models of the cerebral cortex. We propose a strategy for building multilevel computer models that integrate elements from compartmental neuron models, microcircuit representations of neuronal populations, and activity propagation in large-scale neuronal networks. As a working example, we simulate activity propagation in the primate visual cortex, with the aim of relating neuronal activity to cortical activation patterns and onset response latencies to the structure of the underlying anatomical network. This computational approach provides new insights into the functional anatomy of the visual cortex.

16.1. INTRODUCTION

Computational neuroanatomy is concerned with the mathematical formulation of structure–function relationships at the levels of neurons, neuronal populations, and brain systems, with the aim to gain systematic and comprehensive insights into the mechanisms governing brain functions. During the second half of the 20th century, more and more refined morphological and electrophysiological investigations have provided us with a wealth of insights into cellular and synaptic mechanisms of single neurons. Extant examples of resulting concepts are the cable theory of passive dendrites *(1)*, the Hodgkin-Huxley formulation of the relationship between action potential, ionic conductances and membrane particles *(2)*, and the compartmental modeling

From: *Computational Neuroanatomy: Principles and Methods*
Edited by: G. A. Ascoli © Humana Press Inc., Totowa, NJ

approach to complex branched neurons *(3)*. We are now capable of simulating the detailed electrophysiological behavior of single neurons based on full morphological reconstructions and realistic distributions of a dozen of different membrane conductances (e.g., *[4]*). Computational models of neuronal morphology provide insights into the physiological consequences of neuronal size differences as they occur in ontogeny *(5)* and phylogeny *(6)*. Models of intracellular processes address the flow of calcium and other ions as they affect the release of transmitters, the modulation of currents, or the activity of complex signaling pathways *(7–10)*. By contrast, large-scale network models have most commonly addressed the principal computations and dynamics of comparatively simple networks with regular or random connectivity, e.g., *(11–13)* or the implementation of high-level cognitive functions in abstract neural architecture *(14,15)*. They tend to ignore, however, the specific anatomy of cortical networks, which is neither uniform nor random *(16)*, and which puts constraints on global activation patterns and information processing in this complex structure *(16,17)*.

Despite this dichotomy of bottom-up and top-down approaches, neuronal, systems and mental phenomena are inextricably intertwined. Thus, we need to explore the brain mechanisms that relate between detailed anatomy and global functions. It was not before the 1990s, however, that systematic and comprehensive collations of real brain wiring, combined with multivariate statistical approaches, raised a fresh interest in the global anatomical organization of the brain and the functional constraints revealed by its architecture. Based on anatomical connectivity data such investigations have demonstrated, for example: (*i*) the hierarchical organization of the visual system *(18,19)*; (*ii*) its global division into a parietal stream concerned with spatial orientation and a temporal stream specialized in object recognition *(20,21)*, and the differentiation of prefrontal cortex into orbitomedial and lateral areas *(22,23)*; (*iii*) the "small world connectivity" properties of anatomical and functional cortical networks *(21,24)*; and (*iv*) the predictive value of inter-area connectivity for activity propagation in cortical networks *(17)*. The disentangling of the brain's wiring and the functional annotation of its circuit diagrams may turn out to be of similar importance as the sequencing and annotation of the human genome to molecular biologists, or the wiring diagram of a man-made machine to a service technician *(25)*. Functionally annotated wiring diagrams will be required at several levels of detail; patients, clinicians, and neuroscientists with their different view points and questions require information on brain mechanisms that range from a global sketch of interactions between major brain systems *(26)* via the task- and time-dependent activation within distributed networks *(27)* to the detailed mechanisms of interacting processing steps *(28)*. As a specific example, functional brain imaging studies have shown spatially distinct activation patterns under different task conditions; to predict and influence such activation patterns, however, we need to know the functional impact of anatomical connections, the relative contribution of excitation and inhibition, the role of background and anticipatory activity, the sequence of information processing, and the learning of regularities, etc. (see, e.g., *[29–31]*).

This brings us back to the issue of the relationships between different levels of organization ranging from the molecular and cellular to ensemble, area, and systems. Mainstream thinking in neuroscience tends to classify scientists and their work according to the organizational level that they predominantly address. Thus, the categories of molecular, cellular, systems, and behavioral approaches dominate the division into sections of learned societies, job descriptions, conference presentations, and publications in general neuroscience journals. Although it is recognized that these categories do not define independent domains, it requires a substantial departure from predominant habits to target their conceptual integration and to address the question of how one level constrains or informs another. Furthermore, the tendency to categorization and focusing on a single level of organization is not unique to experimental approaches, where it could be argued that it results from the dependence on specialized and expensive equipment. Counterintuitively, this categorization is also noticeable in theoretical and computational approaches that combine relatively homogeneous components in mainstream approaches with a certain conceptual consensus *(32)*. Thus, there are different computational tools for molecular, cellular, network modeling, or statistical analyses, which reinforce the formation of communities that make use of only one subset. The Hodgkin-Huxley formalism is an exceptional example of a multilevel concept that provides an explanation of how the properties of membrane components determine the electrical behavior of entire neurons. A similarly integrative concept is required that would link neuronal activities within a spatially distributed network to functional interactions and global activation patterns in the brain

Cross-level integration is not a simple task: as far as the relationships between phenomena at different levels are known, they are nonlinear and highly complex as, for example, the relationship between neuronal activity and tissue perfusion. Thus, it may be difficult to express them in mathematical terms that are simultaneously simple and comprehensive. A first step towards the specification of such relationships is the detailed investigation of the influence that parameters at one level have on variables at another. Here, the problem is that many crucial variables can only be studied to a limited extent or are not amenable to experimental observation due to technical or ethical limitations. Computational approaches have the advantage that the parameters can be freely manipulated and that all variables of the model are open to inspection. In addition, computational models are, in principle, fully reproducible and can be reanalyzed at later times in the light of further insights. What is needed, then, are computational tools that support and thereby foster multilevel investigations. Ideally, such tools would allow flexible implementation of any kind of mathematical description and, simultaneously, be intuitive and specific enough to represent the keystones of neuroscientific concepts. This twofold ideal is to some extent contradictory as reflected by the division of simulation tools into general-purpose numerical integrators (such as XPP) and dedicated biochemical (GEPASI) or neural simulation systems (such as NEURON or GENESIS; for a comprehensive list see http://www.hirnforschung.net/cneuro/). In fact, there are important considerations of model complexity, temporal resolution, and conceptual compatibility that complicate an integration of models, but which are not insurmountable obstacles. From a technical point of view, the development of flexible object-oriented simulation environments implementing a unified description language for neuroscience objects are desirable *(33)*. In addition, the conceptual issues of multilevel integration have to be addressed. In a previous paper *(9)*, we provided a framework and examples of an integration of biochemical and biophysical modeling using both general-purpose and dedicated simulation environments. Here, we advance the integration of detailed neuronal models with more abstract area and network models for investigation of brain mechanisms underlying activation patterns in the cerebral cortex.

16.2. MODELS OF NEURONS AND NEURONAL POPULATIONS

16.2.1. Neuron Models

Neurons have highly complex morphological, electrophysiological, and biochemical properties as reflected in the many morphological cell types, the repertoire of activity patterns, or the multiple intracellular signaling pathways. Out of these complex phenomena, neuron models represent an extract of features that are thought to be of particular significance. What exactly are the most significant features is a matter of debate. Perhaps the most fundamental distinction in neuronal modeling relates to the divergent intentions of reconstructing their full behavior vs distilling their computational principles. This distinction has produced two different approaches in the history of neuronal modeling: biophysical models and formal (computational) models *(34)*. Further reasons for using different models and different software implementations include detail of available data, ease of implementation, run-time efficiency, and availability of supporting software modules for input, output, and analysis. Clearly, no single implementation is ideal in all respects so that there is scope for variety.

Biophysical models are models of the physical reality in the usual sense. One prominent biophysical model has already been mentioned, the quantitative description of membrane currents and its application to excitable processes in neurons by Hodgkin and Huxley *(35)*. With growing computational resources and increasing amounts of empirical data over the last decades, biophysical neuron models have become more and more complex. Some neuronal models retain the degree of morphological detail obtained with 3D reconstruction techniques following intracellular dye filling *(4)*, or attempt to extract parameters that allow a full realistic reconstruction of the type of neuron *(36)*. Most biophysical neuron models focus on electrophysiological features, which they capture in various forms and in more or less detail *(37,38)*. In our own work, we have built a variety of biophysical neuron models from simplified representations of voltage fluctuations at the soma *(39)* via realistic dendritic trees and spatially distributed conductances *(6)* to interacting biochemical pathways in a single dendritic spine *(9,40)*.

Dedicated simulation software tools, such as GENESIS and NEURON, are well-established means of implementing biophysical neuron models. Standard implementations using such software consist of one or more neuronal compartments representing soma and dendritic segments. Each compartment has passive membrane properties expressed by the compartment's conductance, transmembrane resistance, and axial resistance connecting it with any proximal and distant compartments. In addition, active membrane properties can be added, usually modeled as voltage-dependent ionic conductances. These use variations on the original formulations by Hodgkin and Huxley *(35)* and Connor and Stevens *(41)*, although other formulations are available *(38)*. The axon is for most purposes regarded as a simple delay-line and, therefore, not represented explicitly.

Formal models of neurons and neuronal networks are based on the computational paradigm in neuroscience. This states that the function of the brain and the nervous system is information processing, which can be described by mathematical algorithms. Formal models address the question of how a particular algorithm is implemented in neuronal biophysics. On the level of single neurons, it has been asked how information is coded and processed *(42–44)*. The viewpoint of formal modeling also led to extremely abstract neuron models. In their seminal work, McCulloch and Pitts *(11)* made the most extreme sketch of a neuron by reducing it to a simple threshold gate. Between this extreme and the sophisticated versions of biophysical neuron models, various compromises have been proposed. Examples are the reduction of real neurons to "the simplest type ... that can represent a given neuron in computational form": the "canonical" neuron model *(45)*; the two-compartment model of Pinsky and Rinzel *(46)*, or the leaky integrate-and-fire (I&F) unit *(34,38)*.

In extension of their original purpose, the neuronal simulation tools mentioned above can also be used to implement simpler neuron models. For example, a neuron can be represented as a single passive membrane compartment affected by synaptic input to excitatory and inhibitory synapses. Instead of using active conductance models for simulating action potentials, a simple threshold detector may be used to generate events that would be relayed to the postsynaptic cells with the appropriate conduction delays. Even post-spike polarization of the membrane potential can be modeled by feeding back the action potential event to a synapse on the neuron itself. Such an implementation approximates the I&F unit within the framework of standard neuronal simulator software. The question, of course, arises whether such a simplified compartment model should still refer to the units of real membrane voltage, synaptic conductance, etc., or use some other convenient scale, such as voltage above resting level or graded membrane activation between zero and one.

16.2.2. Neuronal Population and Area Models

Models of neuronal cell populations rely on simplifications to make large networks of neurons amenable to mathematical treatment or to numerical simulation within a reasonable amount of real time. One common approach of neuronal population modeling sketches the single cell by a simplified neuron model. Such models can be justified by the universality hypothesis, a common assumption of statistical physics theories and confirmed in many fields of physics (for example, phase transitions in solid state physics). It states that under certain conditions, the collective behavior of many particle systems (the neurons are considered as the particles) is insensitive to the details of the single particle model. From the variety of studies, we only pick a few examples: as a biophysical model for the collective behavior of cells in hippocampal slices, complex cable models have been proposed that are able to reproduce, for example, collective bursting, oscillations, and synchronization *(47)*. By manipulation of the model, one can single out mechanisms underlying the collective behavior. Over the last 20 years, computational modeling studies proposed a variety of algorithms that can be implemented in neural substrate. Perhaps the first study of this kind was the one by McCulloch and Pitts *(11)*, who used their binary neuron model to show at an abstract level that the nervous system is capable of implementing a Turing machine. Networks of similar binary neurons have been shown to perform associative computational functions, as originally proposed by Hebb *(48)*. These simple neural associative memories could even be treated analytically *(49–52)*. Synchronization of periodic firing in neuronal networks has been proposed as a coding mechanism for information processing *(53)*.

Another common way of coping with large networks of neurons is to lump together the properties of a set of neurons into a single component. In this kind of model, a unit represents not a single cell, but a microcircuit, a cortical column, or even a cortical area. An extreme example is the biophysical model for propagation of epileptiform activity in cortical networks *(17)*. The network model was built from a systematic collation of experimental tracing data and represented cortical areas as simple binary threshold units. Lansner and Fransen *(54)* proposed cortical memory models using the full connectivity of Hopfield networks. Full connectivity is unrealistic if the units are identified with single cells, but becomes plausible if they represent cortical columns as in their model.

All network models mentioned above treat network connectivity as if it was static. This view assumes that the timescale of synaptic plasticity is much slower than that of the neuronal dynamics. Under this assumption the neuronal and synaptic dynamics decouple, in other words: neuronal behavior can be studied in a frozen synaptic structure. This traditional view of neural computation, motivated by the postulate that synaptic plasticity is relevant to long-term memory, is a prerequisite for the analytical treatment of neuronal population dynamics *(49)*. Of course, the dynamics depend on the properties of the fixed synaptic connectivity structure. If the synaptic matrix is symmetric, i.e., if each synaptic contact is reciprocated with a synapse of the same strength, then the dynamics has only fixed point attractors, and all network states converge to stable patterns. In asymmetric networks, dynamical states can also converge to periodic attractors, which are pattern sequences similar to the synfire chains described by Abeles *(55)*.

Synaptic connectivity can be altered by learning processes. For associative memory, Hebb *(48)* proposed a plausible and, as it turned out later, very efficient synaptic learning rule. It prescribes an increase of synaptic strength if pre- and postsynaptic activities coincide (see *[51]* or a discussion of local learning rules). Hebb's learning rule had considerable impact in computational approaches, but it took decades until its existence was confirmed experimentally *(56)*. In fully connected networks and networks with symmetrically diluted synaptic contacts, the original Hebb rule leads to symmetric synaptic connectivity. Modified Hebbian rules have been proposed where not coincidence, but a certain delay between pre- and postsynaptic activity during learning causes optimal synaptic increase *(57)*. These rules result in asymmetric connectivity and can be used for learning of pattern sequences.

Recent investigations suggest that synaptic plasticity is not limited to long-term processes. Neuronal connections also show short-term plasticity, both as depression and as facilitation (see for instance *[58]*). Modeling studies have demonstrated that the coupling of fast synaptic and neuronal dynamics can enrich the repertoire of computations in visual cortex, for example by band-pass filtering and phase shifting *(59)*. However, it is still justified to study the decoupled neuronal dynamics when only influences of synaptic contact structure and long-term plasticity are concerned.

16.2.3. Interfacing Different Models

An important assessment, whether or not a particular model simplification can be justified, is to compare and to link together models with different simplifications, or with different degrees of simplification. Here, we give some examples for such an interfacing of models.

When implementing synchronized network oscillations in a formal model, it is important to know to what extent one may simplify the model. Wennekers et al. *(60)* found that in a random network, the collective oscillatory behavior disappears if the spike mechanism in I&F units is further simplified, for instance, to a spike rate model. Also, finite size effects were found to be important, suggesting that analytical results calculated for an infinite network size do not necessarily apply to small networks. The biophysical model of hippocampal slices by Traub et al. *(47)* consisted of complex cable model neurons with 18 compartments. Pinsky and Rinzel *(46)* derived a model of two-compartment point neurons that reproduced all the relevant biophysical properties, i.e., spiking, bursting, and time-locked γ activity by AMPA receptor-mediated synaptic coupling. The reduced biophysically constrained Pinsky and Rinzel neuron model was suitable to formal modeling exploring cortical synaptic memory under more realistic conditions *(61)*.

Linking simplified neuron and population models introduces further assumptions. For example, reduced dendritic complexity may affect the location, impact, and interactions of dendritic synapses. When reducing the number of neurons in a population model, one has to be aware of a potential increase in behavioral uniformity, which may affect the conclusions on network dynamics. For this reason, large-scale realistic simulations may be required *(62)*. Productivity and insights would be greatly enhanced in the long run, if previous implementations could be reused and be interfaced with others and if the components of a network model could be substituted individually in a modular manner (see *[33]*). For example, Lansner and Fransen refined their original model replacing the single-unit model of a cortical column *(54)* by a circuit of Hodgkin–Huxley type models *(63)*. Clearly, such modularity facilitates the evaluation of effects incurred by more detailed, more simplified, or alternative models for the respective components, and it will help to refine our concepts of brain function.

16.3. MULTILEVEL MODELING OF VISUAL CORTEX

As a practical example of the considerations above, we consider a model for exploring activity propagation in the visual cortex. In particular, we were interested in the mechanisms that determine the onset response latencies in the areas of the visual cortical network to a flash of light as studied experimentally by Schmolesky et al. *(64)*. We realized that for this we had to combine detailed information on spike timing with simplified models of cortical microcircuitry and with large-scale networks defined by anatomical connectivity. These components reflect very different types of data and are difficult to reconcile, both in terms of concepts and implementation. For example, we do not have sufficient information and capacity to implement each visual area in anatomical and physiological detail; on the other hand, we need detailed mechanisms that respond to a set of spike times from single neuron recordings in the thalamus. In our

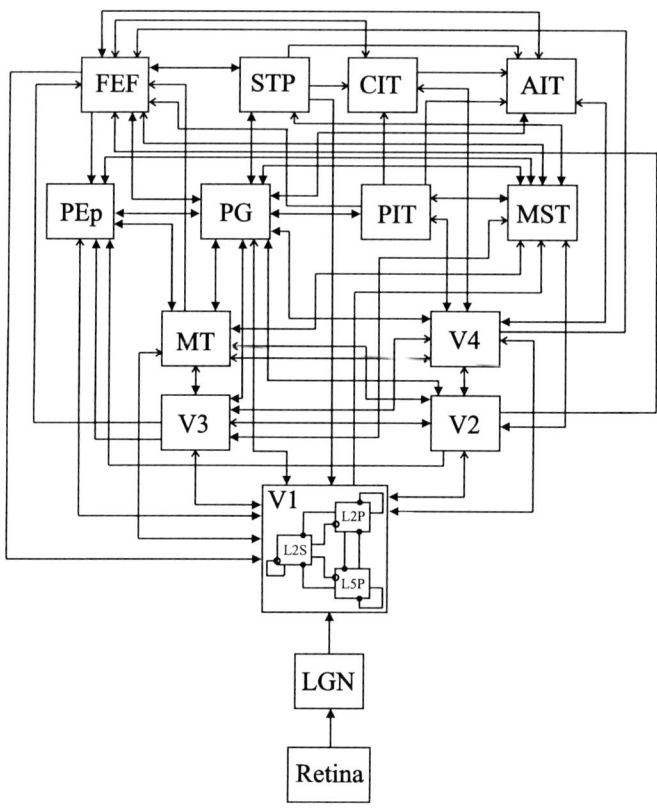

Fig. 1. Schematic drawing of the simulated visual network. Activity propagated from the retina via the thalamic LGN to V1. Area V1 was modeled as a microcircuit consisting of three interconnected neuronal populations indicated in the block diagram: V1–L2P, population of excitatory pyramidal and spiny stellate cells in layers 2–4 of area V1; V1–L2S, population of smooth inhibitory cells in V1; V1–L5P, population of layer 5–6 pyramidal cells in V1. Filled circles represent excitatory synapses, open circles inhibitory synapses. The remaining visual areas are indicated by squares with arrowheads for excitatory inter-area synapses. The laminar specificity is indicated only when terminating on a layer 4 unit (feedforward and lateral interconnections between MT and V4) by the ↑ arrowhead. Interlaminar excitatory and self-inhibitory actions of the cortical units are not shown. V2, V3, V4, visual areas 2, 3, and 4, respectively; MT, middle temporal area (= V5); MST, medial superior temporal area; STP, superior temporal polysensory area; PEp, caudal and medial superior parietal lobule; PG, caudal inferior parietal lobule; AIT, CIT, PIT, anterior, central, and posterior inferotemporal areas, respectively; FEF, frontal eye field.

multilevel approach, we combined a biophysical microcircuit model (neurons with Hodgkin–Huxley mechanisms) of primary visual cortex (V1) with a computational model (I&F units) of the remaining visual cortical network.

Altogether, our model represented 15 visual system components based on some morphological and electrophysiological properties of the cells, layers, areas, and their interconnections (see Fig. 1). Areas were represented in GENESIS at different levels of detail: the simplest representation was the I&F unit for areas PEp (caudal and medial superior) and PG (caudal inferior) of the parietal lobule. These were modeled as a single passive compartment, but extended with an inhibitory autapse (synapse onto itself) for resetting its membrane potential to a low value after suprathreshold activation. A more detailed area model comprised three such I&F units, representing supragranular, granular and infragranular layers of the visual areas 2, 3, and 4 (V2, V3, V4, respectively), middle temporal area (MT), medial superior temporal area (MST), superior temporal polysensory area (STP), anterior, central, and posterior inferotemporal areas (AIT, CIT, and PIT, respectively), and frontal eye field (FEF). A "canonical" microcircuit model represented three excitatory and inhibitory neuronal populations within primary visual area V1. This microcircuit implementation received its input from the model of a cell in the magnocellular part of the thalamic lateral geniculate nucleus (LGN), which in turn responded to pulse input thought to originate in the retina in response to a visual stimulus. Delay lines and excitatory and inhibitory synapse objects transmitted information between the models of areas, layers, and cells. The connectivity between the units was taken from systematical collations of experimental tracing studies in the macaque monkey brain (*[18]* through the CoCoMac database [www.cocomac.org]). The following sections describe the components of the model in detail.

16.3.1. Stimulus Representation

In accordance with the experimental paradigm of Schmolesky et al. *(64)*, we approximated the retinal response to onset and offset of a light flash of 500 ms duration by two 40-ms pulses generating depolarizing current injections into the model of a visual thalamic neuron. The latter was derived from the implementation of a fast spiking cell in area V1 with Na$^+$ and K$^+$ conductances (see V1–L25; Subheading 16.3.2.), whose characteristics provided a good starting point for tuning. Additional background activity was approximated by uniformly random 50 Hz synaptic input to the proximal dendritic compartment. Altogether, the input parameters were tuned such that the average firing rate of the thalamic model (see Fig. 2; LGN) matched histograms from experimental single-unit recordings in the magnocellular (M) layers of the LGN of anesthetized monkeys as shown in Figure 1A of Schmolesky et al. *(64)*. The omission of the parvocellular stream seems justified by our focus on onset response latencies, which are mediated by the magnocellular stream under most circumstances *(64–66)*.

16.3.2. Microcircuit Representation of Primary Visual Cortex

V1 was represented by a canonical microcircuit model, which Douglas and Martin *(67)* proposed as the basic building block of (neo)cortical microcircuitry. This model consisted of three interconnected multicompartmental neuron models, each taken as representative of one or more cell populations: infragranular pyramidal cells (V1-L5P), inhibitory interneurons (V1-L2S), and a combination of supragranular pyramidal cells and layer IV spiny stellate cells (V1-L2P). The dimensions of the compartments representing each cell were taken from Figure 14 in the paper by Douglas and Martin *(67)* (see Table 1).

For all cells, the following passive parameters were used: ra = 2.0 Ohm•m *(68)*, rm = 1.0 Ω•m^2 *(67)*, cm = 0.02 F•m^{-2} *(68)*, Vm = –0.05 V *(67)*, with the exception of the fast spiking cell, where cm = 0.01 Farad•m^{-2}. In addition, each compartment was furnished with the described voltage-gated *(68)* and synaptic conductances *(67,69)*. Since the voltage-gated conductances were reported as being similar to those of Traub et al.

Table 1
Dimensions of Compartments in Cells of the Microcircuit Model (67)

Compartment	Dimension	V1-L5P	V1-L2P	V1-L2S
Soma	diameter	27.3	18.2	25.0
	length	27.3	18.2	25.0
Proximal dendrite	diameter	52.3	15.9	13.6
	length	183.3	144.4	144.4
Medial dendrite	diameter	25.0	4.5	—
	length	127.8	100.0	—
Distal dendrite	diameter	4.6	1.8	4.5
	length	1133.2	100.0	150.0

All units are µm.

Table 2
Conductance Densities in S•m^{-2} of Voltage-Gated and Synaptic Channels in the Cells of the Microcircuit Model (67,68)

Conductance	V1-L5P	V1-L2P	V1-L2S
NaF (soma)	4000	4000	7000
KDR (soma)	800	800	4000
KA (soma)	20	20	—
KAHP (soma)	150	150	—
Ca (soma)	5	5	—
GABA$_A$ (soma)	10	5	5
GABA$_A$ (prox. dend.)	10	5	5
GABA$_B$ (prox. dend.)	5	5	5
ex_syn (prox. dend.)	5	10	10
AMPA (prox. dend.)	80	80	80

(47), we used the already existing GENESIS implementation of the channels from this study. The pyramidal cells had a fast sodium conductance (NaF), a delayed rectifier potassium conductance (KDR), a transient potassium conductance (KA), a calcium-dependent potassium conductance (KAHP), and a high-threshold calcium conductance (Ca). To model the interactions between the Ca and KAHP conductances, calcium concentrations were simulated in a shell of 0.2 µm diameter beneath the somatic cell membrane. These concentrations were then utilized as a parameter for the gating of the KAHP-conductance. The inhibitory interneurons were modeled as possessing only the NaF and KDR voltage-gated conductances. The conductance densities for calculating the maximum conductance values are given in Table 2.

The three cells of the microcircuit model were connected as shown in the block representing V1 in Figure 1. Since each cell model effectively represented an entire cell population, the interactions within a population were modeled by feedback connections of the cell model to itself. This approach ignores the intrinsic and connec-

Table 3
Weights of Synapses Connecting Source Cells (1st Column) to Target Cells (1st Row)

Source\Target	V1-L5P	V1-L2P	V1-L2S
V1-L5P	0.5	0.5	0.5
V1-L2P	0.5	0.5	0.5
V1-L2S	1	1	1

Table 4
Synaptic Parameters of the Microcircuit Model According to Douglas and Martin (67) and Suarez et al. (69)

Synaptic	E_{rev} (V)	τ_1 (ms)	τ_2 (ms)
Conductance			
$GABA_A$	−0.06	10.0	10.0
$GABA_B$	−0.08	80.0	40.0
AMPA	0.0	4.5	1.8
ex_syn	0.0	4.5	1.8

tional heterogeneity of the neuronal populations, but it can be easily extended to account for additional features (see [69]).

Inhibitory synaptic input originating from the inhibitory interneurons was mediated by γ-aminobutyric acid (GABA)$_A$ (soma and proximal dendrite) and GABA$_B$ (proximal dendrite) synapse models in all cells, whereas the excitatory input from both superficial excitatory neurons and deep pyramidal cells were mediated by AMPA-type glutamate synapse objects attached to the proximal dendrites. The effective conductance resulting from a synaptic input was calculated as the product of the time-dependent conductance and the synaptic weight for each connection as given in Table 3.

The excitatory synaptic input from sources outside of area V1 (LGN and nonprimary visual cortical areas in the network model) was mediated by the ex_syn synapse models located on the proximal dendrites of all microcircuit cells with a total synaptic weight of 1.

All synaptic conductances were modeled as dual-exponential functions describing the evolution of the conductance in time after activation (70). Their time constants and reversal potentials are shown in Table 4.

When the membrane potential in the somatic compartment of a cell crossed the threshold of 0 mV, a spike generator object triggered synaptic activation of the postsynaptic neurons with the appropriate delay. Synaptic delays of all axonal connections between cells in V1 were set to 2 ms as specified in the original publication (68).

16.3.3. Network Implementation

Further visual areas were implemented by three I&F units representing supragranular, granular, and infragranular layers of areas within the visual cortical network described by Felleman and Van Essen (18). In distinction to their map of visual cortex, we did not differentiate between dorsal and ventral parts of inferotemporal areas. In addition, two areas peripheral to the visual cortex, PG and PEp of the parietal cortex, were represented by single I&F units.

Each I&F unit consisted of a single passive compartment model of unitary dimensions with a resting membrane potential (Em) of 0.0 V, a membrane capacitance (Cm) of 0.5 nF, and a membrane resistance (Rm) of 40 MΩ. In contrast to the cells in the microcircuit model, action potentials were not modeled explicitly in the I&F units. The compartment sent continuous membrane potential output to a spike generator with a threshold for spike generation of 0.5 V and an absolute refractory period of 3 ms. Each compartment was targeted by an excitatory and an inhibitory synapse. In the current model, all the projections between layers or areas were excitatory, since they generally originate from pyramidal neurons. The excitatory synapse models represented fast AMPA-mediated activation approximated by the following parameters: $E_{rev} = 1.0$ V, $g_{max} = 0.8$ μS, $\tau_1 = 2.3$ ms, and $\tau_2 = 0.1$ ms. The inhibitory synapse model implemented slow self-inhibition of an area and thus used the parameters of the $GABA_B$ synapse models in the microcircuit model ($E_{rev} = 0.0$ V, $g_{max} = 1.0$ μS, $\tau_1 = 40$ ms, and $\tau_2 = 80$ ms, delay = 0.0 s, weight = 1.0). Thereby, the membrane potential of the unit was suppressed to a low value following its suprathreshold activation.

The connectivity within the visual cortex model was derived from several sources. Connections between the layers of the three-tiered areas adopted the basic scheme described by Callaway *(71)*: the granular layer 4 unit projected to the units representing supragranular layers 2/3 and infragranular layers 5/6. The supragranular unit, in turn, projected to the infragranular unit, as well as back to the granular unit. Similarly, the infragranular unit connected to the supragranular unit and to the granular layer 4 unit. Information on inter-area projections, including laminar information, was obtained from *(18)* as collated in the CoCoMac database (www.cocomac.org). Connection strengths were normalized such that the sum of all excitatory inputs to a unit equaled a weight of one, with equal contributions by all its intrinsic and extrinsic afferents. The delays between layers and/or areas were simplified using the same fixed value of 15 ms, which roughly approximates experimental observations *(72)* (see, however, *[56,73]*). Since the network was quite large and complicated, we implemented tools within GENESIS that generated the units and their interconnections automatically from CoCoMac output.

The simulation time step for numerical integration using the standard exponential Euler algorithm was 10 μs. Total simulated time was 1.5 s. This comprised a 0.5-s period preceding the visual stimulus for dissipation of artifacts resulting from initialization of objects and parameters (omitted from the displays). Visual stimulation was taken to start at time t = 0 and to last for 0.5 s as in the experiments of Schmolesky et al. *(64)*. Due to appropriate tuning of its input, the LGN model showed corresponding onset and offset responses for propagation to the visual cortical network.

16.4. RESULTS

The simulated network showed stimulus-correlated propagation of activity from LGN to V1 and all other areas. Histograms, of spike time distributions over 100 runs with random background activity of the LGN model (Fig. 2), confirmed that the firing

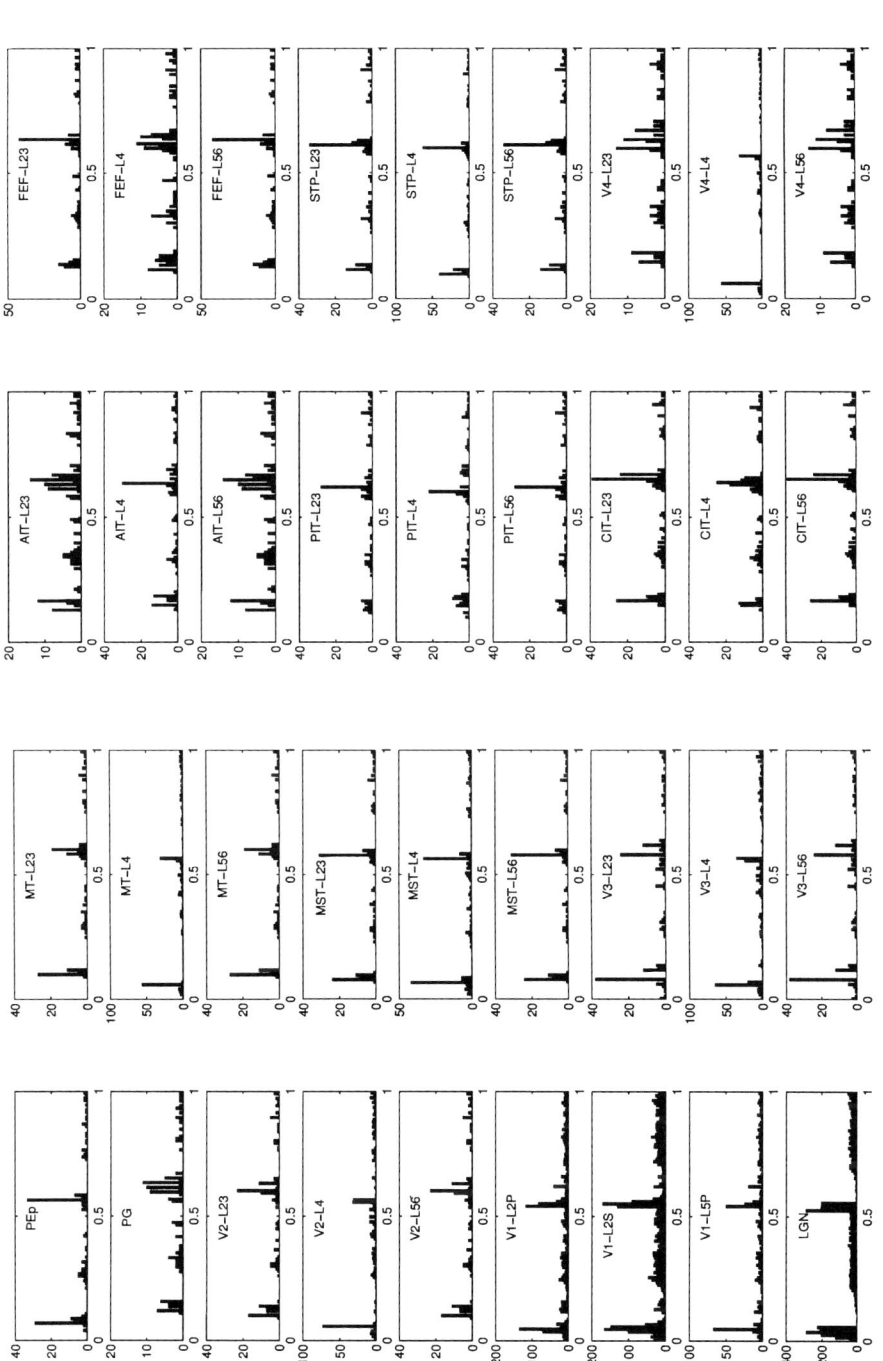

Fig. 2. Histograms of temporal distributions of simulated action potentials in LGN and all units of the 13 cortical areas to visual stimuli lasting from 0–0.5 s (abscissa shows time in s). Acronyms L23, L4, and L56 refer to supragranular, granular, and infragranular layers, respectively. Responses were averaged in bins of 10 ms width over 100 simulations with random seeds. The distribution in LGN was optimized to fit experimentally recorded average frequencies observed in the magnocellular layers of LGN to a corresponding visual stimulus (see Fig. 1 in [64]).

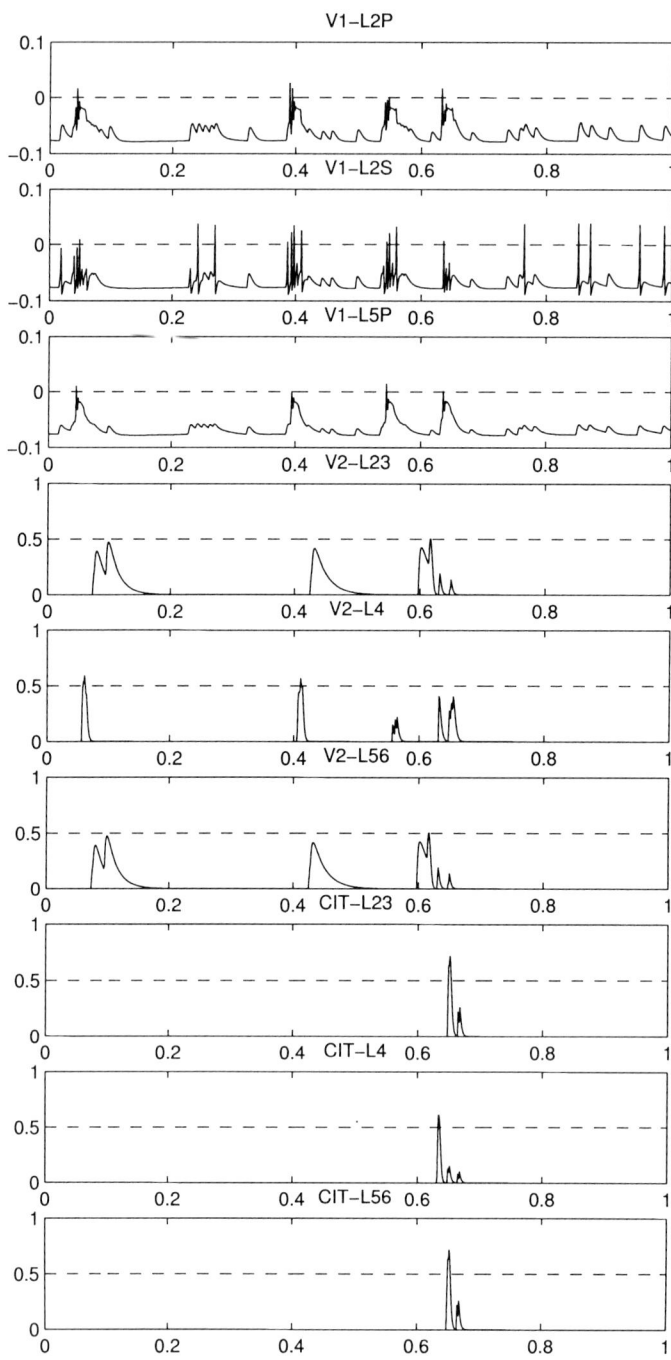

Fig. 3. Sample of simulated voltage traces of the three microcircuit cells in V1, and of the three I&F units (layers) in visual areas V2 and CIT to a visual stimulus lasting from 0–0.5 s. Abscissa shows time in s, ordinate gives membrane potential in V. The hashed horizontal lines indicate the action potential detection threshold. Note that V1 components display realistic membrane potentials, whereas the area models have activation levels in the range of 0–1 and no action potentials. For explanation of abbreviations see legends to Figures 1 and 2.

pattern of the thalamic unit corresponded to the single unit recordings from the magnocellular part of the LGN *(64)*. Subsequently, all cells in the V1 microcircuit responded with activity that had characteristics of both magno- and parvocellular stream neurons when compared to Fig. 1C,D in *64*. The remaining layers and areas of the network showed a considerable variation in the latencies and distributions of their responses. The peaks in the histograms of layer 4 responses occurred before those in the supra- and infragranular layers, with the exception of inferotemporal areas and FEF, where they were nearly simultaneous. Units with direct input from the V1 microcircuit showed activity preceding this peak triggered by ongoing activity in V1, which was not strictly stimulus-coupled.

More detailed inspection of a single run of the simulation showed realistic membrane potential fluctuations in the cells of the V1 microcircuit. Irregular excitatory postsynaptic potentials (PSPs) occasionally summed up and triggered action potentials (Fig. 3). Slow inhibitory PSPs caused by $GABA_B$ synaptic activation resulted in hyperpolarized periods. The inhibitory population unit (V1-L2S) typically repolarized very quickly and fired more frequently. Supra- and infragranular pyramidal population units (V1-L2P and V1-L5P, respectively) showed broader membrane potential depolarizations with decremental fluctuations that led to occasional bursts in the former and single action potentials in the latter. The difference between these two types of cells was explained by the smaller size and higher input resistance, as well as the smaller conductance of the $GABA_A$ inhibitory input of the supragranular model neuron. Altogether, these features were compatible with the observations by Douglas and Martin *(67)* concerned with the first response in V1 neurons to stimulation of thalamocortical afferents.

The I&F units had simplified membrane potential fluctuations, which were fewer, broader, and did not produce action potentials in accordance with the built-in properties of these units (Fig. 3, V2 and CIT). Note that these units did not always respond to the on period of the simulated visual stimulus (0–0.5 s) as shown in this sample of CIT potentials.

Computational modeling allows investigation of variables beyond those that would be simultaneously amenable to experimental investigation. Figure 4 juxtaposes spike events, as well as excitatory and inhibitory synaptic currents and conductances from the same run in the I&F unit representing the granular layer of area V2. Although only two spike doublets occurred during stimulus presentation, the slow self-inhibitory conductance still led to large inhibitory currents when the unit was depolarized by subsequent excitatory events. In this example, the second spike doublet thereby effectively cancelled a stimulus offset response in V2-L4. This plot also gives an impression of differential occurrence and time course of somatic action potentials and excitatory and inhibitory synaptic events.

As a crude marker of activity propagation in this visual network, we measured onset response latencies defined as the period between stimulus onset (t = 0.0 s) and the first subsequent spike in the respective unit (see Fig. 5). The sequence of latencies showed a general feedforward pattern of activity propagation. Median onset response latencies were compatible with a division of visual areas into four groups: (*i*) LGN and V1; (*ii*) the granular layers of V2, V3, V4, MT, and MST, i.e., laminar units directly activated from V1; (*iii*) PEp, STP, and extragranular layers of areas in the second group, except V4; and (*iv*) inferotemporal areas, PG, and extragranular layers of area V4. Particularly

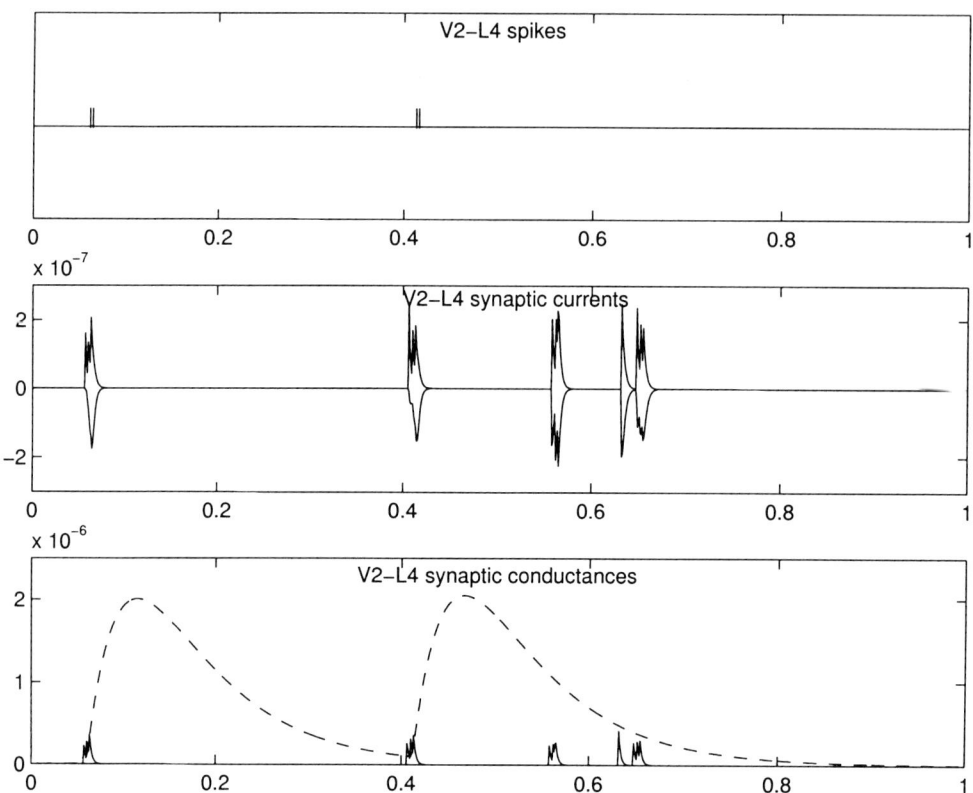

Fig. 4. Sample of simulated spike times, excitatory and self-inhibitory synaptic currents and conductances in layer 4 of area V2 to a visual stimulus lasting from 0–0.5 s. Abscissa shows time in s, ordinates are in Ampere for currents, and Siemens for conductances. Excitatory and inhibitory currents are drawn as positive and negative values, respectively. Excitatory and inhibitory conductances are indicated by full and hashed lines, respectively.

intriguing were the simultaneity of responses across all layers of inferotemporal areas and the huge discrepancy in area V4 between onset latencies of granular and extragranular layers.

The absolute latency values covaried, of course, with the interunit delays (not shown). The sequence of activation among the areas, by contrast, was relatively independent of the interunit delays. What we noted, however, was that the sequence of laminar activation within an area was sensitive to variations of interunit delays. To investigate this issue more thoroughly, we defined a formal index, which we refer to as the "Hierarchical Activation Index" (HAI):

$$HAI = \frac{lat(\hat{L4})}{lat(\hat{L4}) + lat(L4)}$$

where $lat(L4)$ is the mean latency of the first layer 4 spike, and $lat(\hat{L4})$ is the mean of the onset latencies in the supra- and infragranular layers of a given area. A HAI value > 0.5 corresponds to activation of layer 4 before the extragranular layers, which is a

Multilevel Neuron and Network Modeling

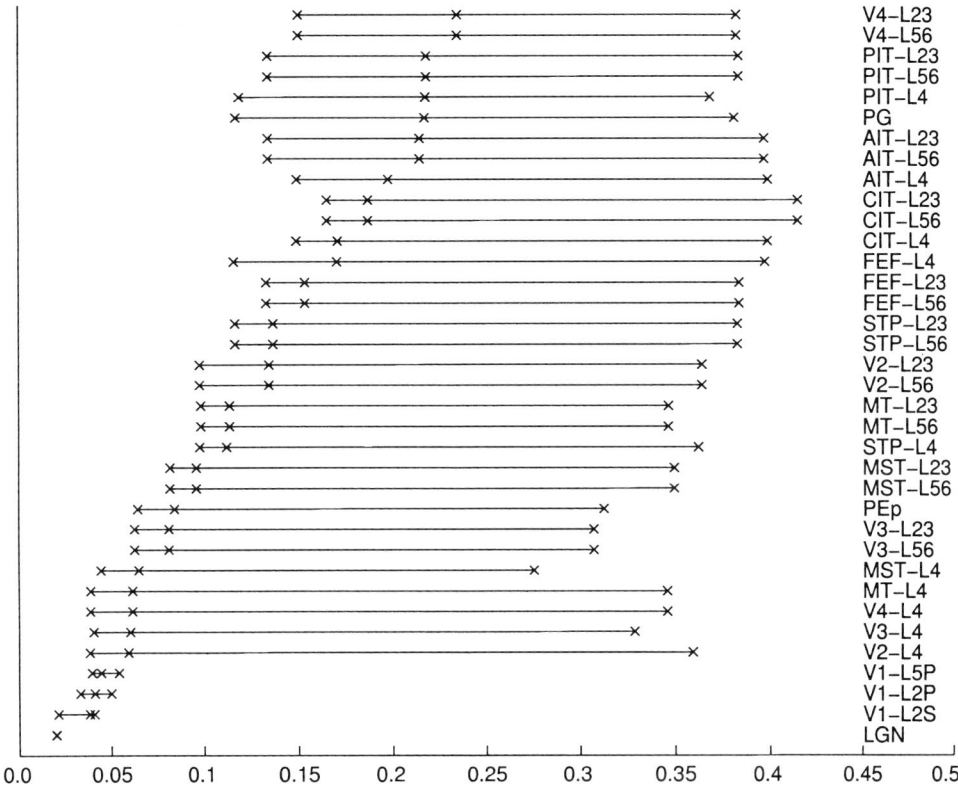

Fig. 5. Distribution of onset response latencies in the simulated visual network in 100 runs with random background activity in LGN. Abscissa shows latencies in s; crosses indicate 10, 50, and 90 percentiles. Delays between I&F units representing layers or areas were set uniformly to 15 ms. Note that the sequence of activation in layered areas generally had feedforward characteristics. Exceptions were the posterior inferotemporal area showing near simultaneous activation of all layers, and the frontal eye field, where the sequence of median latencies indicated feedback activation.

feedforward activation pattern. Conversely, a feedback pattern of activation with longer latencies in layer 4 than in the extragranular layers would result in an HAI < 0.5. Figure 6 shows that the interunit delay of 15 ms resulted almost unanimously in feedforward patterns of area activation. This pattern was most prominent in areas V4, MST, and MT, whereas inferotemporal areas AIT and PIT, as well as FEF, responded almost simultaneously with all layers. With varying interunit delays, we obtained some remarkable alterations in the laminar activation sequences: decreasing interunit delays diminished and reversed the feedforward activation sequence of V4; the opposite occurred in FEF. While the HAI of some areas (V2, V3, MT) varied in a complicated manner, the simultaneous activation of inferotemporal areas was largely unaffected by the interunit delays.

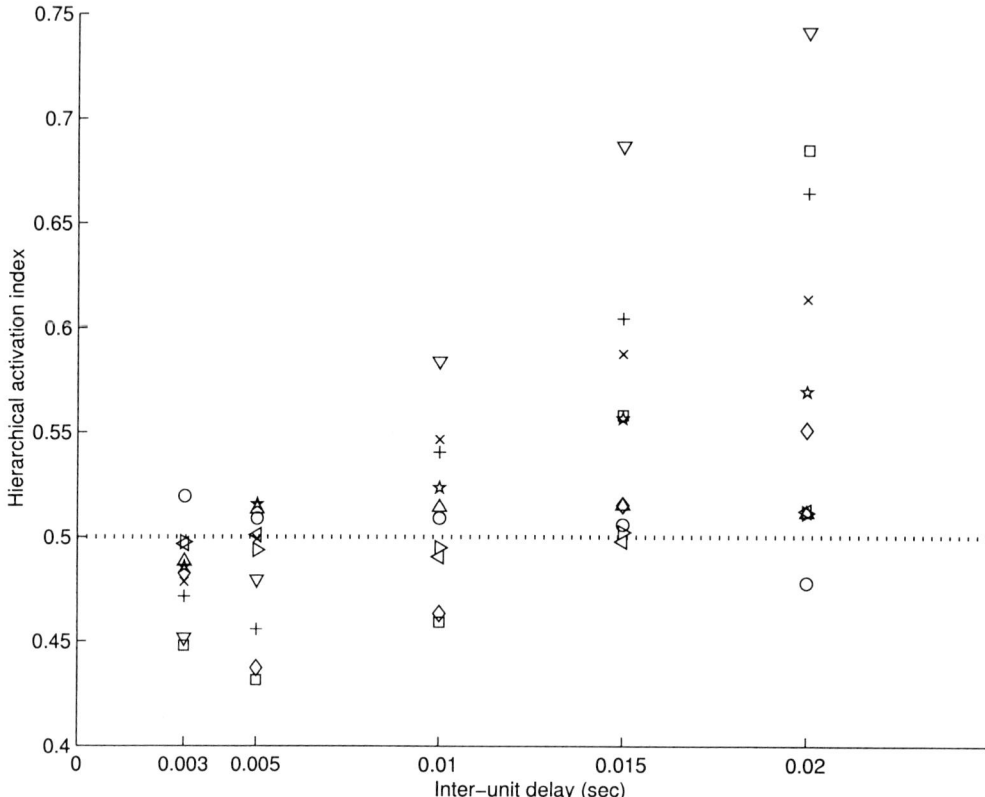

Fig. 6. Distribution of hierarchical activation indices of the 10 three-tiered cortical area models in simulations with five different delays between I&F units (both layers and areas). The HAI scores average onset latencies in extragranular layers over the sum of average onset latencies in granular and extragranular layers. Activation of granular layer 4 before extragranular layers leads to HAI values larger than 0.5, thus indicating feedforward activation. Correspondingly, HAI values smaller than 0.5 indicate feedback activation where the activation of extragranular layers precedes activation of granular layer 4. The values for an interunit delay of 15 ms correspond to histograms in Figure 2 and to median (instead of mean) latencies shown in Figure 5. Note that varying delays had differential effects on the activation characteristics, in particular of areas V4, FEF, and inferotemporal areas. ☐ = V2; ◊ = V3; + = MT; × = MST; ☆ = STP; ∇ = V4; ◁ = AIT; ▷ = PIT; Δ = CIT; O = FEF.

16.5. DISCUSSION

We have integrated several types of cell, layer, and area representations into a functioning network model of the visual system for exploration of patterns of activity propagation induced by visual input. A rich repertoire of responses was obtained that raises hypotheses in need of further investigation.

1. The microcircuit implementation was originally developed to represent the first response of cells in the cat V1 to afferent stimulation *(67)*. In our more extended simulations, the cells show voltage fluctuation that capture elementary features of inhibitory smooth and excitatory pyramidal neurons in the cerebral cortex. Nevertheless, this implementation is too simplified to represent the variety of known cortical cell types (e.g., regular spiking vs

intrinsically bursting cells) or the intricate microcircuitry of are V1 in primates. Thus, it is useful to recognize that the canonical microcircuit concept is an abstract basic scheme that has to be adapted and cast into concrete implementations, which will vary with the precise situation modeled. This view is supported by the various versions of the canonical microcircuit and their sketchy descriptions in the literature *(67–69,74)*. The present study shows that a microcircuit implementation can be integrated as a module within a formal network model. It helps to analyze mechanisms of onset response latencies, and it could be a starting point for more realistic representations of information processing in V1.

2. Typically, visual cortical areas were activated in a sequence that roughly corresponded to the number of stations from LGN as shown in a previous network simulation *(75)*. The number of stations through which a signal has to pass to reach a particular station was referred to as the hodology of the system *(76)*. Hodology explains near simultaneous onset of activity in areas V2, V3, MT, and MST as shown in *(75)* and in Figure 5. Here, we demonstrate that relative onset latencies can be reproduced by a much simpler network model than described previously *(75)*, where each cortical area was implemented as six layers, each with 100 inhibitory and 100 excitatory I&F units. The experimental observations of similar onset latencies in FEF and longer ones in V4 require additional explanations. In this context, it is remarkable that the longer latencies of V4, as well as the more subtle latency differences between areas V3, MST, and V2 *(64,77)* were very well captured by the onset latencies of extragranular layers in our simulation (Fig. 5). This raises the question to what extent the experimental sampling procedure influenced previously reported onset latencies. Some clues as to differential laminar activation have been provided by systematic recordings throughout the depth of visual cortical areas (see *[65,77,78]*). The latency values in our network were obtained using a single (magnocellular) thalamic input. Nevertheless, they displayed some characteristics of both the magnocellular and the parvocellular streams *(64)*. This points to the possibility that differential latency values in the two streams are not simply a consequence of sequential arrival of the separate thalamocortical inputs, but that the latency values in the different areas are substantially shaped by the layout of the cortical network with specific diverging and converging pathways.

3. The visual cortex is hierarchically organized *(18)*, but nevertheless its exact hierarchical structure is indeterminate *(19)*. Our simulations indicate that this anatomical indeterminacy may have a functional counterpart. Not only did we find that the latency differences between granular and extragranular layers varied enormously among areas V2, V3, and V4 despite the same number of processing steps; in addition, the laminar activation sequence switched between feedforward and feedback activation characteristics as a function of varying interunit delays. Since every area may be activated by different routes with different laminar preferences, it can be predicted that the sequence of laminar activation within an area depends on several factors including laminar pre-activation, the functional impact of various afferents, the relative speed of processing in the afferent pathways, and the stimulus statistics (luminance, chromatic contrast, movement, coherence). Thus, not only is the anatomical hierarchy of the visual system indeterminate, but so is its functional hierarchy, i.e., the sequence of areas ordered by their HAI values of relative granular and extragranular activation, dependent on the fine structure of the activity propagation in the entire network.

4. Although several studies reported data on onset response latencies in the visual system *(64,72,77,78)*, the factors that influence them could be scrutinized more thoroughly by adequate experiments. There are two contrasting philosophies, which are both necessary for a comprehensive systems analysis: to measure response latencies throughout the visual system (*i*) using different stimuli optimized for the stimulus preferences (e.g., size, con-

trast, color, movement) of the respective areas (e.g., *[64]*); and (*ii*) using exactly the same stimuli for all areas despite the probability that some areas may not respond very well (e.g., *[77]*). It will be interesting to see whether different stimuli lead to different functional hierarchies as we obtained through variation of interunit delays. It will also be important to compare recordings of field potentials, multi- and single-unit activity, as well as synaptic and metabolic activity through functional imaging. Such experiments will provide information on the relationships between the various processes *(29–31)*, reveal the functional organization of information processing in the visual system *(75)*, and identify the temporal limitations of cognitive processes (see *[79]*). Such data would help to refine models of visual processing, in particular as these start to combine features of cognitive performance, network mechanisms, and neuronal processing.

It is well to keep in mind that a model, be it informal, biophysical, or computational, is always a model of a certain aspect, but never a model of every aspect of the original. Any known model removes the largest part of the complexity of the phenomenon that it intends to model. While this may seem to be a fundamental shortcoming, it is indeed the essence of modeling to extract only those features that are relevant to a certain context or question. The focusing on the essential mechanisms (as opposed to indiscriminate simplification) facilitates an understanding of the phenomenon, which may be cast into mathematical terms and be used for predictions. Whether or not the model still applies in a different context than the original one, however, needs to be investigated with recourse to the original data. In a domain of complex phenomena, such as neuroscience, models focusing on certain aspects are inevitable and not just first steps toward one unified theory. Nevertheless, there is a relationship between the empirical phenomenon that one intends to describe and the complexity of useful models: if the model is too simple, then it fails to describe the phenomenon under investigation; if it is too complicated, then it does not help our understanding. A good model falls between the two extremes and is embedded in sufficient empirical data to constrain its free parameters. After more than a century of quantitative neuroscience and with modern techniques of data production and collation, multilevel models become more and more relevant to this field.

ACKNOWLEDGMENTS

This work was supported by DFG LIS 4 - 554 95(2) Düsseldorf to R.K. and an EU Neuroinformatics fellowship to P.N. We thank Christian Bendicks and Andreas Schwanz for implementing graphics tools aiding the GENESIS simulations.

REFERENCES

1. Rall W, Agmon-Snir H. Cable theory for dendritic neurons. In: Methods in Neuronal Modeling. From Ions to Networks. (Koch C, Segev I, eds.), MIT, Cambridge, MA, 1998, pp. 27–92.
2. Häusser M. The Hodgkin-Huxley theory of the action potential. Nat Neurosci 2000; **3 (Suppl)**:1165.
3. Segev I, Burke RE. Compartmental models of complex neurons. In: Methods in Neuronal Modeling. From Ions to Networks. (Koch C, Segev I, eds.). MIT, Cambridge, MA, 1998, pp. 93–136.
4. De Schutter E, Bower JM. An active membrane model of the cerebellar Purkinje cell I. Simulation of current clamps in slice. J Neurophysiol 1994; **71**:375–400.

5. van Pelt J, Dityatev AE, Uylings HB. Natural variability in the number of dendritic segments: model-based inferences about branching during neurite outgrowth. J Comp Neurol 1997; **387**:325–340.
6. Kötter R, Feizelmeier M. Species-dependence and relationship of morphological and electrophysiological properties in nigral compacta neurons. Prog Neurobiol 1998; **54**:619–632.
7. De Schutter E, Smolen P. Calcium dynamics in large neuronal models. In: Methods in Neuronal Modeling. From Ions to Networks. (Koch C, Segev I, eds.). MIT, Cambridge, MA, 1998, pp. 211–250.
8. Bhalla US, Iyengar R. Emergent properties of networks of biological signaling pathways. Science 1999; **283**:381–387.
9. Kötter R, Schirok D. Towards an integration of biochemical and biophysical models of neuronal information processing: a case study in the nigro-striatal system. Rev Neurosci 1999; **10**:247–266.
10. Peng YY, Wang KS. A four-compartment model for Ca^{2+} dynamics: an interpretation of Ca^{2+} decay after repetitive firing of intact nerve terminals. J Comput Neurosci 2000; **8**:275–298.
11. McCulloch W, Pitts W. A logical calculus of the ideas immanent in nervous activity. Bull Math Biophys 1943; **5**:51–69.
12. Hopfield JJ. Neural networks and physical systems with emergent collective computational abilities. Proc Natl Acad Sci USA 1982; **79**:2554–2558.
13. Kohonen T. Self-organized formation of topologically correct feature maps. Biol Cybern 1982; **43**:59–69.
14. McClelland JL, Rumelhart DE. Explorations in Parallel Distributed Processing. MIT Press, Cambridge, MA, 1988.
15. Murre JM. Interaction of cortex and hippocampus in a model of amnesia and semantic dementia. Rev Neurosci 1999; **10**:267–278.
16. Watts DJ, Strogatz SH. Collective dynamics of 'small-world' networks. Nature 1998; **393**:440–442.
17. Kötter R, Sommer FT. Global relationship between structural connectivity and activity propagation in the cerebral cortex. Philos Trans R Soc Lond B 2000; **355**:127–134.
18. Felleman DJ, Van Essen DC. Distributed hierarchical processing in the primate cerebral cortex. Cereb Cortex 1991; **1**:1–47.
19. Hilgetag CC, O'Neill MA, Young MP. Indeterminate organization of the visual system. Science 1996; **271**:776–777.
20. Young MP. Objective analysis of the topological organization of the primate cortical visual system. Nature 1992; **358**:152–154.
21. Stephan KE, Hilgetag CC, Burns GAPC, O'Neill MA, Young MP, Kötter R. Computational analysis of functional connectivity between areas of primate cerebral cortex. Philos Trans R Soc Lond B 2000; **355**:111–126.
22. Northoff G, Richter A, Gessner M, et al. Functional dissociation between medial and lateral prefrontal cortical spatiotemporal activation in negative and positive emotions: a combined fMRI/MEG study. Cereb Cortex 2000; **10**:93–107.
23. Kötter R, Hilgetag CC, Stephan KE. Connectional characteristics of areas in Walker's map of primate prefrontal cortex. Neurocomputing; **38–40**:741–746.
24. Hilgetag CC, Burns GAPC, O'Neill MA, Scannell JW, Young MP. Anatomical connectivity defines the organization of clusters of cortical areas in the macaque monkey and the cat. Philos Trans R Soc Lond B Biol Sci 2000; **355**:91–110.
25. Kötter R. Neuroscience databases: Tools for exploring brain structure-function relationships. Philos Trans R Soc Lond B; **356**:1111–1120.

26. Scannell JW, Young MP. The connectional organization of neural systems in the cat cerebral cortex. Curr Biol 1993; **3**:191–200.
27. Grasby PM, Frith CD, Friston KJ, et al. A graded task approach to the functional mapping of brain areas implicated in auditory-verbal memory. Brain 1994; **117**:1271–1282.
28. Rao RP, Ballard DH. Predictive coding in the visual cortex: a functional interpretation of some extra-classical receptive-field effects. Nat Neurosci 1999; **2**:79–87.
29. Arbib MA, Bischoff A, Fagg AH, Grafton ST. Synthetic PET: analyzing large-scale properties of neural networks. Hum Brain Map 1995; **2**:225–233.
30. Wright JJ. EEG simulation: variation of spectral envelope, pulse synchrony and approximately 40 Hz oscillation. Biol Cybern 1997; **76**:181–194.
31. Horwitz B, Tagamets MA, McIntosh AR. Neural modeling, functional brain imaging, and cognition. Trends Cogn Sci 1999; **3**:91–98.
32. Koch C, Segev I. Methods in Neuronal Modeling. From Ions to Network. MIT Press, Cambridge, MA, 1998.
33. Goddard NH, Hucka M, Howell F, Cornelis H, Shankar K, Beeman D. Towards NeuroML: model description methods for collaborative modelling in neuroscience. Philos Trans R Soc Lond B; **356**:1209–1228.
34. Softky W, Koch C. Single-cell models. In: The Handbook of Brain Theory and Neural Networks. (Arbib MA, ed.). MIT Press, Cambridge, MA, 1995, pp. 879–884.
35. Hodgkin AL, Huxley AF. A quantitative description of membrane current and its application to conduction and excitation in nerve. J Physiol 1952; **117**:500–544.
36. Ascoli GA, Krichmar JL, Nasuto SJ, Senft SL. Generation, description, and storage of dendritic morphology data. Philos Trans R Soc Lond B Biol Sci; **356**:1131–1145.
37. McKenna T, Davis J, Zornetzer SF. Single Neuron Computation. Academic, Boston, 1992.
38. Koch C. Biophysics of Computation. Oxford University, New York, 1999.
39. Kötter R, Wickens J. Interactions of glutamate and dopamine in a computational model of the striatum. J Comput Neurosci 1995; **2**:195–214.
40. Kötter R. Postsynaptic integration of glutamatergic and dopaminergic signals in the striatum. Prog Neurobiol 1994; **44**:163–196.
41. Connor JA, Stevens CF. Voltage clamp studies of a transient outward membrane current in gastropod neural somata. J Physiol 1971; **213**:21–30.
42. Pfaffelhuber E. Sensory coding and the economy of nerve pulses. Notes Biomath 1974; **4**:467–483.
43. Eckhorn R, Grusser OJ, Kroller J, Pellnitz K, Popel B. Efficiency of different neuronal codes: information transfer calculations for three different neuronal systems. Biol Cybern 1976; **22**:49–60.
44. Bialek W, Rieke F, de Ruyter van Steveninck R. R, Warland D. Reading a neural code. Science 1991; **252**:1854–1857.
45. Shepherd GM. Canonical neurons and their computational organization. In: Single Neuron Computation. (McKenna T, Davis J, Zornetzer SF, eds.). Academic, Boston, 1992, pp. 27–60.
46. Pinsky PF, Rinzel J. Intrinsic and network rhythmogenesis in a reduced Traub model for CA3 neurons. J Comput Neurosci 1994; **1**:39–60.
47. Traub RD, Wong RK, Miles R, Michelson H. A model of a CA3 hippocampal pyramidal neuron incorporating voltage-clamp data on intrinsic conductances. J Neurophysiol 1991; **66**:635–650.
48. Hebb DO. The Organization of Behavior. Wiley & Sons, New York, 1949.
49. Willshaw DJ, Buneman OP, Longuet-Higgins HC. Non-holographic associative memory. Nature 1969; **222**:960–962.
50. Amit D, Gutfreund H. Statistical mechanics of neural networks near saturation. Ann Physiol 1987; **173**:30–67.
51. Palm G, Sommer FT. Associative data storage and retrieval in neural nets. In: Models of

Neural Networks III (Domany E, Van Hemmen JL, Schulten K, eds.). Springer, New York, 1995, pp. 79–118.
52. Schwenker F, Sommer FT, Palm G. Iterative retrieval of sparsely coded associative memory patterns. Neural Networks 1996; **9**:445–455.
53. Singer W. Synchronization of cortical activity and its putative role in information processing and learning. Ann Rev Neurosci 1993; **55**:349–374.
54. Lansner A, Fransen E. Improving the realism of attractor models by using cortical columns as functional units. In: The Neurobiology of Computation. (Bower JM, ed.). Kluwer, Boston, 1995, pp. 251–256.
55. Abeles M. Corticonics: Neural Circuits of the Cerebral Cortex. Cambridge, Cambridge University Press, 1991.
56. Bliss TV, Collingridge GL. A synaptic model of memory: long-term potentiation in the hippocampus. Nature 1993; **361**:31–39.
57. Herz A, Sulzer B, Kuehn R, Van Hemmen JL. The Hebb rule: storing static and dynamic objects in an associative neural network. Europhys Lett 1988; **7**:663–669.
58. Markram H, Tsodyks M. Redistribution of synaptic efficacy between neocortical pyramidal neurons. Nature 1996; **382**:807–810.
59. Chance FS, Nelson SB, Abbott LF. Synaptic depression and the temporal response characteristics of V1 cells. J Neurosci 1998; **18**:4785–4799.
60. Wennekers T, Sommer FT, Palm G. Iterative retrieval in associative memories by threshold control of different neural models. In: Supercomputing in Brain Research: From Tomography to Neural Networks (Herrmann HJ, Wolf DE, Pöppel E, eds.). World Scientific, Singapore, 1995.
61. Sommer FT, Wennekers T. Modelling studies on the computational function of fast temporal structure in cortical circuit activity. J Physiol 2000; **94**:473–488.
62. De Schutter E, Dyhrfjeld-Johnsen J, Maex R. A realistic cerebellar network simulation of mossy fiber induced Purkinje cell activity. Soc Neurosci Abstr 1998; **24**:665.
63. Fransen E, Lansner A. 1998; A model of cortical associative memory based on a horizontal network of connected columns. Network **9**:235–264.
64. Schmolesky MT, Wang Y, Hanes DP, et al. Signal timing across the macaque visual system. J Neurophysiol 1998; **79**:3272–3278.
65. Maunsell JH, Gibson JR. Visual response latencies in striate cortex of the macaque monkey. J Neurophysiol 1992; **68**:1332–1344.
66. Maunsell JH, Ghose GM, Assad JA, McAdams CJ, Boudreau CE, Noerager BD. Visual response latencies of magnocellular and parvocellular LGN neurons in macaque monkeys. Vis Neurosci 1999; **16**:1–14.
67. Douglas RJ, Martin KA. A functional microcircuit for cat visual cortex. J Physiol 1991; **440**:735–769.
68. Bush PC, Douglas RJ. Synchronization of bursting action potential discharge in a model network of neocortical neurons. Neural Comput 1991; **3**:19–30.
69. Suarez H, Koch C, Douglas R. Modeling direction selectivity of simple cells in striate visual cortex within the framework of the canonical microcircuit. J Neurosci 1995; **15**:6700–6719.
70. Wilson MA, Bower JM. The simulation of large-scale neural networks. In: Methods in Neuronal Modeling. From Synapses to Network. (Koch C, Segev I, eds.). MIT Press, Cambridge, MA, 1989, pp. 291-333.
71. Callaway EM. Local circuits in primary visual cortex of the macaque monkey. Ann Rev Neurosci 1998; **21**:47–74.

72. Nowak LG, Bullier J. The timing of information transfer in the visual system. In: Cerebral Cortex (Rockland KS, Kaas JH, Peters A, eds.). Plenum, New York, 1997, pp. 205–241.
73. Kötter R, Scannell JW, Hilgetag CC, Lerzynski G., Stephan KE, Young MP. Response latencies to flashed stimuli reveal differential processing in dorsal and ventral visual streams. Soc Neurosci Abstr 1999; **25:**2061.
74. Douglas RJ, Martin KAC. Exploring cortical microcircuits: a combined anatomical, physiological, and computational approach. In: Single Neuron Computation (McKenna T, Davis J, Zornetzer SF, eds.). Academic, Boston, 1992, pp. 381–412.
75. Petroni F, Panzeri S, Hilgetag CC, Kötter R, Young MP. Simultaneity of responses in a hierarchical visual network. Neuro Report 2001; **12:**2753–2759.
76. Young MP, Scannell JW, Burns GAPC. The Analysis of Cortical Connectivity. Springer, Heidelberg, 1994.
77. Schroeder CE, Mehta AD, Givre SJ. A spatiotemporal profile of visual system activation revealed by current source density analysis in the awake macaque. Cereb Cortex 1998; **8:**575–592.
78. Mehta AD, Ulbert I, Schroeder CE. Intermodal selective attention in monkeys. I: distribution and timing of effects across visual areas. Cereb Cortex 2000; **10:**343–358.
79. Thorpe S, Fize D, Marlot C. Speed of processing in the human visual system. Nature 1996; **381:**520–522.

17
Quantitative Neurotoxicity

David S. Lester, Joseph P. Hanig, and P. Scott Pine

ABSTRACT

Conventional neurotoxicity assessment requires selected behavioral testing and histological analysis. While behavior has been shown to identify 80–90% of adverse neuronal effects, it does not provide positive data regarding mechanism and site of action. Histology has its limitations, in that it is laborious, destroys the intrinsic tissue integrity, and relies on appropriate stain selection. Attempts to quantitate histology have been limited. Some biomedical imaging procedures have the potential of identifying regions of change due to neuronal insult. Technologies such as magnetic resonance imaging, positron emission tomography, and various microscopy applications all are capably of identifying affected regions under specific circumstances. They provide data in 2D and 3D digitized formats, which provide ease in analysis of collected data. Some of these approaches are capable of identifying intrinsic biochemical components, avoiding the potential for artifacts using histology. For clinical purposes, automated analysis is commercially available. This chapter will address the application of some of these imaging modalities to the topic of neurotoxicity and discuss quantitative analysis that are being developed for uniplanar and volumetric analyses of tissue specimens. Limitations, such as sensitivity, spatial resolution, etc., will be discussed in terms of interpretation of the data.

17.1. INTRODUCTION

Neurotoxicity generally refers to the induction of a toxic event in the nervous system. This can occur in either the central or the peripheral nervous system. Neurotoxicity is usually considered to be associated with cell death. However, it is much more complex than toxicity responses in other organ systems. When is something neurotoxic? What is the threshold of response that defines neurotoxicity? These issues are difficult to define and will not be attempted in this chapter. However, we would like to present a number of scenarios for the reader's consideration. If a drug causes a reduction of 20% in total serotonin, is this a neurotoxic response or simply exemplifies the pharmacological effect? If 10% of a particular neuronal cell type is destroyed, yet there is no measurable behavioral change, is this effect considered neurotoxic? What is a more convincing indicator of neurotoxicity, a change in behavior or a histopathologic

From: *Computational Neuroanatomy: Principles and Methods*
Edited by: G. A. Ascoli © Humana Press Inc., Totowa, NJ

effect? These complex questions do not have straightforward answers and have been argued by neurotoxicologists for many years. It should be noted that in the pharmaceutical industry and drug regulatory arena, neuronal cell death is considered to represent neurotoxicity and can have a significant impact on the drug development and regulatory process (1). However, an effect such as cell death of specific neuronal types may not be as drastic as the effect of disorder. It has been shown using imaging technologies that stroke can cause massive damage to the brain (2). The loss of some neuronal cells due to the neuroprotective pharmacotherapy (3), may not be as significant as the damage caused by the stroke. These issues demonstrate that classical procedures for detection of neurotoxicity may not provide adequate answers for both the basic and the applied neuroscientist.

The initial approach for detecting neurotoxicity is by monitoring behavioral changes in the treated specimen. Tests for activities such as locomotor, startle responses, and the more sophisticated cognitive tests are routinely used (4). In addition, easily applied neurological tests such as gait, righting reflex, and front paw extension are used as part of the cage-side observation (5). A standardized functional observational behavioral battery (FOB) is used for environmental agents (6). In addition, there are proposed FOBs for drugs and toxins (7,8). This comprehensive screen analyzes a number of fundamental behavioral phenomena. It does not deal with cognitive changes. Cage-side observations, sometimes called Tier 1 testing, are easy to apply and can often provide significant information (9). The limitation of the more behavioral changes detected using more sophisticated tests is that they take extended periods of time and require large numbers of animals in order to obtain statistical significance. Also, the appropriate behavioral tests must be selected (7,8). Responses can often be correlated with specific regions of the central nervous system. Upon identification of a behavioral change considered to be indicative of an adverse effect, additional studies are needed to determine structural or functional changes that contribute to the defect.

In the basic research arena, neurotoxicity can be determined by a number of what could be referred to as "biomarkers". These are endpoints that fall into a variety of categories. The diversity of biomarkers that have been used for detecting neurotoxicity demonstrates the complexity of neurotoxicity (10,11). All of these various endpoints have their limitations, however, and some of the markers or approaches are more accepted. An example is the glial fibrillary acidic protein, GFAP. Increased synthesis of this protein, or even increased messenger RNA, is considered to be a strong indicator of neurotoxicity. In the case of GFAP, its activation is correlated with activation of glial cells. When there is some toxic insult to neurons, this is usually associated with activation of the associated glial cells. The order and timing of the process is not fully understood. GFAP is usually detected using immunohistochemical procedures (12). Several stains have been shown to demonstrate neurotoxicity. There are numerous well-documented problems relating to use of stains (13,14). Since the process includes fixation, sectioning, and staining, each of these procedures can lead to artifacts. The fixation procedure can change the conformation of the intrinsic components of the section in question. Often, dehydration is required, and thus, the cellular integrity is significantly modified. Sectioning is usually limited to one of three planes, sagittal, horizontal, or coronal. In all cases, the integrity of the tissue or sample structure is destroyed. Therefore, the appropriate orientation of the tissue must be chosen in advance in order to recognize and identify the potential lesion sites. If the site of action of the neurotoxin is unknown, then a detailed histological analysis requires sectioning and subsequent analysis of every slide. In addition, the appropriate stain must be selected. This often requires prior knowledge of the nature of the neurotoxic insult. If the site of action of the neurotoxin is unknown, a detailed histological analysis requires sectioning and subsequent analysis of every slide from the entire sample.

In addition, there are fluorescence indicators for physiological activities such as calcium influx and efflux *(15)*. A sustained increase in intracellular neuronal calcium is considered to be indicative of neurotoxicity. Rapidly induced dendritic outgrowths from neurons have been cited as a structural marker that has been shown to correlate with neurotoxicity *(16)*. These markers are not conclusive. The two examples cited here might also be indicators of a physiological process, since neuronal dendritic outgrowths form upon differentiation of neurons *(16)*. There may also be increased levels of intracellular calcium for extended periods that relate to a physiological process, for example, long term potentiation *(17)*.

In general, it is difficult to distinguish between pharmacological and pathological responses, although reversibility of the effect is often used as a criterion. As discussed previously, neuronal cell death is considered to be indicative of neurotoxicity, but care must be taken in preparation of tissue for analyses using histological approaches.

17.2. IMAGING

The difficulties in determining neurotoxicity using conventional methods are extensive and not easily solved. One approach would be to increase the number of animals and/or analytical approaches. However, in the present research climate of animal welfare, it is desirable to reduce the number of animals used in testing. Toxicity tests such as the LD50 are hardly if ever used anymore. Also, the increase in numbers of research platforms means more animals and more time. This translates into significantly increased costs, an often undesirable requirement. An approach that will be discussed in the remainder of this chapter involves employing imaging technologies.

Biomedical imaging has had an enormous impact on the clinical sciences. Terms such as magnetic resonance imaging (MRI) and computer-assisted tomographic scan (CTscan) have become part of our everyday language. In 1998, over 7 million MRI scans were performed in the U.S., which is a remarkable accomplishment considering that the technology was developed and first used in the early 1980s. The driving force for development of imaging technology has been its clinical applications for disease diagnosis and progression. Hence, the technology has advanced in conjunction with therapy in human use. In recent years, it has been recognized that these technologies could also prove useful in animal research *(18,19)*. Subsequently, academic institutions have begun establishing animal imaging facilities for research studies that employ these technologies. Numerous MRI scanners are commercially available directed towards animal use. More recently, a commercially available positron emission tomography (PET) system has become available. The need for development of scanners specifically for animal studies has been due to the lack of spatial resolution of scanners used for humans.

Interestingly, while pharmacological applications of biomedical imaging have been quite extensive, there has been very limited use of these technologies in toxicology research. Although the potential applications in imaging methodologies to toxicology research are numerous, the transition from clinical to preclinical has been slow. One of the major reasons for development of human imaging approaches was the need to be able to visualize the brain in a noninvasive manner. Thus, it is surprising that imaging has not yet been widely applied to preclinical neurotoxicology. While there have been some applications at the macroscopic level examining such phenomenon as stroke, the application for the type of data that is required for toxicological evaluation in the drug development process is limited.

Considering the cost of imaging systems, why would someone want to use this technology? One of the most powerful features of imaging is that multiple parameters can be obtained from the same animal or patient in a noninvasive manner over extended periods of time. MRI has a number of extensions that allow the measurement of numerous parameters, for example, traditional MRI measures the distribution of protons in a sample. As water is the predominant proton source in all living things, MRI tends to measure the distribution of water concentration in various biological environments. This provides an image of the structural detail (20). Using an extension of MRI, such as functional MRI (fMRI), one monitors blood oxygen levels in response to some functional or behavioral challenge. Thus, the functional activation of the brain can be measured or characterized, and this is then superimposed over the structural data (21). Another example is that of proton MRI in conjunction with PET of a labeled radiotracer (e.g., fluorodeoxyglucose, a tracer for glucose metabolism), which can be imaged simultaneously, supplying structural and functional data (22).

Imaging provides unique information. Utilizing different imaging modalities, data can be obtained relating to anatomy, physiology, genetics, metabolism, as well as functional activity, all in a noninvasive manner. A number of these modalities can be monitored simultaneously.

As all the data is obtained digitally, it is possible to do quantitative analyses of the acquired data. One of the most convincing examples is the ability of PET to measure distribution and localized concentrations of radiolabeled drugs in the brain. This permits an estimation of the rate and total accumulation of drugs in the brain (23). There is no other way of doing this in clinical trials. Using this approach, it has been shown that blood concentration levels of a drug are not representative of concentrations of the drug in the brain (24). PET provides real-time pharmacodynamic and pharmacokinetic information on the intact animal or patient (25). Recently, there has been an increased interest in the ability to do volumetric analysis of tissues in vivo. An example is the change in size of the hippocampus during the progression of Alzheimer's dementia (26). This has been considered to be a potential clinical endpoint. Some examples of this type of approach in neurotoxicity will be discussed in this chapter.

A very powerful application of imaging involves the ability to establish a test protocol in animals and then apply a similar protocol to monitor structural and/or functional changes in clinical trials or diagnostics. This provides the opportunity to evaluate the worth of animal models. Clinical neurological measures are difficult to capture in animal studies. How can we determine whether a rat has a headache or is feeling nausea?

Using imaging, it may be possible to develop approaches such that changes in function can be correlated between animals and humans. Many of these applications of imaging may facilitate the development of biomarkers. The ability to do multiparametric analysis noninvasively provides the opportunity for identifying a much broader number of biomarkers, in which their relevance to the intact organ can be realized. Imaging approaches are very important in the diagnosis of disease states and its progression. With suitable validation, it may be possible to develop imaging data for these surrogate markers to a level better than the available clinical endpoint. The example of multiple sclerosis (MS) will be discussed in this chapter.

17.3. IMAGING AND TOXICOLOGY

When referring to the use of imaging in toxicology, there are not many examples. This is surprising considering the impact of imaging on the clinical sciences. Much of the concern is that the data obtained using innovative technologies may not be easily interpretable. For example, if there is a change in the grey scale contrast in a specific region of the image, what does this signify? This issue can be correlated with the difficulties in interpretation of classical histological analyses. As was described earlier in this chapter, the interpretation of a neurotoxic insult is complex and often not clearly understood. This concern relating to the significance of some change in the image is not unique to biomedical imaging technologies. Even using traditional histopathological techniques, there is often misinterpretation and sometimes there may be incorrect conclusions relating to some difference in "shading" of the stained section *(14)*. We will deal with the various imaging modalities and give examples of how they can or have been used for neurotoxicity and their potential for quantitative assessment.

17.3.1. Magnetic Resonance Imaging

MRI has significant advantages as an imaging modality. It can be used to visualize soft tissue noninvasively. The image contrast obtained as a result of biochemical and biophysical distributions of biochemical entities is entirely intrinsic. The signal may come from a number of intrinsic sources; however, the proton signal is the optimal choice. As opposed to traditional videomicroscopy where the data is collected as pixels (2D), in MRI, it is 3D. The technology is continually improving, such that faster and more sensitive instruments are being developed. MRI instruments are expensive; yet, with appropriate applications, the savings in time and the ability to reduce the number of experimental samples will ultimately make MRI an economical tool for research.

MRI was developed in order to investigate changes in brain structure due to some diseased or pathological state. The signal for proton MRI of the brain is water, which constitutes the major biochemical component of all biological tissue. Considering the relative lack of sensitivity of the MRI technique (usually mM concentrations are needed), it is not surprising that proton MRI is the most applied mode for MRI image acquisition *(20)*. In addition, changes in water content can reflect profound alterations in the molecular, cellular, and tissue components detectable using MRI.

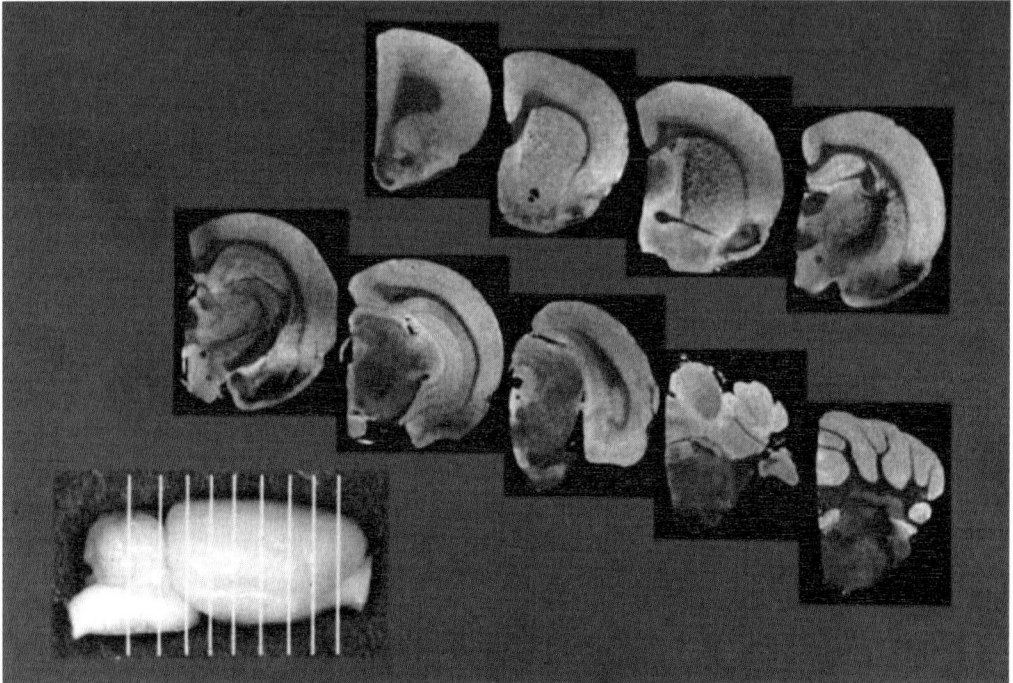

Fig. 1. Computer-generated virtual coronal sections of a rat brain. A 3D volumetric MRM image dataset of a rat brain hemisphere was acquired. Sections were computer-generated along the coronal plane in order to demonstrate the 3D nature of the technology and its ability to rapidly do macrohistological analyses of the sample in a nondestructive manner. A photographic image of the intact image is shown demonstrating the sites of sectioning. Voxel resolution is 47 μm. Details of the scanning procedure are provided in (29). A color version of this figure is available in the CD-ROM.

17.3.2. MRI as a Tool for Preclinical Neurotoxicology

The shortcomings of conventional neurohistological procedures have been previously described. During the last decade, the potential of MRI as a tool for analysis of animal anatomy and physiology has been a topic of research. While MRI has been developed primarily as a clinical tool, its potential in the preclinical arena has not received equal attention. This is in part due to the limited resolution of clinical MRI. Resolution (1–3 mm) is not useful for analysis of the rat or mouse brain. A modification of MRI, MR microscopy (MRM) or micro MRI (μMRI) or high-resolution MRI, has been investigated for its potential as an analytical tool for exploring animal histopathology (27). This methodology differs from classical MRI in that high magnetic field strengths and strong gradients are utilized in the acquisition of data (28). These conditions could not be used in a clinical scanner due to technological issues and safety concerns in humans, but their utilization for excised tissue or even immobilized living animals provides great potential. We have used this approach in collaboration with the Center for In Vivo Microscopy at Duke University Medical Center in order to assess its potential as a tool for neurohistopathology.

Domoic acid. The initial study performed involved the analysis of the action of a known excitotoxin, domoic acid, or the "red tide toxin." Excised rat brains were extracted, and hemispheres were scanned for extended periods of time (8 h), and 3D data sets were obtained. This toxin was a most appropriate initial treatment to study, as the resulting pathology of this toxin, as is the case for other glutamatergic excitotoxin, is difficult to ascertain. The same dose in two different rats may or may not induce convulsions or measurable histopathological effects. In addition, the locus or loci of histopathology is not consistent. Certainly, the main target is the amygdalla, however, there have been reports of lesions in a variety of regions of the cortex and the hippocampus, as well as a number of other brain regions. Hence, traditional histology is tedious and requires analysis of a large number of sections from brain specimens. In contrast, MRM provides the complete 3D data set which can then be examined by preparing "virtual" or computer-generated histological sections (Fig. 1). An additional advantage is that due to the 3D nature of the sample, virtual sections can be generated in all 3 planes *(29,30)*. The MRM data set is then analyzed by inspection, much in the same way as standard pathology. High-resolution detail, as obtained using optical microscopy, was not possible with MRM, however, regions of potential histopathology were readily identifiable *(29,30)*. When a MRM-identified region with different contrast was identified, 1 to 2 sequential virtual sections before and after the identified lesion were examined to see if the lesion had a volume. Generally, if the contrast difference was localized in one slice (47 µm) only, it was dismissed as an imaging artifact. Using MRM, numerous lesions were identified in a number of distinct rat brain regions *(30)*. These fixed scanned brains were then sectioned in the region of the MRM-identified lesion and subsequently stained for identification and verification that the contrast change was truly a pathological lesion (two of the lesions are shown in Fig. 2). In order to identify all of the lesions selected from the MRM scans, a total of 7 different stains were required (not presented here). Some of these were not traditional stains. Thus, considering the time required for sectioning, staining, and analysis, the 8-h MRM scan as an alternative is not that unreasonable.

Interestingly, while the spatial resolution (pixel size) of the MRM technique was greater than 45 µm, cell bodies in rat brain are significantly smaller in diameter, yet the neuronal cell layers were visible. It is proposed that the MRM cannot detect these single cells, but rather it visualizes layers of cell bodies and their surrounding microenvironment. It is assumed that when a cell changes its physiological state, it is not an isolated event. Rather, the microenvironment surrounding the cell is also affected. Thus, the MRM data was convincing beyond expectations and demonstrated outstanding utility and versatility.

Unilateral 6-hydroxydopamine injection. A second system, considered as an animal model that has shown promise as a model for Parkinson's disease, was analyzed in order to develop some quantitative analytical approaches. The reagent inducing the neurotoxic response, 6-hydropxydopamine (6-OHDOPA), is injected using a fine needle (31-gauge) directly into the substantia nigra of one hemisphere of the sedated rat. The advantage of this system is that the animal serves as its own control, that is, one hemisphere is treated, and the other is the control. This treatment is known to destroy dopaminergic terminals and tracts emanating from this brain region, thereby

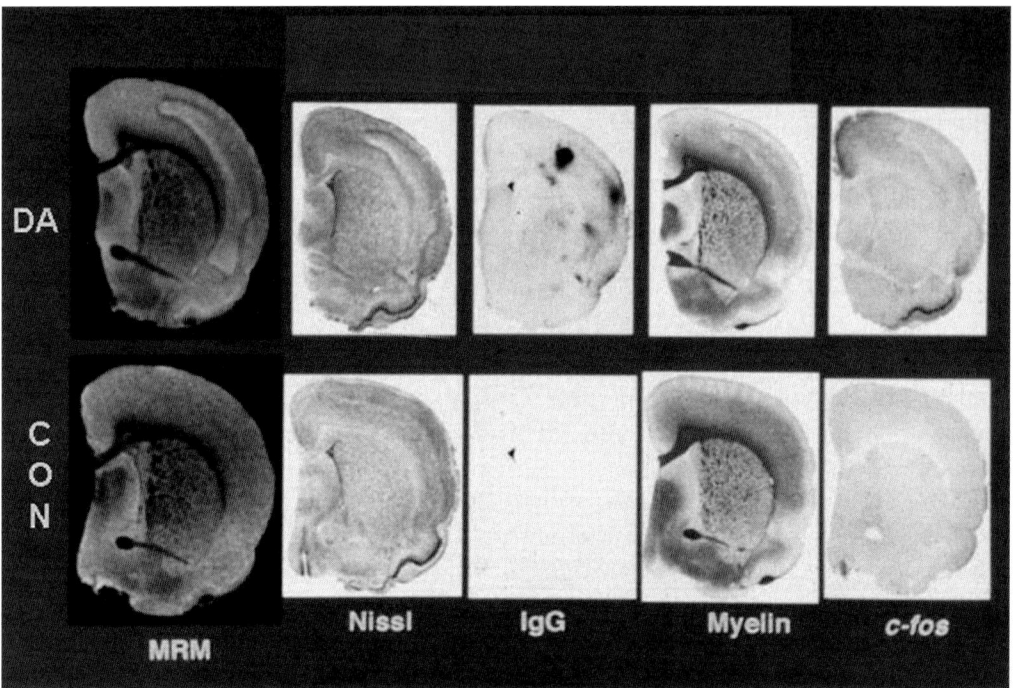

Fig. 2. Standard histological stains verify potential lesion sites identified in MRM scanned rat brain hemisphere. A rat brain from an animal 48 h post-domoic acid injection is scanned using MRM. A slice with potential lesions is identified. This is compared to a section from a control rat in a similar region of the brain. Sections are cut from the scanned brain around the identified site and then stained using conventional histological stains. These stains verify that there is significant neuronal loss in the amygdala, typical for domoic acid treatment. A total of 7 stains were required to identify all of the potential lesion sites observed in the MRM scans. In this figure, two lesions are identifiable, in the cortex and the cortical cell layer of the amygdalla. The stains necessary to identify these pathologies are included. A color version of this figure is available in the CD-ROM.

resulting in a Parkinsonian's type of response in the animal. It is generally considered that the fine 31-gauge needle causes relatively little structural damage to the brain. An MRM image of the injected brain shows the relatively small site of injection (Fig. 3A). Using image analysis, the brain can be made opaque such that the complete needle tract can be seen in comparison with the 3D image of the brain (Fig. 3B). The severity of the physical damage caused by the needle is realized in the coronal and sagittal computer-generated sections of the brain (Fig. 3C,D, respectively). There is considerable needle damage to the hippocampus. In addition, it provides an indication of the accuracy of the needle injection at the desired site, the substantia nigra. An important advantage of MRM can be seen in Figure 3C and D, where the brain dataset can be oriented such that the virtual sections provide a single slice view of the needle tract. The scanning parameters are listed in detail in the relevant publication (29). This would be very difficult to do using histological approaches.

Quantitative measures were made of the 6-OHDOPA brains. The two hemispheres served as a control side and an experimental (injected) side, respectively. The process

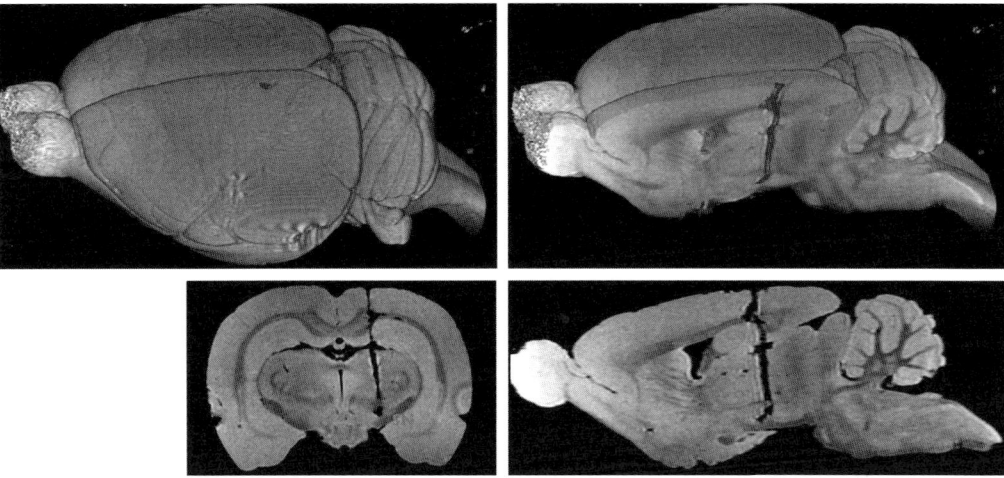

Fig. 3. MRM scanned rat brain with unilateral injection of 6-OHDOPA. A rat receives a unilateral injection with 6-hydroxydopamine in the substantia nigra of one hemisphere. The intact brain is scanned using MRM. The site of injection is visualized (top,left), and the needle tract can be identified in relation to the intact brain (top,right). Coronal (bottom,left) and sagittal (bottom,right) virtual sections are computer generated demonstrating the path of the needle tract and the significant structural damage caused by this treatment. A color version of this figure is available in the CD-ROM.

of quantitative analysis that allows the comparison of the volume of one hemisphere to the other is described in the following image analytical procedure sequence:

1. Fast spin echo (FSE) MRM scans of rat brains were made at Duke University Center of In Vivo Microscopy.
2. 3D datasets were read into the image analysis program, Alice v4.4 (Parexel International, Corp.).
3. The 3D datasets were reoriented to provide symmetrical views of the left and right hemispheres and digitally resectioned to create a series of 343 (~) slices.
4. Within each slice, the exterior boundaries of left and right hemispheres were determined using the program's shrink function, an edge detection algorithm. Shrinking is a process by which a defined region of interest can be decreased or increased in area or volume while maintaining the original shape of the defined region. This is done in order to "fit" one section or volume to another.
5. The area within each hemisphere corresponding to low signal (i.e., "empty space") was determined using the range function for pixels having a value between 0 and 2000 relative units. These ventricles were filled with the solution in which the sample was immersed, a fixation mixture that has no signal.
6. For each slice, the areas of low signal within left and right hemispheres were subtracted from the original area, and the results were expressed as the percentage of change for the left or the right (Figs. 4 and 5).
7. Finally, the relative changes for each hemisphere were compared for each slice and plotted as a function of slice number (Fig. 5).
8. Two regions varied by more than 2%. In the first region, slices 61–65, the differences were due to convolutions in the surface of the brain being seen as surface in one hemisphere and enclosed space in the other. The second region included the needle tract.

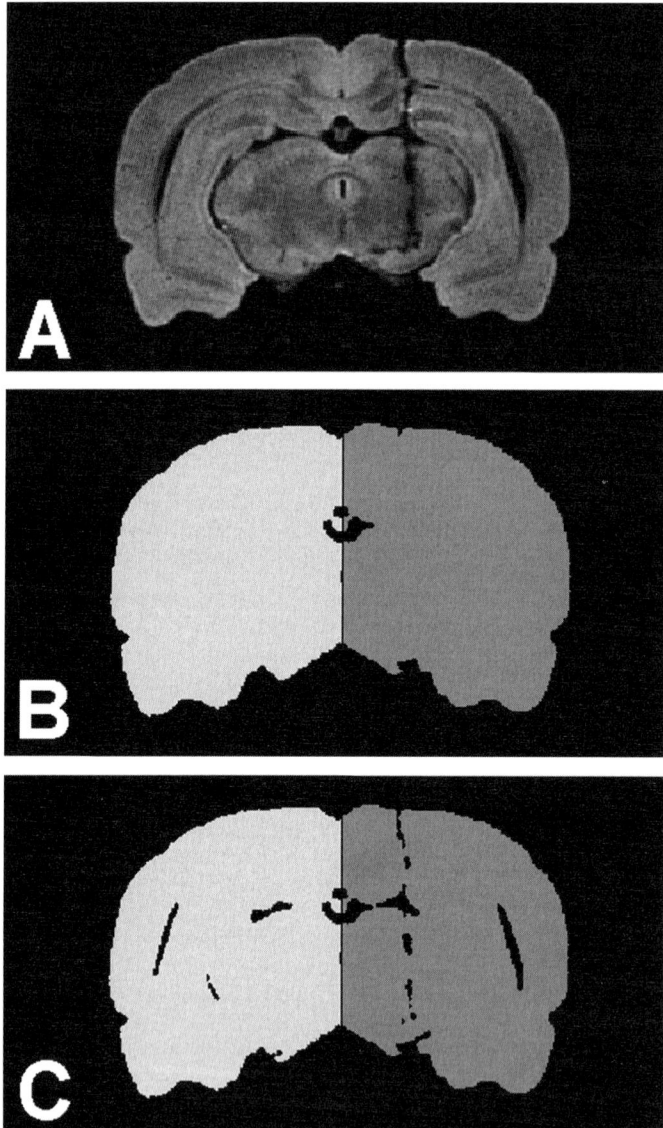

Fig. 4. Differential image processing of MRM rat brain scan. **(A)** MRM image of slice number 165 (see Fig. 3C). **(B)** Area occupied by the right and left hemispheres, light grey and medium grey, respectively. **(C)** Area occupied by right and left hemispheres after low-contrast pixels are subtracted.

These analyses provide insights that have previously not been available. The diversity of structural defects that can be monitored includes cellular changes due to necrosis or gliosis as well as volumetric alterations. Using MRI procedures, it is possible to quantitate and provide volumetric data.

Volumetric analyses of the relative differences in contrast intensity between the treated and untreated hemisphere provides insight into structural changes that are caused by the treatment (Fig. 6C). This graphical visualization of the action of

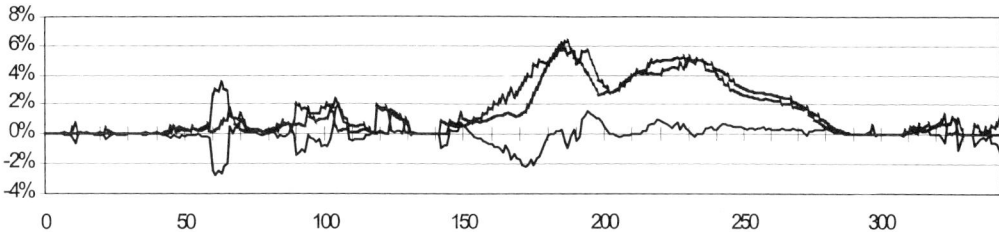

Fig. 5. Relative change in brain hemisphere area per axial slice due to pixel exclusion based on intensity range. The percentage of the original area of the hemisphere that remains after excluding pixels from the low range of MRM contrast (right hemisphere, medium grey; left hemisphere, light grey). The relative difference in the percentage of change between right and left hemispheres per axial slice (black line).

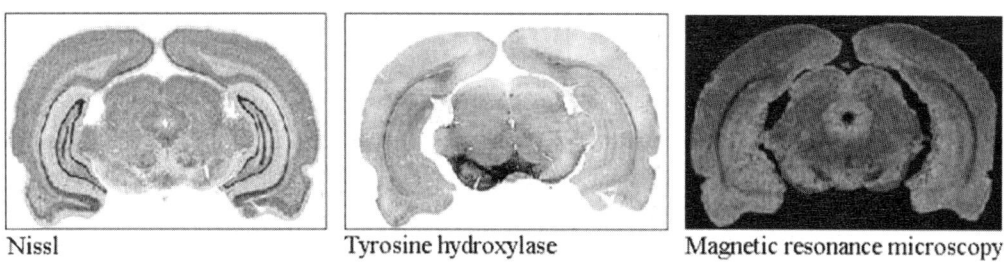

Nissl   Tyrosine hydroxylase   Magnetic resonance microscopy

Fig. 6. Comparison of conventional histological staining to MRM imaging for the detection of 6-OHDOPA-induced neurotoxicity in rat brain. A section has been selected that demonstrates the damage caused by the treatment with the toxin. Sections were cut from this region of the brain. The conventional Nissl-stained section from this region does not show significant differences at this level of magnification (left). A second section from this region has been stained with tyrosine hydroxylase, which is an established marker for dopaminergic neurons. As can be seen, there is a significant depletion of staining in the substantia nigra in the treated hemisphere (center). The MRM section has been pseudocolored in order to visualize subtle contrast differences (right). The left and right hemispheres have diffuse regions of contrast differences indicating that the unilateral treatment has significant effects. A color version of this figure is available in the CD-ROM.

the toxic insult demonstrates the "global" biochemical changes that occur, as expressed and measured by the changes in distribution and concentration of water molecules. This is one of the strengths of MRI, that it is particularly sensitive to changes in water properties, which is an excellent indicator of macroscopic changes in neuronal functioning.

These brains were sectioned, and subsequent sections were stained using Nissl reagent, a standard neurohistological stain, as well as tyrosine hydroxylase (Fig. 6A,B, respectively) *(31)*. The tyrosine hydroxylase stain demonstrates the destruction of the dompaminergic terminals in the injected hemisphere. Quantitative evaluation of MRM contrast is not unlike the qualitative evaluation of the various histochemical stains. It should be noted that digital images of stained histological sections can be obtained, and commercial software is capable of certain quantititative analyses. There are distinct differences in intensity between the injected and control hemispheres (Fig. 6C). The regions are not clearly defined in the MRM section which would suggest that there may be more global effects of the 6-OHDOPA injection affecting more diffuse regions of the brain than is detected using standard histological stains. In order to quantitate these volumes, it is necessary to develop appropriate standards and procedures, such that potential artifacts can be identified and discarded. Appropriate software programs are being developed by a number of groups.

While these studies are all done in ex vivo fixed animal brains, the technology is developing such that similar studies will be possible in living animals. There are significant technical difficulties that need to be overcome for this type of research. For example, imaging scan times cannot be as lengthy as for fixed tissue, which will reduce image quality and resolution. The present application of ex vivo imaging is very practical and can provide considerable insights into the potential sites of neurotoxicity. With the described approach above, it provides examples that demonstrate how much more informative and insightful imaging is when compared to existing approaches.

17.3.3. MRI as a Tool for Detecting Multiple Sclerosis

Axonal loss is considered to be a large component of the pathological substrate for MS. There is a general reduction in the central white matter, which correlates with the progression of MS (32). Up until approximately 10 yr ago, MS was diagnosed using a variety of neurological tests. This approach proved to be inadequate for the field of drug development. Using MRI, this structural change, specifically, the characteristic loss of axons, can be visualized as distinct lesions occurring in a number of different brain regions, as well as in the spinal cord (33). MRI provides the ability to noninvasively estimate the volume of these lesions and their progression over time. Many physicians consider these lesions to be representative of the progression of MS. A correlative study with conventional MRI, histopathology and clinical phenotyping of the spinal cord of MS patients was performed (34). There was a strong correlation between areas scored by the neuropathologist and lesions identified using low-field MRI. High-field MRIs of postmortem spinal cord tissue demonstrated a strong correlation to the subsequently performed histopathological analyses, in particular areas of demyelination. The authors concluded that MRI revealed a great range of abnormalities in spinal cord MS, which related to disease course during life. This study clearly demonstrated the correlation between the lesions in the spinal cord as identified by MRI and the classical histopathology.

Neurological evaluations include ambulation/leg function, arm/hand function and cognition (34). MRI volumetric evaluation of identified lesions shows that there is a significant correlation between these neurological measures and the total volume of these lesions (35). The lesions identified are asymmetrical in the central nervous system and the periphery (limbs) due to the diffuse axonal injury (36). These lesions are considered to be associated with inflammatory disease such as acute disseminated encephalomyelitis. These lesions increase in volume and size over time. Another volumetric approach is simply to measure changes in the volume of the brain with the progression of the disease (37). With the progression of the disease, there is a reduction in the brain volume, due in large part to the decrease in parenchyma. Measuring the total parenchymal (grey matter) volume and comparing it to brain volume is a powerful method of monitoring the progress of the disease. Short-term ventricular changes have also been shown to be a good indicator of the progress of MS. There is a decrease in ventricular volume with the progression of the disease that can be easily determined using serial MRI *(33)*.

Thus, there is significant evidence that MRI is a useful procedure for monitoring structural changes in neuronal tissue that are associated with the progression of MS *(38)*. Unfortunately, the therapies for MS that have been developed using these approaches have not proven too successful *(36,37)*. This raises the issue that the MS lesions identified by MRI are not necessarily indicative of the disease itself, but may correlate with the disease progress or be a particular endpoint. These lesions are considered to be a result of some inflammatory response, not necessarily representative of the actual disease factors *(36,37)*. An additional MR approach that may help in the monitoring of MS progression is MR spectroscopy (MRS) *(39)*. MRS imaging provides spectral information in a noninvasive manner, which is indicative of many of the biochemical activities in the brain. It has the ability of monitoring the biochemical changes caused by a number of brain pathologies related to MS, including inflammation (increase of choline), recent demyelination (increase in lipids and choline), axonal dysfunction (decrease of N-acetyl-aspartate), and gliosis (increase of myoinositol) *(40)*. Some of these biomarkers may detect early onset of the lesions, as well as some of the activities occurring leading up to lesion development. It may be that the combination of MRS imaging and MRI will eventually provide a better way of quantitatively monitoring the disease pathology of MS.

17.3.4. Midinfrared Spectral Imaging of Brain Sections

Infrared (IR) spectroscopy is used in the industrial setting more than any other spectroscopic technique. However, the impact of this technology in biological and biomedical sciences is relatively limited. The technique measures the properties of vibrational functional groups of intrinsic molecules. Most of the biochemical components making up a cell have some specific vibrational properties which can be measured. Some have multiple vibrational properties, for example, phospholipids may have the phosphate head group, and the CH3 and CH2 portions of the fatty acid chain *(40)*. A limitation of IR spectroscopy of biological components is that many of these intrinsic components cannot be identified or quantitated in complex mixtures. Often, multiple principal component analyses are necessary to such a degree that the significance of the data is questionable. Another disadvantage is that, when mixtures are analyzed using spectroscopy, many changes may be "averaged" out and therefore not quantifiable. There has been a considerable body of work published on the mid-IR (1200–5000 nm) properties of lipids and proteins *(40)* describing the specific wavenumber corresponding to a vibrational property. Using this information, it is possible to develop spectral fingerprints for the component of interest. Over the past 5 yr, an innovative approach to IR spectroscopy has provided a potential tool for biological and biomedical issues. The work of Lewis, Levin, and Treado and colleagues has led to the development of an instrument that makes IR spectroscopic imaging possible *(41)*. This has often been called chemical imaging. This procedure uses an array to capture the data, and each pixel records an IR spectrum for the sample contained within the pixel. This approach has been coupled to a microscope such that high-resolution images of IR components can be visualized (see *[41]* for details). Some of the work that has been described using this technique has been done using brain specimens. Thin (8–10 µm) frozen sections are cut on a cryostat and layered on calcium or barium chloride disks and air-dried. The samples must be free of water, as this has a very strong absorption band in the mid-IR. Sections are not subjected to any further processing. One of these studies examined the action of an antineoplastic agent, cytarabine, which is known to have neurotoxic effects. This drug causes the loss of Purkinje cells in the cerebellum. Using chemical imaging, it has been shown that the subsequent spaces ("holes") left by the macrophage removal of dead Purkinje cells can be visualized without any histological stains *(42)*. In addition, the data can be analyzed using a variety of quantitative approaches. The spectral data can be graphed as changes in peak frequency of specific biochemical or biophysical components, such as lipid CH2 symmetric stretching. The data can be graphed as variability in biochemical composition over the sample. Statistical analysis can be made as well and provide unique insights *(43)*. For example, scatter plots can be prepared, in which intensities of both lipid and protein absorbance bands from all pixels are represented in what is typically called "feature space." These are populations with common protein and lipid intensities that define specific regions in the brain tissue sample. Each of the feature space populations can then be mapped back to "image space", where these intensities can be visualized in relation to the image. This allows identification of specific regions in the image with higher or lower intensities. Using this approach, it was shown that the stoichiometry of lipids and protein in specific cerebellar cells, the Purkinje cells, differ in animals treated with an antineoplastic that is known to have neurotoxic effects. This provides a rapid, quantitative approach for identifying necrotic cells.

There are a variety of ways of presenting the data in terms of visualization. We used a 3D imaging software, Voxel View (Vital Images), to examine a data set of a section of the mouse cerebellum *(43)*. While each image is 2D, if we incorporate the spectral information, this provides a third dimension (Fig. 7A). By altering contrast, it is possible to highlight specific brain components (Fig. 7B). We have selected wavenumbers specific for protein (3350 cm^{-1}) and lipid (2927 cm^{-1}) signals (Fig. 7C). As can be seen, the regions of highest protein density are distinct from those of highest lipid density. This technique has been used to compare sections from a wild-type mouse brain, and that of a mutant mouse brain which is known to have a lipid disorder. It is possible to visualize the differences in distribution of these products in the absence of any histological stain. In fact, the spectra obtained, depending on the array detector used, is capable of detecting over 30 different peaks of biochemical components in the mid-IR signal, each attributable to a specific biochemical and or biophysical property. Thus, a single IR spectroscopic array scan can provide more than 30 different stains of the same section. Of course, the ability to obtain an image for that component depends on the intensity of the signal. Many of these components cannot be distinguished in the healthy specimen. However, comparison to a pathological or intoxicated specimen, using the mathematical approaches available, may confer the capability of identifying specific components that are altered. This could lead to the development of specific biomarkers. While the studies described here are using transmission microscopy, this technique is being developed for performance in the reflectance mode for in vivo use.

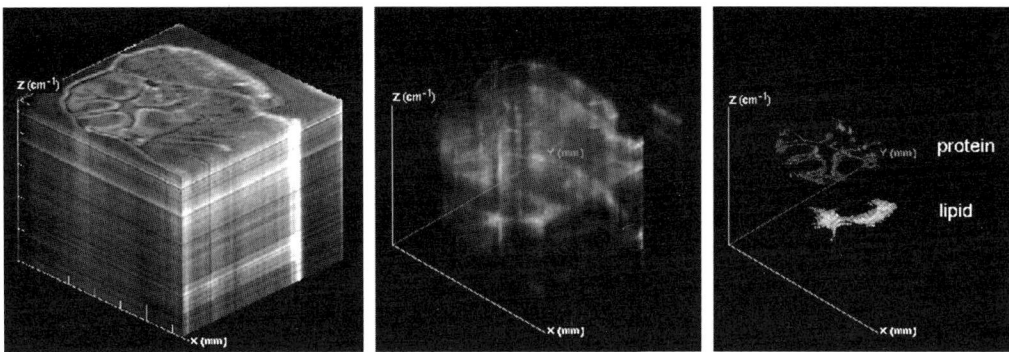

Fig. 7. 3D representation of mid-IR microspectroscopic dataset for a mouse cerebellum. A mid-IR spectral image has collected for a 10-μm thick section of a mouse cerebellum. The third dimension is the spectroscopic data. This can be visualized as a stack (left). The contrast can be adjusted to highlight special structural features of the section (center). The lower threshold has been increased above the level of "noise". Through contrast enhancement, spectral images for the protein and lipid fractions can be identified, highlighted, and identified demonstrating the differential distribution of lipid and protein component gradients (right). The protein absorbance wavenumber is highlighted in red, and the lipid absorbance wavenumber is highlighted in green. A color version of this figure is available in the CD-ROM.

17.4. CONCLUSIONS

Traditional approaches to neurotoxicity screening are relatively successful. Approximately, 80 to 90% of neuronal adverse effects are identified in preclinical studies using behavioral and histopathological approaches. However, this information is not readily transferred to clinical trials due to the reasons discussed in this chapter. Imaging provides the ability of identifying neurological changes in animal studies and then transferring these observations such that appropriate measurements can be made in the clinic. In addition, as the image data is intrinsically digital, quantitative analyses are readily performed. However, imaging does not necessarily solve many of the difficulties in determining neurotoxicity described earlier in this manuscript. There is still the difficulty in distinguishing between pharmacological and toxicological changes. Further controlled studies are necessary in order to further elucidate this difficulty.

ACKNOWLEDGMENTS

We would like to thank Jan Johannessen, Nathan Appel, Marielle Delnomdedieu, Allan Johnson, Neil Lewis, and Linda Kidder for their contributions to portions of the data presented in this manuscript.

REFERENCES

1. Sette WF. Complexity of neurotoxicological assessment. Neurotoxicol Teratol 1987; **9**:411–416.
2. Baird AE, Lovblad KO, Dashe JF, et al. Clinical correlations of diffusion and perfusion lesion volumes in acute ischemic stroke. Cerebrovasc Dis 2000; **10**:441–448.
3. Wozniak DF, Dikranian K, Ishimaru MJ, et al. Disseminated corticolimbic neuronal degeneration induced in rat brain by MK-801: potential relevance to Alzheimer's disease. Neurobiol Dis 1998; **5**:305–322.

4. Moser VC. Applications of a neurobehavioral screening battery. J Am Coll Toxicol 1992; **10**:661–669.
5. Moser VC, Cheek BM, MacPhail RC. A multidisciplinary approach to toxicological screening. III. Neurobehavioral toxicity. J Toxicol Environ Health 1995; **45**:173–210.
6. Baird SJS., Catalano PJ, Ryan LM, Evans JS. Evaluation of effect profiles: functional observational battery outcomes. Fund App Toxicol. 1997; **40**:37–51.
7. Kulig BM. Comprehensive neurotoxicity assessment. Environ Health Perspect 1996; **104**:317–322.
8. Sobotka TJ, Ekelman KB, Slikker W, Raffaele K, Hattan DG. Food and Drug Administration proposed guidelines for neurotoxicological testing of food chemicals. Neurotoxicology 1996; **17**:825–836.
9. Tilson HA, MacPhail RC, Crofton KM. Defining neurotoxicity in a decision-making context. Neurotoxicology 1995; **16**:363–375.
10. Barone S, Das KP, Lassiter TL, White LD. Vulnerable processes of nervous system development: a review of markers and methods. Neurotoxicology 2000; **21**:15–36.
11. Costa LG Biochemical and molecular neurotoxicology: relevance to biomarker development, neurotoxicity testing and risk assessment. Toxicol Lett 1998; **102–103**:417–421.
12. O'Callaghan JP. Biochemical analysis of glial fibrillary acidic protein as a quantitative approach to neurotoxicity assessment: advantages, disadvantages and application to the assessment of NMDA receptor antagonist-induced neurotoxicity. Psychoparm Bull 1994; **30**:549–554.
13. Werner M, Chott A, Fabiano A, Battifora H. Effect of formalin tissue fixation and processing on immunohistochemistry. Am J Surg Pathol 2000; **24**:1016–1019.
14. Fix AS, Garman RH. Practical aspects of neuropathology: a technical guide for working with the nervous system. Toxicol Pathol 2000; **28**:122–131.
15. Lev-Ram V, Makings LR, Keitz PF, Kao JP, Tsien RY. Long-term depression in cerebellar Purkinje neurons results from coincidence of nitric oxide and depolarization-induced Ca^{2+} transients. Neuron 1995; **15**:407–415
16. Rasouly D, Rahamun E, Lester DS, Matsuda Y, Lazarovici P. Staurosporine neurotropic effects in PC12 cells are independent of protein kinase C. Mol Pharmacol 1992; **42**:35–43.
17. Lester DS, Bramham CM. The mechanism of activation of protein kinase C in long-term neuronal processes. Invited Minireview Cellular Signalling 1993; **5**:695–708.
18. Johnson GA, Benveniste H, Engelhardt RT, Qui H, Hedlund LW. Magnetic resonance microscopy in basic studies of brain structure and function. Ann NY Acad Sci 1997; **820**:139–148.
19. Ben-Horin N, Hazvi S, Bendel P, Schul R. The ontogeny of a neurotoxic lesion in rat brain revealed by combined MRI and histology. Brain Res 1996; **178**:97–104.
20. Rajan SS. MRI: A Conceptual Overview. Springer, New York, 1998.
21. Kleinschmidt A, Frahm J. Linking cerebral blood oxygenation to human brain function. Current issues for human neuroscience by magnetic resonance neuroimaging. Adv Exp Med Biol 1997; **413**:221–233
22. Schmidt ME. The future of imaging in drug discovery. J Clin Pharmacol 1999; **39**:45S–50S.
23. Fowler JS, Volkow ND, Ding YS, et al. Positron emisiion tomography studies of dopamine-enhancing drugs. J Clin Pharmacol 1999; **39**:13S–16S.
24. Fowler JS, Volkow ND, Ding YS, Want GJ. PET and the study of drug action in the human brain. Pharmaceutical News 1995; **5**:11–16.
25. Salazar DE, Fischman AJ. Central nervous system pharmacokinetics of psychiatric drugs. J Clin Pharmacol 1999; **39**:10S–13S.
26. Wolf H, Grunwald M, Kruggel F, et al. Hippocampal volume discriminates between normal cognition; questionable and mild dementia in the elderly. Neurobiol Aging 2001; **22**:177–186.

27. Hedlund LW, Johnson GA, Maronpot RR. Magnetic resonance microscopy – A new tool for the toxicologic pathologist. Toxicol Pathol 1996; **24**:36–44.
28 Johnson GA, Benveniste H, Black RD, Hedlund LW, Maronpot RR, Smith BR. Histology by magnetic resonance microscopy. Magn Reson Q 1993; **9**:1–30.
29. Lester DS, Lyon RC, McGregor GN, et al. 3-Dimensional visualization of excitotoxic lesions in rat brain using magnetic resonance imaging microscopy. Neuroreport 1999; **10**:737–741.
30. Lester DS, Pine PS, Delnomdediu M, Johannessen JN, Johnson GA. Virtual neuropathology: 3-dimensional visualization and quantitation of lesions due to toxic insult. Toxicol Pathol 1999; **28**:100–104.
31. Delnomdedieu M, Appel NM, Pine PS, Hayakawa T, Lester DS, Johnson GA. MR microscopy of contrast-structure in a rat model of Parkinson's diseas. Proc ISMRM 1999; **7**:450.
32. Redmond IT, Barbosa S, Blumhardt LD, Roberst N. Short-term ventricular volume changes on serial MRI in multiple sclerosis. Acta Neurol Scand 2000; **102**:99–105.
33. Lycklama a Nijeholt GJ, Bergers E, Kamphorst W, et al. Post-mortem high-resolution MRI of the spinal cord in multiple sclerosis. A correlative study with conventional MRI, histopathology and clinical phenotype. Brain 2001; **124**:154–166.
34. Kalkers NF, Bergers L, de Groot V, et al. Concurrent validity of the MS functional composite using MRI as a biological disease marker. Neurology 2001; **56**:215–219.
35. Friese SA, Bitzer M, Freudenstein D, Voigt K, Kuker W. Classification of acquired lesions of the corpus callusom with MRI. Neuroradiology 2000; **42**:795–802.
36. Filippi M, Rovaris M, Iannucci G, Mennea S, Sormani MP, Comi G. Whole brain volume changes in patients with progressive MS treated with cladribine. Neurology 2000; **55**:1714–1718.
37. Achiron A, Barak Y. Multiple sclerosis –from probable to definite diagnosis: a 7-year prospective study. Arch Neurol 2000; **57**:974–979.
38. Wiendl H, Neuhaus O, Kappos L, Hohlfeld R. Multiple sclerosis. Current review of failed and discontinued clinical trials of drug treatment. Nervenarzt. 2000; **71**:597–601.
39. Viala K, Stievenart JL, Cabanis EA, Lyon-Caen O, Tourbah A. Magnetic resonance spectroscopy in multiple sclerosis. Rev Neurol (Paris) 2000; **156**:1078–1086.
40. Mendelsohn R, Mantsch HH. Fourier transform infrared studies of lipid-protein interaction. In: Progress in Protein Lipid Interactions 2. (Watts A, DePont J, eds.). Elsevier Science Publishers, Holland, 1986, pp.103–146.
41. Lewis EN, Treado PJ, Reeder RC, et al. Fourier transform spectroscopic imaging using an infrared focal plane array detector. Anal Chem 1995; **67**:3377–3381.
42. Lewis EN, Kidder LH, Levin IW, Kalasinsky VF, Hanig JP, Lester DS. Applications of fourier transform infrared imaging microscopy in neurotoxicity. Ann NY Acad Sci 1997; **820**:234–247.
43. Lester DS, Kidder LH, Levin IR, Lewis EN. Infrared microspectroscopic imaging of the cerebellum of normal and drug treated rats. Cell Mol Biol 1998; **44**:29–38.

18
How the Brain Develops and How It Functions
Application of Neuroanatomical Data of the Developing Human Cerebral Cortex to Computational Models

William Rodman Shankle, Junko Hara, James H. Fallon, and Benjamin Harrison Landing

ABSTRACT

For many models of the cerebral cortex, particularly developmental ones, knowledge of cortical structure is essential to a proper understanding of its functional capacities. We have studied the microscopic neuroanatomic changes of the postnatal human cerebral cortex during its development from birth to 72 mo. The microscopic structural changes we have identified to date are complex, yet well organized, and mathematically describable. The discoveries to date arising from analyses of the Conel data include:

1. That the total number of cortical neurons increases by 1/3 from term birth to 3 mo, then decreases back to the birth value by 15 mo, then increases by approximately 70% above the birth value from 15 to 72 mo.
2. Based on 35 cortical areas, the mean number of neurons under 1 mm^2 of cortical surface extending the depth of the cortex decreases by 50% from term birth to 15 mo, then from 15 to 72 mo, increases by 70% above the value at 15 mo. Both of the previous findings provided the first evidence for postnatal mammalian (human) neocortical neurogenesis. These findings have received subsequent support from studies demonstrating cortical neurogenesis in adult macaque monkeys.
3. That changes in total cortical neuron number from birth to 72 mo inversely correlate strongly ($\rho = -0.73$) with the number of new behaviors acquired during this time. The correlation appears strongest when there is a time delay, suggesting that changes in cortical neuron number precede the appearance of newly acquired behaviors.
4. That each of 35 cortical areas analyzed show characteristic increases and decreases in neuron number in a wave-like fashion from birth to 72 mo, suggesting local regulatory control of both neuronal cell death and neurogenesis. The changes in neuron number appear to follow gradients that correspond to functional cortical systems, including frontal (motor, dorsolateral prefrontal, and orbitofrontal, separately), visual (ventral and dorsal streams, separately), and auditory systems.
5. That within any given cortical area from birth to 72 mo, there are functionally related shifts in the relative numbers of neurons in the six cortical layers. Since each of the six cortical layers has a specific function with specific communication to other layers, the

neocortex can create 720 variations (6!) in its function just by changing the relative power of the six cortical layers in all possible permutations. The data clearly show that only a few of these permutations are actually used. It appears that the function of secondary association neocortical areas or higher are most developed when layers III and VI have the most neurons, and that the function of primary sensory, 1st order association, or transitional (i.e., cingulate) neocortical areas are most developed when layers III and IV have the most neurons. Layer III is primarily responsible for long distance cortico-cortical communication; layer IV is primarily responsible for receiving thalamic sensory information from the environment plus feedforward cortico-cortical communication; and layer VI is primarily responsible for sending cortical information back to the thalamus and receiving feedback cortico-cortical information.

In this chapter, we present the data and studies that formed the basis for the above discoveries in postnatal human cerebral cortex from birth to 72 mo. These data have particular relevance to those interested in building computational models of cortical development and provide a basis for concomitant cortical electrophysiological and behavioral developmental changes. Computational models incorporating such knowledge may provide a mechanistic understanding of how the brain develops and how it functions.

18.1. INTRODUCTION: THE CONEL DATA

From 1939 to 1967, J.L. Conel published eight volumes, one for each age point of the postnatal development of the human cerebral cortex from birth to 72 mo *(1–8)*. Each volume reported the microscopic neuroanatomic data derived from measurements made on 5 to 9 brains. To produce these volumes, Conel made more than 4 million individual measurements of the microscopic neuroanatomic features of the developing postnatal human cerebral cortex. Table 1 summarizes the cases Conel studied.

Conel reported these data in three ways. First, he described the state of development of each feature of each layer of each cytoarchitectonic area analyzed. Second, he reported the data in tables, in which each datum represented the mean value of 30 independent measurements per microscopic neuroanatomic feature per cortical layer per cytoarchitectonic area per brain per age point (150 to 270 measurements per datum). Third, he produced photomicrographs for each staining method used in each cytoarchitectonic area studied. He also produced camera lucida drawings of the Golgi-stained material for each cytoarchitectonic area studied. Table 2 summarizes the number of cytoarchitectonic areas Conel studied for each microscopic neuroanatomic feature reported in Conel's data tables.

18.1.1. The Microscopic Neuroanatomic Features

The six microscopic, neuroanatomic features Conel measured were as follows:

1. Layer width: cortical layer thickness (in mm) for left hemisphere gyri.
2. Neuron packing density: numbers of neurons per unit of cortical volume (0.001 mm^3).
3. Somal width: midrange cell width (in µm).
4. Somal height: midrange cell height (in µm).
5. Large fiber density: numbers of Cajal or Golgi-Cox stain-positive large fibers (mostly

Table 1
Description of the Brains Studied by Conel with Adult Brain Weight for Reference

Age (mo)	Age Range, +/- mo.	No. of Brains	No. (fraction) of Males	Mean Brain Wt. (g)	Brain Wt. (std⁴)	W^1	W^2 (g)	W^3
0	0.25	6	4 (0.67)	356	10	0.26	–	–
1	0.25	5	?	413	?	0.31	57	0.16
3	0.5	6	4 (0.67)	575	98	0.43	162	0.39
6	0.5	7	6 (0.86)	761	101	0.56	86	0.15
15	3.0	9	6 (0.67)	950	113	0.70	189	0.25
24	2.75	7	4 (0.57)	1013	99	0.75	63	0.07
48	11.0	8	3 (0.38)	1151	94	0.85	138	0.14
72	8.5	6	4 (0.67)	1248	45	0.92	97	0.08
216⁵	–	–	–	1350		1.0	102	0.08

¹W: Fraction of adult weight for brain weight at the stated age.
²W: Brain weight change in grams from previous age point.
³W: Fractional change in brain weight relative to brain weight at previous age point.
⁴std, Standard deviation.
⁵Accepted brain weights for 18-year-old humans (Blinkov and Glezer, 1968), but not part of the Conel study.

large proximal dendrites, but may include some large diameter axons) per unit area (0.005 mm²) in microscopic fields of the left hemisphere.
6. Myelinated fiber density: number of Weigert stain-positive large myelinated axons per unit area (0.005 mm²) in microscopic fields of the right hemisphere.

18.1.3. Cortical Layers and Fiber Systems

Conel made measurements of the microscopic neuroanatomic features in each of the six cortical layers and for two fiber systems (subcortical and vertical exogenous). Conel described the subcortical fibers as "subcortical arcuate or short association fibers connecting adjacent gyri". He described the vertical exogenous fibers as, "those nerve fibers (axons and dendrites) in the (microscopic) field which have arisen from cells located in other more or less distant parts of the brain ... In the core of the gyrus, they are rather uniformly distributed, but immediately under the cortex, they radiate, ending mostly in layers V and VI. Some continue into layer III, and a few extend as far as II" *(1)*.

18.1.4. Cytoarchitectonic Regions

Depending upon the microscopic neuroanatomic feature measured, Conel made layer-specific measurements in 39 to 43 Von Economo-classified cytoarchitectonic areas. These cytoarchitectonic areas comprise about 73% of the entire surface of the human cerebral cortex. Conel did not measure values for some of the cytoarchitectonic areas found in orbitofrontal, anterior cingulate, and infralimbic regions (14 Von Economo areas: FG, FH, FI, FJ, FK, FL, FM, FN, TD, TH, PA, PC, LB, and LF). Table 3 provides a mapping of the Von Economo and Brodmann classifications of the cytoar-

Table 2
Number of Cytoarchitectonic Areas Studied by Microscopic Neuroanatomic Feature and by Cortical Layer or Fiber System

Feature	Cortical Layer or Fiber System								Total
	I	II	III	IV	V	VI	S^1	V^2	
Layer Width	43	43	41	43	41	41	—	—	252
Neuron dns^3	46	44	44	46	43	43	—	—	266
Somal Width	—	43	43	43	43	43	—	—	215
Somal Height	—	43	43	43	43	43	—	—	215
Cajal dns^4	49	48	49	49	49	49	48	49	390
Weight dns^5	49	48	48	48	49	49	49	49	389
Total	187	269	268	272	268	268	97	98	1727

^{1}Subcortical arcuate or short association fibers.
^{2}Vertical projection fibers coming from the central white matter cores of gyri.
^{3}Neuron packing density.
^{4}Large fiber density (Cajal stain), which are mostly large proximal dendrites and some axons.
^{5}Myelinated fiber density (Weigert stain), which are all axons.

chitectonic areas of the cerebral cortex. Figure 1 illustrates graphically the cytoarchitectonic areas measured by Conel.

18.2. ANALYSES OF DEVELOPMENTAL CHANGES: BIRTH TO 72 MONTHS

18.2.1. Global Analysis of Microscopic Neuroanatomic Changes

Method. Correspondence analysis (CA) is a powerful analytical method that allows one to ask questions about the similarity of development of different cortical areas. We used CA to analyze Conel's original uncorrected data, to determine whether there were similarities among the development of the microscopic features, the layers, the cortical areas, and the developmental stages (age points). The methods and results are fully described elsewhere *(9),* but will be summarized here.

CA is a technique that allows one to obtain a multidimensional representation, in Euclidean space, of the similarities among the rows and columns of a data matrix. In the Euclidean space, rows (and columns) that are more similar to each other are placed closer together than rows (and columns) that are less similar. The Conel data consist of a matrix with 1727 rows and 8 columns. Each row is a profile of the values of a specific microscopic feature, layer, and cortical area at each of the 8 age points for which Conel provided data. Data were available from 39 to 43 cortical (cytoarchitectonic) areas

Table 3
Mapping of Von Economo to Brodmann Area Classification of Cytoarchitectonic Areas of the Cerebral Cortex[1]

Von Economo Area	Von Economo Abbreviation	Brodmann Area	T	P	U	H	S
FAy head, hand, par lob, trunk	FAgc, Fagh, FAgl, FAgp, FAgt	4		X	X		
FB GFM post	FBFM	6, 8, 9			X	X	X
FB GFS post	FBFS	6			X		
FC Bm GFI post	FCFI	44			X		
FC GFS mid	FCFS	8			X	X	
FDR GFI ant, mid	FDFIa, FDFIm	45				X	
FDdelta GFM ant	FDFMa	46				X	
FDdelta GFM mid	FDFMm	8, 9			X	X	X
FDm GFS ant	FDFS	9				X	X
Frontal pole	FEPF	10					X
Frontal orbital	FFGO	11					X
Insula precentralis	IA	14	X				
Insular postcentral	IB	13	X				
Cingulate ant agranular	LA	24	X			X	
Cingulate post granular	LC	23, 31				X	
Retrosplen granular	LD	30	X				
Retrosplen agranular	LE	29	X				
Occip peristriate	OA	19			X		
Occip parastriate	OB	18			X		
Occip Striate, OC	OC	17		X			
Postcentral oralis, head, hand, leg, trunk	PB, PBc, PBh, PBl, PBt	3		X			
Postcentral intermed, head, hand, leg, trunk	PC, PCc, PCh, PCl, PCt	1		X			
Parietal sup lobule	PE	7			X	X	
Parietal inf ant	PF	40		X		X	
Parietal inf post	PG	39				X	
Occip temp intermed	PH	37				X	
Temporal superior	TA	22			X	X	
Temp transv ant post	TB	42			X		
Temporal transv int	TC	41	X				
Temporal mid and inf	TE	21				X	
Fusiform gyrus	TF	36				X	
Temporal pole	TG	38				X	

[1]T, Transitional neocortex; P, primary motor or sensory neocortex; U, unimodal association cortex; H, heteromodal association cortex; S, supramodal association cortex. [2]Terminology from Benson, 1994, *(19)*.

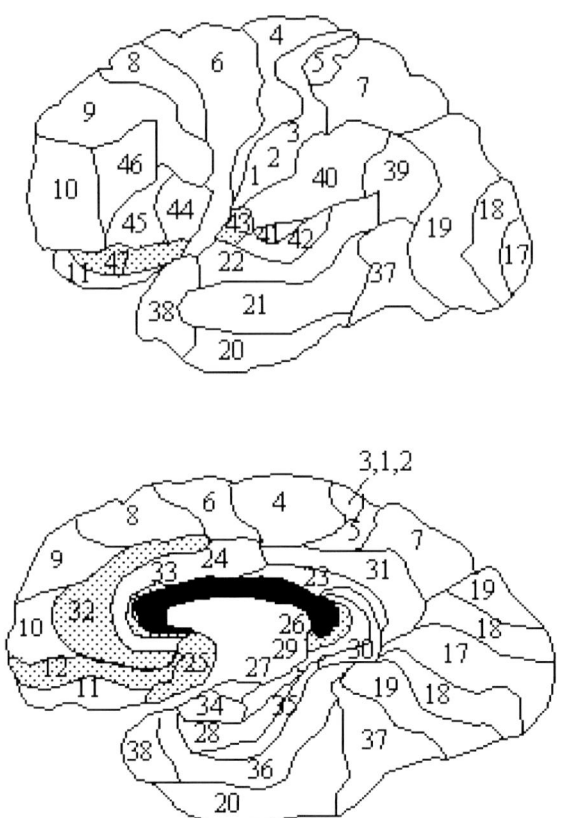

Fig. 1. Map of the cytoarchitectonic areas (Brodman classification) of the postnatal human cerebral cortex studied by Conel. Conel studied 39 to 43 cytoarchitectonic areas depending upon the microscopic neuroanatomic feature measured. The areas not studied by Conel are stippled. A color version of this figure is included in the CD-ROM.

depending upon the microscopic feature studied. The columns represent the developmental stages (age points of the data).

The steps of the CA are as follows. First, the raw 1727 row by 8 column data matrix, A, with cell entries, a_{ij}, is normalized by computing a new matrix, H, with the ijth cell entry given by $h_{ij} = a_{ij}/\sqrt{(a_{i.} a_{.j})}$, where a_{ij} is the original cell frequency, $a_{i.}$ is the total sum for row i, and $a_{.j}$ is the total sum for column j. Second, the normalized matrix is analyzed by singular value decomposition into its triple product, UDV^T, where U contains row scores, V^T contains column scores, and D is a diagonal matrix of singular values. Third, the singular vectors of the U and V^T matrices are used to compute maximally discriminating scores (i.e., optimal scores, canonical scores, variates) for the rows and columns of A. The rescaling formulae for the optimal scores are: $X_i = U_i\sqrt{(a_{..}/a_{i.})}$ and $Y_j = V_j\sqrt{(a_{..}/a_{.j})}$.

Since correspondence analysis scales the profiles solely in terms of overall shape similarity, the resulting row scores remain invariant under multiplication of the data of a given profile by a constant. Thus, two profiles that have the same "shape" will be scaled as identical regardless of how much they differ in absolute magnitude. For this reason, it did not matter whether we used Conel's uncorrected values or values that

A Cortical Layer.

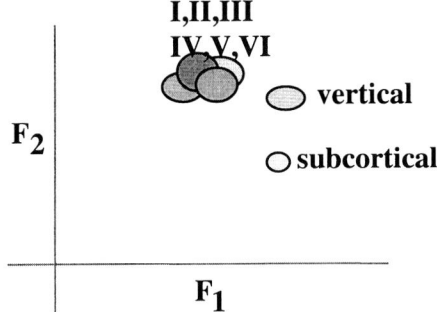

Fig. 2A. Ninety-nine confidence ellipses of the first two correspondence analysis factors for all data within each layer or fiber system. For each cortical layer or fiber system, its confidence ellipse circumscribes the area in which 99% of the values of the first and second correspondence analysis factors, F1 and F2, reside. The extremely small sizes of these ellipses indicate a high degree of similarity among the relative changes of the microscopic features in the different BA within each ellipse. The strong overlap of the ellipses for the six cortical layers also indicates a high degree of similarity among the relative changes in the microscopic features within the different BA across the six cortical layers. The separation of the vertical and subcortical fiber system ellipses from the cortical layer ellipses indicates that different relative changes are occurring during development within these fiber systems.

B Microscopic feature.

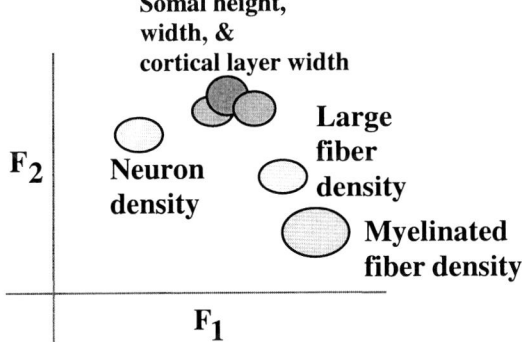

Fig. 2B. Ninety-nine confidence ellipses of the first two correspondence analysis factors for all data within each microscopic neuroanatomic feature. For each microscopic neuroanatomic feature, its confidence ellipse circumscribes the area in which 99% of the values of the first and second correspondence analysis factors, F1 and F2, reside. The extremely small sizes of these ellipses (except that for myelinated fiber density) indicate a high degree of similarity among the relative changes of a given microscopic feature across the cortical layers and cytoarchitectonic areas studied. The strong overlap of the ellipses for somal height, somal width, and cortical layer width indicates that these features are highly similar in their relative developmental changes. The larger confidence ellipse for myelinated fiber density indicates that either cortical layer or cytoarchitectonic area or both has some influence on the relative changes in human cortical myelination.

were corrected for shrinkage and stereologic error. In addition to scaling the profiles in terms of similarity, correspondence analysis scales the eight developmental stages, providing a representation of the relative similarity of the age-specific data among them that will quantify the amount of change from one stage to another.

To examine for structural influences on postnatal human cortical development, we first grouped the factor scores produced by CA for each profile according to cytoarchitectonic area, cortical layer, and microscopic neuroanatomic feature. Next, we computed the 99% confidence ellipses of the mean values of the first two CA factors for each of the three structural groupings.

Finding: CA factor score summary data. The first 2 CA factors explained 71.3% (56.7% and 14.6%, respectively) of the variance of the entire Conel data (representing 4 million measures). Contrary to a belief by some that the Conel data are of poor quality *(10)*, this analysis provides mathematical proof that these data have very little noise and are of very high quality.

Finding: 99% confidence ellipses by cortical layer/fiber system. Figure 2A shows the 99% confidence ellipses of the first two CA factors for the data within each cortical layer or fiber system. The confidence ellipses of all the cortical layers almost entirely overlap, indicating that their relative developmental changes are essentially the same. Their very small 99% confidence ellipses mean that neither cytoarchitectonic area nor microscopic neuroanatomic feature has any effect on the relative developmental changes that occur within each cortical layer. The 99% confidence ellipses for the two fiber systems do not overlap those of the cortical layers. Their complete separation, from each other as well, indicates that the relative changes in the subcortical or short association (usually U fibers) and vertical projection fiber systems develop in different ways from each other as well as differently from the cortical layers.

Finding: 99% confidence ellipses by microscopic neuroanatomic feature. Figure 2B shows the 99% confidence ellipses of the means of the first two CA factors for the data within each microscopic neuroanatomic feature. The data for each confidence ellipse are the relative developmental changes of the given microscopic feature within the cortical layers and cytoarchitectonic areas studied. The ellipses for cortical layer thickness and neuronal somal dimensions strongly overlap, while those for neuron packing density, myelinated axon density, and proximal dendrite density are fully separated. This means that the relative developmental changes in the data for each microscopic neuroanatomic feature provide independent information about cortical development. Because the ellipses for cortical layer thickness and somal dimensions overlap, they appear to be similar in terms of their relative change from birth to 72 mo.

Except for myelinated fiber density, the very small 99% confidence ellipses mean that the data within them are all very similar. Since the data within each ellipse consist of the relative developmental changes of the given microscopic feature for all cortical layers and cytoarchitectonic areas studied, neither cortical layer nor cytoarchitectonic area significantly influences the relative changes in these microscopic neuroanatomic features. The only exception is myelinated fiber density, whose larger confidence ellipse indicates that either cytoarchitectonic area or cortical layer or both have some effect on the relative changes in postnatal human cortical myelination.

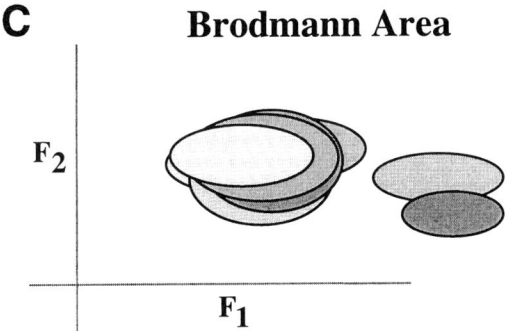

Fig. 2C. Ninety-nine percent confidence ellipses of the first two correspondence analysis factors for all data within each of 43 cytoarchitectonic areas (Brodmann). The two outlying ellipses both come from the hippocampal formation; the remaining 41 cytoarchitectonic areas are neocortical. For each cytoarchitectonic area, its confidence ellipse circumscribes the area in which 99% of the values of the first and second correspondence analysis factors, F1 and F2, reside. The relatively larger sizes of these ellipses, compared to those in Figures 2a and 2b, indicate that the relative developmental changes in the microscopic features within the cortical layers of each area are more variable. Inspection of Figure 2a and b suggests that this greater variability is largely due to the microscopic features since they occupy different parts of the plot of F1 vs F2. The strong degree of overlap for all 41 neocortical areas studied indicates there is no specific influence of any given neocortical area on the relative developmental changes in its microscopic features within its layers. However, the ellipses for two areas within the hippocampal formation (presubiculum and perirhinal cortex), a more primitive three-layered cytoarchitectonic area shows a clear separation from those of the neocortical areas. This separation indicates that the microscopic structural development of archicortex is fundamentally different from that of neocortex.

Finding: 99% confidence ellipses by cytoarchitectonic area. Figure 2C shows the 99% confidence ellipses of the means of the first two CA factors for the data within each cytoarchitectonic area. The data within each confidence ellipse consist of the relative developmental changes of the microscopic features within each of the cortical layers for a given cytoarchitectonic area. The results are striking. The ellipses for all neocortical areas (those with six cortical layers) strongly overlap, indicating that there is very little effect of the specific neocortical area on the relative developmental changes in its microscopic neuroanatomic features across its cortical layers.

However, the sizes of the confidence ellipses for each cytoarchitectonic area are much larger than those for either the cortical layer or the microscopic neuroanatomic feature (compare Fig. 2A,B,C). This means that the relative developmental changes of either the different microscopic neuroanatomic features or the different cortical layers or both, introduce greater variability in the relative developmental changes within each cytoarchitectonic area. Inspection of Figure 2A and B suggests that the larger confidence ellipse sizes of the cytoarchitectonic areas in Figure 2C is largely due to the greater differences in the relative developmental changes of the microscopic neuroanatomic features studied (they occupy different parts of the plot space of F1 and F2).

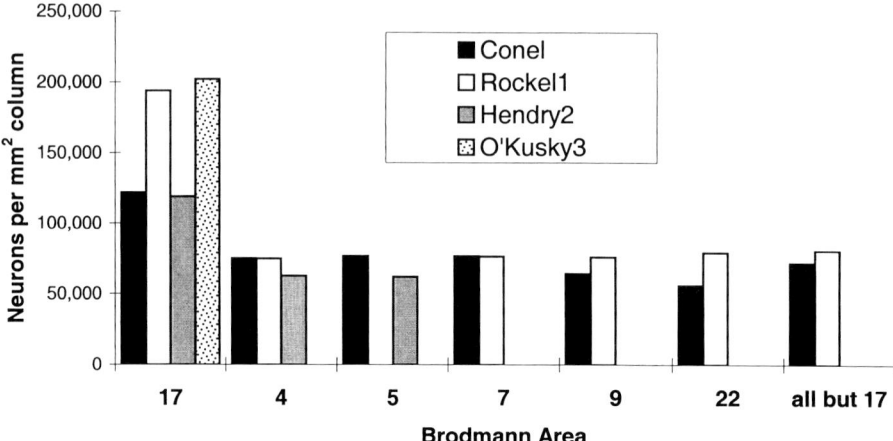

Fig. 3. Comparison of data from Conel vs data from more contemporary authors on the numbers of neurons under 1 mm^2 of cortical surface extending the full depth of the cortex. All data are corrected for shrinkage due to the methods of tissue preparation as well as for stereological error when necessary *(11)*. Comparable data were available for five Brodmann cytoarchitectonic areas (4, 5, 7, 9, 17, and 22). Except for BA 17, primary visual cortex, the values derived from Conel's data are highly similar to those obtained by authors using more contemporary methods of estimating neuron numbers. For BA 17, the data of Conel and Hendry are highly similar while the data of Rockel and O'Kusky give higher values. This most likely represents a sampling bias in BA 17, where some parts of BA 17 have very high numbers of neurons in the granule cell layer. Conel's data for BA 17 neuron number per 1 mm^2 column are about twice the values for other cortical areas he measured, which is generally agreed to be the case. This is one of the pieces of evidence supporting the high quality of Conel's data.

The only two cytoarchitectonic areas whose confidence ellipses did not overlap with the rest were non-neocortical. They were the presubiculum (Von Economo area HC, Brodmann areas [BA] 27 and 35) and perirhinal cortex (Von Economo area HD, BA 27 and 34). These two periarchicortical areas are both part of the hippocampal formation, have only several cortical layers, and are evolutionarily more primitive than neocortex.

In conclusion, models of a particular cytoarchitectonic area should include parameters reflecting effects of cortical layer and/or microscopic, neuroanatomic feature on relative cortical development.

18.2.2. Numbers of Neurons per 1 mm^2 Column per Cytoarchitectonic Area

Method. The numbers of neurons in columns of 1 mm^2 cross-sectional area (neurons per mm^2 column) extending the full depth of the cortex were computed for each of 35 cytoarchitectonic areas at each age point of the Conel data from 0 to 72 mo. The data were corrected for shrinkage due to tissue preparation and for stereologic counting errors using methods fully described elsewhere *(11)*.

Figure 3 shows that the corrected data agree with those of contemporary authors who looked at comparable cytoarchitectonic areas *(12–14)*.

Finding: local control of proliferation and pruning of human postnatal cortical neuron number. Figure 4 presents the corrected mean values for the pyramidal,

How the Brain Develops and Functions

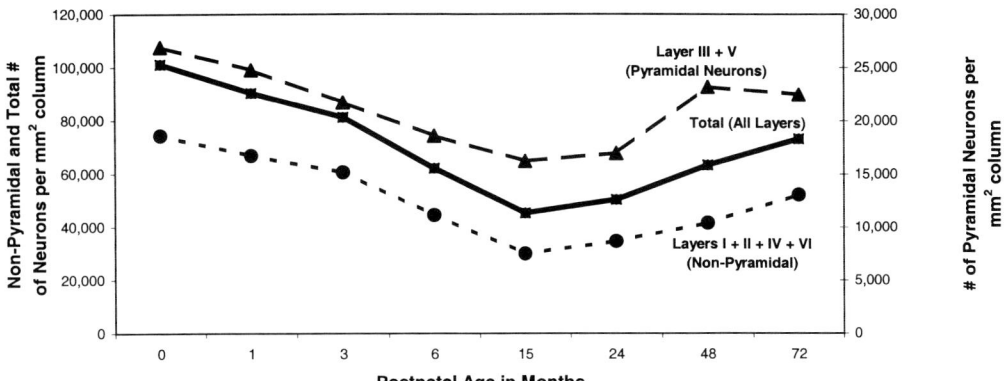

Fig. 4. Mean neuron number per 1 mm² column for pyramidal cell vs non-pyramidal cell neuronal layers. Layers III and V are almost entirely comprised of pyramidal neurons while layers I, II, IV and VI have varying proportions of non-pyramidal neurons. The purpose of this figure is to show that both pyramidal and non-pyramidal neurons have the same mean patterns of increases and decreases. Within the limits of resolution of the present data, it appears that all classes of neurons participate in postnatal neurogenesis, not just granule cell neurons. Because these data are based on 150 to 270 repeated measures per data point, and the standard error of the mean values presented is proportional to $1/\sqrt{N}$, the error bars will be within 6 to 9% of the mean values.

nonpyramidal, and total numbers of neurons per mm² column from birth to 72 mo. The data show a 54% decline from the value at birth to 15 mo, followed by a 70% increase from the value at 15 mo to 72 mo. Given the 1.3-fold increase in human cortical surface from 15 to 72 mo *(15)*, total cortical neuron number at 72 mo would be estimated by this method to be 2.2 times the value at 15 mo. The age at which the smallest (nadir) number of neurons per mm² column occurs varies for different cytoarchitectonic areas and ranges from 6–24 mo (mean nadir age = 15.8 mos). Ninety-five percent of the decline to the nadir neuron number per mm² column can be explained by the concomitant 2.1-fold increase in cortical surface *(15)*. This means that neuronal death can only account for about 5% of the observed changes in neuron numbers per mm² column. However, neuron death may be more prevalent if there is concomitant neurogenesis to minimize net reduction in neuron number per mm² column.

Figure 5A and B show the absolute and normalized (by the area's birth value) numbers of neurons per mm² column for each of 35 cytoarchitectonic areas. Figure 5B shows that, prior to the nadir neuron number per mm² column, no cytoarchitectonic area rises above 120% of its birth value. At the nadir, no cytoarchitectonic area drops below 30% of its birth value. After the nadir, no cytoarchitectonic area rises above 107% of its birth value. Thus, there appears to be an envelope limiting the maximum increase and decrease in neuron number. That the mechanism(s) limiting unconstrained neurogenesis and neuronal death probably operate over a limited distance is suggested by the data of Table 4.

Table 4 shows that, for the 35 cytoarchitectonic areas studied, 15 patterns of increase and decrease in neuron number per mm² column occur from birth to 72 mo. With so many wave-like patterns, a global control of neurogenesis and neuronal death seems

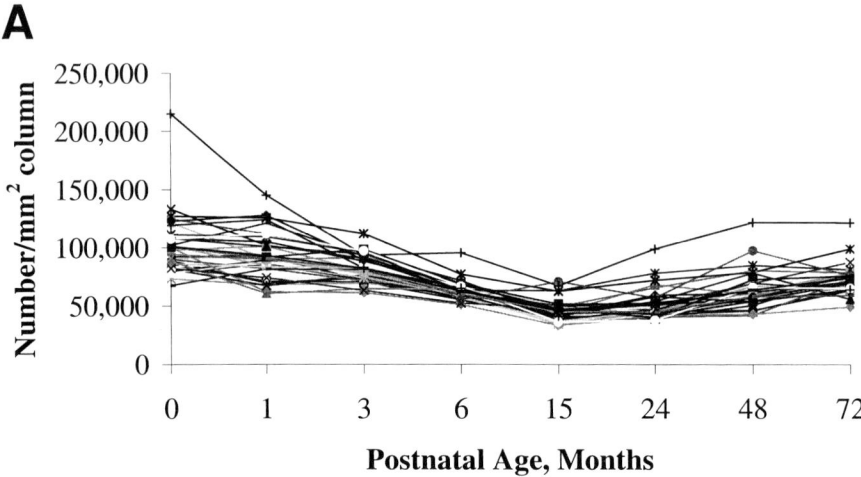

Fig. 5A. Neuron number per 1 mm² column for 35 areas of the human cerebral cortex. These data were corrected for shrinkage due to the methods of tissue preparation used by Conel as well as for stereological error *(11)*. One can see a common pattern of rises and falls in neuron number, with most cortical areas having a nadir value at 15 mo. However, detailed analysis of these data identified 15 patterns of increases and decreases among the 35 cortical areas. A color version of this figure is included in the CD-ROM.

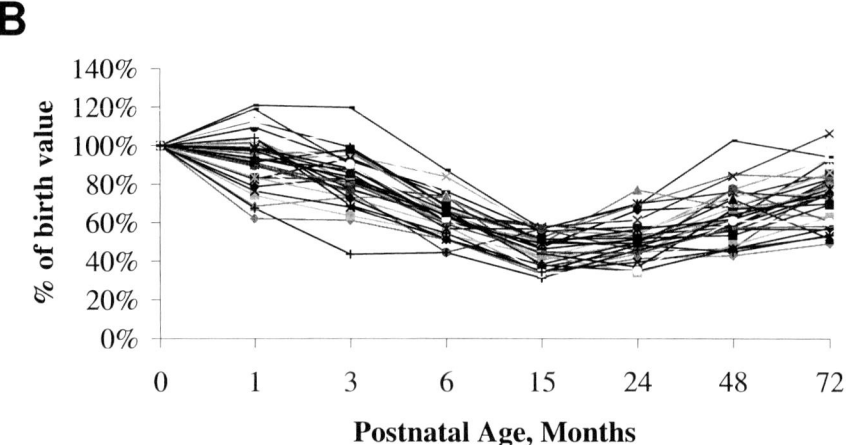

Fig. 5B. Same data as Figure 5A, except that the values of each of the 35 cytoarchitectonic areas are normalized by their birth values. This normalized changes in neuron number per 1 mm² column more clearly show the variability in the patterns of increases and decreases in neuron numbers across different cytoarchitectonic areas. The majority of areas still reach their nadir values at 15 mo. Note also that all areas increase their numbers of neurons from 15 to 72 mo. A color version of this figure is included in the CD-ROM.

unlikely. One mechanism for constraining neurogenesis and neuronal death over a limited distance would be to have the neuronal fate-determining factors be released from cells involved in either neurogenesis or neuronal death. Concentration gradients of these factors would fall off sharply with distance, thus allowing for the observed vari-

Table 4
Cytoarchitectonic Areas with Similar Patterns of Change in Neuron Number per mm² Column from Birth to 72 mo

Cytoarchitectonic Areas		Postnatal age periods of change (months)						
Brodmann	Von Economo	No. of changes	0-1-3	1-3-6	3-6-15	6-15-24	15-24-48	24-48-72
23; 31; 40;	LC; PF; FAGh	1					∨	
8; 45; 42; 41; 38; 24; 29	FCFS; FDFIa; TB; TC; TG; LA; LE	1				∨		
6; 8; 9; 6; 44; 45; 8; 9; 10; 7	FBFM; FBFS; FCFI; FDFIm; FDFm; ; FEPFPE	2	∧			∨		
39	PG	2	∧				∨	
17	OC	2			∧	∨		
19	OA	2			∨			∧
9;22	FDFS; TA	2				∨		∧
46	FDFMa	2		∧		∨		
4; 4; 4; 11; 37	FAGp; FAGl; FAGc; FFGO; PH	3	∨	∧		∨		
4	FAGt	3	∨	∧			∨	
21; 14	TE; IA	3	∧			∨		∧
13	IB	4	∨	∧		∨		∧
36	TF	4	∧			∨	∧	∨
30	LD	4			∧	∨	∧	∨
18	OB	4			∨	∧	∨	∧

∨, Neuron number per mm² column first decreases then increases.
∧, Neuron number per mm² column first increases then decreases.

ety of wave-like patterns of changes in neuron numbers. If the activities of the neurogenesis and neuronal death factors are inversely related, they would also permit the observed envelopes limiting maximal changes in neuron number.

18.2.3. Numbers of Neurons per Cytoarchitectonic Area

Methods. The methods to compute the number of neurons for each of 35 cytoarchitectonic areas included those used to derive neuron number per mm² column *(11)* plus those used to derive each of their cortical surface areas for each of the 8 age points *(15)*. These data were also corrected for shrinkage and stereological error.

Finding: pruning and proliferation are coordinated within functional cortical systems. Figure 6A and B shows the absolute and normalized (by each area's birth value) numbers of neurons per cytoarchitectonic area from birth to 72 mo. Similar to the find-

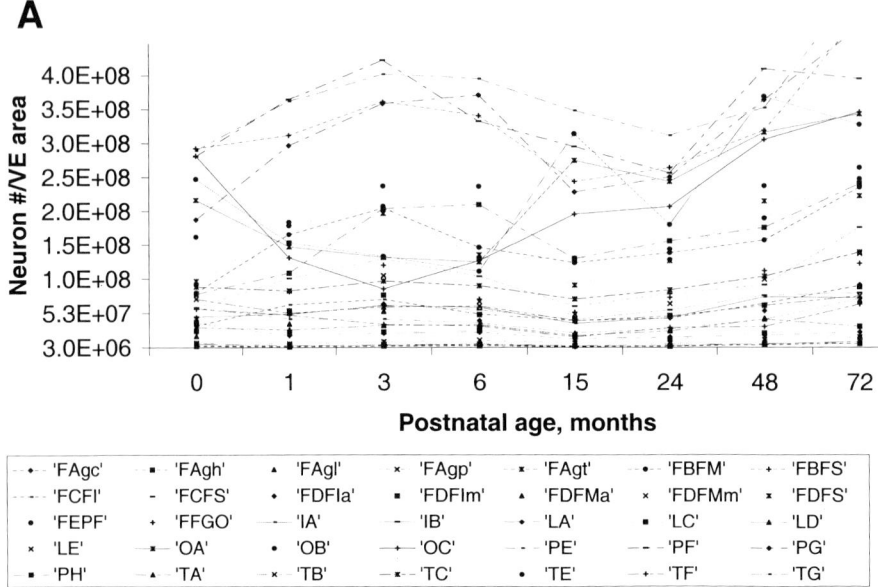

Fig. 6A. Numbers of neurons in each of 35 cytoarchitectonic areas (Von Economo classification). These data were corrected for shrinkage due to the methods of tissue preparation used by Conel as well as for stereological error *(11)*. They are based on measurements taken from the left hemisphere. Estimates of cortical surface area are described in *(15)*. The number of neurons in each cytoarchitectonic area was computed by multiplying the age-specific values of its cortical surface area by its number of neurons per 1 mm^2 column. The much greater spread of these data compared to those of Figure 5 are partly due to cytoarchitectonic differences in their cortical surface areas. A color version of this figure is included in the CD-ROM

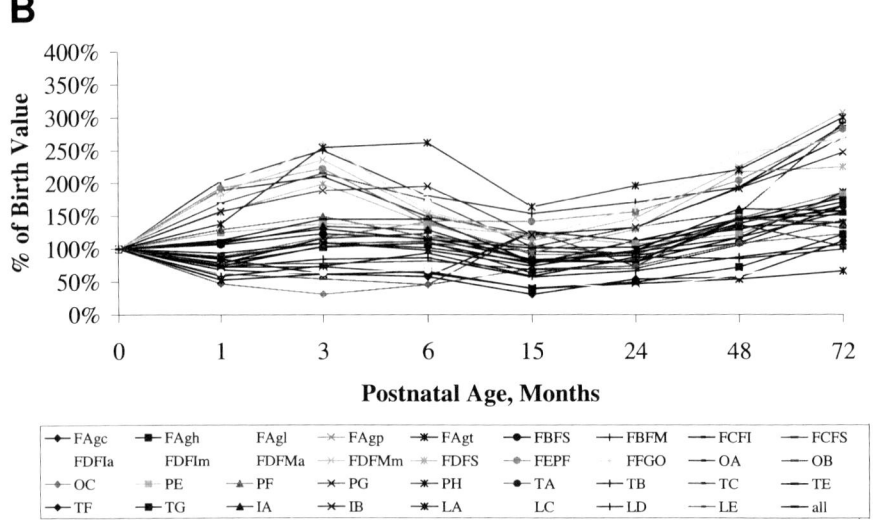

Fig. 6B. Same data as in Figure 6A but the values of each cytoarchitectonic area are normalized by their birth value. The largest postnatal increase in cortical neuron number for all areas is 350% of the birth value for area FDFMa (BA 46). All cortical areas show a postnatal increase in neuron number. A color version of this figure is included in the CD-ROM.

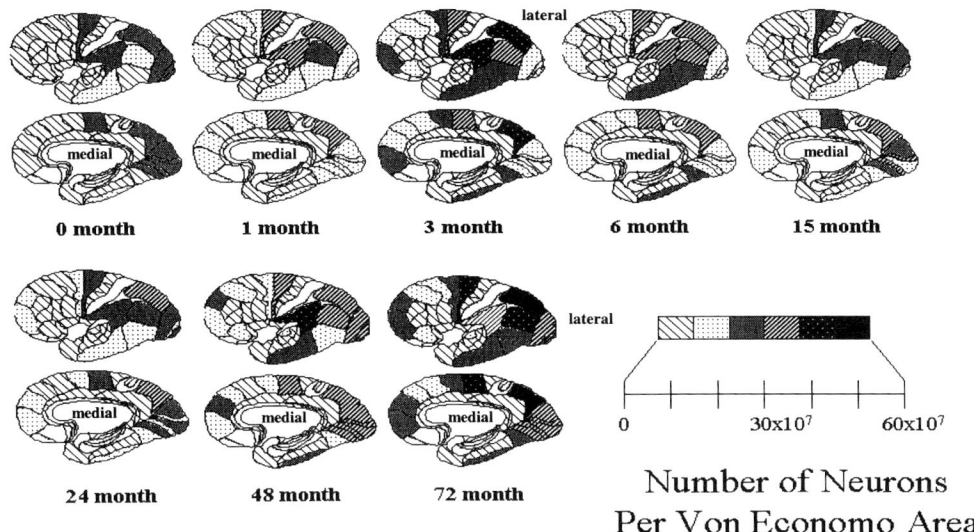

Fig. 7. Lateral and medial views to help visualize the developmental changes in cortical neuron number for each of the 35 Von Economo areas studied. A grey scale represents the lowest numbers of neurons in light cross-hatch and the highest numbers of neurons in black. Lobar patterns of similar developmental changes in cortical neuron numbers are clearly seen. Gradients of neuron number from lateral to medial do not appear to be striking, while frontal, temporal, and cingulate lobes have generally lower numbers of neurons than the parietal and occipital lobes. Note that supplementary motor area is the frontal area with the highest neuron number (gray stipple), and that the posterior superior parietal lobule has the most neurons of all areas (black). A color version of this figure is included in the CD-ROM.

ings in neuron number per mm^2 column, all cytoarchitectonic areas show a series of increases and decreases in cortical neuron number from birth to 72 mo. Also, all cortical areas increase their neuron number to at least 1.5 times their nadir value, with a mean postnadir increase of 2.5 times the mean nadir value.

Figure 7 displays these changes on lateral and medial surface maps of the cerebral cortex for each age point from birth to 72 mo. The color/grey scale represents the number of cortical neurons per cytoarchitectonic area. Adjacent cytoarchitectonic areas show more similar patterns of change in neuron number than nonadjacent ones. Also, gradients of change in neuron number occur according to functional cortical systems, including frontal (motor, dorsolateral prefrontal and orbitofrontal, separately), visual (ventral and dorsal streams, separately), and auditory systems.

18.2.4. Total Numbers of Neurons in the Cerebral Cortex

Methods. Total postnatal human cortical neuron numbers for each of the 8 age points were derived using two estimation methods fully described elsewhere *(15)*. One method (averaging method) used total cortical surface multiplied by mean neuron number per mm^2 column at each age point. The other method (summing method) summed, for each age point, the neuron numbers of the 35 cytoarchitectonic areas and divided this value

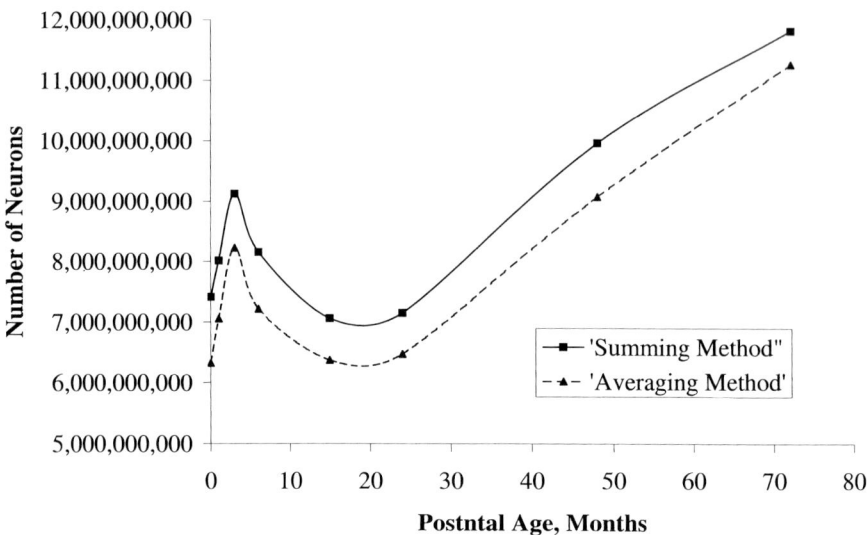

Fig. 8. Total postnatal human cortical neuron numbers for each of the 8 age points were derived using two independent estimation methods fully described elsewhere *(15)*. One method (averaging method) used total cortical surface multiplied by mean neuron number per mm^2 column at each age point. The other method (summing method) summed, for each age point, the neuron numbers of the 35 cytoarchitectonic areas and divided this value by the fraction of the total cortical surface area they occupied. The two methods of estimating total cortical neuron number per age point showed close agreement. The striking conclusion is that there are at least two waves of postnatal neurogenesis from birth to 72 mo. From birth to 3 mo, there is an approximate 1/3 increase, and from 24 to 72 mo there is an approximate 3/4 increase in total cortical neuron number.

by the fraction of the total cortical surface area they occupied. The two methods of estimating total cortical neuron number per age point showed close agreement (Fig. 8).

Finding: approximate doubling of total cortical neuron number from 15 to 72 mo. What was not seen with the data on neuron number per mm^2 column is the approximate 33% increase in total cortical neuron number from birth to 3 mo. This occurs in spite of a decrease in neuron number per mm^2 column from birth to 3 mo, because it is outweighed by such a large concomitant increase in cortical surface area that the total cortical neuron number increases.

Using methods similar to Conel, Rabinowicz found that from gestational age 32 wk to term birth, there is an approximate 70% decline in neuron number for the five cytoarchitectonic areas studied *(16)*. This means that total cortical neuron number at 32 wk gestation is approximately equal to that at 72 mo. Since total cortical neuron number at 72 mo is 10% or more below that of reported adult values *(17)*, there appear to be at least two full cycles of increases and decreases in total cortical neuron number from 32 wk gestation to adulthood (finer-grained age point data may reveal additional cycles).

Although not shown here, inspection of the camera lucida drawings of the Golgi-stained material of any cytoarchitectonic area Conel studied from birth to 72 mo show a characteristic pruning and proliferation of neuronal dendritic arbors *(1–8)*. The present data on neuron numbers indicate that one must also add proliferation and pruning of neuron numbers to this concept.

How the Brain Develops and Functions

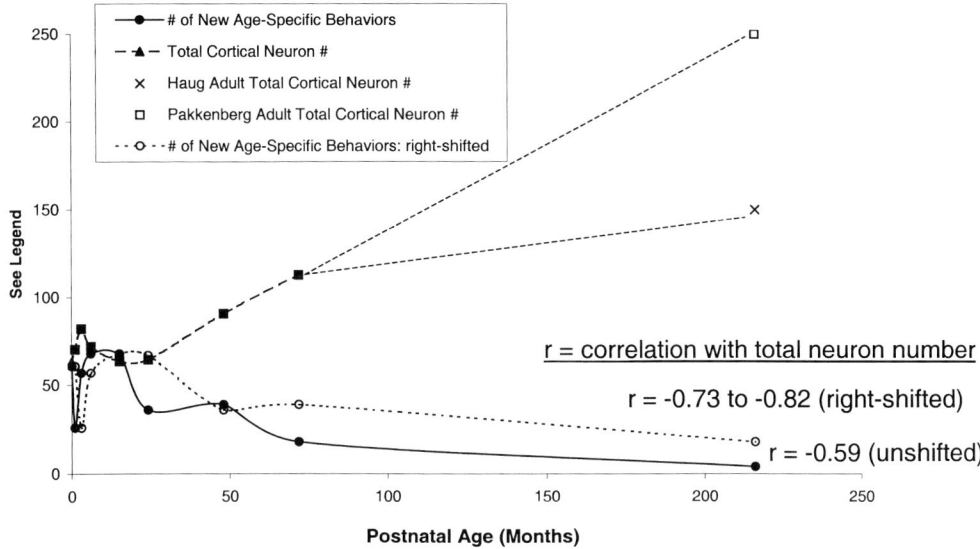

Fig. 9. Inverse relation between total cortical neuron number and number of newly acquired age-specific behaviors. Data on the mean age of acquisition of over 400 behaviors were used to compute the number of such behaviors acquired near the 8 age points Conel studied. The correlation coefficient was –0.59. However, when the behavioral data were shifted one age-point to the right (i.e., the number of newly acquired behaviors at 0 mo was shifted to 1 mo, and the same was done for 1->3, 3->6, 6->15, 15->24, 24->48, 48->72, and 72->100 mo), the correlation coefficient increased to between –0.73 to –0.82, depending upon the total cortical neuron number value used at 100 mo (the minimum and maximum reported values from the contemporary literature were used *(12,17)*. This higher correlation suggests the hypothesis that a delay is required before changes in neuron number can be translated into production of new behavioral abilities. The logistic growth of the brain (faster growth early and greatly reduced growth later) may compensate for the relatively larger shifts in the number of months between the age points of 6, 15, 24, 48, and 72 mos.

18.2.5. Relationship of Total Cortical Neuron Numbers to Acquiring New Behaviors

Methods. Data on over 400 human behaviors acquired at specific ages were obtained from a 10-yr review of the pediatric literature *(18)*. We then plotted total cortical neuron number vs the number of new age-specific behaviors at each of the 8 age points from birth to 72 mo and computed their correlation coefficient (Fig. 9).

Finding: changes in cortical neuron number probably precede acquiring new behaviors. Figure 9 shows a strong inverse correlation ($\rho = -0.59$) between total cortical neuron number and the number of newly acquired behaviors at corresponding age points. However, this correlation increases even further to between –0.73 and –0.82 when the data on age-specific behaviors are shifted by one age point (i.e., the number of newly acquired behaviors at 0, 1, 3, 6, 15, 24, 48, and 72 mo are shifted to correspond to 1, 3, 6, 15, 24, 48, 72, and 100 mo. The total cortical neuron number values used at 100 mo represented the largest and smallest reported values from contemporary

Table 5
Number of Neurons per Layer per Cytoarchitectonic Area[a]

		Postnatal age in months							
		Layers are ranked in descending order of the # of neurons per layer per column.							
VE	BA	0	1	3	6	15	24	48	72
FAGhand	4	3,2,6,1,4,5	3,6,2,4,1,5	3,6,2,4,1,5	3,6,4,2,1,5	3,6,2,4,5,1	3,2,6,4,5,1	3,6,2,4,5,1	3,6,2,4,1,5
FAGhead	4	3,6,1,2,4,5	3,6,2,4,1,5	3,6,2,4,5,1	3,2,6,4,1,5	3,2,4,6,5,1	3,2,6,4,1,5	3,6,2,4,1,5	3,6,2,4,1,5
FAGleg	4	3,4,6,1,5,2	3,6,4,2,1,5	3,6,4,2,1,5	3,6,4,2,1,5	6,3,2,4,5,1	6,3,2,4,5,1	3,6,2,4,1,5	6,3,2,4,5,1
FAGpar	4	3,6,4,1,5,2	3,4,6,1,2,5	3,6,4,2,1,5	3,2,6,4,1,5	3,2,4,6,1,5	3,6,2,4,1,5	3,6,2,5,4,1	6,3,2,4,1,5
FAGtrunk	4	2,3,4,1,6,5	3,6,2,4,1,5	3,6,2,4,1,5	3,2,6,4,5,1	3,2,6,4,1,5	3,6,2,4,5,1	3,6,2,4,5,1	6,3,2,4,1,5
FBFM	6	3,6,4,2,1,5	3,4,6,1,2,5	3,6,2,4,1,5	3,2,6,4,1,5	3,6,2,4,5,1	4,6,3,2,1,5	3,6,2,4,5,1	6,3,2,4,1,5
FBFS	6	3,4,6,1,2,5	3,4,6,2,1,5	3,6,4,2,1,5	3,2,6,4,1,5	3,6,2,4,5,1	3,6,2,4,1,5	3,6,2,4,5,1	6,3,2,4,5,1
FCFI	44	4,3,6,1,2,5	4,6,3,2,5,1	6,2,4,3,1,5	3,4,2,6,1,5	6,3,2,4,5,1	6,3,2,4,1,5	3,6,2,4,5,1	6,3,4,2,1,5
FCFS	8	6,2,3,4,1,5	3,4,6,2,1,5	3,2,4,6,1,5	3,2,6,4,1,5	6,3,2,4,1,5	6,3,2,4,1,5	3,6,4,2,5,1	6,3,2,4,1,5
FDFIa	45	4,6,3,2,1,5	4,6,3,2,1,5	4,6,2,3,1,5	2,3,4,6,1,5	3,4,6,2,5,1	6,3,4,2,5,1	6,3,4,2,5,1	6,4,3,2,1,5
FDFIm	45	6,4,3,1,2,5	4,6,3,2,1,5	4,6,3,2,1,5	3,4,2,6,1,5	3,2,6,4,5,1	6,3,2,4,5,1	6,3,2,4,5,1	6,3,4,2,5,1
FDFMa	46	4,2,6,3,1,5	4,3,6,2,1,5	4,2,3,6,1,5	4,3,2,6,1,5	4,6,3,2,5,1	6,4,3,2,5,1	6,3,4,2,5,1	6,3,4,2,1,5
FDFMm	9	6,4,3,2,1,5	3,4,6,2,1,5	4,3,6,2,1,5	3,2,6,4,1,5	6,3,2,4,1,5	6,3,4,2,1,5	3,6,4,2,5,1	6,3,2,4,1,5
FDFS	9	?	4,3,6,2,1,5	4,2,6,3,1,5	3,6,4,2,5,1	2,3,6,4,5,1	6,3,4,2,1,5	6,3,4,2,5,1	6,3,4,2,1,5
FEPF	10	6,4,2,3,1,5	4,3,6,2,1,5	4,2,3,6,1,5	6,4,2,3,1,5	3,6,4,2,1,5	3,6,4,2,5,1	3,6,4,2,5,1	6,3,4,2,5,1
FFGO	47	4,6,3,2,1,5	4,3,6,2,1,5	4,6,3,2,1,5	4,3,2,6,1,5	6,3,4,2,5,1	6,3,4,2,5,1	6,3,4,2,1,5	6,3,2,4,1,5
IA	14	3,6,4,1,5,2	3,2,6,4,1,5	3,4,6,2,1,5	3,2,6,4,1,5	3,6,2,4,5,1	3,6,4,2,1,5	3,6,2,4,5,1	3,6,4,2,5,1
IB	13	2,4,6,3,1,5	2,4,6,3,1,5	4,3,2,6,1,5	4,2,3,6,1,5	3,6,4,2,5,1	3,6,2,4,5,1	6,3,2,4,5,1	3,2,4,6,1,5
LA	24	?	?	?	6,3,2,4,1,5	3,6,2,4,5,1	6,3,2,5,1,4	6,3,2,5,1,4	6,3,2,5,4,1
LC	31	3,2,6,4,1,5	4,2,3,6,1,5	4,2,6,3,1,5	4,3,2,6,1,5	3,6,4,2,5,1	2,6,3,4,5,1	6,4,3,2,5,1	3,6,2,4,1,5
LD	30	?	3,6,1,4,2,5	3,2,6,1,4,5	3,4,2,6,1,5	3,6,4,2,5,1	3,6,4,2,5,1	3,6,4,2,5,1	6,3,4,2,5,1
LE	29	?	2,1,6,3,5,4	2,4,3,1,6,5	2,6,4,3,1,5	4,6,3,2,5,1	4,6,3,2,5,1	3,4,2,6,5,1	3,6,4,2,1,5
OA	19	4,6,2,3,1,5	4,2,6,3,1,5	4,2,3,6,1,5	4,3,2,6,5,1	3,4,6,2,5,1	6,3,4,2,5,1	3,4,6,2,5,1	4,3,6,2,5,1
OB	18	4,6,2,3,1,5	4,2,6,3,1,5	4,3,2,6,1,5	3,2,6,4,1,5	3,4,2,6,5,1	3,4,2,6,1,5	3,4,6,2,5,1	3,4,6,2,1,5
OC	17	4,3,6,2,1,5	?	4,6,2,3,1,5	4,3,2,6,1,5	4,3,2,6,5,1	4,3,6,2,5,1	4,3,6,2,5,1	4,6,3,2,1,5
PE	7	2,4,3,6,1,5	2,4,3,6,1,5	4,2,6,3,1,5	4,3,2,6,5,1	3,4,6,2,5,1	3,6,4,2,5,1	3,6,2,4,5,1	3,6,2,4,1,5
PF	40	2,4,6,3,1,5	4,2,3,6,1,5	2,4,6,3,1,5	2,4,3,6,5,1	3,6,2,4,5,1	6,3,2,4,1,5	3,6,4,2,5,1	3,4,6,2,1,5
PG	39	2,4,3,6,1,5	2,4,3,6,1,5	2,4,6,3,1,5	4,2,3,6,5,1	3,6,4,2,5,1	3,6,4,2,1,5	3,4,6,2,5,1	3,6,4,2,5,1
PH	37	4,2,3,6,5,1	4,2,3,6,1,5	2,4,6,3,1,5	4,3,2,6,5,1	3,4,6,2,5,1	6,3,4,2,1,5	4,3,6,2,5,1	4,3,6,2,5,1
TA	22	4,3,6,2,1,5	2,4,3,6,1,5	2,4,3,6,1,5	2,4,3,6,1,5	3,6,2,4,5,1	3,6,4,2,5,1	6,3,2,4,1,5	6,3,2,4,5,1
TB	42	2,4,3,6,5,1	2,4,6,3,1,5	2,4,6,3,1,5	2,4,3,6,1,5	3,2,6,4,5,1	3,2,4,6,1,5	3,4,6,2,5,1	3,4,2,6,1,5
TC	41	4,2,3,1,6,5	4,2,3,6,1,5	4,2,3,6,1,5	3,4,2,6,1,5	3,4,2,6,5,1	4,3,2,6,5,1	3,4,2,6,5,1	3,4,5,2,6,1
TE	21	2,4,3,6,1,5	2,4,3,6,5,1	4,2,6,3,1,5	2,4,3,6,1,5	3,6,4,2,5,1	3,6,4,2,5,1	6,3,4,2,5,1	6,4,3,2,5,1
TF	36	2,4,3,6,5,1	4,2,3,6,1,5	4,2,6,3,5,1	4,2,3,6,5,1	3,4,6,2,5,1	3,6,4,2,5,1	3,6,4,2,5,1	6,3,4,2,5,1
TG	38	4,6,3,1,5,2	4,6,2,3,1,5	4,2,6,3,1,5	3,2,6,4,1,5	3,6,2,4,5,1	3,6,4,2,5,1	3,6,4,2,5,1	6,3,4,2,5,1

[a]Color version is included in the CD-ROM. Color coding of the two layers with the largest numbers of neurons per 1 mm2 column for each cytoarchitectonic area at each age-point. The following color key is used:

"?" = cells in which the data for all six layers were not available.
"Red" = cells in which layers III and VI have the most neurons.
"Green" = cells in which layers II and IV have the most neurons.
"Violet" = cells in which layers IV and VI have the most neurons.
"Magenta" = cells in which layers II and VI have the most neurons.
"Light Blue" = cells in which layers III and IV have the most neurons.
"Dark Blue" = cells in which layers II and III have the most neurons.
"Yellow" = cells in which layers I and II have the most neurons.

Of the 15 possible ways of combining any two of the six layers (without regard to order), only 7 ways are used at all, and of these 7, only 4 are used frequently. Layers III and VI (red), layers II and IV (green), and layers III and IV (light blue) are the dominant permutations. The interpretation of these data presume that, for a given cortical area, the number of neurons in a layer is a rough measure of its overall computational power relative to the numbers of neurons in the other layers.

From this perspective, one can see that early on, primary motor cortex emphasizes the processing of long cortico-cortical (layer III), cortico-cortical feedback and cortico-thalamic (layer VI) information. Such integration would be useful in coordinating the survival of essential movements of the newborn child. As another example, one can see that temporal and parietal association cortical areas emphasize the processing of short cortico-cortical (layer II), cortico-cortical feedforward and thalamo-cortical (layer IV) information, but at 15 months or later, shift to emphasizing the processing of long cortico-cortical (layer III), authors for adult human cortex [12,17]). Even though shifting the data on the number of new behaviors one age point to the right shifts each data point by a different number of months, the logistic growth of the human brain, with very rapid development early on and much slower development later, suggests that such a shift may not be unreasonable (e.g., the natural logarithms of 1, 3, 6, 15, 24, 48, 72, and 100 mo give relatively uniformly separated values of 0, 1.1, 1.8, 2.7, 3.2, 3.9, 4.3, and 4.6, respectively). The relatively higher inverse correlation by right-shifting the number of newly acquired age-specific behaviors suggests that changes in cortical neuron number precede the appearance of new behaviors. The strength of this structural–behavioral correlation suggests that models of cortical cytoarchitectonic areas need to incorporate a parameter reflecting the influence of its neuron number to better understand how the functions of these cortical circuits change over development.

18.2.6. Numbers of Neurons per Layer per Cytoarchitectonic Area

Methods. The numbers of neurons per layer per mm^2 column per cytoarchitectonic area from birth to 72 mo were computed using the methods previously described *(11)*. Sufficient data were available for 35 cytoarchitectonic areas. For each cytoarchitectonic area at each age point, the layers were rank ordered from the highest to lowest neuron number per layer per mm^2 column. Table 5 shows the rank ordered data, color/grey scale-coded according to the two layers with the most neurons at each cytoarchitectonic area and age point.

Finding: layer II. At birth, layer II has the largest or second largest number of neurons in 14 of 35 areas, most notably all temporal and parietal association cortices, posterior granular cingulate, and insular cortices (BA 31, 13), the truncal part of primary motor cortex (BA 4), the superior part of the frontal eye fields (BA 8), and the anterior part of BA 46.

In contrast, by 72 mo, layer II ranks in the top two in only one cytoarchitectonic area (posterior granular insula, BA 13). This change in relative neuron number in layer II represents a marked de-emphasis of local (short distance) corticocortical signal processing as the postnatal cortex develops.

Finding: layer IV. At birth, layer IV has the largest or second largest number of neurons in in 24 of 35 areas. The exceptions include: primary motor cortex (except leg area), medial supplementary area (BA 6), superior frontal eye fields (BA 8), anterior agranular insula (BA 14), and cingulate cortex (BA 29, 30, 31).

Table 5 (*continued*) cortico-cortical feedback and cortico-thalamic (layer VI) information (similar to what primary motor and other cortical areas do as development proceeds from birth to 72 months). In general, primary sensory cortices emphasize layers long cortico-cortical (layer III), cortico-cortical feedforward and thalamo-cortical (layer IV) information processing, which is sensible considering their importance in receiving sensory information from the outside world to layer IV.

Because each cortical layer has a specific function, these data suggest that the function of each cortical area relates to the relative number of processing units (neurons) it has in each of the cortical layers. Since these relative numbers change in each cortical area during postnatal development, the function of these cortical areas presumably also changes accordingly.

In contrast, by 72 mo, layer IV ranks in the top two in only 6 of 35 areas. This change in relative neuron number in layer IV represents a marked de-emphasis of thalamocortical signal processing (environmental inputs) as the postnatal cortex develops.

Finding: layer III. At birth, layer III has the largest or second largest number of neurons in 14 or 15 of 35 cytoarchitectonic areas, particularly those associated with primary or secondary sensory, primary motor, or limbic processing (BA 17, 22, 14, 4, 6, 44, and 31). Layer III is not in the top two ranks for most association cortices at birth.

In contrast, by 72 mo, layer III ranks in the top two in all but three areas (it ranks 3rd in BA 17, anterior 45, and 21). This marked change in relative neuron number in layer III indicates an increasing emphasis on long distance corticocortical signal processing as the postnatal brain develops. Such emphasis would allow different sensory, motor, and other modalities of information to interact to produce richer more complex cortical functions (behaviors).

Finding: layer VI. At birth, layer VI has the largest or second largest number of neurons in 13 to 16 of 35 cytoarchitectonic areas, particularly those of the frontal lobe (all prefrontal and orbitofrontal association cortices, plus some primary, supplementary, and frontal eye field motor areas).

In contrast, by 72 mo, layer VI ranks in the top two in all but 7 cytoarchitectonic areas (BA 18, 19, 37, 40, 41, 42, and 13). This change in relative neuron number in layer VI indicates an increasing emphasis on corticothalamic signal processing as the postnatal brain develops. Such an emphasis may allow the cortex to influence environmental sensory inputs as they reach the thalamus before they are sent to the cortex.

Finding: motor cortex (BA 4, 6, 8, and 44). At birth, layer III neuron numbers are largest or second largest in all motor cortical areas except BA 8 (its layers II and VI occupy the top two ranks). Along with BA 14 (insular agranular cortex), motor cortex is the earliest area of the cerebral cortex to have layers III and VI with the most neurons. Some motor areas achieve this pattern by birth (BA 4 (head and lobulus parietalis parts) and medial 6). The majority of the frontal, cingulate, insular, temporal, and parietal association cortical areas also develop this pattern of layers III and VI having the most neurons, but at a later postnatal age (typically 15 to 24 mo) than the motor cortical areas.

The early appearance of layers III and VI having the most neurons in motor cortical areas suggests that the motor system must achieve complex functions early for the human organism to survive the postnatal environment. From this perspective, the association cortices provide additional refinements that come in at later ages (at about the age of the terrible twos!).

Finding: frontal association cortex. At birth, layers IV and VI have the most neurons in all frontal association cortices except the anterior part of BA 46, in which layers II and IV have the most neurons.

In contrast, usually by 24 mo and always by 72 mo, layers III and VI have the most neurons in all but one frontal association cortex (anterior part of BA 45, in which layers IV and VI have the most neurons). This shift in relative neuron number from layers IV and VI to layers III and VI in frontal association cortex by 24 mo indicates increasing emphasis on long corticocortical association signal processing, which provides the opportunity for cortical circuits to integrate more complicated (multimodal) types of information.

Finding: temporal and parietal association cortex. At birth, with the exception of temporopolar cortex (BA 38) and auditory belt association cortex (BA 22), all other temporal and parietal association cortices have layers II and IV with the most neurons.

In contrast, by 15 mo, no temporal or parietal cytoarchitectonic area has layers II and IV in the top two ranks. Instead, these areas develop such that layer III occupies one of the top two ranks along with either layer IV or VI. The only exception is BA 21, in which layers IV and VI occupy the top two ranks. When layers III and IV occupy the top two ranks, the cytoarchitectonic area has a primary sensory function (i.e., BA 41, 42) or is an intermediate association cortex between two lobes (i.e., BA 37, 40). This marked change in relative neuron number from layers II and IV to layers III and IV indicates an increasing emphasis of long corticocortical processing by 15 mo. The shift from layers II and IV to layers III and VI by 15 mo indicates that cortical interaction with thalamic input is also increasingly important as the temporal–parietal association cortex matures and produces functions that are more complex.

Finding: primary sensory, occipital, and transitional cortices (BA 41, 42, 17, 18, 19, 37, and 40). These cytoarchitectonic areas deviate from the motor and association cortical pattern of layers III and VI, ultimately developing the most neurons. Instead, they preserve the number of layer IV neurons in one of the top two ranks from birth to 72 mo, reflecting a continued emphasis on thalamic sensory input for these cortical areas.

Summary. The observed shifts in the relative laminar numbers of neurons in the different cytoarchitectonic areas from birth to 72 mo appear to have functional significance in terms of our current understanding of the functions of the different cortical layers. In general, development of higher order association cortices (those which integrate different types of sensory information) coincide with an increasing emphasis on layers III and VI signal processing from birth to 72 mo. Such increasing emphasis serves to integrate more disparate types of cortical information and to interact with thalamic input before it comes to the cortex. Other types of cortex, such as primary sensory, some unimodal association cortex (i.e., occipital cortex, which integrates only visual features), and transitional cortex (i.e., cortical areas lying between two or more lobes of the brain, whose architecture is designed to integrate their different types of information), deviate from this pattern by preserving a relatively large number of layer IV neurons from birth to 72 mo at least, presumably reflecting continued emphasis on signal processing of environmental (thalamic) input.

18.3. CONCLUSIONS

We have presented microscopic neuroanatomic data of the majority of cytoarchitectonic areas of the postnatal human cerebral cortex from birth to 72 mo. For each measure analyzed, there are consistent findings, indicating underlying rules governing the structuring of the developing human cerebral cortex. The tabular data we have put together can be used to construct computational models at a laminar level of detail of the development and/or function of virtually any cytoarchitectonic area in the human cerebral cortex. Furthermore, we have shown that even at a very coarse level (total cortical neuron number from birth to 72 mo), there is a strong correlation of the number of neurons with acquisition of new behaviors. Such a finding is consonant with the view that neuron number is quite relevant to cortical function.

At a finer level (relative laminar numbers of neurons per cytoarchitectonic area per age point), the data show a sensible developmental pattern to first permit structuring of simple cortical functions to be followed later by more complex ones. In particular, for the association cortical areas, from birth to 6 mo, layers that integrate relatively more homogeneous or unprocessed information have the most neurons (i.e., layers II and IV). From 15 to 72 mo, layers that integrate relatively more heterogeneous and more processed information develop the most neurons (i.e., layers III and VI). Such a finding is consonant with the view that cortical laminar ratios of the numbers of neurons in each cytoarchitectonic area and possibly between cytoarchitectonic areas help govern the functional capacities of cortical areas and the circuits they participate in.

REFERENCES

1. Conel JL. Postnatal Development of the Human Cerebral Cortex, Vol. 1. The Cortex of the Newborn. Harvard University Press, Cambridge, MA, 1939.
2. Conel JL. Postnatal Development of the Human Cerebral Cortex, Vol. 2. The Cortex of the One-Month Infant. Harvard University Press, Cambridge, MA, 1941.
3. Conel JL. Postnatal Development of the Human Cerebral Cortex, Vol. 3. The Cortex of the Three-Month Infant. Harvard University Press, Cambridge, MA, 1947.
4. Conel JL. Postnatal Development of the Human Cerebral Cortex, Vol. 4. The Cortex of the Six-Month Infant. Harvard University Press, Cambridge, MA, 1951.
5. Conel JL. Postnatal Development of the Human Cerebral Cortex, Vol. 5. The Cortex of the Fifteen-Month Infant. Harvard University Press, Cambridge, MA, 1955.
6. Conel JL. Postnatal Development of the Human Cerebral Cortex, Vol. 6. The Cortex of the Twenty-four-Month Infant. Harvard University Press, Cambridge, MA, 1959.
7. Conel JL. Postnatal Development of the Human Cerebral Cortex, Vol. 7. The Cortex of the Forty-eight-Month Infant. Harvard University Press, Cambridge, MA, 1963.
8. Conel JL. Postnatal Development of the Human Cerebral Cortex, Vol. 8. The Cortex of the Seventy-two-Month Infant. Harvard University Press, Cambridge, MA, 1967.
9. Shankle WR, Romney AK, Landing BH, Hara JH. Developmental patterns in the cytoarchitecture of the human cerebral cortex from birth to six years examined by correspondence analysis. Proc Nat Acad Sci USA 1998; **95**:4023–4028.
10. Korr H, Schmitz C. Facts and fictions regarding post-natal neurogenesis in the developing human cerebral cortex. J Theor Biol 1999; **200**:291–297.
11. Shankle WR, Landing BH, Rafii MS, Schiano AVR, Chen JM, Hara J. Numbers of neurons per column in the developing human cerebral cortex from birth to 72 months: evidence for an apparent post-natal increase in neuron numbers. J Theor Biol 1998; **191**:115–140.
12. Rockel AJ, Hiorns RW, Powell TP. The basic uniformity in structure of the neocortex. Brain 1980; **103**:221–244.
13. O'Kusky J, Colonnier M. A laminar analysis of the number of neurons, glia, and synapses in the visual cortex (area 17) of adult macaque monkeys. J Comp Neurol 1982; **210**:278–290.
14. Hendry S, Schwark H, Jones EG, Fan J. Numbers and proportions of GABA-immunoreactive neurons in different areas of monkey cerebral cortex. J Neurosci 1987; **7**:1503–1519.

15. Shankle WR, Rafii MS, Landing BH, Fallon JH. Approximate Doubling of the Numbers of Neurons in the Postnatal Human Cerbral Cortex and in 35 specific cytoarchitectural areas from birth to 72 months. Pediatr Dev Pathol 1999; **2**:244–259.
16. Rabinowicz T, De Courten-Myers GM, Petetot JMC, Xi G, De Los Reyes E. Human cortex development: estimates of neuronal numbers indicate major loss late during gestation. J Neuropath Exp Neurol 1996; **55**:320–328.
17. Pakkenberg B, Jorgen H, Gundersen G. Neocortical neuron number in humans: effect of sex and age. J Comp Neurol (1997) **384**:312–320.
18. Landing BH, Shankle WR, Hara J, Brannock J, Fallon JH. The development of structure and function in the postnatal human cerebral cortex from birth to 72 months: changes in thickness of layers II and III co-relate to the onset of new age-specific behaviors. Pediatr Pathol 2002; in press.
19. Benson DF. The Neurology of Thinking. Oxford University Press, New York, 1994.

19
Towards Virtual Brains

Alexei Samsonovich and Giorgio A. Ascoli

ABSTRACT

In this chapter, we argue that computers are potentially capable of human-level cognition, including experiences involving emotions, creativity, fantasy, humor, etc. In addition, we maintain that computational neuroanatomy will play a key role towards the computer generation of minds by investigating the roots of the structure–activity–function relationships in the nervous system. Using the hippocampus as a working example, we outline a long-term strategy to (*i*) implement an anatomically and biophysically accurate "bottom-up" model reproducing the known neurobiological activity; (*ii*) use the model to investigate and implement simple functional properties such as spatial mapping and path-finding; and (*iii*) insert a higher level "top-down" cognitive capacity to instantiate the concept of *agency* in the context of a generalized memory indexing theory. Finally, we briefly discuss the potential consequences of the computer generation of functionally complete virtual brains for neuroscience research, information technology, and human society.

19.1. A BRIEF HISTORY OF COMPUTERS

What will the twentieth century be remembered for in a thousand years from now? Here is a guess: the emergence of computers[1]. Apparently, the impact of computers on our society was underestimated at the very beginning. At their birth, computers (literally meaning "calculators") were designated for calculating ballistic trajectories in artillery. Here are several definitions of a computer borrowed from modern online dictionaries:

1. A machine for performing calculations automatically (WordNet® 1.6, © 1997 Princeton University).
2. A device that computes (The American Heritage® Dictionary of the English Language, Third EditionCopyright © 1996, 1992 by Houghton Mifflin Co.).
3. A machine that stores a set of instructions (a program) for processing data and then executes those instructions when requested (American Concise Encyclopedia © 1994–1996 by Zane Publishing, Inc. and CLEARVUE/eav, Inc.).

---

[1] No doubt, in such a long perspective as the next millennium, the language will undergo substantial changes, and the word "computer" might not even be appropriate to use at that time in connection with future descendants.

From: *Computational Neuroanatomy: Principles and Methods*
Edited by: G. A. Ascoli © Humana Press Inc., Totowa, NJ

4. A machine that can be programmed to manipulate symbols (The Free On-line Dictionary of Computing, © 1993–2001 Denis Howe).

The idea of using tools for storing and/or processing information encoded symbolically is probably as ancient as the human civilization itself (e.g., the abacus is roughly 7000 years old); however, the concept of information was not understood mathematically before the work of Claude Shannon *(1)*. Similarly, foundations of logic go back to the times of Plato (427–347 B.C.) and Aristotle (384–322 B.C.), but it was only in 1847–1859 A.D. that George Boole developed logic into a mathematical discipline *(2)*. The first blueprint of a general-purpose universal mechanical computer was designed by Charles Babbage in 1834 (the computer was nearly built in 1871). Only in 1936 did Alan Turing turn the idea of a universal general-purpose computer into a precise mathematical concept *(3)*.

Electronic computers emerged as a physical reality in late 1930s–early 1940s, following the onset of the electronic communication age. Here are some of the milestones. In 1906, Lee de Forest invented a vacuum triode by modifying the Edison's lamp patented in 1880. In 1940, Alan Turing's team built a computer on electromechanical relays. Konrad Zuse in 1941 and John Von Neumann in 1942 created the first programmable computers. In 1942, John Atanasoff and Clifford Berry constructed the first electronic calculator. In 1946, Presper Eckert and John Mauchly created ENIAC, the first general-purpose all-electronic computer. In 1947, John Bardeen and Walter Brattain discovered the transistor effect. In 1952, Nathaniel Rochester created IBM 701. FORTRAN has been used since 1956, and LISP since 1958. In 1958, Jack Kilby built a silicon-integrated circuit. In 1960, transistors began to be used in computers, replacing tubes. In 1971, the Intel Corporation introduced a microprocessor.

Now, we live in a world that is impossible to imagine without PCs, Internet, compact discs, and cellular phones. It seems obvious to us that the foregoing historical steps were inevitable, and from this point of view, it is not very essential who and when was the first. However, for some reason all of this did not happen before the last century. It is during the last 25 years that computer hardware underwent a multithousand-times, if not multimillion-times leap, virtually in its every parameter, including the size and the cost of the elements, the speed, the capacity of random access memory (RAM) and of permanent storage devices, the total number of computers in the world, and the scale of their integration into networks *(4)*. This process of exponential growth still shows no sign of slowing down *(5)*.

At present, at least half of our daily activity depends on information technologies and, eventually, on computers. Computer industry alone accounts for about 10% of the gross national product of the United States. Computers gradually invade all ecological niches in the infrastructure of human civilization, with only a negligible fraction of all computer resources being used for computation per se. Nevertheless, the future impact of computers is still underestimated, because the estimates are based on what computers are capable of, not on what they are potentially capable of. The origin of human blindness lies in the nature of the principal barrier in further computer evolution. This barrier has moved from the "hardware problem" to the "software problem." Stated simply, we just do not know all that computers can be used for, and therefore, we cannot imagine the consequences.

19.2. THE GREAT CHALLENGE

In 1955, John McCarthy, Marvin Minsky, Nathaniel Rochester, and Claude Shannon wrote a proposal for a project that addressed questions of how to formalize common sense and how to base computers on the principles of the human brain *(6)*. The proposed goals were to discover the principles that would allow computers to use natural language, form abstractions and concepts, solve any problems that humans can solve, improve themselves, and eventually become potentially as intelligent as humans are. The proposal coined the term "artificial intelligence" (AI) in its title and was intended for a 2 month 10 person study to be carried out during the summer of 1956 at Dartmouth College. This proposal by McCarthy et al. probably presents the greatest challenge left to us from the last century.

From a dialectic point of view, every progress is at the same time a regress. Since computers started using standard programming languages, constraints and rules were continuously introduced that disciplined programmers' thinking. As the programming languages developed, these constraints became more and more complicated, in order to satisfy programmer needs. On the one hand, software developers' tools grew more powerful, inevitably adopting principles of human cognition (e.g., object-oriented programming). On the other hand, similarly to a tight cage, they restrained the human mind, thus making each new methodological breakthrough harder and harder. During this process, breaking one cage resulted in a necessity to create and move to another, more hardened cage. The initial ultimate freedom and cluelessness in programming a Turing machine was traded for powerful and practically effective chains and ball[2]. People created more and more sophisticated software tools trying to meet their needs; however, instead of gaining freedom from exploiting computers, they only multiplied user needs and discovered that they themselves had turned into computer slaves, to the extent that at some point in the future this circumstance could force them to return back to 1955 and to rethink the entire paradigm from scratch.

Several attempts of this sort have actually been made, among which are artificial neural networks and evolutionary computation; however, none of these approaches resulted in a good general solution to the problem of turning computers into "intelligent" devices. Today computer scientists and engineers entertain the idea of creating conscious virtual agents, but nobody seems to understands precisely what this would mean, or, more generally, what consciousness is. This is not surprising, as many other concepts are similarly not yet precisely understood: for instance, the concepts of existence, compassion, humor, pretend play, voluntary intention, the concept of Self, and even the idea of concepts in general. These issues will inevitably become relevant to computer science, if computers are to learn about common sense. Present computers require humans serving them at every step, because they (computers) lack their own common sense initiative. This in turn seems to be due primarily to their "computer autism"; the lack of the ability to represent and use basic human concepts of the types listed above. Something very fundamental is missing in all computer designs and dynamics, while being present in the human brain.

---

[2]Although each modern general-purpose computer in principle could be operated as a Turing machine, this would be an extremely counterproductive way of using it.

What is the "special" ingredient that is so crucial for human cognition, yet still absent from artificial computational processes? And how could it be implemented in computers? In order to find the answers, we need to understand how our mind works. A possible "top-down" approach would be to use introspection in order to study human mind and then to copy all of its properties into a computer. Early in the twentieth century, however, behaviorists dismissed introspection as a scientific method due to its subjectivity. Behaviorist paradigm, although consistent with contemporary cybernetics, brought nothing good in return relative to the understanding of human mind. Thus, a cognitive revolution was necessary in the middle of the century in order to change the dominant paradigm of scientific psychological thinking from behaviorism to cognitive psychology grounded in neuroscience, thus selecting a "bottom-up" approach.

19.3. IMPLEMENTING THE BRAIN

In essence, the cognitive psychological paradigm can be characterized as focusing on information processing in the brain. Methods of analysis involve connectionist and other computer modeling of brain structures and functions. This approach became extremely productive in the second half of the century, when both new brain imaging techniques and new computational tools became available. As a result, brains are now viewed as giant information-processing machines. From this point of view, a thinking machine could be conceived and, in principle, created by understanding the basic information processing functions of the brain, and then transferring them from a protein-and-ion-based computer into another, silicon or GaAs-based computer. In this scenario, the straightforward bottom-up approach to artificial cognition would be to analyze the structure and fundamental dynamics of the brain, to copy them step-by-step into a computer, and to watch higher cognitive functions "emerge".

The key problem of bottom-up modeling strategies is that of defining the appropriate level of detail in the model. What aspects of real brain anatomy and physiology should be reproduced in virtual brains? Are subneuronal dynamics (such as protein regulation) necessary to capture the computing ability of single neurons? Or, should neurons themselves be considered the elementary computing blocks of the brain? Or else, should virtual brains be based not on realistic neurons at all, but rather on blocks representing cell assemblies? The task at hand is thus (*i*) to understand what "bottom level" details are essential for providing the right functionality at the top, and (*ii*) to simplify these details up to a maximal extent, while maintaining the upper-level functionality.

As an initial assumption, we consider neurons as the starting point of a sound bottom-up strategy. This choice is based on the conviction that the action potentials exchanged among neurons are, at present, the best candidates for information carrying signals in the nervous system. However, this initial choice can be modified in the future by the addition of lower level details or by the simplification of neuronal level dynamics into larger components. Thus, assuming that the spiking dynamics of neurons are relevant to cognitive functionality, the first step is to design a statistical model of a neuron capturing the geometrical and electrophysiological characteristics that shape its activity. Much of the current work in computational neuroscience is focusing precisely on this goal (see also part I of this book).

At the network level, we assume that information is processed and carried out by a spatiotemporal pattern of activation of the constituting neurons. Until a less conservative principle or set of principles linking information processing to network activity is discovered, the second step of the bottom-up strategy should, thus, consist of understanding and reproducing the key elements of network composition, connectivity, and spatial arrangement affecting the spatiotemporal activity patters of the network. Unfortunately, however, the precise spatiotemporal activation of a region of the brain cannot be currently recorded at the level of single neurons. Thus, the construction of a model reproducing such an activity must be guided by indirect verification, i.e., by attempting to reproduce related network characteristics that can be measured routinely. Several theories and empirical evidence link cognitive functionality to the presence of rhythmic activity *(7)*. Thus, a reasonable strategy is to attempt the analysis and synthesis of the network properties underlying the emergence of observed rhythms. In this attempt, it should be kept in mind that the reproduction of the rhythms is not the direct goal. The solution should be compatible with the constraints developed in the first step, i.e., using neuronal models whose morphology and physiology result in realistic spiking behavior.

An example of this strategy is the generation of an anatomically and physiologically realistic hippocampal model. Such a model would be based on neurons with accurate shape and biochemical machinery in order to reproduce known single-cell electrophysiology. The neurons should be interconnected based on the available neuroanatomical knowledge. This knowledge (e.g., average number and position of synapses for each morphological class) is far from complete, but the model would be additionally constrained by the attempt to reproduce the emergence of intrinsic oscillations in the hippocampus, specifically those believed to be functionally relevant *(8)*. In this exercise, neuroanatomy plays a fundamental role, because much of the available experimental data can only be interpreted in a precise neuroanatomical framework. For example, the hippocampal model should reproduce characteristic rhythmic activities as a whole, but also upon "virtual slicing" in different planes and under different bathing conditions.

The amount of electrophysiological and anatomical data available for the rodent hippocampus and the great complexity of the system are such that, if a model could reproduce all of the observed properties and behaviors of the natural network, one could assume that some of the properties that are not known would be reproduced as well. Obviously, there would be no guarantee or "mathematical proof", but the continuous accumulation of data and the comparison of experimental and modeled behavior would provide additional support or necessary corrections.

A working model of this sort could allow neuroscientists to test a large number of hypotheses with "virtual experiments", which are impossible to perform in real life, either in principle or, because of technical limitations, for ethical reasons. Intracellular activity could be continuously recorded from every neuron during the propagation of spatiotemporal patterns. Although a complete analysis of such a huge amount of data would certainly require new computational and statistical tools, even the partial examination of the simulation results could foster intuition and facilitate the development of new hypotheses. The same experiment could be repeated multiple times under completely controlled conditions to characterize the influence of every single parameter on the neuronal patterns. The effect of hypothetical drugs (blocking or enhancing specific elements or properties of the network) could be tested, and only the most promising candidates would be commissioned for wet experiments.

19.4. FROM STRUCTURE TO FUNCTION

In order to use an anatomically and biophysically accurate model of a brain region for the study and implementation of cognitive functionality, it is first necessary to understand the principles of neural representation in the real brain structure as well as in the model. In the case of the hippocampus, a relatively well characterized cognitive function is that of supporting spatial navigation *(9–11)*. The neural code underlying hippocampal involvement in spatial navigation is based on the activity of place cells, which are neurons that fire when the animal is in a specific location of the environment *(9,12)*. Although much is known about hippocampal anatomy and physiology and its relationship with the representation of space, the hippocampus cannot be considered an autonomous computational unit. Hippocampal functionality in navigational tasks is intimately related to that of other areas, including the visual system *(13)*, thalamus *(14)*, subiculum *(15)*, septum *(10,16)*, retrosplenial *(15,17)*, and parietal *(18,19)* cortices, just to mention a few.

The development of a realistic and detailed anatomical and physiological model of the whole brain is going to take many decades. However, the realistic hippocampal models could be interfaced with "black box" components representing functionally related areas. These components would be based not on detailed neurobiological knowledge, but on existing AI models of computer vision, natural language parsing, associative memory, automated reasoning and planning, proprioception and motion control, etc. For the detailed hippocampal network model, the system of black-box AI components would represent the "external world", providing both input and output of information. The whole "brain" could then be embodied in a (real or simulated) robot to interact with the (real or simulated) environment *(20,21)*. The advantage of using a purely virtual embodiment is that all simulation parameters would be perfectly controlled, thus allowing an exhaustive and deterministic behavioral analysis. The use of a physically embodied robot, on the other hand, would allow the more faithful and complete reproduction of real-world experience, which includes too many "noisy" details to be simulated. A comprehensive approach would pursue both real and simulated embodiment in parallel to exploit the complementary advantages of both strategies.

The hybrid implementation of an embodied functional "skeleton" of the brain, together with a single initial component, which includes a realistic level of anatomical and physiological details, would be an invaluable research tool to connect the model parameters (electrophysiology and network connectivity) with the emergent characteristics (behavioral and cognitive functions). To investigate hippocampal function in spatial navigation, the robot could be allowed to navigate and explore mazes similar to those used in rodent behavioral experiments or more complicated human-level mazes, such as metropolitan street maps. During the exploration of the environment, the hippocampal network would receive visual and proprioceptive input from the appropriate brain modules *(13–15)*. The implementation of Hebbian and nonHebbian plasticity mechanisms *(22,23)* would allow the emergence of realistic place cells in the hippocampal network upon interaction with the environment. With the aid of information on the current "goal" (e.g., the memorized location of a reward), from subcortical nuclei and the hypothalamic value system *(24,25)*, place cells should naturally acquire the functionality of spatial navigation support and directional representation.

Needless to say, there is a multitude of mechanisms that could underlie each of these functional processes. Each combination of mechanisms would have to be tested, comparing the network activity with known properties of the real hippocampus. The model could eventually bridge ongoing neurobiological research, which employs multicellular recording to partially characterize the neuronal activity on rats exploring simple environments, with machine learning theories, which have extensively characterized the properties of different algorithms that solve spatial navigation tasks.

19.5. IMPLEMENTING THE MIND

The strategy outlined above might be difficult to extend to those "highest" levels of cognition for which current AI models seem to be missing some "special ingredient". The materialist position holds that, once all the aspects of brain anatomy and physiology are correctly implemented, there is nothing "left out" and no additional "special ingredient" to be added (26). Whether or not such an implemented model would itself feel conscious is a question beyond the boundaries of science. What matters is that, if the model reproduces all the structural and functional aspects of the brain, it would have to reproduce all of the observable behavior as well. The problem is, however, that we cannot develop such a "complete" model without an initial knowledge of the cognitive functionality the model is supposed to reproduce. In contrast to the above example of spatial navigation, the semantics of neural representation (i.e., the mapping between the spatiotemporal activity patterns and the cognitive function) in higher brain areas, such as the prefrontal and orbitofrontal cortices, the amygdala, the cerebellum, and the hippocampus itself, for a variety of higher cognitive functions, are largely unknown. The very cognitive functions carried out by these brain regions appear hard to define precisely.

Specific outstanding problems in this context concern autobiographical episodic memory, metacognition (higher-order thoughts), representation of others' minds ("theory of mind"), selected neurological disorders, including schizophrenia, multiple personality, autism, and the "forbidden" topic[3] : consciousness. Remarkably, all these topics have one element in common: they all involve the concept of an agent *(27)*. There is no doubt that much human brain information processing deals with abstract representations of agents, such as instances of the self and others (as opposed to representations of physical bodies of these agents and other things). This sort of representation and information processing could be the key building block missing at all levels of the structure of cognitive neuropsychology. This could be the answer to the question regarding the principal difference between brains and computers as we know them today. If so, then implementing the right sense of agency (based on a human-like theory

---

[3]Scientists learned to substitute words "conscious–unconscious" by all possible means, because each of the substitutes very soon became a "bad word" on its own: "explicit-implicit", "aware–unaware", "attended–unattended", "supraliminal–subliminal", "nocuous–innocuous", and "volatile–nonvolatile".

of mind) in a computer might be the next major historical step in the progress of humanity.

How is theory of mind implemented in the human brain, and how can it be implemented in a computer model? One possibility is that representations of others' minds exist as a system of general concepts instantiated in the brain, so that they can be used by an individual for analysis of behavior of the Self and other agents. Alternatively, or additionally, humans could use mechanisms of their own first-person-experience (in other words, perform mental simulations) in order to understand other minds. What is the rule used to assign appropriate representations to various instances of agents, and which brain areas are involved in this representation and assignment? Several lines of evidence suggest that, together with other higher areas, the hippocampal formation could play an important role in this context.

In rodents, the hippocampus is implicated in representing the animal self-location in space *(12)*. However, the rodent hippocampus can also distinguish between behavioral paradigms and other generalized contexts. For example, an observed place field pattern can be momentarily altered by switching from a random search task to a directed search task in the same environment *(28)*. In primates, hippocampal activity patterns could represent the current focus of attention (detected by eye position) rather than the current location of the animal's body *(9)*. In humans, the semantics of the hippocampal activity patterns is even more complicated; in addition to spatial correlates, hippocampal cell firing is related to recent and remote episodic memory retrieval *(29)*, perception of time, natural language processing, and many other elements of cognition. Although such a broad spectrum of cognitive functionality seems to be difficult to encompass with a single clear definition of hippocampal function, a possibly unifying idea is to view the human hippocampus as a generalized context indexing device *(30)*. According to this theory, episodic memories are labeled by their generalized contexts. Among the parameters that together constitute a "generalized context" of an episode are the time, the spatial location, and the gist of the event, as well as the subject to whom the experience is attributed (the mental perspective). How does the hippocampus index episodes? Each experienced generalized context could be represented by a pattern of activity stored in the hippocampus. This activity pattern would be associated with the details of the experience, which are stored in the neocortex.

The number of life experiences and the richness and similarity of different generalized contexts could pose a problem of storage capacity, if each context had to be represented by a unique pattern of neuronal activity. However, to move mentally from one context to another similar one, following remembered context relations, could be much easier than to recall a specified context from scratch. Therefore, remembering a reference to a particular episode might amount to the ability to relate the context of that episode to contexts of other remembered episodes, as well as of currently ongoing events. In other words, retrieving a memory would consist of finding a "path" from one context to another. Individual steps of this path involve dropping or accepting various assumptions, beliefs, rules, conditions, etc., that apply to the entire state of mind representing a given episode. If the hippocampus relates generalized contexts to particular episodic memories, then the problem of memory retrieval becomes naturally related to the problem of navigation in a graph of all previously experienced contexts. This could explain the unification of the spatial and memory functions in the hippocampus.

We can now further expand the concept of generalized contexts into the hypothesis that human hippocampal activity patterns encode various instances of the Self as an agent, e.g., instances associated with different spatial locations and, generally speaking, different generalized contexts. This hypothesis can be illustrated with a particular model in which various instances of the Self, such as "I-Now", "I-Past", "I-Imagined", "I-Pretend-Play", "He-Now", etc., are encoded by patterns of activity in two areas: the hippocampus and the prefrontal cortex *(31)*. In this model, the hippocampal representations are "allocentric", i.e., a given event (and the corresponding instance of the Self) are always associated with one and the same pattern of hippocampal activity, in the same way as "September 11, 2001" always stands for one and the same historical day. In contrast, prefrontal representations are "egocentric", in the sense that, e.g., "I-Now" is always encoded by one and the same pattern, regardless of the moment of time referred to as "now". Other instances of "I" are assigned patterns in the prefrontal cortex based on their current relations to "I-Now". In this model, an experience labeled "I-Now" cannot be forgotten and subsequently retrieved as "I-Past" without the help of the hippocampus, because of a "context shifting" problem. In order to initiate this retrieval process from the prefrontal cortex, one needs to activate a representation of the desired "I-Past" there and to "synchronize" it with the hippocampal activity. Reactivation of the hippocampal pattern then results in retrieval of the remembered experience in a new context.

This model requires that the space underlying various instances of the Self (i.e., the space of generalized contexts, which reduces to a 2D map of an environment in the rodent navigation case) be represented outside of the hippocampus. Similarly, particular associations between general facts and locations in this space (the semantic knowledge), as well as details of experience, are also stored outside of the hippocampus proper, presumably in various parts of the neocortex. The exclusive role of the hippocampus would be to bind experience by the sense of agency, which is semantically reduced to its simplest form: a pattern labeling a specific instance of the Self or "I" *(32)*. Therefore, hippocampal loss or damage should disrupt autobiographical memories without affecting the semantic memory system, which is consistent with cases of human hippocampal amnesia *(29)*. A direct test of this model would consist of the reactivation of specific hippocampal patterns to elicit the associated behavioral and introspective correlates.

19.6. CONCLUDING REMARKS

Our main points can be summarized as the following: (*i*) in order to build functional virtual brains, we need to know neuroanatomy and neurophysiology as much as we need to know cognitive science; (*ii*) in order to understand the architectural principles underlying cognitive functions in the brain, it is necessary to generate virtual neuronal networks with the right structure, connectivity, and dynamics, using computational neuroanatomy; (*iii*) a key step in the construction of virtual brains, as well as smart computers of the future, will be the understanding and implementation of human-like representations of agency; and (*iv*) A practical strategy to advance towards virtual brains is based on a "hybrid" neuroanatomical and algorithmic approach, in which a realistic neural network is placed in a context of simplified AI brain modules.

The availability of a complete model of brain anatomy and physiology would be of tremendous value to develop scientific intuitions and ideas and to foster education in neuroscience and cognitive psychology. The incremental development of the model would boost our understanding of the relationship between brain composition and connectivity, the spatiotemporal pattern of electrical activity, and the resulting cognitive functionality. This is basically the application of the standard scientific search for structural–functional relationships to the brain–mind problem.

Finally, a great obstacle on the way towards the computer generation of functionally and structurally realistic virtual brains is represented by the dynamic process of development. At birth, the human brain possesses only coarse neuroanatomical features of the adult brain, and its cognitive functionality is minimal. Throughout postnatal development, the structural complexity of the brain, particularly in higher areas such as the cortex *(33)*, undergoes an explosive increase, which is dependent on the interaction with the environment. Higher cognitive functions, such as episodic memory and consciousness, develop in parallel upon continuous sensory and motor experience in the real world *(34)*. In order to create a virtual brain that truly reproduces the function of the mammalian nervous system, it will be necessary to model its structure not in three but in four dimensions (space and time). The development of neuroanatomy in the model will have to rely upon interactions with the environment, thus requiring (virtual and/or real) embodiment even during the process of ontogeny. The optimal mixture of initial structural elements and growth rules in the developing virtual brain will have to be simple enough to maximize the role of self-organization, yet complex enough to produce the capacity to learn from its own experience and from the interaction with other (real or virtual) agents.

Once the architectural and developmental principles of brain anatomy that are sufficient and necessary to implement cognitive functionality were understood, the model could be progressively simplified by reducing its bottom level statistical complexity, while preserving higher cognitive functions. The exact location and shape of neurons or their intracellular machinery might be essential for some purposes but not for others. The level of viable simplification will thus depend on what aspects of the functionality are to be reproduced and on whether the implementation is limited to specific subregions of the brain. In the end, this level of understanding and its implementation in "virtual brains" will lead to new computational paradigms and to the next generation of computing machines. The details of neuroanatomy might not be implemented in these final functional models, but the neuroanatomical principles will be. At the present stage, neuroanatomical details must be included in the model, because we do not yet have a complete understanding of the neuroanatomical principles.

The future implementation of virtual brains is going to raise ethical and epistemological issues quite different from those discussed today in the scientific community. Should the machines implementing the models, or even the models themselves, be given rights? Should direct interfaces between machine-implemented virtual brains and real human brains be encouraged or discouraged? How about "cognitive augmentation" by the addition of nonhuman-based computational modules performing tasks in which machines already outperform us? These questions, which touch the very definition of human being, will derive directly from today's neuroscience, computer science, and cognitive psychology research.

REFERENCES

1. Shannon CE, Weaver W. The Mathematical Theory of Communication. University of Illinois Press, Urbana, IL, 1949.
2. Boole G. The Mathematical Analysis of Logic: Being an Essay Towards a Calculus of Deductive Reasoning. Macmillian, Barclay, and Macmillian, Cambridge, 1847.
3. Turing AM. On computable numbers, with an application to the Entscheidungs problem. Proc Lond Math Soc 2nd Series, 1936; **42**:230–265.
4. Polsson K. Chronology of Events in the History of Microcomputers from 1947 to 2000. Published online at (http://www.islandnet.com/~kpolsson/comphist/) © 1995–2000 Ken Polsson.
5. Kurzweil R. The Law of Accelerating Returns. Published online at (http://www.kurzweilai.net/articles/art0134.html) © 2001 Raymond Kurzweil.
6. McCarthy J, Minsky ML, Rochester N, Shannon CE. A proposal for the Dartmouth summer research project on artificial intelligence. 1955. Published online in 1996 by John McCarthy at (http://www-formal.stanford.edu/jmc/history/dartmouth/dartmouth.html).
7. Engel AK, Singer W. Temporal binding and the neural correlates of sensory awareness. Trends Cogn Sci 2001; **5**:16–25.
8. Klimesch W. Memory processes, brain oscillations and EEG synchronization. Int J Psychophysiol 1996; **24**:61–100.
9. Rolls ET. Spatial view cells and the representation of place in the primate hippocampus. Hippocampus 1999; **9**:467–480.
10. Ono T, Nishijo H. Active spatial information processing in the septohippocampal system. Hippocampus 1999; **9**:458–466.
11. Maguire EA, Burgess N, O'Keefe J. Human spatial navigation: cognitive maps, sexual dimorphism, and neural substrates. Curr Opin Neurobiol 1999; **9**:171–177.
12. Bures J, Fenton AA, Kaminsky Y, Zinyuk L. Place cells and place navigation. Proc Natl Acad Sci USA 1997; **94**:343–350.
13. Wylie DR, Glover RG, Aitchison JD. Optic flow input to the hippocampal formation from the accessory optic system. J Neurosci 1999; **19**:5514–5527.
14. Taube JS. Head direction cells and the neurophysiological basis for a sense of direction. Prog Neurobiol 1998; **55**:225–256.
15. O'Mara SM, Commins S, Anderson M, Gigg J. The subiculum: a review of form, physiology and function. Prog Neurobiol 2001; **64**:129–155.
16. Brandner C, Schenk F. Septal lesions impair the acquisition of a cued place navigation task: attentional or memory deficit? Neurobiol Learn Mem 1998; **69**:106–125.
17. Maguire EA. The retrosplenial contribution to human navigation: a review of lesion and neuroimaging findings. Scand J Psychol 2001; **42**:225–238.
18. Berthoz A. Parietal and hippocampal contribution to topokinetic and topographic memory. Philos Trans R Soc Lond B Biol Sci 1997; **352**:1437–1448.
19. Arbib MA. From visual affordances in monkey parietal cortex to hippocampoparietal interactions underlying rat navigation. Philos Trans R Soc Lond B Biol Sci 1997; **352**:1429–1436.
20. Almassy N, Edelman GM, Sporns O. Behavioral constraints in the development of neuronal properties: a cortical model embedded in a real-world device. Cereb Cortex 1998; **8**:346–361.

21. Weng J, McClelland J, Pentland A, et al. Artificial intelligence. Autonomous mental development by robots and animals. Science 2001; **291**:599–600.
22. Lechner HA, Byrne JH. New perspectives on classical conditioning: a synthesis of Hebbian and non-Hebbian mechanisms. Neuron 1998; **20**:355-358.
23. Atkins PW. What happens when we relearn part of what we previously knew? Predictions and constraints for models of long-term memory. Psychol Res 2001; **65**:202–215.
24. Ciompi L. Affects as central organising and integrating factors. A new psychosocial/biological model of the psyche. Br J Psychiatry 1991; **159**:97-105.
25. Patton PE, McNaughton B. Connection matrix of the hippocampal formation: I. The dentate gyrus. Hippocampus 1994; **5**:245–286.
26. Dennett DC. The Fantasy of First-Person Science (a written version of a debate with David Chalmers, held at Northwestern University, Evanston, IL, February 15, 2001, supplemented by an email debate with Alvin Goldman). Published online at (http://ase.tufts.edu/cogstud/papers/chalmersdeb3dft.htm), March 2001.
27. Baron-Cohen S. Mindblindness: An Essay on Autism and Theory of Mind. MIT Press, Cambridge, MA, 1995.
28. Markus EJ, Qin YL, Leonard B, Skaggs WE, McNaughton BL, Barnes CA. Interactions between location and task affect the spatial and directional firing of hippocampal neurons. J Neurosci 1995; **15**:7079–7094.
29. Nadel L, Samsonovich A, Ryan L, Moscovitch M. Multiple trace theory of human memory: computational, neuroimaging, and neuropsychological results. Hippocampus 2000; **10**:352–368.
30. Wheeler MA, Stuss DT, Tulving E. Toward a theory of episodic memory: the frontal lobes and autonoetic consciousness. Psychol Bull 1997; **121**:331–354.
31. Samsonovich AV, Nadel L, Moscovitch M. A theory-of-mind connectionist model of episodic memory consolidation. 2000; J Neurosci Suppl **26**:1498.
32. Ascoli GA. Association, abstraction, and the emergence of the Self. Noetic J 1999; **2**:9–20.
33. Pallas SL. Intrinsic and extrinsic factors that shape neocortical specification. Trends Neurosci 2001; **24**:417–423.
34. Siegel DJ. Memory: an overview, with emphasis on developmental, interpersonal, and neurobiological aspects. J Am Acad Child Adolesc Psychiatry 2001; **40**:997–1011.

Index

A

Abacus, 426
Absolute latency values, 374
Acetyltransferase, 193
Active models
   neuronal modeling, 107–109
Activity-independent mechanisms
   column development, 351–354
AD, 113–120, 172
ADC maps, 276–278, 280f
Adjacency Rule, 76
Adult brain weight
   Conel data, 403f
Adult maps
   visual cortex columnar structures, 338
Adult neuronal morphology
   model, 9–10
Afferent. *See also* Thalamocortical afferent (TCA)
   basal forebrain
      differential distribution, 184f
   connectivity patterns, 302f
   ingrowing
      construction method, 261–262
   navigation elements, 255
   neuron density cloud
      representation, 158f
   restricted localization
      basal forebrain corticopetal system, 183
Algorithm
   Borg-Graham's, 144
   branch growth, 33
   computational
      local *vs.* global, 112–113
   elastic net
      visual cortex columnar structures, 343–350
   FDP, 78–80, 79f
   Floyd's, 305
   genetic, 78f
   growth
      axons, 264–266
   iterative, 53
   Kohonen, 343–344
   L-Neuron, 53–55
   modified Hillman, 53–55
   Monte Carlo growth, 31–32
   stochastic
      digital dendrites, 50–51
   STRETCH, 77f
Alice software, 391
Alzheimer's Disease (AD), 113–120, 172
Amaral's collection, 67f
AMPA receptor-mediated, 365
Anatomical model
   dentate gyrus, 16f
Anatomical representation, 135–140
Anatomical tract-tracing techniques, 337
Angular deviation, 39–41
Angular measures
   RMS error, 41–42
Anisotropic diffusion, 271
Anisotropy
   relative, 281
   water diffusion, 275

f: Figure
t: Table

Anisotropy map, 280–282
Apical
   daughter ratio, 59f
   initial tree diameter, 58f
   tree type parameters, 63t
Apical tree, 62f
Apoptosis, 383
Apparent diffusion constant (ADC) maps, 276–278
   examples, 280f
Approximations, 262–263
Arborization, 255–262, 256
ArborVitae, 51, 55, 112, 251–252, 256
   implementation, 13
   observed TCA behavior, 258
   terminal arborization behavior, 260
Area models, 363–365
Aristotle, 426
Artificial intelligence, 427
Artificially generated neurons, 112–113
Associative memory, 364
Atanasofff, John, 426
Attention, 4, 171
Auditory systems
   sensory map transformations, 200–201
Author
   definition, 97
AutoCad (.dxf)
   L-Neuron, 55
Average degree on connectivity, 299–301
Axodendritic synapses, 261
Axon
   CGRP, 183
   competition, 230–238
   guidance, 219–220
   innervating competition, 220
   large-scale optimization, 75–76
Axonal functions, 5
Axonal growth
   regulation, 13
Axonal morphogenesis
   early growth, 219
Axonal navigation
   dendritic morphology, 12–13
   voxel substrates, 245–266

Axonal pathways
   method for specifying simulated, 245–263
Axonal structure, 49
Axosomatic synapses, 261
Azimuthal orientation, 41
Azimuth angles, 41

B

Babbage, Charles, 426
Bardeen, John, 426
Basal
   daughter ratio, 59f
   diameter
      scatter plot, 117f
   initial tree diameter, 58f
   parameters
      L-Neuron parameter, 60t
   path
      scatter plot, 117f
   tree, 62f
   tree type parameters, 63t
Basal forebrain, 12
   afferents differential distribution, 184f
   CGRP-containing axons, 185f
   cholinergic dendritic segments, 185f
   cholinergic neurons, 173f, 176f
   compartments, 172
   composition, 171
   corticopetal system
      afferent restricted localization, 183
      animals, 188
      chemically identified cell inhomogeneous distribution, 177–179
      computational anatomical analysis, 171–194
      connections probability, 183–186
      data acquisitions, 188–189
      data analysis, 190–194
      hodologically identified neural populations, 174–177
      merging datafiles, 186
      neuron selection for dendritic tracing, 189–190

regionally selective dendritic
orientation, 179–183
tissue processing, 188
noncholinergic neurons, 175f
Base
definition, 97
BDA, 201
Behavioral approach, 360–361
Behaviors
cortical neuron numbers, 417–419
Berry, Clifford, 426
Bifurcations
number, 50
Binary format (.vol)
L-Neuron, 55
Binocularly-deprived, 352
Biocytin, 27
Biological branching
axons, 264
Biological Modeling Framework
(BMF), 93–94
extensible architecture
highly modular, 94–95
Biological navigation
axons, 264
Biomarkers, 384
Biophysical microcircuit model, 366
Biophysical models, 362
Biotinylated dextran amine (BDA), 201
Bjaalie, Jan, 12
BMF, 93–94
extensible architecture
highly modular, 94–95
BMF Core Plugins, 93f
Boole, George, 426
Borg-Graham's algorithm, 144
Bounded-resource philosophical
critique, 72
Brain
computing, 3–19
connectivity
computational methods for analysis, 295–330
functional mapping, 15
graph theoretical perspective, 299

context
rendered data, 256f
development, 402–422
birth to 72 months, 404–421
3D volumetric MRM, 388f
function, 402–422
implementing, 428–430
microscopic neuroanatomic changes
global analysis, 404–410
neuroanatomical model, 18
sections
midinfrared spectral imaging,
395–396
structure
combinatorial network
optimization, 71
virtual, 425–435
Brain stem
3D reconstruction, 205f
nuclei
local coordinate systems, 205–207
Branch angles
frame of reference for measuring,
38–39
Branch diameter, 31f
Branch growth algorithm, 33
Branch order
distributions model, 34
vs. taper rate PK, 65f
Branch parameter, 226–230
map, 224f
scattergram, 225f
Branch point, 39–41
Branch probability
growth curves, 223f
Branch taper
histogram, 30f
Brattain, Walter, 426
Brownian motion, 272–273
Buchs' Toolbox, 145
Build pane, 87–90
Burke's model, 8
Bursting cell, 114f

C

CA. *See* Correspondence analysis (CA)
Caenorhabditis elegans
    ganglion placement optimization, 76–80
Caenorhabditis elegans ganglia, 72, 78f, 79f
Cage-side observation, 384
Cajal's neuron doctrine, 3
Cajal's qualitative laws of protoplasmic economy, 72
Cajal's theory, 5
Calbindin, 193
Calcitonin-gene-related-peptide (CGRP)
    axons, 183, 185f
Calcium flow, 360
    fluorescence indicators, 385
Calretinin, 193
Canonical microcircuit, 15
Canonical neuron model, 363
CANTOR network processor, 322–324
CANTOR system
    hierarchical analysis, 328
CA1 pyramidal cells, 117–118
    case study, 51–68
CA3 pyramidal cells, 117–118
    computer representation, 143f
    examples, 114f
Catacomb, 128
Cat a-motoneuron dendrites
    2D drawing, 44f
    features, 29
Cell biological mechanisms, 226–230
Cell deaths, 383
Cell matrix
    size, 45–46
Cellular approach, 360–361
Cellular detail
    merge simulated, 252
CellViewer, 111
Centers of mass (COM), 38
Central Dogma of genetics, 77–78
Cercal sensory system
    cricket, 154
        functional organization, 155f
    physiological characteristics of neurons, 154–156

Cerebellum
    3D reconstruction, 205f
    mid-IR microspectroscopic dataset, 397f
    Purkinje cells, 396
Cerebral cortex
    cytoarchitectonic areas
        classification, 405t
        maps, 406f
    neuron number, 412f
Cerebro-cerebellar systems
    sensory map transformations, 200–201
Cerebro-pontine projection, 200–201
CGRP
    axons, 183, 185f
Chaotic organization, 199
Characteristic path length, 305, 307–309
Chemically identified cell inhomogeneous distribution, 177–179
    differential density 3D scatter plot, 177, 178f
    isorelational surface rendering, 177–178, 178f
Cholinergic cells
    primary dendrites, 179
Cholinergic corticopetal neurons, 172
Cholinergic neurons
    distribution, 187f
Climbing fiber type contact, 183
Climbing type synapses, 185
Cluster
    analysis, 316–318
        example, 318
    index, 307–309
    vs. random, 308–309
CoCoMac database
    web site, 370
Cognition
    understanding, 4
Cognitive tests, 384
Color map, 282–283
Columnar structures
    visual cortex, 337–354
Column development
    activity-independent mechanisms, 351–354

COM, 38
Combinatorial network optimization
  brain structure, 71
Combined approaches, 318–320
Compartmentalization
  dendritic structure, 128
Compartmental models, 5, 359–360
Compartment-based neuronal models
  branched tubules, 248
Compartment dimensions
  microcircuit model, 368t
Competition
  axons, 230–238
  dendritic trees, 220
  developmental phases, 220
  innervating axons, 220
  neuronal morphogenesis, 219–239
Complex branched neurons, 360
Complex dendritic morphology
  spread of, 108
Component placement optimization, 73, 76f
Compression-tension, 75
Computational algorithms
  local *vs.* global, 112–113
Computational implementation, 130–136
Computational modeling
  viable methodology, 117
Computational model study
  discussion, 35–37
Computational neuroanatomy, 3
  network modeling, 359–378
  neuron modeling, 359–378
  role, 12
Computational Neuroanatomy Group, 112
Computational properties
  challenges, 152
Computational studies
  successful, 4
Computational tools
  neuronal structure-activity analysis, 107–120
Computational treatment
  experimental data, 297–298
Computed tomography (CT), 385–387

Computer
  defined, 425
  growing power, 4
  hardware
    plateau, 4
  history, 425–426
  model, 432
  simulation
    nervous system development and function, 220–221
Computing brain, 3–19
Concentration pool
  definition, 98
Conceptual hypothesis testing, 320–328
Conductance densities
  microcircuit model, 368t
Conel data, 402–404
  adult brain weight, 403f
  cortical layers, 403
  cytoarchitectonic regions, 403–404, 404t, 410f
  fiber systems, 403
  microscopic neuroanatomic features, 402–403
Confidence ellipses
  CA
    cortical layer/fiber system, 408
    cytoarchitectonic area, 409–410
    microscopic neuroanatomic feature, 408
  correspondence analysis factors, 407f
Connectedness, 304–305
Connections
  probability
    basal forebrain corticopetal system, 183–186
Connectivity
  average degree, 299–301
  experimental identification, 297
  global patterns
    hierarchical cluster tree, 320f
  indices, 301–303
  nerve connections
    development, 219–239
  neural
    description, 297–299

patterns, 302f
statistical exploration, 310–320
vertex, 305–306
Connect pane, 87–90
Continuity
mathematical formulation, 341–343
Convex hulls, 38
CORBA/IIOP, 100
Core plugins, 93f
Correlation measurements, 50
Correspondence analysis (CA), 404–410
cerebral cortex
neuron number, 415–416
confidence ellipses
cortical layer/fiber system, 408
cytoarchitectonic area, 409–410
factors, 407f
microscopic neuroanatomic
feature, 408
cytoarchitectonic area
neuron numbers, 410–413
factor score summary data, 408
Cortex
simulated data, 259f, 260f–261f
Cortical delta oscillations, 186
Cortical layers
Conel data, 403
Cortical maps
activity, 339
Cortical neuron
number, 402
numbers
new behaviors, 417–419
Cortical visual system
connectivity data, 300f
Coverage uniformity
mathematical formulation, 341–343
Cricket
cercal sensory system, 11–12, 154
neural maps, 160–161
Crook, Sharon, 163–164
Crossing over
geometry, 183
Cross-level integration, 361
CT, 385–387
Cvapp, 57, 145

Cytoarchitectonic regions
Conel data, 403–404, 404t, 410f
neuron number, 413–415, 413t, 414f
per layer, 418f, 419–421
Cytochrome oxidase staining, 337

D

Data
definition, 97
Databases
interacting, 99–101
Datafiles
merging, 193–194
Daughter branch
database for choosing diameters, 33f
diameters
determining, 33
directions, 41
Daughter diameter ratio frequency
distribution, 59f
De Forest, Lee, 426
Dendrites
central axis, 39
isolated neurons
two-dimensional analysis, 28–35
large-scale optimization, 75–76
morphogenesis
early growth, 219
spatial orientation, 66–67
synaptic information, 27
Dendritica, 145
Dendritic arbor, 254
optimization analysis, 77f
Dendritic branching, 106
angles
calculate, 39f
mean orientation, 180f
patterns, 221
process
continuous time modeling, 226
modeling, 221–225
Dendritic functions, 5
Dendritic growth
model, 9–10
results, 226
regulation, 13

Dendritic length, 50
Dendritic material
  log-log plot, 31f
Dendritic morphology, 6–7, 11
  axonal navigation, 12–13
  development, 221–226
  research strategy, 9f
  variations, 109
Dendritic processes
  2D orientation, 192
  mean 3D vector, 192
Dendritic structure, 49
  anatomical representation, 10
  compartmentalization, 128
Dendritic territories
  overlap, 38
Dendritic trees
  competition, 220
  3D simulation, 42
  frequency distribution, 224f
  map, 224f
  morphology of, 28
Dendrogram, 52f, 57
Density cloud representation
  primary afferent neuron, 158f
Dentate gyrus
  anatomical model, 16f
2-deoxyglucose, 337
Diameter, 305
Differential density 3D scatter plot, 190–191
Diffusion
  restricted, 271
  water molecules, 272
Diffusion ellipsoid
  characterized, 278f
Diffusion MRI sequence
  example, 276f
Diffusion MRI technique
  measurement and calculation, 275–283
Diffusion process, 272–274
Diffusion tensor imaging (DTI), 13
  background, 271–275
  based color map, 283f
    parcellation of white matter, 283f
  description, 271
  future directions, 287
  principle and applications, 271–287
Diffusion-weighted images
  brain regions, 273f
  examples, 277f
  human brain, 272f
Digital dendrites
  stochastic algorithm, 50–51
Digital reconstructions, 50
Digraph, 298
  paths and cycles, 303–304
  small world, 308f
Directed graph
  example, 304f
Directionality
  water diffusion, 271
Disjoint paths, 305–306
Distance matrix, 305
  example data, 306f
Distribution morphometrics, 110
Domoic acid, 389
Dormant growth cones, 229
Drugs
  FOBs, 384
DTI. *See* Diffusion tensor imaging (DTI)
Duke-Southampton format (.swc)
  L-Neuron, 52–57
dxf (AutoCad)
  L-Neuron, 55
Dynamic animation, 164
DynamicAtlas, 163–164
Dynamic pattern prediction
  neural map, 163f

E

Eckert, Presper, 426
Edge, 305–306
Edge-disjoint
  directed paths, 305–306
Efferent connectivity patterns, 302f
Elastic net (EN), 340
  algorithm
    visual cortex columnar structures, 343–350

cylindrical feature
  one spatial dimension, 349f
developmental stages, 345f
ocular dominance, 347–348
  maps, 346f
orientation, 348–350
Electrode physiology, 337
Electrophysiological simulations, 5
  anatomically accurate, 50
  software packages, 10
Electrotonic Workbench
  web site, 108
Elliott and Shadbolt model, 232
Ellipsoid
  parameters, 275f
Elongation process
  modeling, 226
Emergent parameters
  apical and basal, 63t
EN. *See* Elastic net (EN)
ENIAC, 426
Eph family, 351
Epileptic seizures, 5
Epileptiform activity, 364
EPSPs, 108
Errors
  computer implementations, 130
Eutectic, 55, 143
Excitatory inputs
  prediction, 165f
    primary sensory interneuron
      prediction, 167f
Excitatory post-synaptic potentials
  (EPSPs), 108
Excitatory synaptic currents, 373
EXtensible Markup Language (XML), 84
Extreme anisotropy, 281
Eye saccadic control, 4

F

Factor analysis, 313–316
Fast spin echo (FSE)
  MRM, 391
FDP algorithm, 78–80
Ferret
  neural activity perturbed patterns, 353

Fiber orientation
  water diffusion, 275
Fiber systems
  Conel data, 403
Fiber tracking
  approaches, 285f
  examples, 284f
Figure-ground segregation, 4
Finkels model, 131
Firing types
  different current injections, 116f
  examples, 114f
Floyd's algorithm, 305
Fluid-dynamical model, 75
FOB, 384
Force-directed placement (FDP)
  algorithm, 78–80, 79f
Forebrain. *See* Basal forebrain
Foreign databases, 99–100
Formal description
  neuronal networks, 298–299
Formal models, 363
FORTRAN, 426
Fractional anisotropy, 281
Frontal association cortex, 420–421
Front paw extension, 384
FSE
  MRM, 391
Functional cluster analyses, 326–327
  result, 327f
Functional mapping
  brain connectivity, 15
Functional observational behavioral
  battery (FOB), 384
Fundamental parameters of form, 28

G

GABA, 369
Gait, 384
Gamma-aminobutyric acid (GABA), 369
Ganglion placement optimization
  Caenorhabditis elegans, 76–80
Gaussian centers, 346
GenAlg, 78–79, 78f
Gene expression models, 353–354
Gene network, 353–354

General-purpose numerical
    integrators, 361
GENESIS, 10, 84, 95, 111, 127, 362, 368
    Menschik model, 130–132, 130f
    neural simulator format, 113
    web site, 102
Genetic algorithm, 78f
Geometry
    crossing over, 183
GFAP, 384
Glial fibrillary acidic protein (GFAP), 384
Global connectivity patterns
    hierarchical cluster tree, 320f
Global distortions, 350–351
Goodhill model
    cortical maps
        global structure, 348f
Gradient, 275–283
    examples, 277f
Graph. *See also* Digraph
    directed
        example, 304f
    examples, 323f
    nondirected, 298
    random, 306–307
    theoretical analysis, 299–310
    theoretical measures
        example, 304f
    theoretical perspective
        brain connectivity, 299
Graphical formats
    L-Neuron, 55
Graphical user interface (GUI), 153
Growth algorithms
    axons, 264–266
GUI, 153

H

Hebbian learning rules, 230–232, 339, 346
    modified, 231–232
Hierarchical activation indices
    distribution, 376f
Hierarchical analyses, 327–328
Hierarchical cluster tree
    direct connectivity, 319f
    global connectivity patterns, 320f

High-resolution MRI, 388–390
Hillman, Dean, 7–9
Hillman algorithm
    modified
        L-Neuron, 53–55
Hillman's seven fundamental
    parameters, 28
Hippocampal model, 429
Hippocampal pyramidal cell
    remodeling, 67f
Hippocampus
    function, 430–431
    needle damage, 390
    self-location in space, 432
    structure, 430–431
Histochemical data
    brain, 247
Histology
    definition, 97
Hodgkin-Huxley formulation, 361, 365
    multi-level network modeling, 359–360
Hodologically identified neural
    populations, 174–177
    isodensity surface rendering, 174–176
    overlap analysis, 174
Hopefield networks, 364
Horseradish peroxidase (HRP), 27
HRP, 27
HTTP, 100
6-hydroxydopamine (6-OHDOPA)
    MRM, 391f
    MRM *vs.* conventional histological
        staining, 393f
    quantitative measures, 389–394

I

Image-combining microscopy
    sensory map transformations, 202–204
Imaging
    toxicology, 387–397
Imaging/histology
    definition, 97
Imaging technique, 13, 385–387
Implementation
    ArborVitae, 13

Information
    top-down, 14
Ingrowing afferents
    construction method, 261–262
Inhibitory interneurons
    role, 4
Inhibitory synaptic currents, 373
Innervating axons
    competition, 220
Innervation
    coexistence of stable states, 237f
Input-integration-output, 5
Integration method CVODE, 132–133
Integrative aims
    neuroscientists, 247–248
Interdaughter angle, 41
Interfacing models, 365
Interstitial neurons, 179
Intracellular calcium levels, 219
Intracellular mechanisms
    definition, 97
Intracellular processes
    models, 360
In voxo tissue culture, 252–255
Ionic pump
    definition, 98
Ischemia
    water diffusion, 275
Isodensity surface mapping, 191
Isodensity surface rendering
    hodologically identified neural
        populations, 174–176
Isomap, 312
Isorelational surface rendering, 191–192
    chemically identified cell
        inhomogeneous distribution,
        177–178
Iterative algorithm, 53

J

Jacobs, Gwen, 11
Java applet, 111–112
    Cvapp, 57, 145

K

Kainic acid (KA)
    lesioned cells, 117–120
        differences, 119t, 121t
Kilby, Jack, 426
Knowledge-based computational
        neuroanatomy project, 15
Knowledge integration
    issue, 14
Kohonen algorithm, 343–344

L

Labeling techniques, 27
Lamellipodia, 219
Large-scale optimization
    dendrites and axons, 75–76
Lateral geniculate nucleus (LGN),
        338, 352
Layered framework
    modeler's workspace, 93
Leaky integrate-and-fire unit, 363
Learning processes, 364
Lemniscal nuclei
    3D reconstruction, 210f, 211f
    pathways, 200–201
LGN, 338, 352
Ligand-activated channel
    definition, 98
LLE, 312
L-Measure, 9, 51, 110
L-Neuron, 8–9
    algorithms, 53–55
    description, 49
    Duke-Southampton format (.swc),
        52–57
    generation and description, 49–68
    graphical formats, 55
    modified Hillman algorithm, 53–55
    parameter
        basal parameters, 60t
    persistency of vision (.pov), 55
    Ray Dream, 55
    virtual reality markup language
        (.wrl), 55

L-Neuron Hillman/PK algorithm flow chart, 54f
Local connectivity indices, 301–303
Local distortions, 350–351
Locally linear embedding (LLE), 312
Locomotor responses
    tests, 384
Loyez myelin stain
    future directions, 266
Lumbosacral motoneurons, 43–44
L-Viewer, 55, 57
Lyndenmayer rewrite rules, 53–55

M

Macaque
    ocular dominance stripes, 347
Magnetic field gradient, 275–283
Magnetic resonance imaging modification (MRM), 388, 390f
    differential image processing, 392f
Magnetic resonance imaging (MRI), 17, 385–388
    brain fiber pathways, 246–247
    diffusion sequence
        example, 276f
    diffusion technique
        measurement and calculation, 275–283
    and DTI, 271–272
    MS, 394–395
    neuroanatomical studies, 271–287
    preclinical neurotoxicology, 388–390
    toxicology, 387–388
    vs. videomicroscopy, 387
Magnetic resonance spectroscopy (MRS), 394–395
Map development
    visual cortex columnar structures, 338–339
Mapping chaotic optimization landscapes, 78–79
Mathematical formulation
    continuity, 341–343
    coverage uniformity, 341–343
Mathematical modeling process
    diagram, 128f

MatLab, 128
MatLab R11, 191
Mauchly, John, 426
McCarthy, John, 427
MDS, 311
Mean 3D vector
    dendritic processes, 192
Membrane capacitance, 370
Membrane potential
    spatial distribution, 137f
Memory, 171
Memory encoding, 4
Menschik model, 130–132
    GENESIS, 130–132
    NEURON, 130–132, 130f
Mental phenomena, 360
Method
    definition, 97
Michaelis-Menten function, 233
Michaelis-Menten steady state, 131
Microcircuit cells
    voltage traces, 372f
Microcircuit implementation, 376
Microcircuit model
    compartment dimensions, 368t
    conductance densities, 368t
    synaptic parameters, 369t
Microcircuit representation
    primary visual cortex (V1), 367–368
Micro3D program, 193
Micro MRI, 388–390
Microprocessor
    introduction, 426
Microscopic neuroanatomic features
    Conel data, 402–403
MicroTrace
    GUI, 203f
Microtubule-associated proteins, 229–230
Midinfrared spectral imaging
    brain sections, 395–396
Migliore model, 132, 137f
    histogram, 138f
Miller, John, 163–164
Mind
    implementing, 431–433

Mind-brain science, 72
Minimal spanning tree
  illustration, 75f
Minsky, Marvin, 427
MODECLUS procedure, 317
Model
  definition, 97
Model-based approaches
  need for, 84–86
Model-based formulation, 341
Modeler's workspace, 83–102
  default templates, 95–96
  elements, 86–92
  example, 91–92
  representation language, 95–99
  user interface, 87
    prototype, 88f
    underlying architecture, 92–95
  web site, 102
  workspace database, 87
Modeler's workspace directory (MWD),
    88, 101
Model-free formulation, 341–342
Model sensory system
  for studying ensemble encoding,
    154–159
Modified Hebbian learning rules,
    230–232
Modified Hillman algorithm
  L-Neuron, 53–55
Modified MRI, 388–390
Molecular approach, 360–361
Molecular guidance cue models,
    353–354
Monte Carlo growth algorithm, 31–32
Monte Carlo method, 8, 29
Moore's Law, 4
Morphological data
  measuring, 109–111
Morphologically distinct growth
    phases, 257
Morphological parameters
  3D scatter plot, 7f
Morphological simulation
  case study, 51–68

Morphological variability
  computational studies, 107–120
  experimental studies, 106–107
Morphology
  influences physiology hypothesis
    testing, 113–120
  motoneuron
    parsimonious description, 8
Motivation, 171
Motoneuron
  dendrites
    2D drawing, 44f
    3D structure, 41f
    features, 29
  estimation of volume fraction, 45f
  morphology
    parsimonious description, 8
  quantitate 3D morphology, 36f
  virtual, 51–68
Motor cortex, 420
Mouse atlas, 248–251
MRI. *See* Magnetic resonance imaging
    (MRI)
MRM, 388–392
MR microscopy (MRM), 388–390
MRS, 394–395
MS
  MRI, 394–395
  MRS, 394–395
Multidimensional scaling (MDS), 311
Multilevel modeling
  visual cortex, 365–370
Multi-level neuron modeling
  computational neuroanatomy,
    359–378
Multiple correspondence analysis, 316
Multiple Sclerosis (MS)
  MRI, 394–395
  MRS, 394–395
MWD, 88, 101
Myelin slice data, 249f
  sagittal and horizontal, 250f

N

Navigation, 255
NEOSIM, 84, 128
Nernst
  definition, 98
Nerve connections
  development, 219–239
Nervous system
  descriptive geography, 72
  development and function
    computer simulation, 220–221
  generative grammar, 72
  key substrate, 3
  model-based studies, 83–102
Network data
  volumetric data, 246–247
Network dynamics
  statistical measures, 322
Network implementation, 369–370
Network levels, 4
Network optimization theory, 73–75
Neural circuits
  refinement, 230–238
Neural computation
  traditional view, 364
Neural connectivity
  description, 297–299
Neural databases and models
  web site, 298
Neural ensembles
  probabilistic representations, 156–158
Neural fluid mechanics, 75
Neural maps
  cricket cercal system, 160–161
  dynamic pattern prediction, 163f
  predictions of spatial patterns, 161f
  stimulus parameters
    functional representation, 159f
Neural optimization
  functional role, 80
Neural oscillations, 4
Neural simulation packages, 91
Neural Tracing System (NTS), 143
Neurite branching, 229–230
Neurite elongation, 226–230

Neuroanatomical model
  brain, 18
Neuroanatomical variability, 105
Neuroanatomy
  archives, 111–112
  optimal-wiring models, 71–80
  potential, 5
Neurobiological models, 132
  aspect, 128
NeuroGenerator, 145
Neuroinformatic knowledge base
  web site, 207
Neurological rehabilitation, 5
Neurological tests, 384
Neurolucida, 55, 188–189, 193
  software, 181
  system, 143
NeuroML, 96
Neuromorphology
  influences neurophysiology
    hypothesis, 109–110
NEURON, 10, 84, 108, 111, 127, 362
  Menschik model, 130–132, 130f
Neuron
  anatomical characteristics, 156–158
  classified, 6–7
  compartmental model, 227f
    results, 228f
  3D anatomical reconstruction, 156
  definition, 97
  dendritic orientation
    regional differences, 179–180
  dendritic tracing selection, 189–190
  3D reconstruction, 157f
  ensembles
    transfer of information, 166–168
  internal structure, 14
  mapped, 12
  to networks, 11–13
  part
    definition, 97
  three dimensions, 37–46
  ultrastructural investigation, 5–6
  virtual
    creating, 51

Neuronal activity
  computational studies, 127–146
  models, 353–354
  vs. neuronal shape, 105–121
Neuronal anatomy
  definition, 97
Neuronal electrophysiology
  anatomically accurate simulations
    practical aspects, 127–146
Neuronal ensembles
  predicting emergent properties, 151–168
Neuronal models, 362–363
  and simulation, 107–109
Neuronal morphogenesis
  competition, 219–239
Neuronal morphology
  computational models, 360
  generation and description, 49–68
  visualization, 57
Neuronal networks
  challenges, 152
Neuronal population models, 363–365
Neuronal response, 105
Neuronal shape, 49
  vs. neuronal activity, 105–121
Neuronal simulation tools, 363
Neuronal structure-activity analysis
  computational tools, 107–120
Neuronal tracing, 50
Neuron_Morpho, 143
Neuropeptide Y (NPY)
  neurons, 186
    distribution, 187f
Neuropil, 43
Neuroscholar, 15
Neuroscience
  key substrate, 3
Neuroscientists
  challenges, 152
NEUROSYS, 153
  general applications, 168
  to study emergent properties, 160–166
  web site, 151
NeuroToolBox, 145

Neurotoxicity
  detection, 384
NeuroTrace, 145
Nevins/Claiborne, 55
New behaviors
  cortical neuron numbers, 417–419
Nissl, 249f, 393f
  stained volumetric data
    future directions, 266
NMDS, 311–314
Nondeterministic polynomial-time
    complete (NP-complete), 73
  combinatorial optimization
    problem, 343
  theory, 73
Nondeterministic polynomial-time hard
    (NP-hard), 73, 75
Nondirected graphs, 298
Noninvasive imaging techniques
  resolution, 17
Nonmetric multidimensional scaling
    (NMDS), 311–314
  analysis
    example, 313
  representation
    of dataset, 314f
NP-complete, 73
  combinatorial optimization
    problem, 343
  theory, 73
NP-hard, 73, 75
NPY
  neurons, 186
    distribution, 187f
NTS, 143
Numbers of neurons
  stereological methods, 43

O

Object-oriented style
  advantages, 98–99
Object recognition, 4
Occipital cortex, 421
Ocular dominance
  EN, 347–348
  orientation maps, 350

Ocular dominance column formation, 338
Ocular dominance maps, 352
 EN, 346f
6-OHDOPA
 quantitative measures, 389–394
Onset response latencies, 373, 377
 distribution, 375f
Optical imaging, 337
Optic nerve
 stimulating cuff, 353
Optimal cluster arrangements
 connectivity data, 326f
Optimal-wiring models
 neuroanatomy, 71–80
Optimization analyses, 322–328
Optimization mechanisms, 75–80
Organizational level, 360–361
Orientation
 EN, 348–350
 independent visualization
  technique, 279
   anisotropy, 280–282
 maps, 352–353
  ocular dominance, 350
 water movement, 273–274, 274f
Outliers, 117f, 118f
Overlap analysis, 193
 hodologically identified neural
  populations, 174

P

Pain
 mechanism, 5
Parent branches
 scatter plots, 30f
Parietal association cortex, 421
Parkinson's disease, 172
Parsimonious description
 motoneuron morphology, 8
Parvalbumin, 193
Passive models
 neuronal modeling, 107–109
Path distance distributions
 model, 34
Pathological states, 5

Paths and cycles
 digraph, 303–304
Pathway development, 263
Patterns
 afferent connectivity, 302f
 dendritic branching, 221
 dynamic prediction
  neural map, 163f
 efferent connectivity, 302f
 global connectivity
  hierarchical cluster tree, 320f
 neural activity perturbed
  ferret, 353
 predictions of spatial
  neural map, 161f
 similarity indices
  primate visual areas, 302f
 spatio-temporal
  activity ensemble of sensory
   neurons, 161–163
  activity within neural ensembles,
   163–166
PCA, 313–316
Pearson's correlation, 312
Persistency of vision (.pov)
 L-Neuron, 55
PET, 17, 385–387
PHA-L, 201
Phaseolus vulgaris-leucoagglutinin
 (PHA-L), 201
Physiological rhythms, 5
Pinksy and Rinzel
 two-compartment model, 363, 365
PK. *See* Terminal length (PK)
Plateau potential cell, 114f
Plato, 426
Polar histogram
 regionally selective dendritic
  orientation, 181–182
 vs. vector representation, 181, 182f
Pontine projections
 anterograde axonal tracing, 202f
Pontocerebellar projection neurons
 distribution, 206f
Positron emission tomography (PET),
 17, 385–387

Postsynaptic cell, 230
Postsynaptic potentials (PSPs), 373
Potassium conductance, 368
pov (persistency of vision)
    L-Neuron, 55
Preclinical neurotoxicology
    MRI, 388–390
Predicting spatio-temporal patterns of activity
    within neural ensembles, 163–166
Primary afferent neuron
    density cloud representation, 158f
Primary dendrites
    cholinergic cells, 179
Primary sensory cortex, 421
Primary sensory interneuron, 154–156
    excitatory input
        prediction, 167f
    spatial relationship, 165f
Primary visual cortex (V1), 337
    microcircuit representation, 367–368
Primate visual areas
    pattern similarity indices, 302f
Principal components analysis (PCA), 313–316
Probability density cloud, 158
Programmed cell death, 383
Propagation techniques, 285
Proton density
    human brain, 272f
Proximal dendritic trees
    photomontage, 43f
PSPs, 373
Purkinje Cell Inspector, 92
Purkinje cells
    cerebellum, 396
    virtual, 51–68
Pyramidal cells
    case study, 51–68
    computer representation, 143f
    examples, 114f
    vs. non-pyramidal cell
        neuron numbers, 411f

Q

Quadratic assignment problem, 73
Quantitative dendritic morphology approaches, 27–46
Quantitative neurotoxicity, 383–397

R

Rabbit retina
    optimization analysis, 77f
Rall's equation, 53
Random
    vs. clustered, 308–309
Random graphs, 306–307
Randomized organization, 199
Ray Dream
    L-Neuron, 55
Reachability matrix, 304–305
Red tide toxin, 389
Reference
    definition, 97
Regionally selective dendritic orientation, 179–183
    2D dendritic stick analysis, 181–182
    mean 3D vector, 179–183
Relative anisotropy, 281
Restricted diffusion, 271
Retina
    optimization analysis, 77f
Retinotectal map formation, 343
Retinotopy distortions, 350–351
Righting reflex, 384
Robustness
    observation, 78
Rochester, Nathaniel, 427
Rooted binary topological tree
    example, 222f
Root mean square (RMS) error
    angular measures, 41–42
Rule of 1/2, 132
Rule of 1/3, 135–139

S

SAS software, 317
Save wire, 80
Save wire neuroanatomy optimization, 75
Scalar morphometrics, 110
Scale-free attributes, 309
Schizophrenia, 172
Search pane, 87–90
Search pane
  prototype, 89f
Self-inhibitory synaptic currents
  spike times, 374f
Self-organizing map (SOM), 340
Sensory information
  model, 154–159
Sensory map transformations
  architecture, 199–213
  auditory systems, 200–201
  brain stem nuclei
    local coordinate systems, 205–207
  cerebro-cerebellar systems, 200–201
  3D reconstruction, 204–205
  image-combining microscopy, 202–204
  labeled axons and cells distribution, 207–213
    density gradient analysis, 211–212
    3D reconstructions slicing, 207–208, 208f, 209f
    spatial overlap, 212–213
    stereoimaging, 212
    surface modeling, 209–211
  neural tracing techniques, 201–202
Sensory receptor neurons, 154
Shannon, Claude, 426, 427
Shape
  water movement, 273–274, 274f
Shared target-derived resources, 231–232
Shepard scatterplot, 315f
Shift
  water movement, 273–274, 274f
Sholl analysis, 50, 55
Sib deviation, 41
Sib vector
  calculating, 41
Signaling pathways, 360
Silicon-integrated circuit, 426
Simulated action potentials
  temporal distributions
    histograms, 371f
Simulated CA3 pyramidal cell
  response, 130f
Simulated visual network
  schematic drawing, 366f
Simulation tools, 361
  web site, 361
Single neuron morphology
  computer simulations, 13
Site
  definition, 97
Site browser
  description, 91
  prototype, 90f
Six degrees of separation, 307
6-hydroxydopamine (6-OHDOPA)
  MRM, 391f
  MRM vs. conventional histological staining, 393f
  quantitative measures, 389–394
Size
  water movement, 273–274, 274f
Slicing
  virtual, 429
Small-world attributes, 307–309
Small world digraphs, 308f
SOM, 340
Somatic functions, 5
Somatic membrane potential, 129f
Somatofugal COM axis, 39–41
Spatial distribution
  membrane potential, 137f
Spatial orientation
  dendrites, 66–67
Spatial patterns
  predictions
    neural map, 161f
Spatio-temporal patterns
  activity
    ensemble of sensory neurons, 161–163
  predicting
    of activity
      within neural ensembles, 163–166

Spike events, 373
Spiking, 111
Spiking cell
   examples, 114f
Sprague-Dawley rats, 188
Startle responses
   tests, 384
STATISTICA, 317
   manual, 311
Statistical exploration
   connectivity, 310–320
Steiner tree, 73
   illustration, 75f
Stem diameter frequency distribution, 58f
Stereological methods
   numbers of neurons, 43
Stereopsis, 4
Stimulus representation, 367
Stochastic algorithm
   digital dendrites, 50–51
Stochastic model, 221–226
Stochastic sampling, 8
STRETCH, 75
   algorithm, 77f
Stroke studies
   example, 280f
   trace image, 279–280
Structural cluster analyses, 324–326
Structurally realistic models
   data evaluation, 85–86
   functional assessment, 85–86
   role, 85
Structurally realistic neuronal model, 83
Substantia innominata, 185
Surf-Hippo, 128
   simulation software, 144
swc (Duke-Southampton format)
   L-Neuron, 52–57
Synapse
   axodendritic, 261
   axosomatic, 261
   climbing type, 185
   formation, 219–220
   rearrangement, 220
   weights
      connecting source cells to target cells, 369t

Synaptic boutons
   estimation of volume fraction, 45f
Synaptic information
   dendrites, 27
Synaptic normalization, 230–232
Systems approach, 360–361

T

Taper rate and PK, 61–65
   vs. branch order, 65f
Target-derived neurotrophic factor, 232
Taylor approximation, 132–133
TCA, 252–262, 266
Template-driven search interface, 100–101
Template hierarchy
   modeler's workspace, 96–98
Temporal cortex, 421
Tensarama, 76, 78–79, 79f
Tensegrity, 75
Tensor analysis
   brain fiber pathways, 246–247
Terminal arborization behavior
   ArborVitae, 260
Terminal length (PK), 53–54
   branch order
      vs. taper rate PK, 65f
   L-Neuron Hillman/PK algorithm flow chart, 54f
   taper rate and PK, 61–65
   taper rate PK
      vs. branch order, 65f
Terminal segments
   frequency distribution, 224f
   map, 224f
Terminating
   scatter plots, 30f
Tetrodotoxin (TTX), 339
Thalamic projection cells, 254
Thalamocortical afferent (TCA), 252–262
   pathway
      future directions, 266
Thalamocortical pathway
   simulated data, 257f

Thalamus
    simulated data, 258f
Theory of mind, 431
Thermal motion (Brownian motion), 272–273
3D anatomical reconstruction
    neurons, 156
3D-based DTI techniques
    and applications, 283–287
3D brain fiber reconstruction
    examples, 286f
3D dendrites
    building, 38–42
3D reconstruction
    sensory map transformations, 204–205
3D subcellular detail, 246
3D volumetric scans
    brains, 245
Tissue preparation, 50
Top-down
    information, 14
Topographic organization, 199
Toxicology
    imaging, 387–397
Toxins
    FOBs, 384
Trace image, 279
    stroke studies, 279–280
Tract reconstruction
    examples, 286
Transistors, 426
Transitional cortex, 421
Translational motion (diffusion)
    water molecules, 272
Transmembrane mechanisms
    definition, 97
Transmembrane Mechanism template, 100
Traub's model, 117
Traveling Salesman Problem (TSP), 343
Tree type parameters
    apical, 63t
    basal, 63t
Trigeminocerebellar projection neurons
    distribution, 206f
TSP, 343
Tubulin decay, 227f

Tubulin dynamics, 229–230
Tubulin polymerization, 226–229
Turing, Alan, 426
Turing machine, 363
T1-weighted image
    postmortem anatomical preparation, 283f
T2-weighted image
    example, 282f
Two-compartment model
    Pinksy and Rinzel, 363, 365
2D dendritic stick analysis
    regionally selective dendritic orientation, 181–182
2D DTI data analysis
    brain study application, 279–283
2-deoxyglucose, 337
2D orientation
    dendritic processes, 192
Tyrosine hydroxylase, 393f

U

Ultrastructural investigation
    neurons, 5–6
Unilateral 6-hydroxydopamine (6-OHDOPA)
    injection, 389–394
User interface
    elements, 87–90
    modeler's workspace, 87

V

Vacuum triode
    invention, 426
Van Essen's tension-based model
    cortical folding, 75
Van Ooyen and Willshaw model, 232–238
    description, 232–233
    empirical data, 236–238
    extension, 238
    illustrated, 234f
    results, 235–236
Van Pelt's model, 51
VB, 252–255

Vector representation
  vs. polar histogram, 181, 182f
Ventrobasal complex (VB), 252–255
  slice data, 253f
Vertex connectivity, 305–306
Vertex-disjoint
  directed paths, 305–306
Videomicroscopy
  vs. MRI, 387
Virtual brains, 425–435
Virtual cell
  dendrogram, 62f
Virtual experiments, 429
Virtual motoneurons, 51–68
Virtual NeuroMorphology Electronic Database
  web site, 112
Virtual neurons
  creating, 51
Virtual Purkinje cells, 51–68
Virtual reality markup language (.wrl)
  L-Neuron, 55
Virtual slicing, 429
Visual cognition, 4
Visual cortex
  columnar structures, 337–354
    adult maps, 338
    elastic net algorithm, 343–350
    map development, 338–339
    theoretical models, 339–343
  hierarchically organized, 377
  multilevel modeling, 365–370
Visual cortical areas
  activation, 377
Visual cortical map development, 337
Visualization techniques
  brain study application, 279–283
  for orientation, 282
vol (binary format)
  L-Neuron, 55
Voltage-gated channel
  definition, 97–98
  template, 100
Voltage-sensitive dyes, 337
Volumetric anatomic data, 245–246

Von Economo areas
  neuron number, 415f
Von Neumann, John, 426
Von Neumann computational
  architecture, 72
Voxel, 190
  based atlases
    mouse brain, 248
  future directions, 266
  representational flexibility, 262
  substrates
    axonal navigation, 245–266
VoxelMath, 248, 252
VoxelView, 248, 252, 396

W

Water diffusion
  process
    brain, 274–275
  schematic view, 274f
Waterflow
  branching networks, 76
Water movement
  modes, 273–274, 274f
Wheat germ agglutinin-horseradish
  peroxidase (WGA-HRP), 201
Willshaw model. *See* Van Ooyen
  and Willshaw model
Wirecosts
  distribution, 74f
Wiring principles, 321
Workspace database, 99
  modeler's workspace, 87
Wrl (virtual reality markup language)
  L-Neuron, 55

X

Xgobi
  web site, 116
XML, 84
XPP, 84

Z

Zaborszky, Laszlo, 12
Zeiss Axioscope, 188
Zuse, Konrad, 426

Authors

Giorgio A. Ascoli, PhD

Giorgio Ascoli is head of the Computational Neuroanatomy Group at the Krasnow Institute for Advanced Study and is an Associate Professor in the Department of Psychology at George Mason University. He received a PhD from the Scuola Normale Superiore of Pisa, Italy and experimental training at the National Institutes of Health. His interests include human cognition and the generation of virtual nervous systems. His home page is at: www.krasnow.gmu.edu/ascoli.

David Beeman, PhD

David Beeman is an Adjunct Professor of Electrical and Computer Engineering at the University of Colorado, Boulder where he is developing educational materials for computational neuroscience using the GENESIS simulator. He spent 20 years at Harvey Mudd College engaged in undergraduate teaching and research in computational solid state physics after receiving his PhD in theoretical solid state physics from UCLA in 1967.

Jan G. Bjaalie, MD, PhD

Jan Bjaalie is the founder and head of the Neural Systems and Graphics Computing Laboratory (www.nesys.uio.no) at the Institute of Basic Medical Sciences at the University of Oslo. His main fields of interest include neuroinformatics tools for neuroanatomy and structural and functional organization of visual, somatosensory, and auditory systems.

Sybrand Boer-Iwema, MS

Sybrand Boer-Iwema received an MS in Chemistry from the University of Leiden (Netherlands) in 2001, spending the last semester of his curriculum as a student intern in the Computational Neuroanatomy Group at the Krasnow Institute for Advanced Study. His interests include neuroscience, computational modeling, and biking.

From: Computational Neuroanatomy
Edited by: Giorgio A. Ascoli © Humana Press Inc., Totowa, NJ

James M. Bower, PhD

James M. Bower is a Professor of Computational Biology with appointments at the University of Texas Health Science Center at San Antonio and the University of Texas at San Antonio. Research in his laboratory focuses on anatomical, physiological, and model-based studies of information processing in cortical structures of the mammalian brain. He received his PhD in Neurophysiology from the University of Wisconsin, Madison in 1981.

Derek L. Buhl, PhD

Derek L. Buhl graduated at Rutgers University in 1999 with a BA in Psychology. He is now working on his doctoral degree in Behavioral and Neural Sciences. His current work includes analyzing the differences in hippocampal electrophysiology in numerous types of transgenic and knockout mice.

Robert E. Burke, MD

Robert E. Burke received his MD degree from the University of Rochester School of Medicine and Dentistry in 1961 and clinical training in Internal Medicine and Neurology at the Massachusetts General Hospital. In 1964 he joined the Spinal Cord Section in the National Institute of Neurological Diseases and Blindness at the National Institutes of Health in Bethesda, MD. He has spent his entire career at NIH and is currently Chief of the Laboratory of Neural Control in NINDS. Dr. Burke's research interests center on the structure and function of motor control systems in the spinal cord, with emphasis on the mechanisms of locomotion and on computational approaches to understanding electrophysiological and neuroanatomical data.

Miguel Á. Carreira-Perpiñán, PhD

Miguel Á. Carreira-Perpiñán has university degrees in Computer Science and Physics (Technical University of Madrid, Spain, 1991) and a PhD in Computer Science (University of Sheffield, UK, 2001), on the use of continuous latent variable models for dimensionality reduction and data reconstruction. His current research interests are statistical pattern recognition and computational neuroscience.

Author Biographies

Christopher Cherniak, PhD

Christopher Cherniak has degrees in philosophy from Harvard, Oxford, and University of California, Berkeley. He is a member of the Committee on History and Philosophy of Science at the University of Maryland, College Park.

Attila Csordas, MD

Attila Csordas studied medicine in Hungary. During medical school and after graduation, he worked at the Institute of Microsurgery, Pecs on the topic of neuronal apoptosis. After two years of cardiovascular surgery residency, he returned to neuroscience and for two years was a post-doctoral research fellow at the Center for Molecular and Behavioral Neuroscience, Rutgers University, working on the computational analysis of basal forebrain-cortex connections.

Duncan E. Donohue, BA

Duncan Donohue is a second-year PhD student in the Department of Psychology at George Mason University. He received his undergraduate degree in Biology from the University of Virginia and is interested in the application of computational methods in neuroscience.

Alvaro Duque, PhD

Alvaro Duque received his BS in Electrical Engineering from Columbia University, New York. He then joined the Cellular and Molecular Biodynamics group and the PhD Neuroscience program at Rutgers University, where he received his PhD degree in 2001. Currently he is a post-doctoral fellow at the Section of Neurobiology at the Yale University School of Medicine.

Jonas Dyhrfjeld-Johnsen, MS

Jonas Dyhrfjeld-Johnsen obtained his MS in Biophysics through studies in Denmark and Belgium and is currently working on his PhD in the Computational Systems Neuroscience group at the Centre of Anatomy and Brain Research, Düsseldorf University.

James H. Fallon, PhD

James Fallon is a Professor of Anatomy and Neurobiology at UC Irvine. He recently discovered and published a method of using TGF-a to generate large numbers of new neurons from stem cells in adult rat brains lesioned with 6-OH dopamine. The animals clinically recovered from their parkinsonian behavior.

Geoffrey J. Goodhill, PhD

Geoffrey J. Goodhill has a BSc in Mathematics and Physics from the University of Bristol, an MSc in Artificial Intelligence from the University of Edinburgh, and a PhD in Cognitive Science from the University of Sussex. Following post-doctoral training in computational neuroscience at the University of Edinburgh, Baylor College of Medicine, and the Salk Institute, he joined the faculty of Georgetown University in 1996. He is currently an Associate Professor in the Department of Neuroscience.

Josef P. Hanig, PhD

Josef P. Hanig is the Deputy Director of the Division of Applied Pharmacology Research, Center for Drug Evaluation and Research, Food and Drug Administration, Laurel, MD. He received his PhD from New York Medical College. Dr. Hanig is a fellow of the Academy of Toxicological Sciences and a Diplomat of the American Board of Toxicology.

Junko Hara, PhD

Junko Hara is a research fellow in Computer Science at UC Irvine. She trained in artificial intelligence and electrical and biomedical engineering at the University of Tokushima and Keio University. She has developed computational models of EEG for Alzheimer's Disease, vascular dementia and normal again, as well as developed mathematical models for the developing human cerebral cortex.

Claus C. Hilgetag, PhD

Claus C. Hilgetag studied Biophysics in Berlin and Neuroscience in Edinburgh, Oxford, Newcastle, and Boston. He is an Assistant Professor of Neuroscience at the newly founded International University Bremen. His research focuses on the organization of cerebral connectivity and architecture as well as the neural mechanisms of spatial attention. Further information can be found at: http://www.iu–bremen.de/directory/faculty/01065.

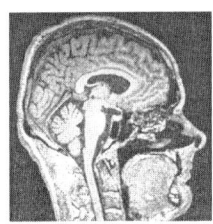

Michael Hucka, PhD

Michael Hucka received his PhD in Computer Science and Engineering in 1998 from the University of Michigan. He has worked in the areas of artificial intelligence, cognitive science, and computational neuroscience. He is currently engaged in developing software to help neurobiologists interact with simulation tools, databases, and other resources.

Gwen A. Jacobs, PhD

Gwen A. Jacobs is an Associate Professor of Neuroscience and Head of the Department of Cell Biology and Neuroscience at Montana State University. She is an active research neuroscientist, focusing on the neural basis of information processing in sensory systems, using experimental and computational approaches. She is actively engaged in neuroinformatics with a focus on developing databases and computational and visualization tools for neuroscientists.

Rolf Kötter, MD

Rolf Kötter studied medicine and computer science in Germany, Britain, and France. He specialized in anatomy and leads the Computational | Systems | Neuroscience group at the Centre of Anatomy and Brain Research, Düsseldorf University. His work includes experimental and computational approaches to elucidate structure-function relationships in the brain.

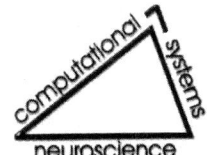

Jeffrey L. Krichmar, PhD

Jeffrey L. Krichmar received a BS in Computer Science in 1983 from the University of Massachusetts at Amherst, an MS in Computer Science from The George Washington University in 1991, and a PhD in Computational Sciences and Informatics from George Mason University in 1997. Currently, he is an Associate Fellow in theoretical neurobiology at The Neurosciences Institute in San Diego, CA. Besides investigating the relationship between neuronal shape and function, his research interests include biologically plausible models of learning and memory and simulating the nervous system in a real-world artifact ("robot") interacting with the environment.

Benjamin Harrison Landing, MD

The late Benjamin H. Landing was an Emeritus Professor of Pediatrics and Pathology at USC and served as director of laboratories and chair of pediatric pathology at Children's Hospital Los Angeles from 1959 to 1988. He was one of the founders of Pediatric Pathology and trained more pediatric pathologists than anyone of his generation. His discoveries include defining the structure of skeletal muscle and liver, discovering GM1 gangliosidosis, and overturning the dogma of no mammalian post-natal neurogenesis. He died in the summer of 2000 from intracerebral hemorrhage and is remembered as a true scientist and scholar.

Maciej T. Lazarewicz, MD

Maciej Lazarewicz is a postdoctoral researcher in the Computational Neuroanatomy Group at the Krasnow Institute for Advanced Study. He received an MD from the Medical Academy of Gdansk (Poland). His research interests are in computational neuroscience: numerical methods, drug and neurological disease mechanisms, visual attention, and consciousness.

Trygve B. Leergaard, MD, PhD

Trygve Leergaard is a postdoctoral research fellow at the Neural Systems and Graphics Computing Laboratory at the Institute of Basic Medical Sciences at the University of Oslo. He is experienced with computerized, three-dimensional anatomical analyses of developing and adult cerebro-cerebellar systems in rat, with a main focus on somatosensory map transformations.

David S. Lester, PhD

David S. Lester is presently the Director of Clinical Technologies at Pharmacia at Peapack, NJ. Previously, he was Senior Science Advisor in the Office of Pharmaceutical Sciences, Center for Drug Evaluation and Research, Food and Drug Administration, Rockville, MD. He earned his PhD from Northwestern University.

William B. Marks, PhD

William B. Marks received his BS in Physics at MIT in 1955 and his PhD at the Biophysics Department of Johns Hopkins University in 1964. From 1956 to 1964 he worked on artificial intelligence with Oliver Selfridge, the spinal cord with Patrick D. Wall, and visual pigments with Edward F. MacNichol. During 1965 to 1973, while on the faculty of Biophysics at JHU, he studied the visual physiology of the optic tectum and methods for simultaneous recording of activity from many individual neurons. He continued the latter work at NIH from 1973 to 1980. He then turned to the neurophysiology of locomotion with Gerald Loeb in 1980–1986, the fractal analysis of neuronal morphology with T. G. Smith in 1986–1988, and modeling of neuronal shape with Bob Burke from 1988 to the present. He has maintained an interest in theoretical and philosophical aspects of neuroscience throughout.

Zekeria Mokhtarzada, BS

Zekeria Mokhtarzada is studying computer science at the University of Maryland; he is also a computer scientist at NIH.

Susumu Mori, PhD

Susumu Mori received his PhD in Biophysics from Johns Hopkins University in 1996 and has been an Assistant Professor since 1997. His current interests include the study of brain white matter anatomy using diffusion tensor imaging.

Zoltan Nadasdy, PhD

Zoltan Nadasdy has received his PhD in neuroscience at Rutgers, The State University of New Jersey. Granted by the Lady Davis fellowship he started his post-doctoral studies at the Interdisciplinary Center for Neural Computation of The Hebrew University in Jerusalem. Currently he is a post-doctoral scholar in biology at Caltech.

Slawomir J. Nasuto, PhD

Slawomir J. Nasuto received a MSc in Mathematics from the University of Marie Curie-Sklodovska in Lublin, Poland and a PhD in Cybernetics from the University of Reading, UK in 1999. He spent a year as a postdoctoral fellow at the Krasnow Institute for Advanced Study at George Mason University, where he worked in the Computational Neuroanatomy Group. He is currently a lecturer in the Department of Cybernetics, at The University of Reading, UK.

Pernille Nielsen, BS

Pernille Nielsen holds a BS in Physics and is currently working on her MS in Biophysics at the University of Copenhagen in collaboration with the Computational Systems Neuroscience group at the Centre of Anatomy and Brain Research at Düsseldorf University.

Uri Nodelman, BS

Uri Nodelman has received a BS in mathematics from the University of Maryland, College Park and is currently a PhD student in Computer Science at Stanford University.

Georg Northoff, MD, PhD

Georg Northoff studied medicine and philosophy in Hamburg, Essen, Bochum, and New York. He holds both an MD in Medicine and a PhD in philosophy. His work as a psychiatrist at the University Hospital Magdeburg focuses on functional imaging studies of psychiatric diseases.

Author Biographies

P. Scott Pine, MA

P. Scott Pine is in the Division of Applied Pharmacology Research, Center for Drug Evaluation and Research, Food and Drug Administration, Laurel, MD. He obtained his BS from Michigan State University and his MA from the State University of New York at Buffalo.

Colin S. Pittendrigh, PhD

Colin (Sandy) Pittendrigh is a Senior Software Engineer in the Center for Computational Biology at Montana State University. He is the chief architect of the NeuroSys System and is interested in building easy-to-use databases and data analysis tools for biologists.

Alexei Samsonovich, PhD

Alexei Samsonovich is a postdoctoral researcher in the Computational Neuroanatomy Group at the Krasnow Institute for Advanced Study. He received a PhD from the University of Arizona at Tucson where he was also a post-doctoral fellow. His interests include artificial consciousness and computational neuroscience.

Ruggero Scorcioni, BS

Ruggero Scorcioni is the software engineer of the Computational Neuroanatomy Group at the Krasnow Institute for Advanced Study while studying for a PhD in the School of Computational Science at George Mason University. He graduated in electronic engineering from the University of Modena, Italy.

Stephen L. Senft, PhD

Stephen Senft is interested in visualization of brain anatomy and activity at the cellular level. He began his study of Neuroscience with Steve George at Amherst College, and pursued his graduate study of Neuroscience first at the University of Oregon and later at Washington University, with intervening study in Dan Alkon's laboratory at the MBL. He obtained his Ph.D. in the Woolsey laboratory, with an analysis of the ingrowth of thalamic afferents to mouse somatosensory ("barrel") cortex. Subsequently he was a co-founder of Vital Images, a company devoted to scientific and medical 3D imaging. Currently he is a lecturer in Neuroscience at Yale University, and a fellow at the Krasnow Institute for Advanced Science at George Mason University.

Kavita Shankar, PhD

Kavita Shankar received both her BS in Human Biology (1984) and MS in Anatomy (1986) from the All India Institute of Medical Sciences, New Delhi, and a PhD in Craniofacial Biology (1993) from the University of Southern California, Los Angeles. She was the recipient of the NIH-National Research Service Award grant for carrying out her doctoral work. She previously worked in UCLA's Laboratory of Neuroimaging. Her current interests include neuroinformatics and neuroimaging.

William Rodman Shankle, MD

William Shankle is a neurologist, statistician, and Professor of Cognitive Science and pharmacology at UC Irvine. He, along with Drs. Landing, Hara, Rafii, and Fallon, provided the first evidence for mammalian (human) cortical neurogenesis after birth, first in 1993 and again in 1998.

Friedrich T. Sommer, PhD

Friedrich T. Sommer received the Physics Diploma from Tübingen University and a PhD from Düsseldorf University. He was a postdoc and research associate at the Universities of Ulm and Tübingen, and a visiting scientist at the Massachusetts Institute of Technology. His research interests include fMRI data analysis, associative memory models, neuronal computations, and large-scale activity spread in the cortex.

Jozsef Somogyi, PhD

Jozsef Somogyi received his Ph.D. from the Hungarian Academy of Sciences. Currently he is a Senior Research Associate at the Center for Neuroscience, The Flinders University of South Australia. His current research focuses on the cellular mechanisms of synaptic transmission.

Olaf Sporns, PhD

Olaf Sporns is an Assistant Professor in the Department of Psychology and Programs in Cognitive and Neural Science at Indiana University, Bloomington, IN. He received a PhD in neuroscience at the Rockefeller University, New York. His main research interest is theoretical and computational modeling of the brain. His work includes the design of neuronal network models that can be interfaced with real-world autonomous behaving systems (robots) and are used to study perceptual categorization, sensorimotor development, value-dependent learning, and the development of neuronal receptive field properties. In addition, he has developed computational methods for analyzing patterns of anatomical and functional connectivity. His homepage is at: php.indiana.edu/~osporns.

Klaas E. Stephan, MD

Klaas Stephan graduated with a degree in medicine from the Heinrich Heine University Düsseldorf. Now working at University Hospital Aachen and Research Centre Jülich, he is also pursuing degrees in Computer Science (Hagen) and Psychology (Newcastle). His research focuses on computational approaches to the analysis of structural, functional, and effective connectivity in the brains of macaque and man.

Arjen van Ooyen, PhD

Arjen van Ooyen is a researcher at the Netherlands Institute for Brain Research. He has a PhD in theoretical neurobiology from the University of Amsterdam. His principal research concerns modeling neural development: neurite outgrowth, axon guidance, and axonal competition. Further information can be found on his website at: www.anc.ed.ac.uk/~arjen.

Jaap van Pelt, PhD

Jaap van Pelt received his PhD in Physics in 1978 at the Free University in Amsterdam. His research group, "Neurons and Networks", at the Netherlands Institute for Brain Research (NIBR) investigates, both experimentally and by theoretical and computational approaches, neuronal morphology and activity-dependent mechanisms in neurite outgrowth and neuronal network formation. Further information can be found on his website at: www.nih.knaw.nl/~jaapvanpelt.

Laszlo Zaborszky, MD, PhD

Laszlo Zaborszky received his MD from Semmelweis University, Hungary in 1969 and his PhD from the Hungarian Academy of Sciences in 1981. Prior joining the Center for Molecular and Behavioral Neuroscience at Rutgers University in 1993, he was an Associate Professor in the Department of Neurology at the University of Virginia, Charlottesville. His current research interest is the functional organization of the basal forebrain using a combination of techniques, including electrophysiology, computational anatomy and functional imaging. Among others he is an author of a monograph on hypothalamic connections (Springer, 1982) and co-editor of a textbook on anatomical tract tracing methods (Plenum, 1989). His homepage is at: www.zlab.rutgers.edu.

CD-ROM Contents

This CD-ROM contains primarily GIF, JPEG, PDF, MPEG, and AVI files for *Computational Neuroanatomy: Principles and Methods* by Giorgio A. Ascoli. These files are mainly color versions of the artwork found in the book. Animation files are included in some chapters. Chapters 3 and 7 on the CD contain PC-executable and UNIX files. Users should note below under "System Requirements."

System Requirements

This self-launching CD-ROM is compatible with both PC and Macintosh operating platforms on which Internet Explorer or Netscape Navigator version 4.0 or later is installed. (Users are required to have a Pentium III processor, 256 MB RAM, and Windows 98 or later for some Chapter 3 materials. Users are required to have a 1.2 Ghz processor, 256 MB RAM, and Linux Redhat 7.1 for some Chapter 7 materials. Most of the software in these two chapters is Mac/PC compatible.) Users should note that their browser's settings should be configured to playback files in MPEG format (Quicktime 4.0 or later is recommended). Users can re-launch the CD-ROM by double-clicking the icon which reads "index.htm" in the "Computational Neuroanatomy" folder.

Limited Warranty and Disclaimer

Humana Press Inc. warrants the CD-ROM contained herein to be free of defects in materials and workmanship for a period of thirty days from the date of the book's purchase. If within this thirty day period Humana Press receives written notification of defects in materials or workmanship, and such notification is determined by Humana Press to be valid, the defective disk will be replaced.

In no event shall Humana Press or the contributors to this CD-ROM be liable for any damages whatsoever arising from the use or inability to use the software or files contained therein.

The authors of this book have used their best efforts in preparing this material. These efforts include the development, research, and testing of the theories and programs to determine their effectiveness. Neither the authors nor the publisher make warranties of any kind, express or implied, with regard to these programs or the documentation contained within this book, including, without limitation, warranties of merchantability or fitness for a particular purpose. No liability is accepted in any event, for any damages including incidental or consequential damages, lost profits, costs of lost data or program material, or otherwise in connection with or arising out of the furnishing, performance, or use of the programs on this CD-ROM.